GENETICS OF
BACTERIAL DIVERSITY

Genetics of Bacterial Diversity

We apologise for a minor printing error on the cover of this book.

The correct authorship is DAVID A HOPWOOD and KEITH F CHATER, as given on the title page.

The error will be corrected in future printings,

Academic Press

GENETICS OF BACTERIAL DIVERSITY

Edited by

David A. Hopwood

John Innes Institute
and AFRC Institute of Plant
Science Research
Norwich NR4 7UH, UK

Keith F. Chater

John Innes Institute
and AFRC Institute of Plant
Science Research
Norwich NR4 7UH, UK

1989

ACADEMIC PRESS

Harcourt Brace Jovanovich, Publishers

London San Diego New York Berkeley
Boston Sydney Tokyo Toronto

ACADEMIC PRESS LIMITED
24/28 Oval Road
London NW1 7DX

United States Edition published by
ACADEMIC PRESS INC.
San Diego, CA 92101

British Library Cataloguing in Publication Data
Hopwood, D. A.
Genetics of bacterial diversity
1. Bacteria. Genetics
I. Title II. Chater, K. F.
589.9′015

ISBN 0-12-355574-4
ISBN 0-12-355575-2 Pbk

Typeset by Paston Press, Loddon, Norfolk
Printed in Great Britain by Alden Press, Oxford

Contributors

N. L. Brown, Department of Biochemistry, University of Bristol, Bristol, BS8 1TD, UK

E. J. Bylina, Department of Applied Biological Sciences, Massachusetts Institute of Technology, Cambridge, MA 02139, USA

K. F. Chater, John Innes Institute and AFRC Institute of Plant Science Research, Norwich, NR4 7UH, UK

M. J. Daniels, The Sainsbury Laboratory, John Innes Institute, Norwich, NR4 7UH, UK

J. Engebrecht, The Agouron Institute, La Jolla, CA 92037, USA

J. Errington, Microbiology Unit, Department of Biochemistry, University of Oxford, Oxford, OX1 3QU, UK

S. Harayama, Department of Medical Biochemistry, Geneva University Medical Centre, Geneva, Switzerland

G. Hinson, Department of Genetics, University of Leicester, Leicester, LE1 7RH, UK

D. A. Hodgson, Department of Biological Sciences, University of Warwick, Coventry, CV4 7AL, UK

P. J. J. Hooykaas, Department of Biochemistry, State University of Leiden, 2333 AL Leiden, The Netherlands

D. A. Hopwood, The John Innes Institute and AFRC Institute of Plant Science Research, Norwich, NR4 7UH, UK

A. W. B. Johnston, John Innes Institute and AFRC Institute of Plant Science Research, Norwich, NR4 7UH, UK

D. Kaiser, Department of Biochemistry, Stanford University, Stanford, CA 94305, USA

C. Kennedy, AFRC Institute of Plant Science Research, Nitrogen Fixation Laboratory, University of Sussex, Brighton, Sussex, BN1 9RQ, UK

L. Kroos, The Biological Laboratories, Harvard University, Cambridge, MA 02138, USA

R. Losick, The Biological Laboratories, Harvard University, Cambridge, MA 02183, USA

P. A. Lund, Department of Biochemistry, University of Bristol, Bristol, BS8 1TD, UK

M. Martin, The Agouron Institute, La Jolla, CA 92037, USA

A. Newton, Department of Molecular Biology, Princeton University, Princeton, NJ 08544, USA

N. Ni Bhriain, Department of Biochemistry, University of Dundee, Dundee, DD1 4HN, UK

M. Roberts, Department of Genetics, University of Leicester, Leicester, LE1 7RH, UK

J. R. Saunders, Department of Microbiology, University of Liverpool, Liverpool, L69 3BX, UK

M. Silverman, The Agouron Institute, La Jolla, CA 92037, USA

R. K. Taylor, University of Tennessee, Medical School Center, Department of Microbiology and Immunology, Memphis, TN 38163, USA

K. N. Timmis, Department of Medical Biochemistry, Geneva University Medical Centre, Geneva, Switzerland

P. H. Williams, Department of Genetics, University of Leicester, Leicester, LE1 7RH, UK

J. P. W. Young, John Innes Institute and AFRC Institute of Plant Science Research, Norwich, NR4 7UH, UK

P. Youngman, Department of Microbiology, University of Pennsylvania, School of Medicine, Philadelphia, PA 19104, USA

D. C. Youvan, Department of Applied Biological Science, Massachusetts Institute of Technology, Cambridge, MA 02139, USA

Preface

E. coli K-12 is a marvellous subject for genetical research. Without the flowering of molecular genetics that followed Lederberg and Tatum's classical discovery of sex in this simple bacterium, it is difficult to imagine that we would know so much about the structure, organization and expression of genes as we do today. (Even the discovery of such marvels of eukaryote genetics as introns, homeoboxes, or oncogenes depended directly on the development of the K-12 strain and its accessory DNA elements for the cloning and analysis of recombinant DNA.) But *E. coli* K-12 does not do everything. It does not degrade complex phenolics, photosynthesize, fix nitrogen or produce antibiotics or light. It does not differentiate to form stalks, spores or fruiting bodies. Nor does the K-12 strain parasitize animals or plants. Other bacteria do, and the application of molecular genetics is leading to an understanding of these phenomena. Such is the subject matter of this book.

For some time, we had felt the time to be ripe for a collection of articles on the genetics of a range of interesting things that bacteria do. But we were daunted by the conviction that such a book would need to contain, as a background, a detailed description of the full range of the "mainstream" topics of bacterial genetics that *E. coli* represents, since there appeared to be no suitable up-to-date textbook in this field. This deficiency was recently remedied by the publication of the book, *Genetics of Bacteria*, edited by Scaife, Leach and Galizzi (Academic Press, 1985); and the comprehensive and specialized compilation, *Escherichia coli and Salmonella typhimurium: Cellular and Molecular Biology*, edited by Neidhardt *et al.* (American Society for Microbiology, 1987). Our book is thus able to concentrate on the rapidly developing field of "non-K-12" bacterial genetics that is largely outside the scope of these other texts.

After an introductory chapter outlining the phylogenetic relationships of bacteria and the range of metabolic, behavioural and developmental phenomena displayed by them, Chapter 2 reviews the genetic processes found in bacteria generally, and Chapter 3 discusses a range of genetic techniques that have been used to analyse the various special systems described in the body of the book. The book is not meant to be exhaustive. Not only is our choice of topics deliberately eclectic, but we have encouraged the authors of most of the chapters to deal with a particular example of each topic in detail, rather than reviewing everything that is known about the phenomenon, such as light production or sporulation, in all bacteria that display it. In this way, we hope that the book will give a strong flavour of the biology of bacteria and the ways in which the techniques of genetics are used to study them, which will be of interest to a wide range of scientists, rather than providing an exhaustive review of any particular topic for the specialist. In this vein, we asked authors to choose, wherever possible, key references and reviews to cite in limited bibliographies, rather than charting every step in the development of the subject.

Chapters 4–20 can be read in any order, even though we have grouped them into four main subject areas. Thus, Chapters 4–9 deal with various special metabolic capabilities characteristic of certain groups of bacteria (light production, photosynthesis, nitrogen fixation, antibiotic production, degradation of aromatic compounds and mercury resistance); Chapters 10–12 describe the developmental processes of cell-cycle associated motility, sporulation, and specialized colonial behaviour; Chapters 13–16 cover four components of bacterial pathogenicity for animals; and Chapters 17–19 deal with pathogenic and symbiotic interactions of bacteria with higher plants. Chapter 20 stands by itself. It is written by a population geneticist and explains, for the rest of us, mainly molecular, geneticists, some of the concepts and the progress now being made in the application of population genetics to bacteria.

As Hodgson points out in Chapter 1, the bacteria studied by geneticists cover only a narrow part of the whole phylogenetic spectrum. The glaring omission in current bacterial genetics is the study of archaebacteria. These prokaryotes grow at extreme pH values, temperatures and salt concentrations and have important metabolic capabilities such as methanogenesis and chemolithotrophy. They are clearly far removed phylogenetically from other groups of bacteria, and show some novel genetic properties; for example, their ribosomes have features intermediate between those of eubacteria and eukaryotes. A start has been made on their genetics and, if there were to be a second edition of this book, it is to be hoped that they would figure prominently in it.

We have deliberately avoided any unnecessary justification for research on bacterial genetics arising from its biotechnological opportunities, real or conjectural, although work on a large proportion of the organisms covered in the book—including of course the pathogenic systems—has in fact been motivated by anthropocentric considerations, and several applications have been discussed in the appropriate chapters. They range from the construction of *Pseudomonas* strains, able to degrade novel hydrocarbons polluting the environment, to the use of *Agrobacterium* as a microbial route to the genetic engineering of plants.

We hope that the book will be of interest to microbiologists wishing to catch up on the genetic basis of some of the classical phenomena of bacteriology, and geneticists unfamiliar with some of the things that bacteria can accomplish.

We hope it will be useful to undergraduates, graduate students and research scientists at many levels. Above all, we hope that it will help to dispel the sometimes tacitly assumed opinion that modern genetics has left prokaryotes behind to concentrate on eukaryotes. Eukaryotes are amazing genetic systems, but so too are bacteria!

We should like to thank the authors of the various chapters for joining with us in producing this book, for preparing their manuscripts promptly and for responding in a friendly way to our editorial suggestions. We are especially indebted to Anne Williams for the considerable secretarial task involved at all stages from the original conception of the book to its completion, and Gina Fullerlove of Academic Press for her efficient and sympathetic handling of the whole project.

David Hopwood
Keith Chater

Norwich, April 1988

Contents

Section **I**

Introductory Chapters—the Diversity of Bacteria and of Bacterial Genetics

Bacteria are typified by the very small size of their cells, which are usually of the order of 1 μm in diameter, and the relative simplicity of their structure—they do not contain mitochondria or chloroplasts, nor is their genetic material contained within a nuclear membrane. The two bacterial kingdoms, eubacteria and archaebacteria, differ radically in their cell wall organization, and substantially in the details of their transcription and translation systems. In these latter features, archaebacteria are more like eukaryotes than eubacteria. This, and their remarkable ability to survive in various extreme environments, are now attracting considerable research interest to the archaebacteria. As yet, however, systems for their genetic analysis are not well advanced, a technical obstacle shared with many of the lesser-known groups of eubacteria whose remarkable biochemical, structural and ecological attributes are described in Chapter 1. The phylogenetic relationships of bacterial groups displayed in Fig. 2 of Chapter 1 reveal that those few bacterial groups for which extensive genetic analysis is already possible all fall in one sector of the phylogeny containing the purple group of Gram-negative bacteria (which includes *Escherichia coli*) and the Gram-positive bacteria.

The availability of systems of genetic analysis has allowed characterization of many basic genetic features of these groups, such as genome size and circularity; the strategies of plasmids, transposons and phages; the various means by which genetic exchange can take place; and the means by which gene expression occurs and is regulated. Chapter 2 shows how this has led to a more fully substantiated view of the generalities of bacterial genetics than could be obtained by the study only of the single *E. coli* paradigm.

In briefly summarizing the overall genetic phenomenology of these bacteria in Chapter 2, and in dealing with special attributes of particular bacteria in later Chapters, it is inevitable that details of the techniques by which the discoveries were made should be omitted. This deficiency is redressed in Chapter 3, which specifically describes the techniques employed in molecular genetic studies of various bacteria, and serves as a reference point and glossary for the technical background of the more specific chapters that make up the main body of the book. It is clear that, while the use of the wonderful *E. coli* cloning and analytical systems provides great benefits in the study of all organisms, the most striking advances are made when recombinant DNA technology and allied techniques are also well-developed in the organism under scrutiny. Fortunately, the persistence of aficianados of different bacteria has resulted in the extension of *E. coli*-derived technology into most purple Gram-negative bacteria via wide host-range plasmids and transposons, and its use in targeted modifications of the genotype of the relevant organism by gene replacement or disruption. It has proved less straightforward to introduce *E. coli*-based technology into Gram-positive bacteria, and there has been much more emphasis on the exploitation of the native genetic elements and genetic features of these organisms in developing molecular genetic techniques for them. As a result, efficient plasmid- and phage-based cloning systems and well-developed transposon mutagenesis for Gram-positive bacteria are available both for the low and the high (G + C) divisions of this group.

Chapter 1

Bacterial Diversity: the Range of Interesting Things that Bacteria Do

D. A. HODGSON

I. INTRODUCTION

The aim of this chapter is to provide a bacteriocentric overview of the role of bacteria in nature. Hence the biotechnological roles of bacteria; both ancient and modern, are excluded, as are "bacteria as model systems". Rather I hope to show that bacteria are so interesting in themselves that there is no need to justify their study by recourse to some principle of anthropocentric utility.

Inevitably, this chapter contains a personal, possibly idiosyncratic, selection of topics. The chapter is organized into four main sections. In Section II, I discuss the diversity of bacteria in terms of the multiplicity of ecological niches that they fill. This is followed by a short discussion of the recent triumph of bacterial phylogeneticists in the assembly of a molecular phylogeny of bacteria using analysis of 16S rRNA and 5S rRNA (Section III). This allows us in Section IV to discuss the evolution of homologous and analogous structures, physiologies and behaviours in bacteria. Finally, there is a discussion of phenomena that might not have been expected to be present in bacteria (Section V).

II. ECOLOGICAL NICHES

Bacteria can be found almost everywhere. This remarkable ubiquity results from and reflects their ability to gain energy and the necessary building blocks of life from diverse sources, and the evolution of novel physiologies and behaviours to allow life in extreme environments.

A. Bacterial Nutrition is Diverse

1. Inorganic sources of carbon, nitrogen and energy can be exploited by some prokaryotes. The ability of bacteria to gain energy and reducing power by the oxidation or reduction of simple minerals or inorganic salts (chemolithotrophy) has been well documented since Winogradsky's report of 1890, but its investigation has largely been limited to biochemistry and physiology. Only recently has the analytical power of genetics begun to be applied to the methane oxidizers, the sulphur oxidizers and the carbon dioxide reducers (the methanogens). This is because many of these bacteria are obligate chemolithotrophs, difficult to culture on solid media and very slow growing. The many advances in gene cloning have allowed the transfer of genetic material to more amenable hosts where molecular analysis can now take place. The discovery that the methanogens belong to a third primary kingdom, the archaebacteria, separate from both the eubacteria and the eukaryotes, may have repercussions for the genetic analysis of these organisms.

Many bacteria can fix elemental nitrogen, an ability not found in eukaryotes. The complexity of the intensively studied gene system that determines this ability (Chapter 6) illustrates the general principle that nutritional simplicity is allied to genetic complexity.

A more limited range of bacteria are able to obtain all their energy for CO_2 reduction from light (Chapter 5). Some other bacteria can use light as an auxiliary energy source: for example, members of the Halobacteriaceae—an archaebacterial group—use bacterial rhodopsin to generate ATP. Both these groups of light-exploiting bacteria exhibit phototaxis; the mechanisms of which have been the subject of a recent review (1).

2. *Some bacteria can utilize toxic compounds.* Most bacteria are heterotrophs—they depend on complex sources of carbon and nitrogen for nutrition and energy. Amongst them, many have characteristics unique to the prokaryotic world. Some can use organic compounds that are toxic to other organisms to generate reducing power and release energy: thus, the abilities of one group, the pseudomonads, to deal with xenobiotics are described in Chapter 8. Plants, fungi and bacteria produce a very diverse range of compounds. So it is not surprising that the bacterial world already contains organisms which can deal with unusual chemicals. For example, *Streptomyces achromogenes* var. *streptozoticus* produces streptozocin (Fig. 1). Within this compound, which contains a dinitrogen bond, there lurks the active moiety of *N*-methyl-*N'*-nitro-*N*-nitroso-guanidine, a most powerful mutagen. Streptozocin biosynthesis may involve nitrous acid, another potent mutagen (2). It has been said that bioactive molecules, produced by bacteria and fungi, are no longer sought only for their potential use but also as a spur to the organic chemist to show what range of molecular structures are possible (3).

3. *Extracellular digestion allows bacteria to dissolve macromolecules.* An important process in many heterotrophs is the extracellular secretion of enzymes which digest insoluble materials, to give products that can be absorbed back into the

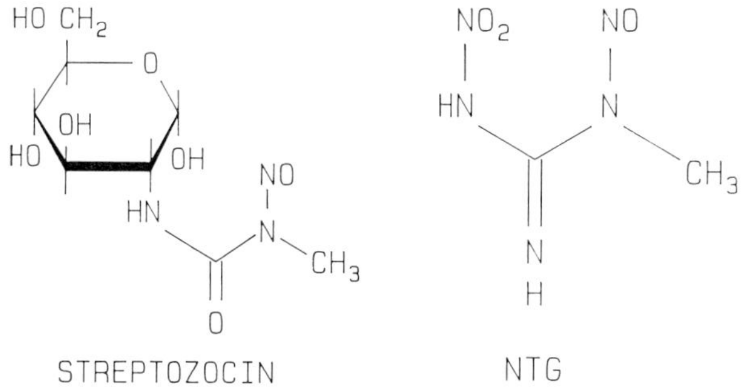

Figure 1. Chemical structures of streptozocin and *N*-methyl-*N'*-nitro-*N*-nitroso-guanidine (NTG).

cell. With Gram-positive bacteria, the secretion process appears relatively simple: a proenzyme is targeted to the cell membrane and released by proteolytic cleavage of an NH_2-terminal signal sequence (4). In the case of Gram-negative cells, the process is complicated by the presence of an outer membrane. Therefore *Escherichia coli* is generally unable to export enzymes beyond the periplasm. However, many Gram-negative plant pathogens secrete pectinases, cellulases and proteases, presumably to macerate plant tissues. The application of molecular genetics to genera such as *Erwinia* (5) and *Xanthomonas* (Chapter 17) promises to reveal the special mechanisms that are responsible.

B. Life in Extreme Environments

1. Life on or in other organisms. Many bacteria are adapted to use other organisms as an ecological niche: so exquisitely in some cases that they have become obligate pathogens and have proved refractory to man's attempts to culture them outside the cell or tissue. Perhaps the most famous example is the leprosy bacillus. Even the secrets of this refractory organism may soon be accessible through the achievements of recombinant DNA technology. *Mycobacterium leprae* genes have been cloned in *E. coli* (6) and recently there is a strong suggestion that many more will be well-expressed in the more closely-related streptomycetes (7). Intracellular parasitism, as in leprosy, provides pathogens with one form of defence against the host's immune system. Other organisms have evolved ways to change their surface antigens. Well-studied examples include phase variation in *Salmonella* and antigenic variation in *Neisseria* (Chapter 13).

A milder kind of host colonization is the induction of crown-gall tumours and "hairy roots" on dicotyledonous plants by *Agrobacterium tumefaciens* and *A. rhizogenes*. These bacteria have evolved a mechanism of genetic colonization which results in the insertion of DNA encoding production of complex amino acids (opines) into the plant. The opines can then act as sole carbon and nitrogen source for the *Agrobacterium* which caused the colonization (Chapter 18).

As well as having pathogenic life styles, bacteria are capable of many symbiotic relationships with eukaryotes. An intensively-studied example is the legume/*Rhizobium* interaction (Chapter 19), but other examples, which are receiving increasing attention, include cyanobacteria in lichens and sulphur oxidizers in gutless worms (8). The origin of chloroplasts and mitochondria from endosymbiotic bacteria has been proposed and extensively discussed by Margulis (9).

2. Life in physicochemically extreme environments. The archaebacterial kingdom has a penchant for extreme environmental niches. Halobacteria can grow in

2.5–4.25 M saline. Other archaebacteria such as the sulphur oxidizer, *Sulfolobus* and the extreme thermophile, *Thermoplasma* live in an extremely acidic environment (pH 1 and lower). This habitat is also shared by the eubacterial sulphur oxidizers, e.g. *Thiobacillus* spp. Another, though less tolerant, group of acidophilic eubacteria are the lactic acid bacteria. Both the sulphur oxidizers and the lactic acid bacteria generate the extreme environments within which they live by the production of acids (sulphuric and lactic respectively) as a waste product. This strategy has given the lactic acid bacteria the leeway to lose many of their biosynthetic capabilities because, presumably, they can be sure of finding the necessary nucleic acid bases, amino acids and growth factors in the environment that they have "poisoned" for their competitors. A number of bacteria (e.g. *Alcaligenes*) can also grow in extremely alkaline environments. How they manage to maintain a proton gradient across their cell membrane is still an open question.

Extreme thermophiles are found in both the eubacteria (*Thermus*) and the archaebacteria (*Thermoplasma*, *Thermococcus*, etc.), where growth at 80°C is commonplace and up to 105°C possible. Some years ago, there was a report of bacteria living at 250°C in deep sea black smokers. However, this observation is disputed from both technical and theoretical points of view (10).

As a final example of a bacterium able to survive in an extreme environment, I include *Deinococcus*. Organisms of this genus, and of a related genus *Deinobacter*, can tolerate extremes of ionizing radiation. However, perhaps surprisingly, they are susceptible to some chemical mutagens. This has rendered them amenable to genetic analysis, and a transformation system has been developed in *Deinococcus radiodurans*. Analysis of the mechanisms of DNA repair has shown that a number are operating concomitantly. Also there appear to be four copies of the chromosome in each cell (11).

III. TOWARDS A PHYLOGENY OF BACTERIA

A. Construction of the Tree

The development of 16S and 5S rRNA oligonucleotide cataloguing and, more recently, of rapid RNA sequencing has allowed progress towards an objective and consistent bacterial phylogeny, well reviewed by Woese (12) and Stackebrandt (13). Ribosomal RNA is a particularly good "molecular clock" because it performs the same function in all organisms and is essential under all physiological conditions. Therefore, sequences have altered with time at a slow enough pace to allow the comparison of genera which have little overall DNA homology.

Figure 2 is a diagram of the proposed universal phylogenetic tree of bacteria, and should be studied in conjunction with Table I. It is obvious at first glance

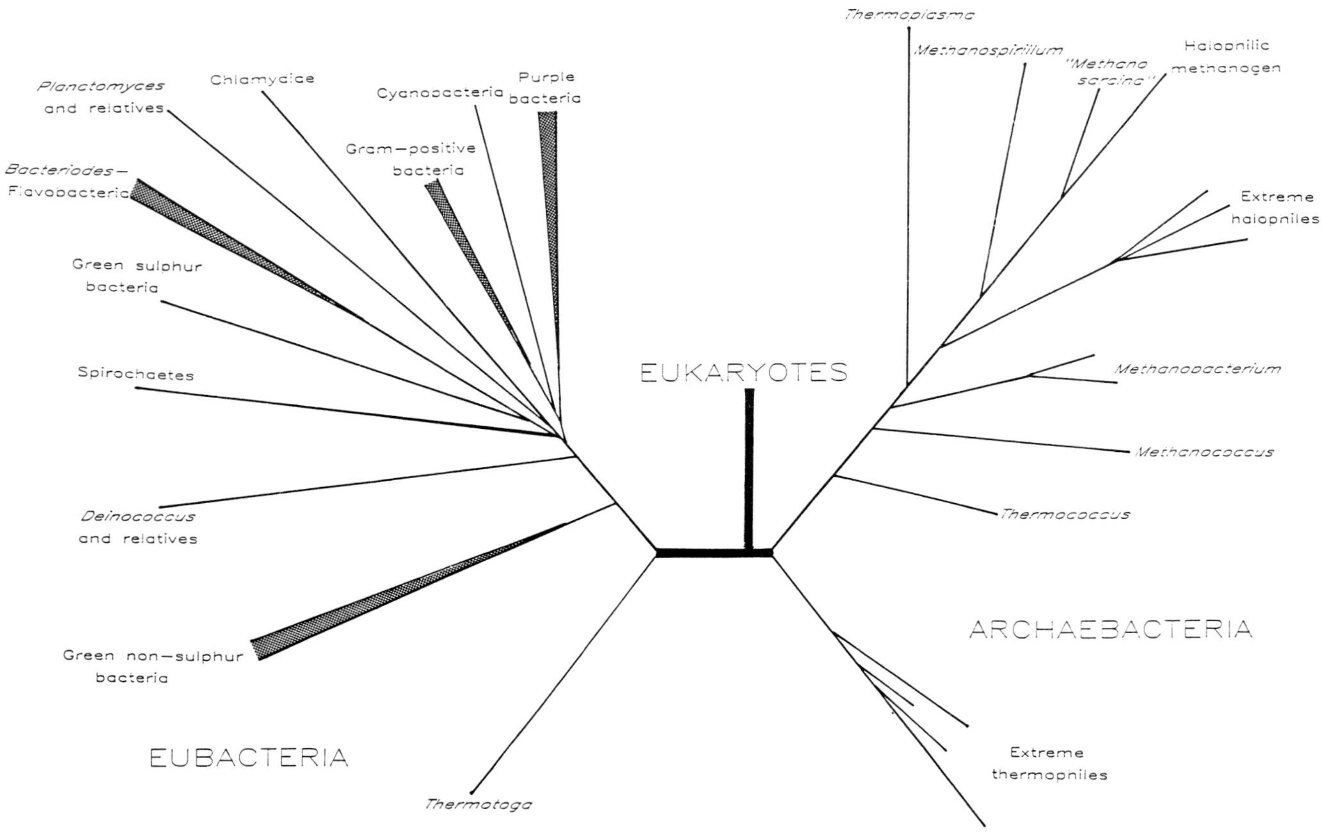

EUKARYOTES
Thermoplasma
Methanospirillum
"Methano sarcina"
Halophilic methanogen
Extreme halophiles
Methanobacterium
Methanococcus
Thermococcus
ARCHAEBACTERIA
Extreme thermophiles
Planctomyces and relatives
Chlamydice
Cyanobacteria
Purple bacteria
Gram—positive bacteria
Bacteriodes—Flavobacteria
Green sulphur bacteria
Spirochaetes
Deinococcus and relatives
Green non—sulphur bacteria
EUBACTERIA
Thermotoga

Figure 2.

Table I

The eubacteria and archaebacteria phyla and their major subdivisions. Taken from Tables 2 and 14 of Woese (12) by permission of the author with additions from ref. 29

Eubacteria and their subdivisions

Purple bacteria
- α subdivision
  - Purple non-sulphur bacteria (*Rhodobacter, Rhodopseudomonas*) rhizobacteria, agrobacteria, rickettsiae, *Nitrobacter, Thiobacillus* (some), *Azospirillum, Caulobacter*
- β subdivision
  - *Rhodocyclus* (some), *Thiobacillus* (some), *Alcaligenes, Bordetella, Spirillum, Nitrosovibrio, Neisseria*
- γ subdivision
  - Enterics (*Acinetobacter, Erwinia, Escherichia, Klebsiella, Salmonella, Serratia, Shigella, Yersinia*), vibrios, fluorescent pseudomonads, purple sulphur bacteria, *Legionella* (some), *Azotobacter, Beggiatoa, Thiobacillus* (some), *Photobacterium, Xanthomonas*
- δ subdivision
  - Sulphur and sulphate reducers (*Desulphovibrio*), myxobacteria, bdellovibrios

Gram-positive eubacteria
- A. High (G + C) species
  - *Actinomyces, Streptomyces, Actinoplanes, Arthrobacter, Micrococcus, Bifidobacterium, Frankia, Mycobacterium, Corynebacterium*
- B. Low (G + C) species
  - *Clostridium, Bacillus, Staphylococcus, Streptococcus*, mycoplasmas, lactic acid bacteria
- C. Photosynthetic species
  - *Heliobacterium*
- D. Species with Gram-negative walls
  - *Megasphaera, Sporomusa*

Cyanobacteria and chloroplasts
- *Oscillatoria, Nostoc, Synechococcus, Prochloron, Anabaena, Anacystis, Calothrix*

Spirochaetes and relatives
- A. Spirochaetes
  - *Spirochaeta, Treponema, Borrelia*
- B. Leptospiras
  - *Leptospira, Leptonema*

Green sulphur bacteria
- *Chlorobium, Chloroherpeton*

Bacteroides, flavobacteria and relatives
- A. Bacteroides group
  - *Bacteroides, Fusobacterium*
- B. Flavobacterium group
  - *Flavobacterium, Cytophaga, Saprospira, Flexibacter*

continued

Figure 2. Universal phylogenetic tree of the bacteria. Line lengths are proportional to evolutionary distances except that the thick black lines represent five times the distance of the thin lines. The figure is compiled from Figs. 4, 11 and 13 of Woese (12) with permission of the author and this reference should be consulted for details of the method of construction of the trees. This figure should be consulted in conjunction with Table I.

Table I
Continued

Planctomyces and relatives
- A. Planctomyces group
  Planctomyces, Pasteuria
- B. Thermophiles
  Isocystis pallida

Chlamydiae
Chlamydia psittaci, C. trachomatis

Radio-resistant micrococci and relatives
- A. Deinococcus group
  Deinococcus radiodurans
- B. Thermophiles
  Thermus aquaticus

Green nonsulphur bacteria and relatives
- A. Chloroflexus group
  Chloroflexus, Herpetosiphon
- B. Thermomicrobium group
  Thermomicrobium roseum

Archaebacterial subdivisions

Extreme halophiles
Halobacterium, Halococcus morrhuae

Methanobacter group
Methanobacterium, Methanobrevibacter, Methanosphaera stadtmaniae, Methanothermus fervidus

Methanococcus group
Methanococcus

"Methanosarcina" group
Methanosarcina barkeri, Methanococcoides methylutens, Methanothrix soehngenii

Methanospirillum group
Methanospirillum hungatei, Methanomicrobium, Methanogenium, Methanoplanus limicola

Thermoplasma group
Thermoplasma acidophilum

Thermococcus group
Thermococcus celer

Extreme thermophiles
Sulfolobus, Thermoproteus tenax, Desulfurococcus mobilis, Pyrodictium occultum

that the bacteria which have received most attention, and indeed to which most of this book is dedicated, namely the Gram-positive and the purple bacteria, form only a narrow part of the phylogenetic tree. It should, however, be noted that some branches of the tree contain very few members.

B. Genetic Exchange Between the Branches of the Tree

It turns out that the bacteria used to study exchange of genetic material between genera (Chapter 2), in the form of plasmids or transposons, are in fact relatively closely related. The ability of the Inc P group plasmids to transfer to a large number of bacteria of very different properties and appearance suggests that all Gram-negative bacteria are tightly affiliated. However, the bacteria in question all fall into the purple α, β and γ groups. When RP4 was transferred to a δ group bacterium (*Myxococcus xanthus*), it could only be maintained if it was inserted into the chromosome (14). The recently identified plasmids that appear to be able to replicate autonomously in both purple Gram-negative and Gram-positive organisms (15), and transposons that can transpose in both bacterial groups (16), may be better examples of promiscuous genetic elements. Plasmids have been identified in bacteria from many other branches of the eubacterial and archaebacterial kingdoms, but their genetic mobility is still essentially unstudied.

The construction of a phylogenetic tree gives rise to some interesting taxonomic problems (Chapter 20). For instance: What is a genus and what is a species in the bacterial world? If we ignore plasmids, phages and transposons, which have evolved to cross taxonomic groups as mentioned above, the ability to exchange and integrate chromosomal DNA via homologous recombination might still be used as a definition of species (albeit one with severe experimental restrictions). Curiously, this ability differs in different branches of the tree. Chromosomal DNA can be exchanged and assimilated by recombination, given enough selective pressure, between what are classed as different genera in the enterics, e.g. *Salmonella* and *Escherichia*. In contrast, within the genus *Streptomyces*, where species separation is very shallow using numerical taxonomic assignment, some species clusters contain strains that show little DNA homology with each other (17). This paradox is yet to be resolved.

C. The Archaebacteria

The identification of the archaebacteria as a separate kingdom in 1977 (12) stimulated a great deal of interest in these otherwise quite disparate bacteria.

Their possession of many unusual features—lipids not previously seen in the bacterial world, and unusual cofactors like cofactor M—was not perhaps surprising considering their unusual life styles (see above). However, the discovery of singular features of the transcriptional/translational machinery of these organisms emphasized their separation from the eubacterial kingdom. A particular singularity was the discovery of introns in *Sulfolobus solfataricus* and *Halobacterium volcanii* (18). However, the bizarre nature of this discovery was mitigated by the discovery of an intron in the coliphage T4 (19). Attempts to develop genetic analysis of both the *Halobacteriaceae* and the *Methanobacteriaceae* have begun. Recent papers have reported bacteriophage transfection and plasmid transformation of *Halobacterium* spp. and low level transformation of chromosomal markers in *Methanococcus voltae*. In addition, Mevarech and Werczberger have reported a natural system of genetic exchange in *H. volcanii* (20).

IV. REINVENTION THROUGHOUT THE PHYLOGENETIC TREE

One problem with developing a phylogeny of bacteria is to take account of the possibility that DNA has been exchanged between bacteria from different genera and different branches of the tree. As discussed above, plasmids and transposons can potentially be exchanged between different branches. Drug resistance genes may have been exchanged between the producers of antibiotics, mostly Gram-positive bacteria, and the purple group of bacteria (though the number of examples on which this rests has recently been shown to be smaller than previously thought; 21). The following discussion is intended to highlight areas where either reinvention or gene transfer may have taken place, based on our present knowledge of bacterial phylogeny.

A. Endospores Have Probably Been Invented Only Once, but Exospores Several Times

Many bacteria produce resting structures and/or specialized dispersal stages. It has now become clear that endospores (Chapter 11) are produced by a phylogenetically closely knit group, namely the low (G + C) Gram-positive bacteria that includes *Thermoactinomyces*, a genus once classified with the Actinomycetes.

Exospore formation or cyst formation—the terminology is confused in different systems—appears to have been reinvented many times. The cyanobacteria, purple bacteria and Gram-positives all contain a variety of examples. The *Myxobacteriales* (purple bacteria) and the *Streptomycetaceae*

(Gram-positive, high (G + C) DNA) inhabit similar ecological niches: both secrete extracellular enzymes that digest insoluble polymeric carbon and nitrogen sources in the soil (see above). It is an advantage in such cases to maintain a high cell density to promote efficient substrate solubilization. The two groups of bacteria have solved the same problems in two different ways. The actinomycetes (like the lower fungi) form a complex network of interlaced cells—a substrate mycelium: an immobile life-style that necessitates a dispersal stage, the production of somewhat resistant exospores (22).

The myxobacterial vegetative cells are fully motile, but cell coordination has given rise to complex behaviours. Swarms of cells behave as "wolf packs" which overrun and digest bacteria and other prey. When food runs out, the swarm cooperates in the formation of a supracellular fruiting body within which resistant exospores are developed (Chapter 12). (In some myxobacteria, the fruiting body is elegantly sculptured into a multilobed structure consisting of sporangioles. This life-style means that two dispersal stages are possible: vegetative cells can glide away; and the sporangioles, each containing a potential swarm, can also be dispersed.) It is clear from the different pathways of development of actinomycete and myxobacterial spores that this is an example of evolutionary reinvention.

B. Multiple Invention of Motile Dispersal Stages and Motility Mechanisms

The evolution of motile dispersal cells has occurred a number of times, presumably independently. Examples include: the swarmer cells of *Caulobacter* (Chapter 10) and the Rhodospirillaceae; the hormogonia of cyanobacteria (23); and motile spores of *Actinoplanes* and related genera. In each case, there is a non-motile "growth" phase (stalk cell, filament and substrate mycelium respectively) from which the motile cells can disperse. It appears safe to suggest that these are independent reinventions considering their occurrence in the phylogenetic tree and the observation that one type—hormogonia—move not by swimming but by gliding (see below). The reader is recommended to read the review by Dow *et al.* (24) about swarmers as metabolic shutdown cells. It would be interesting to see if the *Actinoplanes* spores and cyanobacterial hormogonia are metabolically similar to the swarmer cells.

In addition to gliding and swimming, other forms of motility such as twitching, sliding, darting and swarming, have also been described (25). Here I discuss only swimming and gliding. Examination of Table II will show the intermingling of the two types of motility system. Reinvention or gene transfer must have occurred a number of times. The discovery of the singular anomalies of "gliding" by mycoplasmas (26) and "swimming" by a cyanobacterium (27)—although in the latter case, no flagella could be found—will require

Table II

Occurrence of gliding and swimming throughout the eubacteria. Data taken from Woese (12), Stackebrandt (13) and Bergey's Manual (41)

Branch of phylogenetic tree		Swimming genera	Gliding genera
Purple bacteria	α	More than 10	None identified
	β	More than 10	*Simonsiella* *Vitreoscilla*
	γ	More than 20	*Alysiella* *Beggiatoa* *Leucothrix*
	δ	*Bdellovibrio* *Desulfovibrio* and related sulphur/sulphate reducers	*Cystobacter* *Myxococcus* *Nannocystis* *Sorangium* *Stigmatella*
Gram-positive bacteria	High (G + C)	*Dactylosporangium*, *Ampullariella*, *Actinoplanes*, some *Arthrobacter*	None identified
	Low (G + C)	More than 10 including *Bacillus*, *Chlostridium* and *Spiroplasma* (No flagella observed)	*Mycoplasma*
Cyanobacteria		*Synechococcus* sp. C	*Synechococcus* (*Agmenellum*, *Anacystis* and *Synechococcus*) *Fisherella*, *Oscillatoria* (*Spirulina*)
Green sulphur bacteria		*Chlorobium*	*Chloroherpeton*
Bacteriodes and relatives		*Bacteriodes* *Fusobacterium* *Flavobacterium*	*Saprospira* *Haliscomenobacter* *Flexibacter* *Sporocytophaga* *Cytophaga* Anaerobic "*Bacteroides*"-like glider
Green nonsulphur bacteria		None identified	*Chloroflexus* *Herpetosiphon*

more study before strict comparison can be made to "true" gliders or swimmers.

Flagellum-driven swimming has received a great deal of attention (Chapter 10) and clearly many genes are involved in synthesis and control. Recent research has demonstrated an alternative RNA polymerase sigma subunit mechanism in the control of flagellar and chemotaxis genes in *Bacillus subtilis* and possibly in *E. coli* (28). Gliding has been subjected to genetic analysis in *M. xanthus*, in which two systems were found; one involved in movement of individual cells (the "adventurous" system) and the other in movement of groups of cells (the "social" system). Again many genes are necessary for each system. The possibility that gliding motility has been reinvented a number of times may indicate that the hope for a general mechanism for gliding is a forlorn one. The presence of flagella on some Halobacteriaceae and Methanobacteriaceae implies that flagellum-driven swimming occurred very early in evolution or has been invented more than once.

C. Sulphur Oxidation and Reduction, Unlike Nitrate Oxidation, Appear to Have Been Reinvented Several Times

Nitrate oxidation is apparently limited to the α division of the purple bacteria, whilst ammonia oxidation is generally limited to the β division with one example (*Nitrosococcus*) fitting into the γ division. However, sulphur oxidation, a property that has been used to define a "genus", *Thiobacillus*, is spread throughout the α, β and γ divisions of the purple bacteria (29) and is also found in the archaebacterial kingdom (*Sulfolobus*). Discovery of such a spread of *Thiobacillus* spp. within the phylogenetic tree was not perhaps surprising because the biochemistry of sulphur oxidation appears to be very different in different *Thiobacillus* species (30). A similar story of reinvention appears to apply to sulphate/sulphur reduction. The eubacterial sulphur/sulphate reducers belong to the δ group of the purple bacteria. Again the archaebacterial kingdom contains representation of the same ecological niche: the anaerobic extreme thermophiles are all sulphur-dependent. Interestingly, one species of *Sulfolobus*, *S. acidianus*, can both reduce and oxidize sulphur.

D. Pathogenicity is Particularly Associated With Certain Disparate Groups of Bacteria

The α subdivision of the purple bacteria contains *Rhizobium*, a genus of bacteria capable of invading plant cells and *Rochalimaea* (Rickettsiaceae), a genus capable of invading animal cells. The same subdivision includes the agrobacteria which genetically colonize plant cells. Similar clustering of plant and animal pathogens is seen in the *Enterobacteriaceae*, with *Shigella*, *Yersinia*, *Salmonella* and *Erwinia* (although *Erwinia* spp. are not intracellular pathogens), and the Gram-positive coryneform bacteria.

The recent molecular biological studies of Isberg and colleagues (31) on *Yersinia* have illustrated, perhaps surprisingly, how little additional genetic information, in this case for a single protein of 103 000 daltons, *E. coli* needs to be able to invade an animal cell. Similarly, *E. coli* containing a *Rhizobium nod* operon could induce pseudonodule formation on plant roots, although no bacteria were found in the nodule (32). The same genes in *Agrobacterium* caused the host to induce infection thread formation, but the process of intracellular invasion was aborted before the bacteria were individually budded off within the cell (Chapter 18).

V. SOME UNEXPECTED ATTRIBUTES OF BACTERIA

A. Bacterial Colonies Can Have Very Ordered Structures

Work by Shapiro (33) has indicated that the colonies of apparently simple bacteria like *E. coli* and *Pseudomonas putida* are very structured. He showed, using both the scanning electron microscope and Mu *dlac* transposons with X-gal as a histochemical stain for β-galactosidase, that a colony is composed of localized cell types. The patterns of X-gal staining reflected both clonal features resulting from inherited genetic changes in sectors of the colony, and non-clonal features resulting from cell sorting within the colony. The analysis of colonies resulting from mixed suspensions of cells with transposons inserted in different parts of the genome implied that different cell types were indeed sorted.

At least part of the mechanism of the ordering of bacterial colonies has been uncovered in the highly-ordered colonies of some of the *Actinomycetales* and the *Myxobacteriales* discussed above. Pheromones which are involved in the induction of resting stage formation have been identified in members of these two families and in the cyanobacterium *Cylindrospermum licheniforme* (34).

As well as these examples of "whispers" within a colony, "shouts" between genetically dissimilar strains of the same species have also been heard. The peptide sex pheromones of the streptococci are a particularly good example (35). Pheromone-based interactions between bacteria and plants have recently been uncovered in *Agrobacterium* where acetosyringone activates the virulence genes (see Chapter 18); and in *Rhizobium* spp. where various flavonoids act as inducers and anti-inducers of the *nod* genes (Chapter 19).

B. Not All Cells Within a Colony are Immortal

There are many examples in the bacterial world of the ultimate altruism: some cells die so that others can live. This is perhaps surprising in the bacterial world where usually all cells are totipotent. In the myxobacteria, the majority of the cells that go into a fruiting body lyse and only a minority become spores. It has been presumed that this behaviour supplies nutrients to allow the spores to complete development. A similar phenomenon is observed in *Streptomyces* and other colonial *Actinomycetales* and the mycelial fungi. Here, substrate mycelium lyses and the aerial mycelium grows saprophytically on these lysis products. This has lead to the proposal (Chapter 7) that antibiotics, which are synthesized concurrently with sporulation in all three above examples, are a means of stopping competitors attempting to exploit this resource. This principle of autolysis of "feeder cells" may also apply to the endospore-forming bacteria, in which the mother cell is also sacrificed to the formation of the spore, and again antibiotics are formed. (Alternative generalized roles for antibiotics may be spore germination inhibition, to ensure staggering of germination within a population of spores, and capture and lysis of prey bacteria and fungi and, indeed, perhaps lysis of the feeder cells (36).)

Another example of altruism in *Streptomyces* is seen in the "phage growth limitation" (Pgl^+) phenotype of *Strep. coelicolor* A3(2). When actinophage ϕC31 invades a Pgl^+ cell, it causes cell lysis. However, the resultant phages are modified in some way such that they can no longer infect Pgl^+ cells. Hence the sacrifice of one cell allows protection of all the others in the mycelium. This mechanism is a clear improvement over classical restriction/modification as no protected phage can escape through the net even at a low level (21). A similar "scorched earth" policy (the hypersensitive response) is often seen in plant responses to pathogen invasion.

Nitrogen fixation provides two further examples of altruism within a population. *Rhizobium* bacteroids that fix nitrogen in legume root nodules are all destined to lyse when the nodules senesce; bacteroid formation is irreversible (Chapter 19). However, as many as 10^7 rhizobia may be obtained from a nodule. Therefore a large number of "free loaders" must be present which do not differentiate, and which exploit the nodule as a source of food and shelter. The heterocysts of filamentous cyanobacteria are another site of nitrogen fixation fated to senesce. Synthesis of the nitrogenase enzyme in heterocysts of some strains depends on at least two irreversible gene rearrangements (37) (Chapter 6). A close analogy to this differentiation-specific genome change is presented in the DNA rearrangements involved in antibody formation in mammalian B cells. No evidence has yet been presented to demonstrate that irreversible DNA rearrangements are significantly involved in any of the other examples of programmed death discussed above.

C. Not All Bacteria in a Colony are Genetically Identical

The last example in the previous section illustrated that a subpopulation of cells (heterocysts) are genetically different from the other cells. This observation is not unique in the bacterial kingdoms. Sapienza and Doolittle (38) demonstrated the extremely unstable nature of the chromosomes of the archaebacterial genus *Halobacterium*. They identified a great number of repeated sequences and transposons that were constantly rearranging. It was calculated that siblings of a division have only 80% chance of being genetically identical.

Fluidity of the genome, but by no means as extreme, has been demonstrated in the streptomycetes. Several examples of DNA sequences that can become massively reiterated have been identified. The process of reiteration is often accompanied by extensive deletion and spontaneous loss of the reiteration is also observed (21). This may in part explain the frequent observation of chronic instability of some streptomycete phenotypes.

Somewhat lower degrees of genomic instability are brought about by the movement of various transposable elements and by site-specific recombination events (39).

D. Some Bacterial Cell Walls are not Rigid Exoskeletons

Cytophaga and *Flexibacter* move not by swimming but by gliding (see above). Whilst investigating the mechanism of gliding using these organisms, Lapidus and Berg (40) reported that latex beads could attach to the cell surfaces and were then seen to track backwards and forwards over the cell. Figure 3 indicates diagrammatically what was observed using a video-microscope with two beads attached to a cell. It became obvious that more than one track of movement was possible. The nature of the tracks and the points of attachment of the beads are not known. However, with the advances in knowledge of the genetics of myxococcal gliding, the mechanism may be within our grasp (but note that the mechanisms in *Myxococcus* and *Cytophaga* may be different, as discussed above).

V. CONCLUSION

This chapter was conceived without particular reference to the specific contents of the rest of the book: yet all the systems outlined in the following chapters have found more or less prominent places here. Nevertheless, many of the subjects that I have touched on do not receive further attention in later chapters. This is partly a reflection of the (not unreasonably) anthropocentric

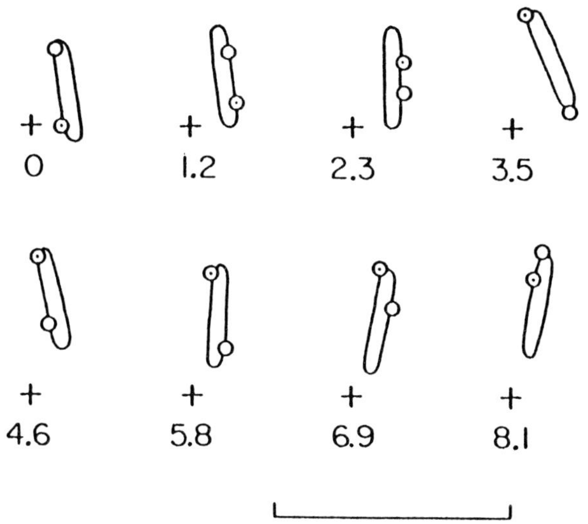

Figure 3. Some bacterial cell walls are not rigid exoskeletons.

Two 0.26 μm diameter polystyrene latex spheres moving on the surface of a *Cytophaga* sp. strain U67 cell, 4.4 μm long, gliding on the top of a slide. Selected frames of a frame-by-frame analysis spanning a time of 8.1 s are shown. The cross is fixed in the frame of reference of the microscope. The numbers are the elapsed time in seconds. The first sphere (marked with a dot) was moving downward at time 0. It looped around the lower pole, travelled up the right side of the cell, crossed upward from right to left over the dorsal surface of the cell, and stopped (at 3.5 s): later it started moving again (at 7.9 s) and moved down the left side of the cell. The second sphere also was moving downward at time 0. It crossed downward from left to right over the dorsal surface of the cell, moved down the right side, passed the first sphere (at 1.9 s), looped around the lower pole, travelled up the left side of the cell, backed up (at 4.6 s), looped around the lower pole once again, travelled up the right side of the cell, and finally looped around the upper pole. In subsequent frames (not shown), both spheres moved down the left side of the cell at a speed of about 1.6 μm s^{-1}. The cell glided upward between 0 and 1.2 s, stopped between 1.2 and 2.3 s, continued to glide upward between 2.3 and 5.2 s, glided downward between 5.2 and 6.9 s and finally glided upward again between 6.9 and 8.1 s. Later on (not shown), it continued to glide upward. The speed of the glide varied from about 1 to 2 μm s^{-1}. Note that the times at which the spheres and the cell stopped or changed direction were not the same. Scale bar = 10 μm.

Reproduced from Lapidus and Berg (40) by permission of the authors.

view of most microbiologists, and partly a reflection of the relative experimental approachability of the different subjects. A book of this kind could not have been written at all a decade ago. I expect that in less than 10 years from now, it will be possible to write a comparable book on the more "difficult" subjects, as the remarkable advances in molecular genetic techniques continue to open up new, almost undreamed of, experimental possibilities. It is to be hoped that the crucial ground of *E. coli* research will not be altogether abandoned for these exciting new territories.

Acknowledgements

I thank the following for their generous help: N. G. Carr, D. R. Gill, J. C. D. Hinton, D. P. Kelly, B. E. B. Moseley, P. R. Norris, G. P. C. Salmond, E. M. H. Wellington and M. Wyman. I thank Mrs. Len Schofield for typing the manuscript.

References

(1) Häder, D.-P. (1987). Photosensory behaviour in prokaryotes. *Microbiol. Rev.* **51**, 1–21.

(2) Hornemann, U. (1981). Nitrogen–nitrogen bond containing antibiotics: biosynthesis of streptozocin. *In* "Antibiotics IV—Biosynthesis" (J. W. Corcoran, ed.), pp. 313–324. Springer-Verlag, New York.

(3) Wolfe, S., Demain, A. L., Jensen, S. E. and Westlake, D. W. S. (1984). Enzymatic approach to syntheses of unnatural Beta-lactams. *Science* **226**, 1386–1392.

(4) Pugsley, A. P. and Schwartz, M. (1985). Export and secretion of proteins by bacteria. *FEMS Microbiol. Rev.* **32**, 3–38.

(5) Ji, J., Hugouvieux-Cotte-Pattat, N. and Robert-Baudouy, J. (1987). Use of Mu-*lac* insertions to study the secretion of pectate lyases by *Erwinia chrysanthemi*. *J. Gen. Microbiol.* **133**, 793–802.

(6) Young, R. A., Mehra, V., Sweetser, D., Buchanan, T., Clark-Curtiss, J., Davis, R. W. and Bloom, B. R. (1985). Genes for the major protein antigens of the leprosy parasite *Mycobacterium leprae*. *Nature* **316**, 450–452.

(7) Kieser, T., Moss, M. T., Dale, J. W. and Hopwood, D. A. (1986). Cloning and expression of *Mycobacterium bovis* BCG DNA in "*Streptomyces lividans*". *J. Bacteriol.* **168**, 72–80.

(8) Jannasch, H. W. (1984). Microbes in the oceanic environment. *In* "Symposium of the Society for General Microbiology, Vol. 36, Part II, The Microbe 1984. II. Prokaryotes and Eukaryotes" (D. P. Kelly and N. G. Carr, eds.), pp. 97–122. Cambridge University Press, Cambridge.

(9) Margulis, L. (1981). "Symbiosis in Cell Evolution". W. H. Freeman and Co., San Francisco.

(10) White, R. H. (1984). Hydrolytic stability of biomolecules at high temperatures and its implication for life at 250°C, *Nature* **310**, 430–432.

(11) Moseley, B. E. B. (1982). Photobiology and radiobiology of *Micrococcus* (*Deinococcus*) *radiodurans*. *Photochem. and Photobiol. Rev.* **7**, 223–274.

(12) Woese, C. R. (1987). Bacterial evolution. *Microbiol. Rev.* **51**, 221–271.

(13) Stackebrandt, E. (1985). Phylogeny and phylogenetic classification of prokaryotes. *In* "Evolution of Prokaryotes" (K. H. Schleifer and E. Stackebrandt, eds.), pp. 309–334, Academic Press, London.

(14) Breton, A. P., Jaoua, S. and Guespin-Michel, J. (1985). Transfer of plasmid RP4 to *Myxococcus xanthus* and evidence for its integration into the chromosome. *J. Bacteriol.* **161**, 523–528.

(15) Lacks, S. A., Lopez, P., Greenberg, B. and Espinosa, M. (1986). Identification and analysis of genes for tetracycline resistance and replication functions in the broad-host-range plasmid pLS1. *J. Mol. Biol.* **192**, 753–765.

(16) Lereclus, D., Mahillon, J., Menou, G. and Lecadet, M.-M. (1986). Identification of Tn*4430*, a transposon of *Bacillus thuringiensis* functional in *Escherichia coli*. *Mol. Gen. Genet.* **204**, 52–57.
(17) Mordarski, M., Goodfellow, M., Williams, S. T. and Sneath, P. H. A. (1985). Evaluation of species groups in the genus *Streptomyces*. *In* "Sixth Int. Symp. on Actinomycetes Biology" (G. Szabó, S. Biró and M. Goodfellow, eds.), pp. 517–525. Akadémiai Kiadó, Budapest.
(18) Daniels, C. J., Gupta, R. and Doolittle, W. F. (1985). Transcription and excision of a large intron in the $tRNA^{Trp}$ gene of an archaebacterium, *Halobacterium volcanii*. *J. Biol. Chem.* **260**, 3132–3134.
(19) Hall, D. H., Povinelli, C. M., Ehrenman, K., Pedersen-Lane, J., Chau, F. and Belfort, M. (1987). Two domains for splicing in the intron of the phage T4 thymidylate synthase (*td*) gene established by non-directed mutagenesis. *Cell* **48**, 63–71.
(20) Charlebois, R. L., Lam, W. L., Cline, S. W. and Doolittle, W. F. (1987). Characterization of pHV2 from *Halobacterium volcanii* and its use in demonstrating transformation of an archaebacterium. *Proc. Acad. Natl. Sci. USA* **84**, 8530–8534.
(21) Chater, K. F., Henderson, D. J., Bibb, M. J. and Hopwood, D. A. (1988). Genome flux in *Streptomyces coelicolor* and other streptomycetes and its possible relevance to the evolution of mobile antibiotic resistance determinants. *In* ref. (39) pp. 7–42.
(22) Chater, K. F. and Merrick, M. J. (1979). Streptomycetes. *In* "Developmental Biology of Prokaryotes" (J. H. Parish, ed.) pp. 93–114. Blackwell Scientific Publications, Oxford.
(23) Castenholz, R. W. (1982). Motility and taxis. *In* "The Biology of Cyanobacteria" (N. G. Carr and B. A. Whitton, eds.), pp. 413–439. University of California Press, Berkeley and Los Angeles.
(24) Dow, C. S., Whittenbury, R. and Carr, N. G. (1983). The 'shut down' or 'growth precursor' cell—an adaptation for survival in a potentially hostile environment. *In* "Symposium of the Society for General Microbiology, Vol. 34. Microbes in Their Natural Environments" (J. H. Slater, R. Whittenbury and J. W. T. Wimpenny, eds.), pp. 187–247. Cambridge University Press, Cambridge.
(25) Henrichsen, J. (1972). Bacterial surface translocation: a survey and a classification. *Bacteriol. Rev.* **36**, 478–503.
(26) Piper, B., Rosengarten, R. and Kirchhoff, H. (1987). The influence of various substances on the gliding motility of *Mycoplasma mobile* 163K. *J. Gen. Microbiol.* **133**, 3193–3198.
(27) Waterbury, J. B., Willey, J. M., Franks, D. G., Valois, F. W. and Watson, S. W. (1985). A cyanobacterium capable of swimming motility. *Science* **230**, 74–76.
(28) Helmann, J. D. and Chamberlin, M. J. (1987). DNA sequence analysis suggests that expression of flagellin and chemotaxis genes in *Escherichia coli* and *Salmonella typhimurium* is controlled by an alternative σ factor. *Proc. Natl. Acad. Sci. USA* **84**, 6422–6424.
(29) Lane, D. J., Stahl, D. A., Olsen, G. J., Heller, D. J. and Pace, N. R. (1985). Phylogenetic analysis of the genera *Thiobacillus* and *Thiomicrospira* by 5S rRNA sequences. *J. Bacteriol.* **163**, 75–81.
(30) Kelly, D. P. (1988). Oxidation of sulphur compounds. *In* "Symposium of the Society for General Microbiology, Vol. 42, The Nitrogen and Sulphur Cycles" (J. A. Cole and S. J. Ferguson, eds.), pp. 65–98. Cambridge University Press, Cambridge.
(31) Isberg, R. R., Voorhis, D. L. and Falkow, S. (1987). Identification of invasin: a protein that allows enteric bacteria to penetrate cultured mammalian cells. *Cell* **50**, 769–778.
(32) Hirsch, A. M., Wilson, K. J., Jones, J. D. G., Bang, M., Walker, V. V. and Ausubel, F. M. (1984). *Rhizobium meliloti* nodulation genes allow *Agrobacterium tumefaciens* and *Escherichia coli* to form pseudonodules on alfalfa. *J. Bacteriol.* **158**, 1133–1143.
(33) Shapiro, J. (1987). Organization of developing *Escherichia coli* colonies viewed by scanning electron microscopy. *J. Bacteriol.* **169**, 142–156.
(34) Stephens, K. (1986). Pheromones among the prokaryotes. *CRC Critical Reviews in Microbiology*, **13**, 309–334.

(35) Clewell, D. B., White, B. A., Ike, Y. and An, F. Y. (1984). Sex pheromones and plasmid transfer in *Streptococcus faecalis*. *In* "Microbial Development" (R. Losick and L. Shapiro, eds.), pp. 133–149. Cold Spring Harbor Laboratory, Cold Spring Harbor, New York.
(36) Varon, M., Tietz, A. and Rosenberg, E. (1986). *Myxococcus xanthus* autocide AMI. *J. Bacteriol.* **167**, 356–361.
(37) Haselkorn, R., Golden, J. W., Lammers, P. J. and Mulligan, M. E. (1986). Developmental rearrangement of cyanobacterial nitrogen-fixation genes. *Trends in Genetics* **2**, 255–259.
(38) Doolittle, W. F. (1985). Genome structure in archaebacteria. *In* "The Bacteria. A Treatise on Structure and Function, Vol. 8. Archaebacteria" (C. R. Woese and R. S. Wolfe, eds.), pp. 545–560. Academic Press, Inc., Orlando.
(39) Kingsman, A. J., Chater, K. F. and Kingsman, S. M. (1988) Eds. "Symposium of the Society for General Microbiology, Vol. 43. Transposition". Cambridge University Press, Cambridge.
(40) Lapidus, I. R. and Berg, H. C. (1982). Gliding motility of *Cytophaga* sp strain U67. *J. Bacteriol.* **151**, 384–398.
(41) Holt, J. G. (1984–1986). Bergey's Manual of Systematic Bacteriology, 9th Edition, Vols. 1 and 2. Williams and Wilkins, Baltimore. Buchanan, R. E. and Gibbons, N. E. (1974). Bergey's Manual of Systematic Bacteriology, 8th Edition. Williams and Wilkins, Baltimore.

Chapter **2**

Diversity of Bacterial Genetics

K. F. CHATER *and* D. A. HOPWOOD

I. INTRODUCTION

Many special attributes of a variety of bacteria are now being illuminated by the power of genetics, and most of this book is devoted to descriptions of such studies. In this chapter, we survey the genetics of bacteria from a different viewpoint and ask to what extent fundamental genetic processes, including genome structure and organization, gene exchange and the regulation of gene expression, are similar in all bacteria. We assume some familiarity with *Escherichia coli* genetics and look beyond it to reveal many ways in which non-*E. coli* bacteria are extending the rich tapestry of prokaryotic molecular biology.

II. THE PROKARYOTIC GENOME

A. How Big are the Genomes of Bacteria?

Estimates of genome size, usually obtained from renaturation kinetics of sheared DNA, range from ca. 600 kb in *Chlamydia trachomatis*, a nutritionally demanding obligate parasite, to ca. 13 000 kb for the cyanobacterium *Calothrix* (1). However, most bacterial genomes probably fall within a factor of two of the size of the *E. coli* chromosome, most accurately determined as 4500 kb with the emergence of the technique of pulsed field gel electrophoresis for separating very large DNA fragments (2).

B. Gross Genome Organization and its Slow Change During Prokaryote Evolution

In those bacteria in which enough genetic mapping has been done, all the genes needed for normal growth and development are linked to each other, implying a single chromosome. Such functions include: uptake and assimilation of commonly-available nutrients; intermediary metabolism; essential macromolecular synthesis; cell maintenance and repair; recombination; cell division; and, where relevant, motility and sporulation. Many bacterial linkage maps are circular and it is generally assumed that chromosome circularity is characteristic of bacteria.

Satisfactory linkage mapping is only possible if the arrangement of the genes is nearly stable. (Physical analysis of the chromosomes of members of the archaebacterial genus *Halobacterium* has revealed an extraordinary degree of instability (Chapter 1) but these chromosomes have not been mapped genetically.) In *Streptomyces*, there is an apparent paradox: especially after exposure to stressful conditions, many strains frequently suffer a variety of huge deletions, of as much as 900 kb in *Streptomyces glaucescen*: (R. Hütter, personal communication) and massive tandem amplification of certain segments of chromosomal DNA, which can make up 10–50% of the total DNA (3); yet reliable linkage maps have been constructed in some of these strains. The paradox is explained by the fact that the rearrangements usually lead to obvious phenotypic deficiencies in properties such as sporulation, production of secondary metabolites or extracellular enzymes, and (perhaps surprisingly) arginine biosynthesis (Chapter 7). Colonies showing such properties are probably at a disadvantage in nature, and are usually selected out in the laboratory by the regular isolation of "healthy" single colonies.

Most bacterial chromosomes appear to have been rather stable over many millions of years. Closely similar linkage maps for different species within genera have been found for *Rhizobium* (4), *Bacillus* (5) and *Streptomyces* (6). Amongst pseudomonads, fundamentally similar linkage maps differ in the accretion of genes for additional specialized functions such as catabolism of aromatic hydrocarbons, resistance to antibiotics and bacteriocin production (7). The maps of *E. coli* and *Salmonella typhimurium* are also very similar to each other, even though there have been changes at most "silent" bases in coding regions throughout the chromosomes since the progenitors of these two organisms diverged perhaps 10^8 years ago (8) (Chapter 1). However, the longer periods of evolutionary separation (perhaps 7×10^8 years: 8) between enterobacteria and pseudomonads (both members of the γ group of purple bacteria) and between them and purple bacteria of group α (e.g. *Rhizobium*) ($>10^9$ years) have obliterated discernible linkage map resemblances. The same is true between the low (G + C) Gram-positive *Bacillus subtilis* (9) and the high (G + C) Gram-positive *Streptomyces coelicolor* (6).

C. The Occurrence and Possible Significance of Clustering or Scattering of Gene Sets

Operons of cotranscribed, physiologically-related genes abound in *E. coli* (10), though some of its sets of physiologically-related genes are scattered while still being coregulated (e.g. the *arg* and *cys* genes: 11). (There are also some operons containing genes for apparently unrelated functions; the *rpsU-dnaG-rpoD* operon has genes involved in protein, DNA and RNA synthesis: 10.) In most other groups of bacteria that have been studied, clustering is not so extensive as in *E. coli* for biosynthetic genes such as those for amino acids; *Pseudomonas* shows little or none (7). However, clustered genes for catabolic pathways occur widely: thus, *Streptomyces* operons specify galactose and glycerol utilization (12, 13). There are also many examples of catabolic operons in pseudomonads (7). (Interestingly, the glycerol utilization genes of *E. coli* are scattered (11) while those of *B. subtilis* are clustered but probably not cotranscribed: 14.) With genes for more idiosyncratic functions, clustering and polycistronic transcription are the rule. The following chapters demonstrate this for genes involved in bioluminescence, photosynthesis, nitrogen fixation, antibiotic biosynthesis, aromatic hydrocarbon degradation, metal resistance, pathogenicity and symbiosis. However, genes for morphological differentiation are generally scattered, often as small operons (Chapters 10–12, and Ref. 15).

What determines clustering vs. scattering? Clustered genes could easily be transferred together, facilitating the lateral spread of a specialized property. Cotranscription simplifies production of the subunits of hetero-oligomeric proteins in stoichiometric amounts and coordination of the levels of consecutively acting enzymes in a pathway, as well as minimizing the number of regulatory elements. Scattering allows relative gene dosage to be varied in response to growth rate since, at fast growth rates, there will be more copies of genes close to the origin of chromosomal replication (perhaps accounting for the close linkage of many genes for ribosome biosynthesis with the replication origin in *E. coli* and *B. subtilis*: 9, 11). Separate transcription also permits different regulatory elements to act on individual genes, which would be helpful when biochemical pathways converge, diverge or form parts of networks. Widely differing ecologies, such as those of *E. coli* and *Pseudomonas*, are likely to be best served by differently orchestrated intermediary metabolisms, resulting in selection for differences in the clustering of particular gene sets.

D. Mobile Genetic Elements Occur in Most Prokaryotes

Plasmids, insertion sequences, transposons and phages have all been studied most extensively in *E. coli*. Are the results generally applicable to all prokaryotes?

1. Plasmids are widespread among bacteria. Plasmids, ranging in complexity from a few to hundreds of kilobase pairs, appear to occur throughout bacteria. Many of them are "cryptic", conferring no detectable phenotype, but others contribute significant phenotypic potential (Table I).

The involvement of plasmids in particular phenotypes has become easier to determine: methods for identifying plasmid DNA have improved; transformation of bacteria with plasmid DNA leading to acquisition of a phenotype is often possible; and transposon-induced mutations and cloned genes used as probes allow the physical localization of genes. These procedures are generally supplanting more indirect criteria often previously applied (16), such as: lack of linkage with chromosomal genes, which can be determined only with a fairly extensive background of chromosomal genetics; infectious transfer in matings; and instability, especially following so-called "curing" treatments. "Curing" evidence is particularly unreliable because many plasmids are not eliminated by these treatments, while some chromosomal segments may be deleted at

Table I
Phenotypic properties attributable to bacterial plasmids

Phenotype	Organism	References
Toxin production	Certain (but not all) *V. cholerae* and *E. coli* strains	Chapter 15
Resistance to antibiotics and heavy metals	Purple bacteria and low (G + C) Gram-positive bacteria	17
Bacteriocin production	Purple bacteria	17
UV resistance	*E. coli*	Appendix B in Ref. 123
Catabolism of stable compounds	*Pseudomonas*	Chapter 8
Nodulation and nitrogen fixation	*Rhizobium*	Chapters 6, 19
Induction of plant tumours and hairy roots by transformation of plants	*Agrobacterium tumefaciens* and *A. rhizogenes*	Chapter 18
Galls on plants	*Pseudomonas savastanoi*	124
Adhesins	*E. coli*	Chapter 14
Iron uptake	*E. coli*	Chapter 16
Haemolysin	*S. faecalis*	Appendix B in Ref. 123
Insecticidal toxin production	*B. thuringiensis*	39
Antibiotic production	*S. coelicolor* and *S. violaceus-ruber*	Chapter 7

frequencies that are naturally high and are increased by treatment with "curing" agents.

2. *Novel mechanisms of plasmid transfer occur in Gram-positive bacteria.* An important aspect of plasmids is their mobility. Many plasmids found in a wide range of bacteria specify their own transfer systems, allowing the dispersal of genes that they carry. Transfer of the well-known plasmids of Gram-negative bacteria depends on surface appendages (pili) to establish copulation. Genetic determination of such pili is complex; for example, the assembly of the pili of IncF plasmids requires a dozen genes out of the whole 33 kb transfer region (17). Among Gram-positive bacteria, things seem to be very different. In *Streptomycetes*, the most-studied of the high (G + C) group, transfer functions generally occupy only a few kilobases of plasmid DNA and so quite small plasmids are conjugative (18). In the 9 kb pIJ101, only one gene seems to be directly involved in transfer (19). Perhaps the reason is that the naturally immobile streptomycetes need no specialized surface appendages to stabilize mating pairs. Indeed, the filamentous habit of streptomycetes may have been responsible for the evolution of mechanisms of intramycelial migration of plasmids: it is postulated that the so-called "spread" functions (requiring two further genes on pIJ101: 19) allow the invasion of the recipient mycelium from the point of initial intercellular transfer (18). Among the low (G + C) Gram-positive bacteria also, broader host-range plasmids, such as pAMβ1 and pIP501, transfer well only on solid surfaces and may resemble *Streptomyces*

plasmids in devoting only a few genes to the mating process (20). Streptococci also have some very narrow-host-range plasmids (such as pAD1) that cause the host to respond to an extracellular signal, or "phermone", produced by the potential recipient; the response is to synthesize an "adhesin" protein which causes cell clumping and hence makes plasmid transfer very efficient, even in liquid medium (21).

3. Novel mechanisms of plasmid replication. Some small plasmids of *Staphylococcus* (also studied in *Bacillus*) and of *Streptomyces* seem to replicate via single-stranded circular intermediates which may accumulate if second-strand synthesis is impaired (22–24). It remains to be seen if a similar situation occurs for some plasmids of Gram-negative bacteria.

In various streptomycetes and in *Borrelia* spp. (spirochaetes that cause relapsing fever), linear plasmids have been discovered (25–27). The most studied is pSLA2, a 17 kb DNA molecule found in *Streptomyces rochei*, which carries proteins covalently attached to the 5′ ends of terminal inverted repeats presumably involved in plasmid replication (25). It is not known whether the giant linear plasmids that occur in *Streptomyces* (e.g. SCP1, of 350 kb: 26; Chapter 7) have similar terminal organization.

4. Study of Gram-positive bacteria extends the range of elements that can be called transposable. Extensive studies of enteric bacteria have revealed several kinds of transposable elements (28–30). Among the Class I elements, insertion sequences (IS elements) are small (mostly 1–2 kb) and only encode functions needed for transposition, which act on the short terminal inverted repeats of the element. However, composite transposons (e.g. Tn*5*, Tn*9*, Tn*10*, Tn*951*) consist of genes with other functions, like antibiotic resistance or catabolic properties, flanked by two identical (or nearly identical) IS elements which can sometimes transpose independently. Class II transposons (e.g. Tn*3*) have short (ca. 38 bp) inverted repeats at the ends of the entire element and, like composite transposons, generally specify additional properties. Transposing phages, such as Mu, use transposition as an essential part of their replication strategy; the ends of Mu DNA carry 22 bp imperfect inverted repeats, with further versions of the sequences present near both ends (31). The unifying feature of all these elements is that their transposition involves recognition of the terminal sequences by the cognate transposase, which then cleaves one or both strands at each end and also introduces a staggered cut at the transposition target site; following transposition, DNA synthesis leads to the duplication of the few base pairs of target DNA between the cut sites, which are then found on either side of the transposed element (28–31). It is not yet clear where the large and complex Tn*7* (29) and the plasmid pUB2380, which exhibits "one-ended transposition" (32), fit into this scheme.

Elements belonging to all the well-characterized classes occur in nonenteric bacteria. IS-like elements have been found, for example, in *Pseudomonas aeruginosa* (33), in low (G + C) Gram-positive bacteria such as *Staphylococcus*

(34) and in archaebacteria (35). They are a heterogeneous group, generally showing little sequence relatedness. Nevertheless, it is interesting that IS*4* of *E. coli* and IS*231* of *Bacillus thuringiensis* possess similar terminal inverted repeats and specify similar proteins (36); and that IS*2*—an abundant element in *E. coli* strains—occurs in the cyanobacterium *Chlorogloeopsis fritschii* (37). In *Streptomyces*, there is some circumstantial evidence pointing to replication of an IS-like element, IS*110*, via a circular intermediate (3). Few composite transposons have been described in nonenteric bacteria: some, specifying antibiotic resistance, have been found in *Staphylococcus aureus* (a particularly intensively-studied organism because of its importance in hospital infections that fail to respond to antibiotic treatment) (34, 38), and the insecticidal toxin genes of some *B. thuringiensis* strains are flanked by IS-like elements (39). Transposons with terminal inverted repeats related to those of Tn*3* occur widely: they have been found in *Pseudomonas* (Tn*501*; Chapter 9), *Streptococcus faecalis* (Tn*917*; 40), *S. aureus* (Tn*551*; 34), *B. thuringiensis* (Tn*4430*; 41) and *Streptomyces fradiae* (Tn*4556*; 42). Mu-like transposing phages have been described in *Pseudomonas* (D3112; 43) and *Vibrio cholerae* (Chapter 15).

Several transposable elements that do not fit neatly into this classification occur in the chromosomes of Gram-positive bacteria. The remarkable conjugative transposon Tn*916* (16.4 kb: 44) mediates conjugation between *Streptococcus* cells, leading to intercellular transposition in the absence of any other plasmid and with no free intermediate being detected. A *Streptomyces* element exists as two integrated copies on the *S. coelicolor* chromosome and is also detectable as a low abundance CCC DNA molecule, the 2.6 kb "minicircle", capable of chromosomal integration when introduced artificially into a new host (3). Another group of actinomycete elements, represented by SLP1 (14 kb) and its relatives (3, 45), are "plasmidogenic": they exist as conjugative plasmids after excision from the chromosome and can reintegrate after transfer to a new host. These elements, and a *Staphylococcus* transposon, Tn*554* (6.691 kb: 34), all contain either no (Tn*554*, SLP1) or poor (Tn*916*, minicircle) terminal inverted repeats; they do not generate target site duplications; and all except Tn*916* exhibit strong insertion site specificity. Such specificity might theoretically involve substantial sequence homology between the element and a target *att* site. However, this appears to be the case only for the SLP1-like elements which all appear to use ca. 47 bp of sequence identity between themselves and their preferred chromosomal target as the site of recombination. It might be fruitful to compare these elements with an *E. coli* chromosomal element (e14, of 14.4 kb) which excises to form a circular molecule capable of site-specific reintegration and which (as judged by the published Southern blots) has perhaps 100 bp of sequence in common between its *att* site and that on the chromosome (46).

5. Bacteriophages. The well-known phages of *E. coli* are very diverse. Not surprisingly, even greater diversity is seen when phages of different groups of bacteria are analysed. In this brief treatment, we confine discussion to the properties of temperate phages and small lytic phages which often impinge strongly on host features of genetic interest, and to those new paradigms of phage biology found in noncoliforms.

Small single-stranded (ss) RNA phages are known only in the purple bacteria. They always use pili as adsorption organs. Many of them—but not Cb5 of *Caulobacter crescentus* (47)—are plasmid-specific. The apparent absence of sex pili from Gram-positive bacteria may explain why no ssRNA phages or other sex-specific phages have been described, but this does not account for the failure to discover, in Gram-positive bacteria, any small ss circular DNA icosahedral phages, such as ϕX174 of *E. coli*, or MAC-1 of *Bdellovibrio bacteriovorus* (48), which are not sex- or pilus-specific.

Temperate phages with at least superficial similarity to coliphage λ are widely distributed. Two phages—ϕ105 of *B. subtilis* (49) and ϕC31 of *Streptomyces* (50)—have been developed as useful cloning vector systems. They are both λ-like in morphology, dimensions, *cos* site-specific cleavage of DNA during packaging, and in forming prophages through specific interactions between phage and chromosomal *att* sites, though there are differences from the λ model in the relative positions of landmarks such as *c*, *att* and *cos* (49, 50) and in the molecular details of the region controlling lysogeny (49, 51). A much larger temperate phage of *B. subtilis*, Spβ (DNA content 126 kb), which is also finding increasing and subtle use as a cloning vector (49) must evidently have features absent from λ. Among these are a system for sequence-specific methylation of vegetative Spβ DNA, and a bacteriocin determinant (49; Chapter 11). The ability of some temperate phages to form plasmid prophages (like P1 in *E. coli*: 52) is not limited to the enteric bacteria: the prophage of ϕSF1 of *Streptomyces fradiae* not only exists as a plasmid, but even has genes specifying conjugal transfer (50).

Two kinds of phages are conspicuously different from any known in enteric bacteria. The *Pseudomonas phaseolicola* phage ϕ6—one of several unusual lipid-containing phages of *Pseudomonas*—is the only known dsRNA phage: each of its particles contains three distinct linear molecules of ca. 3.5, 4.6 and 7.5 kb, together with an RNA-dependent RNA polymerase (53). Phages ϕ29 of *B. subtilis* and Cp-1 of *Streptococcus pneumoniae* replicate as linear DNA molecules with proteins covalently bound to their ends (54, 55); in this they resemble the linear plasmid pSLA2 of *Streptomyces* mentioned above.

III. TRANSFER OF CHROMOSOMAL DNA BETWEEN BACTERIA

A. Plasmid-Mediated Conjugation

Conjugal transfer of chromosomal DNA between bacteria apparently always depends on the action of self-transmissible plasmids, which are said to have chromosome-mobilizing ability (Cma). It was first shown for F-mediated conjugation in *E. coli* where only a section of the donor chromosome is transferred. Even in *E. coli*, mystery still surrounds some aspects of Cma. Thus, the recombination in $F^+ \times F^-$ crosses that is not due to the presence of Hfr variants in the F^+ culture is not well understood, although a transient recombinational interaction of the plasmid with the donor chromosome is suggested, and transient interactions may well account for most other cases of Cma in Gram-negative bacteria (56). In *V. cholerae*, the inconveniently low frequency of Cma conferred by the indigenous P-factor plasmid was increased by inserting transposable elements into the plasmid and chromosome to provide regions of homology (Chapter 15). Amongst low (G + C) Gram-positive bacteria, the Cma of transmissible plasmids has been little studied. However, among the high (G + C) species, SCP1 of *S. coelicolor* can certainly integrate to give highly fertile donor strains (18). SCP1 is a linear molecule (26), raising the question of how it integrates. It would also be interesting to understand the inheritance at high frequency of donor markers on both sides of the integration point in crosses involving some (but not all) donors carrying integrated SCP1. Other *Streptomyces* plasmids can act as efficient sex factors without any evidence of stable interaction with the host chromosome (18).

In exploiting Cma for linkage mapping, the well-known interrupted mating procedure is not universally applicable. For example, in *Streptomyces* and *Rhizobium*, recombinants are selected after many hours of coculture of the two parental strains and analysed for the pattern of inheritance of unselected markers.

B. Transduction

Fine genetic mapping in *E. coli* and *S. typhimurium* has depended heavily on generalized transduction mediated by the P1 and P22 phages which use a nonspecific "headful" packaging mechanism (57, 58). In *B. subtilis*, the giant PBS1 phage can transduce perhaps 200–300 kb and, in the absence of conjugation, was a principal tool in constructing a gross linkage map (59). Generalized transducing phages have been exploited in several other bacteria, both for basic genetic mapping, as with the *Streptomyces venezuelae* phage SV1 (60), and for the transduction of transposon-induced mutations, as with myxophages Mx4 and Mx8 (Chapter 12), staphylococcal phage 80 (61), and *Rhizobium* phages RL38 and RL39 (62). Interestingly, the *Rhizobium* phages are lytic, their most effective use depending on UV inactivation of the phages to avoid killing of the host.

Phage λ provides the classical example of the ability of chromosomally located prophages to excise inaccurately, and acquire adjacent segments of chromosomal DNA, becoming specialized transducing phages. Specialized transduction is not always easy to demonstrate, but has been shown for

Rhizobium meliloti (63), with Spβ for *B. subtilis* (49), and in slightly atypical circumstances with ϕC31 for *Streptomyces* (64). For phage λ (65) and Spβ (49), it has been possible to delete the preferred chromosomal attachment site (or, for Spβ, to use an integration-defective phage mutant: 49) so that a range of secondary sites is occupied by the prophage, allowing the *in vivo* cloning of flanking DNA from several chromosomal regions.

C. Natural Transformation

Diverse bacteria exhibit natural competence for efficient transformation by fragments of chromosomal DNA. This property, which *E. coli* lacks, is determined by specific genes rather than depending on "accidental" DNA uptake, and is clearly a highly evolved process, involving (in *B. subtilis*) a large set of coactivated genes (66). Its molecular details differ between bacterial groups. In the Gram-positive *B. subtilis* and *Streptococcus pneumoniae* DNA uptake is sequence-independent (although its recombination into the chromosome depends on sequence homology). However in the Gram-negative *Haemophilus influenzae*, a special sequence (5′ AAGTGCGGTCA 3′) must be recognized for DNA to be incorporated (67). Such a sequence serves to tag DNA as self because it occurs some 600 times on the *H. influenzae* chromosome, whereas its chance occurrence would be only about once per bacterial genome.

Apart from its use in fine linkage analysis, natural transformation is particularly valuable for other genetic studies. For example, in *S. pneumoniae* (the organism in which study of the transforming principle gave the first clues that genes were made of DNA: 68), the ability to transform with artificially constructed DNA heteroduplex molecules has led to penetrating analyses of mismatch repair (69); and in *B. subtilis*, transformation has been cleverly exploited for rapidly replacing segments of DNA both of chromosomal regions and of various genetic elements, such as Tn*917* and Spβ derivatives, used as molecular genetic tools (Chapter 11).

D. Capsduction in *Rhodobacter capsulatus*

The above three well-known routes of genetic exchange do not represent the complete repertoire of bacterial sex. Cultures of the phototrophic purple nonsulphur bacterium *Rhodobacter capsulatus* can release random fragments of their DNA packaged in particles (the "gene transfer agent") and the particles can deliver this DNA to recipient bacteria. This capsduction is used analytically much like transduction in other bacteria (70). It has proved particularly useful in transferring *in vitro*-manipulated DNA sequences into the *R. capsulatus* chromosome (as in interposon mutagenesis: Chapter 5).

IV. GENE EXPRESSION

The study of gene expression in a range of bacteria has widened and deepened the picture derived from studies of *E. coli*. In this section we illustrate this by comparing *E. coli* with a representative of each of the two major Gram-positive subgroups: those with low (G + C) DNA (*B. subtilis*) and those with high (G + C) DNA (*S. coelicolor*). What is true for all three of these examples might well apply to all eubacteria (but we may nevertheless be in for surprises); in archaebacteria, which do not feature in any of the specialized chapters of this book, further differences are apparent.

A. Initiation of Transcription

RNA polymerase holoenzyme in bacteria as distant from each other as *E. coli*, *B. subtilis*, *S. coelicolor* and the cyanobacterium *Anacystis nidulans* consists of a catalytically-active core with the subunit structure $\beta\beta^1\alpha_2$ and an additional subunit, the sigma factor, which directs the holoenzyme to specific promoter sequences in the DNA (71). In *B. subtilis* (and *Lactobacillus curvatus*, another low (G + C) Gram-positive species) an additional component of the holoenzyme, δ, appears to compensate for the absence from their principal sigma factors of an amino-terminal region present in the *E. coli* protein. In *A. nidulans*, the sigma factor is a nondissociating part of the enzyme rather than an associated factor (71).

One of the most important recent generalizations about the molecular biology of eubacteria is that they usually contain several distinct sigma factors, and therefore a corresponding number of holoenzymes, each capable of recognizing different promoter classes (72). Detailed analysis of sigma factors and the promoters with which they interact very strongly suggests that specific contact between RNA polymerase holoenzymes and most cognate promoters involves direct interaction between particular regions of the sigma factors and two approximately 6 bp DNA sequences separated by about 17 bp. These sequences (Table II), which have centres about two turns of the DNA double helix apart (and therefore lie approximately on the same face of the helix), are the famous "−10" and "−35" sequences (measured from the transcription start site at +1). In broad terms, the majority of genes that are expressed during normal vegetative growth of *E. coli* and *B. subtilis*, and many of those of *S. coelicolor*, have promoters that resemble the consensus sequence for the major holoenzyme of *E. coli*; and sigma factors corresponding in promoter specificity to the major sigma factor (σ^{70}) of *E. coli* have been biochemically demonstrated for each species. In contrast, specialized gene sets that are normally expressed only during some relatively major shift in cell physiology or development are often transcribed from unusual promoters following the synthesis, or activa-

Table II
Sigma factors and cognate promoter consensus sequences

Sigma factor molecular weight in kD or genetic determinant	Consensus promoter recognition sequence[a]	Organism	Known classes of genes controlled
70	TTGACA(17)TATAAT(6) + 1	*E. coli*	Housekeeping genes[b]
	TTGACA(17)TATAAT(6) + 1	*B. subtilis*	Housekeeping genes[b]
35	TTGACA(18)TAGGAT(6) + 1	*S. coelicolor*	Housekeeping genes[c]
32	CTTGAA(13–15)CCCCAT-TA(7) + 1	*E. coli*	Heat shock genes[b]
37	AGG-TT(13–16)GG-ATTG-T(6) + 1	*B. subtilis*	Unknown[b]
49	(Resembles that for 6^{37} of *B. subtilis*)	*S. coelicolor*	Several catabolic genes (*endoH, dagAp3*)[c,d]
32	AAATC(14–15)TA-TG-TT-TA(2) + 1	*B. subtilis*	Unknown[b]
29	TT-AAA(14–17)CATATT(8–10) + 1	*B. subtilis*	Sporulation genes[b]
30	GCAGGA(17)GAATT(?) + 1	*B. subtilis*	Sporulation genes[b,e]
spoIIAC	Unknown	*B. subtilis*	Sporulation genes[b]
whiG	Unknown	*S. coelicolor*	Sporulation genes[f]
28	CTAAA(16)CCGATAT(7) + 1	*B. subtilis*	Motility genes[b]
flbB, flaI	TAAA(15)GCCGATAA(?) + 1	*E. coli*	Motility genes[b]
	TTGGCCC(5)TTGC(9–11) + 1	*C. crescentus*	Motility genes[g]
54	CTGGCAC(5)TTGCA(6–11) + 1	*E. coli*	Nitrogen regulated genes[b]
28	Unknown	*S. coelicolor*	Agarase (*dagAp2*)[d]
Gene 28	T-AGGAGA--A(15–16)TTT-TTT(4–7) + 1	SP01/*B. subtilis*	Phage middle genes[b]
Genes 33/34	CGTTAGA(17–19)GATATT(?) + 1	SP01/*B. subtilis*	Phage late genes[b]
Gene 55	(No−35) TATAAATA(3–6) + 1	T4/*E. coli*	Phage late genes[b]

[a]Numbers indicate the spacing between the −35, −10 and +1 regions. Most of the consensus sequences, apart from those for sigma−70 and sigma−43, are based on small numbers of examples.
[b]Ref. 72.
[c]Refs. 77, 81.
[d]Ref. 125.
[e]R. Losick, personal communication.
[f]C. J. Bruton and K. F. Chater, unpublished.
[g]Chapter 10.

tion, of alternative sigma factors. These shifts include stages of sporulation, heat-shock response, a requirement for motility and chemotaxis, and starvation for readily-available nitrogen (Table II). Some phages provide special examples of transcriptional switching, either by encoding sigma factors which direct the host polymerase to recognize phage promoters (Table II) or by specifying entirely new holoenzymes with very pronounced specificity for specialized phage promoters (e.g. T7 and Sp6: 73).

It is interesting to find that homologous sigma factors and their cognate promoters do not always seem to retain the same uses across even quite narrow taxonomic divisions. For example, promoters of an unusual class are associated with genes for nitrogen assimilation in *Rhizobium*, *Klebsiella* and *E. coli* and are recognized by RNA polymerases containing σ^{54} (a sigma factor that may have evolved separately from all other characterized sigma factors: Chapter 6 and Ref. 71); but promoters with very similar sequence features are associated with genes for catabolism of aromatic hydrocarbons in *Pseudomonas* (Chapter 8) and with motility genes in *Caulobacter* (Chapter 10). The last case is particularly intriguing, because two of these genes specify flagellins partially homologous with those of *B. subtilis* and *S. typhimurium* (74), the genes for which have promoters of a quite different class (Table II; see also Chapter 6).

Significant conservation of RNA polymerase core enzyme and sigma factors is indicated by the ability of *E. coli* core enzyme to interact productively with *B. subtilis* σ^{43} to transcribe T4 phage promoters (75, 76); of *B. subtilis* core enzyme to interact with *E. coli* σ^{70} to transcribe DNA of a *Bacillus* phage (76); and of two different *S. coelicolor* sigma factors to activate transcription from two *B. subtilis* promoters of different classes by *B. subtilis* core enzyme (77). However, the extent of this interchangeability is not unlimited, and additional features of promoters and of particular holoenzymes may influence the efficiency of cross-generic transcription. *B. subtilis* promoter regions are particularly rich in A + T, and *B. subtilis* RNA polymerase appears to be rather exclusively adapted to this feature: transcription from "foreign" promoters, except those of other low (G + C) Gram-positive bacteria, is usually less efficient (75, 76). On the other hand, RNA polymerase from *S. coelicolor* (which has promoter regions much less rich in A + T) is apparently less fastidious, and can transcribe conventional promoters from *E. coli*, *Serratia marcescens* or *Bacillus* (77, 78). Putting these findings together with a number of observations on the extent of expression of *E. coli* genes in relatively high (G + C) Gram-negative bacteria (79) and *vice-versa*, it seems likely that there is no particular transcription barrier between Gram-positive and Gram-negative bacteria, but rather that organisms that contain relatively high (G + C) in their DNA (and at least to some extent in their promoters) are more permissive for the expression of low (G + C) promoters than the reverse.

B. Transcription Termination and mRNA Processing: an Insufficiently Studied Subject

Termination of transcription has not been intensively studied outside *E. coli*. Nevertheless, it is clear that so-called factor-independent termination sequences from *E. coli* can cause termination in *B. subtilis* and *Streptomyces* (80, 81), and that potential hairpin loop structures are found at the 3′ ends of many mRNA species in these organisms, just as in *E. coli*. In a very few cases, such structures have been shown to correspond to terminator sites; more often it remains possible that the hairpin loop is a processing point beyond which exonucleolytic digestion of longer initial transcripts seldom proceeds. In *E. coli* and *S. typhimurium*, the presence of such structures is important in determining mRNA half-life (82). This is a regulatory parameter that has received less attention than it deserves: it has been shown to influence the relative expression of different regions of a photosynthetic operon in *R. capsulatus* (83).

A general feature of factor-independent terminator structures in enteric bacteria and *B. subtilis* is their possession of a run of U residues located 3′ to the stem-loop. Such runs are not characteristic of putative *Streptomyces* terminators (Chapter 7). It is instructive to consider this observation in the light of a discussion (84) of an *E. coli* mutant RNA polymerase that terminates at unusually short U tracts. It is postulated that the wild-type RNA polymerase may carry a stabilizing function for rU-dA base pairs (the weak hydrogen-bonding of which might otherwise lead to abortion of incomplete transcripts at U-rich regions of coding sequences), and that particularly long U runs are needed to mark *bona fide* termination points. Mutational loss of the stabilizing function would then lead to premature termination at short U runs. In the (G + C)-rich *Streptomyces*, runs of U residues are very rare in coding sequences (e.g. TTT for phenylalanine occurs only four times in more than 12 kb of accumulated *Streptomyces* coding sequence, compared with 176 occurrences of its synonym TTC; M. J. Bibb, personal communication). We may therefore postulate that the *Streptomyces* enzyme does not need, or have, a stabilizing function for rU-dA pairs.

There is even less information for nonenteric bacteria about factor-*dependent* terminators and the role of RNA polymerase-associated termination factors such as *rho* and the *nusA* gene product that have been analysed in *E. coli* (85). Since factor-dependent terminators lack features that can readily be recognized by sequence inspection, their discovery is likely to depend on more intensive studies than have yet been attempted.

C. Translation

Eubacterial ribosomes are highly conserved and it is reasonable to expect that many aspects of translation will be common to all eubacteria. Generally, coding sequences in mRNA from *E. coli*, *B. subtilis* and *S. coelicolor* are preceded by a short region (the "ribosome-binding site" or Shine–Dalgarno sequence) complementary to part of the 3′-terminus of 16S rRNA, typically centred 8–15 bases upstream of the start codon (86). These sequences of 16S rRNA from *E. coli*, *B. subtilis* and *Streptomyces* are almost the same, so it is not surprising that the ribosome-binding sites are also similar and that cross-generic recognition can take place. Just as for promoters, there are some limitations on this: more extensive ribosome-binding sites are generally found in low (G + C) Gram-positive bacteria (*Bacillus* and *Staphylococcus*: 87) than in enteric bacteria (86) or *Streptomyces* (81). This probably has functional importance, since *B. subtilis* is more limited than the other two groups in its ability to initiate translation of transcripts of "foreign" bacterial genes (81, 87).

Not all transcripts possess conventional ribosome-binding sites. In several cases, the translational start codon is at or within one or two bases of the 5′ end of the mRNA and sometimes (e.g. the *c*I gene transcribed from P_{RM} (88), and the rRNA methylase gene of *Saccharopolyspora* (formerly *Streptomyces*) *erythraea*: 81) this is associated with a very low level of translation product (though a causal relationship is not proven). However, transcripts lacking untranslated leader sequences can give rise to abundant protein products, as in the case of the aminoglycoside phosphotransferase gene of *S. fradiae* (81). Clearly, translation initiation is a complex mechanism, and conventional ribosome-binding sites are only part, albeit often crucial, of the story. Even for transcripts with such sites, other mRNA-ribosome interactions are important (86).

While translation is nearly always initiated from an AUG codon (for formylmethionyl-tRNA) in *E. coli*, other initiator codons, presumably also recognized by the same tRNA through specialized "wobble" effects on base-pairing, are relatively common in bacteria with different overall (G + C) content: UUG in low (G + C) *B. subtilis* (89), and GUG in high (G + C) *Streptomyces* (81). Codon usage patterns can also differ from those in *E. coli*: inevitably, the third codon position, which is the main position of variation between synonymous codons, is strongly biased in accordance with the overall base composition of the organism (90), to the extent that there may be overall shifts in the use of certain amino acids (91). At the extremes of base composition, certain codons are very rarely used, raising the possibility that they could assume significant regulatory functions. A likely example of this is the possible role of the UUA codon in developmental switching in *S. coelicolor* (Chapter 7). It remains to be seen whether there are equivalent roles of rare (G + C)-rich codons in low (G + C) bacteria.

With the increased availability of DNA sequences, we can expect that other correlations will emerge: for example, the choice of translation termination codon is not independent of codon usage (92).

V. GENE-, PATHWAY- AND REGULON-SPECIFIC REGULATORY MECHANISMS

A. Positive and Negative Regulation of Transcription Initiation

A variety of mechanisms have been described, particularly in *E. coli*, for controlling transcription initiation of individual genes, operons and regulons (in which several genes or operons respond to common regulators: 93, 94). In negative control, typified by the action of the *lacI* gene product on the *lac* operon of *E. coli* (95), repressor proteins bind to operator sequences located near the promoters of structural genes, thereby interfering with the binding or movement of RNA polymerase; expression of the structural gene(s) generally requires the repressor to interact with a low molecular weight inducer, or to cease binding to a low molecular weight corepressor, thereby favouring an allosteric form of the repressor that does not bind to the operator. Alternatively, the repressor may be permanently inactivated, as in the proteolytic cleavage of the λ prophage and *lexA* repressors that follow UV irradiation of *E. coli* (96). Some simple negative control systems have been found in Gram-positive bacteria (e.g. the gluconate utilization operon of *B. subtilis* (97)), but it is not clear how common they are.

Pathway- and regulon-specific positive control systems appear to be more widespread (94). In these, transcription is stimulated by the binding of an activator protein in a suitable configuration to a specific site in the DNA near the promoter. Most commonly, the activator is stabilized in this configuration by binding an intracellular low molecular weight effector (as in the *E. coli* maltose regulon: 98); by protection against rapid proteolytic inactivation, the means by which the *c*III protein of phage λ allows the *c*II activator protein to direct expression of genes needed for prophage establishment (96); or by interacting with a membrane-bound "sensor" protein (99) which is itself activated in response to extracellular signals such as changing osmolarity, phosphate limitation or (in the case of *Agrobacterium*: Chapter 18) plant exudate. It is proposed that in such bipartite sensor-regulator systems, the extracellular (or periplasmic) amino-terminal domain perceives the signal, which causes an allosteric change in the cytoplasmically located carboxy-terminal domain. This domain, which is relatively conserved among sensor proteins, interacts in turn with the amino-terminal domain of the regulator protein (also relatively conserved), to modify it in a way that affects the ability of the carboxy-terminal domain of the regulator to bind to suitable DNA sequences and thus activate transcription (99). In the NtrB–NtrC interaction, regulator activation may be by phosphorylation (99; Chapter 6). There are variations on this bipartite scheme: in the NtrB–NtrC system, the NtrB sensor is entirely cytoplasmic and recognizes (through a further regulatory cascade) intracellular nitrogen stress, and in the *toxR* system of *V. cholerae* a single membrane-bound protein combines the sensory and regulatory functions and interacts directly with target DNA (Chapter 15).

Activator proteins may also be involved in negative regulation as in the classical example of the *araC*-specified protein that regulates arabinose utilization genes in *E. coli* (10). In the presence or absence of arabinose, *araC*-specified protein binds to the *araI* site; but when arabinose is absent, the bound protein interacts with *araC* protein bound at a separate site (*ara*O_2), an interaction that requires a loop to be formed in the DNA and which leads to transcriptional shutdown.

The adoption of positive or negative control for a particular pathway may differ between bacterial groups. For example, the genes for glycerol utilization are negatively controlled by the *glpR* product in *E. coli*, positively controlled by the $glpP_I$ product in *B. subtilis*, and probably subject to both positive and negative control by the *gylR* product in *S. coelicolor* (100). The demand theory (101) suggests that negative control is generally used for genes that are likely to be switched off for many generations. Perhaps the alternation of environments which *E. coli* inhabits (inside or outside the mammalian colon) results in the long-term switching off of certain gene sets (glycerol is essentially absent from the colon), whereas *B. subtilis* and *S. coelicolor* inhabit only a single kind of environment (soil, in which glycerol is frequently available), and many of their metabolic gene sets are constantly in use (100).

The well-known positive regulatory role of the cAMP-cAMP receptor protein complex in activating genes for carbon source utilization in *E. coli* (an important part of the system of catabolite or glucose repression: 102) does not appear to be universal among prokaryotes. In *B. subtilis*, cAMP levels are so low that the system could not work (103) and in *S. coelicolor*, cAMP levels do not vary with the carbon source (104); moreover, the enzymology of glucose phosphorylation, which is important in glucose repression in *E. coli*, is apparently quite different in *S. coelicolor*; and *S. coelicolor* mutants defective in ATP-dependent glucose kinase are relieved from catabolite repression (81).

B. Attenuation

Regulation of gene expression may also be effected through the translation machinery, in response to factors affecting ribosome pausing. Such attenuation (105) systems have been most studied in certain amino acid biosynthetic operons, in which an initial short open reading frame (ORF) encodes a leader peptide which has a specific regulatory function. For example, in the *trp* operons of enteric bacteria, this ORF contains consecutive codons for tryptophan: in conditions of tryptophan starvation, ribosomes stall at these codons and allow the nascent downstream mRNA to adopt a secondary structure which precludes formation of an alternative structure, a strong transcription terminator. Hence, tryptophan starvation allows transcription to proceed into the structural genes. There is an interesting variation on this theme in the *trp* operon of *Bacillus* spp: here there is no leader peptide, but the formation of a terminator structure in the mRNA is controlled by a tryptophan-activated regulatory protein, the product of a separate gene, *mtr* (105).

In addition to this transcriptional attenuation, translational attenuation has also been observed: the transcripts specifying resistance of ribosomes to certain antibiotics (for example, erythromycin-resistance by enzymatic modification of ribosomal RNA, determined by plasmid-borne genes in *Staphylococcus* and *Bacillus*) contain leader regions with the potential to form complex secondary structures (106, 107). Low levels of antibiotic cause ribosome pausing in a short ORF in the leader sequence, which in turn limits the kind of secondary structure that can be formed, and makes available the ribosome-binding site for the expression of the resistance protein.

C. Regulation by Anti-Termination

In phage λ, the transcription of some genes depends on the ability of RNA polymerase to read through upstream terminators. This depends in turn on the product of an upstream gene that changes the state of RNA polymerase as it traverses particular signals in the transcribed DNA sequences, so that the subsequent terminator is not recognized. Since downstream genes can be transcribed only when sufficient antiterminator protein has been made, this provides a mechanism for a controlled cascade of gene expression which is obviously suitable for a rapid developmental process such as phage infection.

Antitermination is also used in the regulation of some metabolic operons (e.g. the *bgl* operon) of *E. coli* (108). It does not appear to have been demonstrated in other bacteria, though it has been proposed that the *strR* gene of *Streptomyces griseus* specifies an antiterminator, whose action would allow readthrough transcription from the streptomycin resistance gene into a gene for streptomycin biosynthesis (109); thus resistance would be established before potentially toxic antibiotic production (see also Chapter 7). It is tempting to speculate that one significance of the multiple transcription initiation sites used for some genes in *B. subtilis* (110) and *Streptomyces* (81) could be to allow some, but not all, transcripts of a particular region to traverse an antitermination-inducing signal in the template.

D. Genetic Regulation by Complementary RNA Species ("Antisense RNA")

Natural examples of the recently discovered phenomenon of antisense RNA regulation, in which the biological activity of an RNA species is prevented by its base-pairing with a complementary RNA sequence, are widespread in *E. coli* (111). They are mostly of the kind where the two complementary RNA species are transcribed from different segments of the DNA. Thus, translation of *ompF* mRNA (encoding a major outer membrane protein) is inhibited by sequestering of its ribosome-binding site by *micF* RNA transcribed from a distant region of the chromosome; and an antisense transcript of the promoter region of the *crp* gene (for cAMP receptor protein) can apparently pair with bases 2–11 of the *crp* mRNA to form a structure sufficiently like the stem of a terminator to cause transcription to be aborted.

In the copy-number control region of plasmid ColE1, a small antisense transcript (RNA II) is specified by the same DNA segment as the sense transcript. It inhibits the initiation of ColE1 DNA synthesis by interacting with the sense RNA molecule (RNA I) that primes the new DNA strand at the origin of replication. An interesting feature is the importance of the RNA secondary structure: both RNAs I and II fold into complex stems and loops, and interactions between the complementary single-stranded loop regions, stabilized by a plasmid-specified protein, are crucial for efficient antisense regulation.

Antisense mechanisms have also been implicated in regulating transposition of Tn*10* and in the lytic development of phage λ (111). Since they appear in such a wide range of elements in *E. coli*, they are likely to be widespread among bacteria. In this book there is at least one set of circumstances that suggest antisense regulation: the diverging and converging overlapping transcripts in and around certain resistance genes associated with antibiotic production in *Streptomyces* (Chapter 7).

E. Regulation of Gene Expression by Changes in DNA Structure

It is sometimes advantageous for particular phenotypes to be expressed only in a fraction of the bacterial population. In these circumstances, reversible covalent changes in DNA structure are responsible for gene activation or inactivation, nearly all involving either recombination events or changes in methylation of particular bases (Table III). Those events involving recombination usually occur at such frequencies where typically one cell in every 10^2–10^4 exhibits the changed DNA, and often site-specific recombination systems are necessary to bring particular genes into contact with particular promoters (112). In the remarkable case of the cyanobacterium *Anabaena* (113), irreversible DNA rearrangements in the terminally differentiated heterocysts are needed to allow them to fulfil their function of nitrogen fixation (Chapters 1 and 6).

In other cases, methylation of residues in GATC-sequences present in certain promoters influences promoter strength, with the result for Tn*5* and Tn*10* that transposase is especially poorly expressed, and transposition is

Table III
Regulation of gene expression by covalent changes in DNA structure

Phenotype involved	Organism	Nature of change in DNA	References
Flagellar type	*Salmonella typhimurium* and *E. coli*	Site-specific DNA inversion	112, 126
Host range of temperate phages	*E. coli* phages Mu, D108 and P1	Site-specific DNA inversion	126
Penicillinase regulation	*Staphylococcus aureus* plasmid pI524	Site-specific DNA inversion	127
Change of pilus type (phase and antigenic variation)	*Neisseria gonorrhoeae*	Gene conversion events and site-specific recombinational rearrangements involving multiple loci	Chapter 13 and Ref. 112
Opacity variation	*Neisseria meningitidis*	Misreplication of a 5 base repeated sequence leading to changes in reading frame at multiple loci	Chapter 13 and Ref. 112
Antigenic variation in relapsing fever	*Borrelia hermsii*	Movement of different coding sequences into "expression site" in a linear plasmid	112
Bacteriophage development	*E. coli* phage λ	Site-specific recombination at *att* site, changing 3′-OH end of *int-xis* transcript, and influencing its stability	128
Nitrogen fixation in heterocysts	*Anabaena*	Site-specific excision of DNA intervening between a gene and its promoter or interrupting a coding sequence	Chapter 6
Regulation of transposition frequency of Tn*5* and Tn*10*	*E. coli*	Methylation at GATC sequences (Dam methylation)	29, 30

potentially kept at a low level. It is proposed that this effect is most marked in fully methylated DNA, but that hemimethylated DNA produced during replication would be more efficiently transcribed. This could stimulate transposition in cells containing, as a result of replication, at least two chromosomes—an advantage for Tn*5* and Tn*10*, since their transposition is lethal to the donor DNA molecular (29, 30). This attractive model may be somewhat undermined by the absence of GATC-methylating (Dam) enzymes from most bacteria other than *E. coli*, and by the natural occurrence of Tn*5* mainly in noncoliform bacteria.

VI. DIFFERENCES BETWEEN PROKARYOTE AND EUKARYOTE GENETICS

The distinction between prokaryotes and eukaryotes is formally defined by the absence of the nuclear membrane. In addition, and despite the diversity of genetic phenomena in bacteria, there are several genetic features that distinguish all eubacteria that have been studied from eukaryotes (but remember that very large areas of the phylogenetic tree have been ignored: Chapter 1). In this chapter we summarize these features. Further information on many of these aspects of eukaryotic genetics can be found in Ref. 114.

A. Chromosome Structure

Generalizing from the small available sample, the typical bacterium seems to have a single, circular chromosome, while the genes of a eukaryote are distributed over a set of several or many separate chromosomes, each of which is a linear structure with a special telomere at both of its free ends. In eukaryotes, the DNA is extensively complexed with histone and nonhistone proteins to form the chromatin with its nucleosome organization, whereas bacterial DNA, though largely covered by histone-like proteins, does not seem to associate with so many structural proteins (115).

Bacterial chromosomes, like those of eukaryotes, must interact with some architectural feature of the cell to ensure accurate partitioning of chromosome copies to daughter cells after DNA replication. Most experimental work (in *E. coli*) on this problem has used plasmids (which can be regarded as minichromosomes) as model systems. The result has been the recognition of *par* (partition) regions. Each consists of a short (A + T)-rich *cis*-acting DNA sequence and two structural genes whose products may complex with the *cis*-acting site and the cell membrane to form a partition complex (116). The system is thus analogous, even if not homologous, to the centromeres of eukaryotic chromosomes which interact with the spindle fibres during nuclear division.

B. Plasmids

As discussed in Section II.D.1, plasmids often carry genes for a variety of specialized, or locally adaptive, phenotypes in a wide range of bacterial groups. In eukaryotes, examples of the same phenomenon are almost vanishingly rare: cobalt resistance in the cellular slime mould *Dictyostelium discoideum* may be the only case so far (117).

C. Merodiploidy

In spite of the diversity of ways in which DNA can be transferred between bacteria (Section III), eubacterial "alternatives to sex" have one feature in common which sets them all apart from sexual and parasexual phenomena in eukaryotes. In the latter, no matter how DNA from two individuals is brought together—and the variety of sexual processes that achieve this in eukaryotes is truly remarkable—fusion of nuclei each containing a complete set of genes to give a diploid condition occurs at some stage in the life cycle, and is followed, sooner or later, by a meiotic or mitotic haploidization with reassortment of genes. In eubacteria, complete diploidy seems not to occur (except as a rare limiting case when plasmid-mediated conjugation goes to completion before the mating cells separate). Instead, merodiploidy is the rule, reassortment of genes taking place in a cell containing a complete set of genes from the recipient and a partial set from the donor. Only when bacterial cells are artificially united by the technique of protoplast fusion (119) is a transient diploid stage regularly produced, providing an opportunity for crossing-over to generate all possible combinations of the genetic differences that distinguished the two parents. (Remarkably, *B. subtilis*, but not *S. coelicolor*, progeny that arise from protoplast fusion often, and as yet inexplicably, contain complete genomes from both parents, with one genome being silent: 115.) A gene exchange system has been discovered in the archaebacterium *Halobacterium volcanii* (119) but it is not known whether it too leads to merodiploidy as in eubacteria.

D. Gene Expression

Several features of transcription differ between eubacteria and eukaryotes, including the nature of promoter sequences and the process of transcription termination. Soon after its initiation, a typical eukaryotic mRNA is capped at the 5′-end by a 7-methyl guanosine residue which provides a site for mRNA-ribosome interaction entirely different from the ribosome-binding sites of eubacterial mRNAs. In eukaryotes there seems to be no equivalent of the hairpin-loop characteristic of factor-independent prokaryotic terminators; instead transcription ends at a characteristic nucleotide sequence, with subsequent nucleolytic processing of the transcript and attachment of a poly-A tract. Presumably because of the presence of the eukaryotic nuclear membrane, transcription and translation are temporally separated rather than being concurrent. The transcription apparatus is itself relatively complex in eukaryotic nuclei, involving three classes of RNA polymerase holoenzyme each with ten or more subunits. They transcribe rRNA (polymerase I), mRNA (polymerase II), tRNA and 5S rRNA (polymerase III). However, there are similarities with eubacterial holoenzyme: all three classes have their four largest subunits in common, and these subunits are immunologically related to subunits of the principal *E. coli* holoenzyme (71). Archaebacterial RNA polymerases are very like the eukaryotic enzymes (71).

E. Operons and Introns

In Section II.C, we reviewed the occurrence of operons containing the cotranscribed structural genes for related steps in the same biochemical or developmental pathway and saw that the details vary for homologous genes in different bacteria and for different sets of genes in the same bacterium. Nevertheless, the organization of at least some genes into operons seems to be characteristic of all prokaryotes that have been investigated (and this includes archaebacteria and chloroplasts). This sets them apart from eukaryotes, in which mature transcripts appear to be uniformly monocistronic. On the other hand, eukaryotic genes are often more complex than those of prokaryotes in containing introns which are spliced out to give the mature transcript. Some introns even have coding capacity for proteins unconnected with the function of the gene in which they are embedded (120).

While all eukaryotes probably have some introns in some of their genes, and higher eukaryotes have many large introns in most of them, there is no report of an intron in any eubacterial gene. Even considering the whole known range of prokaryotic genomes, examples of introns are extremely few (Chapter 1): the thymidilate synthase gene of coliphage T4 provides one example (121), and a tRNA gene of the archaebacterium *H. volcanii* provides a second (122). The T4 intron is an example of the self-splicing group I introns and so it does not imply the existence in the host cell of an enzyme-catalysed splicing machinery.

VII. CLOSING REMARKS

This chapter has been wide-ranging and necessarily selective. For example, we have found no space to deal with many aspects of DNA metabolism—its replication, repair, recombination, modification and restriction. Nevertheless, the chapter should have provided a useful checklist of those aspects of prokaryotic genes which must be kept in mind in interpreting the emerging details of every gene system, both those presented in the remaining chapters of this book, and those omitted or yet to be analysed.

Acknowledgements

We thank Mervyn Bibb, Tobias Kieser, Rich Losick and Phil Youngman for their helpful comments on the manuscript of this Chapter, and Mark Buttner for useful discussions.

References

(1) Riley, M. and Krawiec, S. (1987). Genome organization. *In* "*Escherichia coli* and *Salmonella typhimurium*. Cellular and Molecular Biology" (F. C. Neidhardt, J. L. Ingraham, K. B. Low, B. Magasanik, M. Schaechter and H. E. Umbarger, eds.), pp. 967–981. American Society for Microbiology, Washington, D.C.

(2) Smith, C. L., Econome, J. G., Schutt, A., Klco, S. and Cantor, C. R. (1987). A physical map of the *Escherichia coli* K12 genome. *Science* **236**, 1448–1453.

(3) Chater, K. F., Henderson, D. J., Bibb, M. J. and Hopwood, D. A. (1988). Genome flux in *Streptomyces coelicolor* and other streptomycetes and its possible relevance to the evolution of mobile antibiotic resistance determinants. *In* "Transposition" (A. J. Kingsman, K. F. Chater and S. M. Kingsman, eds.), pp. 7–42. Cambridge University Press, Cambridge.

(4) Beringer, J. E., Johnston, A. W. B. and Kondorosi, A. (1984). Genetic maps of *Rhizobium leguminosarum, R. meliloti, R. phaseoli* and *R. trifolii*. *In* "Genetic Maps 1984" (S. J. O'Brien, ed.), pp. 202–205. Cold Spring Harbor Laboratory, Cold Spring Harbor, New York.

(5) Vary, P. A. and Tao, Y-P. (1988). Development of genetic methods in *Bacillus megaterium*. *In* "Genetics and Biotechnology of Bacilli II" (J. A. Hoch and A. T. Ganesan), Academic Press (In press).

(6) Chater, K. F. and Hopwood, D. A. (1984). *Streptomyces* genetics. *In* "The Biology of the Actinomycetes" (M. Goodfellow, M. Modarski and S. T. Williams, eds.), pp. 229–286. Academic Press, London.

(7) Holloway, B. W. and Morgan, A. F. (1986). Genome organization in *Pseudomonas*. *Ann. Rev. Microbiol.* **40**, 79–105.

(8) Ochman, H. and Wilson, A. C. (1987). Evolution in bacteria: evidence for a universal substitution rate in cellular genomes. *J. Molec. Evol.* **26**, 74–86.

(9) Wilson, G. (1984). Genetic map of *Bacillus subtilis*. *In* "Genetic Maps 1984" (S. J. O'Brien, ed.), pp. 169–185. Cold Spring Harbor Laboratory, Cold Spring Harbor, New York.

(10) Beckwith, J. (1987). The operon: an historical account. *In* "*Escherichia coli* and *Salmonella typhimurium*. Cellular and Molecular Biology" (F. C. Neidhardt, K. B. Low, B. Magasanik, M. Schaechter and H. E. Umbarger, eds.), pp. 1439–1443. American Society for Microbiology, Washington, D.C.

(11) Bachmann, B. (1987). Linkage map of *Escherichia coli* K12, 7th Edition. *In* "*Escherichia coli* and *Salmonella typhimurium*. Cellular and Molecular Biology" (F. C. Neidhardt, K. B. Low, B. Magasanik, M. Schaechter and H. E. Umbarger, eds.), pp. 807–876. American Society for Microbiology, Washington, D.C.

(12) Adams, C. W., Fornwald, J. A., Schmidt, F. J., Rosenberg, M. and Brawner, M. E. (1988). Gene organization and structure of the *Streptomyces lividans gal* operon. *J. Bacteriol.* **170**, 203–212.

(13) Smith, C. P. and Chater, K. F. (1988). Cloning and transcription analysis of the entire glycerol utilization (*gylABX*) operon of *Streptomyces coelicolor* A3(2) and identification of a closely associated transcription unit. *Mol. Gen. Genet.* **211**, 129–137.

(14) Lindgren, V. and Rutberg, L. (1976). Genetic control of the *glp* system in *Bacillus subtilis*. *J. Bacteriol.* **127**, 1047–1057.

(15) Chater, K. F. (1984). Morphological and physiological differentiation in *Streptomyces. In* "Microbial Development" (R. Losick and L. Shapiro, eds.), pp. 89–115. Cold Spring Harbor Laboratory, Cold Spring Harbor, New York.

(16) Hopwood, D. A. (1978). Extrachromosomally determined antibiotic production. *Ann. Rev. Microbiol.* **32**, 373–392.

(17) Willetts, N. (1985). Plasmids. *In* "Genetics of Bacteria" (J. Scaife, D. Leach and A. Galizzi, eds.), pp. 16–19. Academic Press, London.

(18) Hopwood, D. A., Kieser, T., Lydiate, D. and Bibb, M. J. (1986). *Streptomyces* plasmids: their biology and use as cloning vectors. *In* "The Bacteria. A Treatise on Structure and Function. Vol. IX. Antibiotic-producing *Streptomyces*" (S. W. Queener and L. E. Day, eds.), pp. 159–229. Academic Press, Orlando, Florida.

(19) Kendall, K. and Cohen, S. N. (1987). Plasmid transfer in *Streptomyces lividans*: identification of a *kil-kor* system associated with the transfer region of pIJ101. *J. Bacteriol.* **169**, 4177–4183.

(20) Evans, R. P., Winter, R. B. and Macrina, F. L. (1985). Molecular cloning of a pIP501 derivative yields a model replicon for the study of streptococcal conjugation. *J. Gen. Microbiol.* **131**, 145–153

(21) Clewell, D. B., White, B. A., Ike, Y. and An, F. A. (1984). Sex pheromones and plasmid transfer in *Streptococcus faecalis*. *In* "Microbial Development" (R. Losick and L. Shapiro, eds.), pp. 133–149. Cold Spring Harbor Laboratory, Cold Spring Harbor, New York.

(22) te Riele, H., Michel, B. and Ehrlich, S. D. (1986). Are single-stranded circles intermediates in DNA replication? *EMBO J.* **5**, 631–637.

(23) Deng, Z., Kieser, T. and Hopwood, D. A. (1988). "Strong incompatibility" between derivatives of the *Streptomyces* multi-copy plasmid pIJ101. *Mol. Gen. Genet.* (In press).

(24) Schrempf, H. and Pigac, J. (1986). Single-stranded plasmid DNA in *Streptomyces*. p. 41, Abstract Book, 5th International Symposium on the Genetics of Industrial Microorganisms, Split.

(25) Hirochika, H., Nakamura, K. and Sakaguchi, K. (1984). A linear DNA plasmid from *Streptomyces rochei* with an inverted terminal repetition of 614 base pairs. *EMBO J.* **3**, 761–766.

(26) Kinashi, H., Shimaji, M. and Sakai, A. (1987). Giant linear plasmids in *Streptomyces* which code for antibiotic biosynthesis genes. *Nature* **238**, 454–456.

(27) Plasterk, R. H. A., Simon, M. I. and Barbour, A. G. (1985). Transposition of structural genes to an expression sequence on a linear plasmid causes antigenic variation in the bacterium *Borrelia hermsii*. *Nature* **318**, 257–263.

(28) Bennett, P. (1985). Bacterial transposons. *In* "Genetics of Bacteria" (J. Scaife, D. Leach and A. Galizzi, eds.), pp. 97–115. Academic Press, London.

(29) Craig, N. L. and Kleckner, N. (1987). Transposition and site-specific recombination. In "*Escherichia coli* and *Salmonella typhimurium*. Cellular and Molecular Biology" (F. C. Neidhardt, J. L. Ingraham, K. B. Low, B. Magasanik, M. Schaechter and H. E. Umbarger, eds.), pp. 1054–1070. American Society for Microbiology, Washington, D.C.

(30) Berg, D. E., Kazic, T., Phadnis, S. H., Dodson, K. W. and Lodge, J. K. (1988). Mechanism and regulation of transposition. *In* "Transposition" (A. M. Kingsman, K. F. Chater and S. M. Kingsman, eds.), pp. 107–129. Cambridge University Press, Cambridge.

(31) Craigie, R., Mizuuchi, M., Adzuma, K. and Mizuuchi, K. (1988). Mechanisms of the DNA strand transfer step in transposition of Mu DNA. *In* "Transposition" (A. M. Kingsman, K. F. Chater and S. M. Kingsman, eds.), pp. 131–148. Cambridge University Press, Cambridge.

(32) Bennett, P. M., Heritage, I., Comanducci, A. and Dodd, H. M. (1986). Evolution of R-plasmids by replicon fusion. *J. Antimicrobial Chemotherapy* **18**, Suppl. C. 103–111.

(33) Gertmann, E., White, B. N., Berry D. and Kropinski, A. M. (1986). IS222, a new insertion element associated with the genome of *Pseudomonas aeruginosa*. *J. Bacteriol.* **166**, 1134–1136.

(34) Murphy, E. (1988). Transposable elements in *Staphylococcus*. *In* "Transposition" (A. M. Kingsman, K. F. Chater and S. M. Kingsman, eds.), pp. 59–89. Cambridge University Press, Cambridge.

(35) Ebert, K., Hanke, C., Delius, H., Goebel, W. and Pfeifer, F. (1987). A new insertion element, ISH26, from *Halobacterium halobium*. *Mol. Gen. Genet.* **206**, 81–87.

(36) Mahillon, J., Seurinck, J., van Rompuy, L., Delcour, J. and Zabeau, M. (1985). Nucleotide sequence and structural organization of an insertion sequence element (IS231) from *Bacillus thuringiensis* strain berliner 1715. *EMBO J.* **4**, 3895–3899.

(37) Machray, G. C., Vakeria, D., Codd, G. A. and Stewart, W. D. P. (1988). Insertion sequence IS*2* in the cyanobacterium *Chlorogloeopsis fritschii*. *Gene* **67**, 301–305.

(38) Lyon, B. R. and Skurray, R. (1987). Antimicrobial resistance of *Staphylococcus aureus*: genetic basis. *Microbiol. Rev.* **51**, 88–134.

(39) Whiteley, H. R. and Schnepf, H. E. (1986). The molecular biology of parasporal crystal body formation in *Bacillus thuringiensis*. *Ann. Rev. Microbiol.* **40**, 549–576.

(40) Perkins, J. and Youngman, P. J. (1984). A physical and functional analysis of Tn*917*, a *Streptococcus* transposon in the Tn*3* family that functions in *Bacillus*. *Plasmid* **12**, 119–138.

(41) Lereclus, D., Mahillon, J., Menou, G. and Lecadet, M-M. (1986). Identification of Tn*4430*, a transposon of *Bacillus thuringiensis* functional in *Escherichia coli*. *Mol. Gen. Genet.* **204**, 52–57.

(42) Olson, E. R. and Chung, S-T. (1988). Transposon Tn*4556* of *Streptomyces fradiae*: nucleotide sequence of the ends and the target sites. *J. Bacteriol.* **170**, 1955–1957.

(43) Krylov, V. N., Plotnikova, T. G., Kulakov, L. A., Fedorova, T. V. and Eremenko, E. N. (1982). Integration of Mu-like *Pseudomonas aeruginosa* bacteriophage D3112 genome into RP4 plasmid and its transfer by the hybrid plasmid into *Pseudomonas putida* and *Escherichia coli* C600 bacterial cells. *Genetika (USSR)* **18**, 5–12.

(44) Clewell, D. B., Senghas, E., Jones, J. M., Flannagan, S. E., Yamamoto, M. and Gawron-Burke, C. (1988). Transposition in *Streptococcus*: structural and genetic properties of the conjugative transposon Tn*916*. *In* "Transposition" (A. L. Kingsman, K. F. Chater and S. M. Kingsman, eds.), pp. 43–58. Cambridge University Press, Cambridge.

(45) Omer, C. A. and Cohen, S. N. (1986). Structural analysis of plasmid and chromosomal loci involved in site-specific excision and integration of the SLP1 element of *Streptomyces coelicolor*. *J. Bacteriol.* **166**, 999–1006.

(46) Brody, H., Greener, A. and Hill, C. W. (1985). Excision and reintegration of the *Escherichia coli* K12 chromosomal element e14. *J. Bacteriol.* **161**, 1112–1117.

(47) Schmidt, J. M. (1966). Observations on the adsorption of *Caulobacter* bacteriophages containing ribonucleic acid. *J. Gen. Microbiol.* **45**, 347–353.

(48) Roberts, R. C., Keefer, M. A. and Ranu, R. S. (1987). Characterization of *Bdellovibrio bacteriovorus* bacteriophage MAC-1. *J. Gen. Microbiol.* **133**, 3065–3070.

(49) Zahler, S. A. (1988). Temperate bacteriophages of *Bacillus subtilis*. *In* "The Bacteriophages, Vol. I" (R. Calender, ed.), pp. 559–592. Plenum Press, New York.

(50) Chater, K. F. (1986). *Streptomyces* phages and their application to *Streptomyces* genetics. *In* "The Bacteria. Vol. IX. Antibiotic-producing *Streptomyces*" (S. E. Queener and L. E. Day, eds.), pp. 119–158. Academic Press, Orlando, Florida.

(51) Sinclair, R. B. and Bibb, M. J. (1988). The repressor gene (c) of the *Streptomyces* temperate phage ϕC31: nucleotide sequence, analysis and functional cloning. *Mol. Gen. Genet.* (In press).

(52) Sternberg, N. and Hoess, R. (1983). The molecular genetics of bacteriophage P1. *Ann. Rev. Genet.* **17**, 123–154.

(53) Mindich, L. (1978). Bacteriophages that contain lipid. *In* "Comprehensive Virology. Vol. 12. Newly Characterized Protist and Invertebrate Viruses" (H. Fraenkel-Conrat and R. R. Wagner, eds.), pp. 271–335. Plenum Press, New York.

(54) Mellado, R. P. and Salas, M. (1983). Initiation of phage ϕ29 DNA replication by the terminal protein modified at the carboxyl end. *Nucleic Acids Res.* **11**, 7397–7407.

(55) Escarmis, C., Gomez, A., Garcia, E., Ronda, C., Lopez, R. and Salas, M. (1984). Nucleotide sequence at the termini of the DNA of *Streptococcus pneumoniae* phage Cp-1. *Virology* **133**, 166–171.

(56) Holloway, B. W. (1979). Plasmids that mobilize bacterial chromosome. *Plasmid* **2**, 1–19.

(57) Masters, M. (1985). Generalized transduction. *In* "Genetics of Bacteria" (J. Scaife, D. Leach and A. Galizzi, eds.), pp. 197–215. Academic Press, London.

(58) Margolin, P. (1987). Generalized transduction. *In* "*Escherichia coli* and *Salmonella typhimurium*. Cellular and Molecular Biology" (F. C. Neidhardt, J. L. Ingraham, K. B. Low, B. Magasanik, M. Schaechter and H. E. Umbarger, eds.), pp. 1154–1168. American Society for Microbiology, Washington, D.C.

(59) Dubnau, D., Goldthwaite, C., Smith, I. and Marmur, J. (1967). Genetic mapping in *Bacillus subtilis*. *J. Mol. Biol.* **27**, 163–185.

(60) C. Stuttard (1983). Cotransduction of *his* and *trp* loci by phage SV1 in *Streptomyces venezuelae*. *FEMS Microbiol. Lett.* **20**, 467–470.

(61) Schroeder, C. J. and Pattee, P. A. (1984). Transduction analysis of transposon Tn*551* insertion in the *trp-thy* region of the *Staphylococcus aureus* chromosome. *J. Bacteriol.* **157**, 533–537.

(62) Buchanan-Wollaston, V. (1979). Generalized transduction in *Rhizobium leguminosarum*. *J. Gen. Microbiol.* **112**, 135–142.

(63) Svab, Z., Kondorosi, A. and Orosz, L. (1978). Specialized transduction of a cysteine marker by *Rhizobium meliloti* phage 16–3. *J. Gen. Microbiol.* **106**, 321–327.

(64) Lydiate, D. J., Henderson, D. J., Ashby, A. M. and Hopwood, D. A. (1987). Transposable elements of *Streptomyces coelicolor* A3(2). *In* "Genetics of Industrial Microorganisms, Part B" (M. Alačević, D. Hranueli and Z. Toman, eds.), pp. 49–56. Pliva, Zagreb.

(65) Weisberg, R. A. (1987). Specialized transduction. *In* "*Escherichia coli* and *Salmonella typhimurium*. Cellular and Molecular Biology" (F. C. Neidhardt, J. L. Ingraham, K. B. Low, B. Magasanik, M. Schaechter and H. E. Umbarger, eds.), pp. 1169–1176. American Society for Microbiology, Washington, D.C.

(66) Aghion, J., Albano, M., Guillen, N., Hahn, J., Mohan, S., Weinrauch, Y. and Dubnau, D. (1988). Gene interactions in the development of genetic competence. Abstract 39, Program of the 10th International Spores Conference, Woods Hole, Massachusetts.

(67) Smith, H. O. and Danner, D. B. (1981). Genetic transformation. *Ann. Rev. Biochem.* **50**, 41–68.

(68) Avery, O. T., Macleod, C. M. and McCarty, M. (1944). Studies on the chemical nature of the substance inducing transformation of pneumococcal types. I. Induction of transformation by a deoxyribonucleic acid fraction isolated from pneumococcus type III. *J. Exptl. Med.* **79**, 137–157.

(69) Claverys, J-P. and Lacks, S. A. (1986). Heteroduplex deoxyribonucleic acid base mismatch repair in bacteria. *Microbiol. Rev.* **50**, 133–165.

(70) Marrs, B. (1974). Genetic recombination in *Rhodopseudomonas capsulata*. *Proc. Natl. Acad. Sci. USA* **71**, 971–973.

(71) Zillig, W., Schnabel, R., Gropp, F., Reiter, W. D., Stetter, K. and Thomm, M. (1985). The evolution of the transcription apparatus. *In* "Evolution of Prokaryotes" (K. H. Schleifer and E. Stackebrandt, eds.), pp. 45–72. Academic Press, London.

(72) Helmann, J. D. and Chamberlin, M. J. (1988). Structure and function of bacterial sigma factors. *Ann. Rev. Biochem.* **57**, 839–872.

(73) McClure, W. R. (1985). Mechanism and control of transcription initiation in prokaryotes. *Ann. Rev. Biochem.* **4**, 171–204.

(74) Gill, P. R. and Agabian, N. (1982). A comparative structural analysis of the flagellin monomers of *Caulobacter crescentus* indicates that these proteins are encoded by two genes. *J. Bacteriol.* **150**, 925–933.

(75) Whiteley, H. R. and Hemphill, H. E. (1970). The interchangeability of stimulatory factors isolated from three microbial RNA polymerases. *Biochem. Biophys. Res. Commn.* **41**, 647–654.

(76) Shorenstein, R. G. and Losick, R. (1973). Comparative size and properties of the sigma subunits of ribonucleic acid polymerase from *Bacillus subtilis* and *Escherichia coli*. *J. Biol. Chem.* **248**, 6170–6173.

(77) Westpheling, J., Ranes, M. and Losick, R. (1985). RNA polymerase heterogeneity in *Streptomyces coelicolor*. *Nature* **313**, 22–27.

(78) Bibb, M. J. and Cohen, S. N. (1982). Gene expression in *Streptomyces*: construction and application of promoter-probe plasmid vectors in *Streptomyces lividans*. *Mol. Gen. Genet.* **187**, 265–277.

(79) Johnston, A. W. B., Bibb, M. J. and Beringer, J. E. (1978). Tryptophan genes in *Rhizobium*—their organization and their transfer to other bacterial genera. *Mol. Gen. Genet.* **165**, 323–330.

(80) Peschke, U., Beuck, V., Bujard, H., Gentz, R. and LeGrice, S. (1985). Efficient utilization of *Escherichia coli* transcriptional signals in *Bacillus subtilis*. *J. Mol. Biol.* **186**, 547–555.

(81) Hopwood, D. A., Bibb, M. J., Chater, K. F., Janssen, G. R., Malpartida, F. and Smith, C. P. (1986). Regulation of gene expression in antibiotic-producing *Streptomyces*. *In* "Regulation of Gene Expression—25 Years On" (I. Booth and C. Higgins, eds.), pp. 251–276. Cambridge University Press, Cambridge.

(82) Newbury, S. F., Smith, N. H., Robinson, E. C., Hiles, I. D. and Higgins, C. F. (1987). Stabilization of translationally active mRNA by prokaryotic REP sequences. *Cell* **48**, 297–310.

(83) Chen, C-Y. A., Beatty, J. T., Cohen, S. N. and Belasco, J. G. (1988). An intercistronic stem-loop structure functions as an mRNA decay terminator necessary but insufficient for *puf* mRNA stability. *Cell* **52**, 609–619.

(84) Galloway, J. L. and Platt, T. (1985). Control of prokaryotic gene expression by transcription termination. *In* "Regulation of Gene Expression—25 Years On" (I. R. Booth and C. F. Higgins, eds.), pp. 155–178. Cambridge University Press, Cambridge.

(85) Yager, T. D. and Von Hippel, P. H. (1987). Transcript elongation and termination in *Escherichia coli*. *In* "*Escherichia coli* and *Salmonella typhimurium*. Cellular and Molecular Biology" (F. C. Neidhardt, J. L. Ingraham, K. B. Low, B. Magasanik, M. Schaechter and H. E. Umbarger, eds.), pp. 1241–1275. American Society for Microbiology, Washington, D.C.

(86) Gold, L. and Stormo, G. (1987). Translational initiation. *In* "*Escherichia coli* and *Salmonella typhimurium*. Cellular and Molecular Biology" (F. C. Neidhardt, J. L. Ingraham, K. B. Low, B. Magasanik, M. Schaechter and H. E. Umbarger, eds.), pp. 1302–1307. American Society for Microbiology, Washington, D.C.

(87) McClaughlin, J. R., Murray, C. L. and Rabinowitz, J. C. (1981). Unique features of the ribosome binding site sequence of the Gram-positive *Staphylococcus aureus* β-lactamase gene. *J. Biol. Chem.* **256**, 11283–11291.

(88) Walz, A., Pirrotta, V. and Ineichen, K. (1976). Lambda repressor regulates the switch between P_r and P_{rm} promoters. *Nature* **262**, 665–669.

(89) Hayer, P. W. and Rabinowitz, J. C. (1985). Translational specificity in *Bacillus subtilis*. *In* "The Molecular Biology of Bacilli. Vol. 2" (D. A. Dubnau, ed.), pp. 1–32. Academic Press, Orlando, Florida.

(90) Bibb, M. J., Findlay, P. R., and Johnson, M. W. (1984). The relationship between base composition and codon usage in bacterial genes and its use in the simple and reliable identification of protein-coding sequences. *Gene* **30**, 157–166.

(91) Sueoka, N. (1961). Compositional correlation between deoxyribonucleic acid and protein. *Cold Spring Harbor Symp. Quant. Biol.* **26**, 35–43.
(92) Sharp, P. M. and Bulmer, M. (1988). Selective differences among translation termination codons. *Gene* **63**, 141–145.
(93) Neidhardt, F. C. (1987). Multigene systems and regulons. *In* "*Escherichia coli* and *Salmonella typhimurium.* Cellular and Molecular Biology" (F. C. Neidhardt, J. L. Ingraham, K. B. Low, B. Magasanik, M. Schaechter and H. E. Umbarger, eds.), pp. 1313–1317. American Society for Microbiology, Washington, D.C.
(94) Hoopes, B. C. and McClure, W. R. (1987). Strategies in regulation of transcription initiation. *In* "*Escherichia coli* and *Salmonella typhimurium.* Cellular and Molecular Biology" (F. C. Neidhardt, J. L. Ingraham, K. B. Low, B. Magasanik, M. Schaechter and H. E. Umbarger, eds.), pp. 1231–1240. American Society for Microbiology, Washington, D.C.
(95) Beckwith, J. (1987). The lactose operon. *In* "*Escherichia coli* and *Salmonella typhimurium.* Cellular and Molecular Biology" (F. C. Neidhardt, J. L. Ingraham, K. B. Low, B. Magasanik, M. Schaechter and H. E. Umbarger, eds.), pp. 1444–1452. American Society for Microbiology, Washington, D.C.
(96) Gottesman, S. (1987). Regulation by proteolysis. *In* "*Escherichia coli* and *Salmonella typhimurium.* Cellular and Molecular Biology" (F. C. Neidhardt, J. L. Ingraham, K. B. Low, B. Magasanik, M. Schaechter and H. E. Umbarger, eds.), pp. 1308–1312. American Society for Microbiology, Washington, D.C.
(97) Fujita, Y., Miwa, Y. and Fujita, T. (1988). Molecular mechanism of the gluconate-inducible system of the *Bacillus subtilis gnt* operon. Abstract 45 in Program of the 10th International Spores Conference, Woods Hole, Massachusetts.
(98) Schwarz, M. (1987). The maltose regulon. *In* "*Escherichia coli* and *Salmonella typhimurium.* Cellular and Molecular Biology" (F. C. Neidhardt, J. L. Ingraham, K. B. Low, B. Magasanik, M. Schaechter and H. E. Umbarger, eds.), pp. 1482–1502. American Society for Microbiology, Washington, D.C.
(99) Ronson, C. W., Nixon, B. T. and Ausubel, F. M. (1987). Conserved domains in bacterial regulatory proteins that respond to environmental stimuli. *Cell* **49**, 579–581.
(100) Smith, C. P. and Chater, K. F. (1987). Physiology, genetics and molecular biology of glycerol utilization in *Streptomyces coelicolor. In* "Genetics of Industrial Microorganisms, Part B" (M. Alačević, D. Hranueli and Z. Toman, eds.), pp. 7–15, Pliva, Zagreb.
(101) Savageau, M. A. (1977). Design of molecular control mechanisms and the demand for gene expression. *Proc. Natl. Acad. Sci. USA* **74**, 5647–5651.
(102) Magasanik, B. and Neidhardt, F. C. (1987). Regulation of carbon and nitrogen utilization. *In* "*Escherichia coli* and *Salmonella typhimurium.* Cellular and Molecular Biology" (F. C. Neidhardt, J. L. Ingraham, K. B. Low, B. Magasanik, M. Schaechter and H. E. Umbarger, eds.), pp. 1318–1325. American Society for Microbiology, Washington, D.C.
(103) Setlow, P. (1973). Inability to detect cyclic AMP in vegetative or sporulating cells or dormant spores of *Bacillus megaterium. Biochem. Biophys. Res. Commn.* **52**, 365–372.
(104) Hodgson, D. A. (1980). Carbohydrate utilization in *Streptomyces coelicolor* A3(2). Ph.D. Thesis, University of East Anglia, Norwich.
(105) Yanofsky, C. (1988). Transcription attenuation. *J. Biol. Chem.* **263**, 609–612.
(106) Weisblum, B. (1983). Inducible macrolide-lincosamide-streptogramin resistance: a review of the resistance phenotype, its biological diversity, and structural elements which regulate expression. *In* "Gene Function in Prokaryotes" (J. Beckwith, J. Davies and J. A. Gallant, eds.), pp. 91–121. Cold Spring Harbor Laboratory, Cold Spring Harbor, New York.
(107) Dubnau, D. (1984). Translational attenuation: the regulation of bacterial resistance to the macrolide-lincosamide-streptogramin B antibiotics. *Crit. Rev. Biochem.* **16**, 103–132.
(108) Mahadevan, S. and Wright, A. (1987). A bacterial gene involved in transcription termination: regulation at a rho-independent terminator in the *bgl* operon of *E. coli. Cell* **50**, 485–494.

(109) Distler, J., Ebert, A., Mansouri, K., Pissowotzki, K., Stockmann, M. and Piepersberg, W. (1987). Gene cluster for streptomycin biosynthesis in *Streptomyces griseus*: nucleotide sequence of three genes and analysis of transcriptional activity. *Nucleic Acids Res.* **15**, 8041–8056.
(110) Carter, H. L., Wang, L-F., Doi, R. H. and Moran, C. P. (1988). *rpoD* operon promoter used by σ^{H}-RNA polymerase in *Bacillus subtilis*. *J. Bacteriol.* **170**, 1617–1621.
(111) Green, P. J., Pines, O. and Inouye, M. (1986). The role of antisense RNA in gene regulation. *Ann. Rev. Biochem.* **55**, 569–597.
(112) Meyer, T. F. and Haas, R. (1988). Phase and antigenic variation by DNA rearrangements in prokaryotes. *In* "Transposition" (A. M. Kingsman, K. F. Chater and S. M. Kingsman, eds.), pp. 193–219. Cambridge University Press, Cambridge.
(113) Haselkorn, R., Golden, J. W., Lammers, P. J. and Mulligan, M. E. (1987). Rearrangement of *nif* genes during cyanobacterial heterocyst differentiation. *Phil. Trans. R. Soc. Lond. B* **317**, 173–181.
(114) Alberts, B., Bray, D., Lewis, D., Raff, M., Roberts, K. and Watson, J. D. (1983). Molecular Biology of the Cell. Garland Publishing Inc., New York.
(115) Schmid, M. (1988). Structure and function of the bacterial chromosome. *Trends in Biochemical Science* **13**, 131–135.
(116) Austin, S. and Abeles, A. (1985). The partition functions of P1, P7 and F miniplasmids. *In* "Plasmids in Bacteria" (D. R. Helinski *et al.*, eds.), pp. 215–216. Plenum, New York.
(117) Rush, M. G. and Misra, R. (1985). Extrachromosomal DNA in eukaryotes. *Plasmid* **14**, 177–191.
(118) Hopwood, D. A. (1981). Genetic studies with bacterial protoplasts. *Ann. Rev. Microbiol.* **35**, 237–272.
(119) Mevarech, M. and Werczberger, R. (1985). Genetic transfer in *Halobacterium volcanii*. *J. Bacteriol.* **162**, 461–462.
(120) De la Salle, H., Jacq, C. and Slonimski, P. P. (1982). Critical sequences within mitochondrial introns: pleiotropic mRNA maturase and *cis*-dominant signals of the *box* intron controlling reductase and oxidase. *Cell* **28**, 721–732.
(121) Hall, D. H., Povinelli, C. M., Ehrenman, K., Pederson-Lane, J., Chu, F. and Belfort, M. (1987). Two domains for splicing in the intron of the phage T4 thymidylate synthase (*td*) gene established by non-directed mutagenesis. *Cell* **48**, 63–71.
(122) Daniels, C. J., Gupta, R. and Doolittle, W. F. (1985). Transcription and excision of a large intron in the $tRNA^{Trp}$ gene of an archaebacterium, *Halobacterium volcanii*. *J. Biol. Chem.* **260**, 3132–3134.
(123) Bukhari, A. I., Shapiro, J. A. and Adhya, S. L. (1977). DNA Insertion Elements, Plasmids and Episomes. Cold Spring Harbor Laboratory, Cold Spring Harbor, New York.
(124) Comai, L. and Kosuge, T. (1980). Involvement of plasmid deoxyribonucleic acid in indoleacetic acid synthesis in *Pseudomonas savastanoi*. *J. Bacteriol.* **143**, 950–957.
(125) Buttner, M. J., Smith, A. M. and Bibb, M. J. (1988). At least three different RNA polymerase holoenzymes direct transcription of the agarase gene (*dagA*) of *Streptomyces coelicolor* A3(2). *Cell* **52**, 599–607.
(126) Van de Putte, P. (1988). Site-specific inversion in bacteriophage Mu. *In* "Transposition" (A. M. Kingsman, K. F. Chater and S. M. Kingsman, eds.), pp. 183–192. Cambridge University Press, Cambridge.
(127) Murphy, E. and Novick, R. P. (1979). Physical mapping of *Staphylococcus aureus* penicillinase plasmid pI524: characterization of an invertible region. *Mol. Gen. Genet.* **175**, 19–30.
(128) Echols, H. and Guarneros, G. (1983). Control of integration and excision. *In* "Lambda II" (R. W. Hendrix, J. W. Roberts, F. W. Stahl and R. A. Weinberg, eds.), pp. 75–92. Cold Spring Harbor Laboratory, Cold Spring Harbor, New York.

Chapter 3

Cloning and Molecular Analysis of Bacterial Genes

K. F. CHATER *and* D. A. HOPWOOD

I. INTRODUCTION

The contents of this book amply illustrate the power of modern molecular genetics: the ability to isolate and characterize genetic determinants, to manipulate them outside the cell, and to examine the phenotypic consequences of these changes after reintroducing the changed DNA into the host. The basic procedures of molecular genetics are described in numerous excellent textbooks and manuals (1–3), and we need not review them here. Instead, this chapter discusses the cloning systems available for bacteria other than *Escherichia coli*, and the ways in which their DNA, once cloned, can be made to provide information to the experimenter. Many of these techniques are referred to in the specialized chapters that follow.

II. CLONING BACTERIAL DNA

A. *E. coli* can be Helpful in Primary Cloning

It is often possible to isolate genes from other bacteria by first cloning them in *E. coli*. Natural expression signals, within the constraints of heterologous gene expression (Chapter 2), or expression from *E. coli* signals in appropriate lambda or plasmid vectors, may sometimes allow synthesis of sufficient gene product to give a perceptible phenotype, even if this may require a sensitive immunological test to reveal it: the λgt11 system, in which translational fusions to the *E. coli lacZ* gene allow the expression in plaques of foreign proteins that can be detected by antibody reactions, is a particularly powerful one (4). Alternatively, clones can be detected with hybridization probes, usually synthetic oligonucleotides devised from a knowledge of a protein sequence. The task of choosing appropriate synthetic probes can be simplified for organisms with a (G + C) content far from the random value of 50% because such bacteria preferentially use codons with A or T, or alternatively G or C, in the third position (5). The most useful probes are then provided by long "guessomers" of ca. 40–60 bases synthesized assuming an appropriately non-random codon choice, rather than by mixtures of oligonucleotides containing all possible short sequences of DNA that could code for the desired gene product: *Streptomyces* provides several examples (6). Alternatively, success has been reported for somewhat shorter probes with the "neutral" base inosine in the third codon position (7).

Many other cloning strategies depend on painstakingly acquired knowledge of the organism in which the interesting genes normally reside. Particularly valuable approaches have depended on the availability for that organism either of transposon mutagenesis or of procedures to introduce cloned DNA into it for functional tests—usually for the complementation of mutants. These approaches are described next.

B. Wide Host-Range Vectors for Purple Bacteria

The transformation systems available for several kinds of Gram-negative purple bacteria are seldom as efficient as those developed for *E. coli*. Moreover, the cloning of large DNA fragments in cosmid vectors containing the *cos* site, followed by packaging into λ particles, is possible only if *E. coli* is used as an intermediate host, since λ adsorption (to the LamB protein) is *E. coli*-specific. Hence, much of the cloning done in *Vibrio* (Chapter 4), *Rhodobacter* (Chapter 5), *Klebsiella* and *Azotobacter* (Chapter 6), *Pseudomonas* (Chapter 8), *Caulobacter* (Chapter 10), *Xanthomonas* (Chapter 17), *Agrobacterium* (Chapter 18) and *Rhizobium* (Chapter 19) has involved initial establishment of a DNA library in

E. coli on vectors containing the replication and transfer functions of wide host-range transmissible plasmids, with subsequent transfer of the clones to the host of choice. A particularly useful strategy has been to clone large DNA segments in *E. coli* on a shuttle cosmid such as pLAFR1, which carries the replication region of a P-group plasmid (8), and to use them to complement various kinds of mutants, like those defective in nodulation in *Rhizobium*, pathogenicity in *Xanthomonas* and motility in *Caulobacter*. Interestingly, the cloned *lamB* gene can be artificially introduced into other purple bacteria, which can then absorb λ (Chapter 15). This might allow cosmid libraries to be used directly on hosts other than *E. coli*—a strategy that would be advantageous if there were a significant restriction barrier between *E. coli* and the host in question.

C. Transposon Mutagenesis in Gram-negative Bacteria

The potential of transposons as molecular genetic tools was clearly seen early in the history of transposon biology (9) and has been fully realised not only in bacteria, but also in eukaryotes. Transposon delivery systems based on the inability of many *E. coli* vectors to replicate in certain bacteria other than *E. coli* have been widely exploited for genomic mutagenesis in purple bacteria. For example, the *E. coli* transducing phage P1 injects DNA into *Myxococcus xanthus* without producing either lysis or a stable P1 prophage. Thus, infection with a P1::Tn*5* derivative, followed by selection for the antibiotic resistance of the Tn*5* derivative, allowed the assembly of a large set of *M. xanthus* strains carrying Tn*5* insertions (Chapter 12). These strains can be screened for interesting phenotypes that may result from transposon insertion, and the transposon plus flanking sequences can then be cloned from the relevant mutants into *E. coli* with selection for the resistance determinant of the transposon. The flanking sequences, used as hybridization probes, in turn allow the uninterrupted sequence to be isolated and its functional integrity can subsequently be confirmed by complementation of the transposon-induced mutant.

Transposon mutagenesis is also widely used in an analytical mode on more closely defined target sequences. Chapter 19 provides an example. Here, an *E. coli* strain carrying a cloned DNA fragment encoding nodulation functions in a wide host-range plasmid vector was subjected to transposon mutagenesis, and plasmids from the resulting strain collection were introduced into a *Rhizobium* strain lacking a nodulation plasmid, with selection for the resistance determinant of the transposon. The locations of the transposon in those plasmids that conferred various kinds of mutant phenotypes helped to define the relevant genes, and led to construction of a combined physical and functional map.

Table I
Some genes that are widely used as reporters

Reporter gene	Detection	Comments on utility
lacZ from *E. coli* (31)	β-galactosidase activity, detected by cleavage of: (i) 5-bromo-4-chloro-3-indolyl-β-D-galactoside (X-gal); (ii) 4-methylumbelliferyl-β-D-galactoside (MUG); (iii) *O*-nitrophenyl-β-D-galactoside (ONPG).	X-gal or MUG give high sensitivity in plates. X-gal is expensive. MUG gives highly diffusible fluorescent umbelliferone. Many selective procedures for increased or decreased expression of *lacZ* (32). Requirement for subunit interactions has been exploited in looking for extremely rarely expressed promoters. Applicable in most bacteria (but gives various difficulties in *Streptomyces*: 33). Often requires physiological or genetic elimination of host β-galactosidase.
galK from *E. coli* (34)	Galactokinase activity, detected on plates by galactose utilization or by pH change during galactose fermentation. Assayed: (i) immunologically; (ii) by filter-binding of ^{14}C-galactose-1-phosphate; (iii) by phosphorylation of galactose coupled to NADH oxidation.	Moderate sensitivity in plates. Selective procedures for increased or decreased expression. Has been successfully applied in *Streptomyces*. Usually requires physiological or genetic elimination of host galactokinase.
phoA from *E. coli* (35)	Alkaline phosphatase, detected by cleavage of: (i) *p*-nitrophenyl phosphate; (ii) 5-bromo-4-chloro-3-indoyl phosphate (X*P*).	X*P* is expensive but sensitive. Alkaline phosphatase readily exported when 5′-terminally deleted *phoA* gene is fused to export signal sequences. Dimerization needed for activity: dimers cannot form in cytoplasm, hence activity/inactivity of *phoA* protein fusions indicates whether fusion point is on outside or inside of membrane. Widely applicable; often used in conjunction with inorganic phosphate to repress endogenous activity.
luxAB from *Vibrio harveyi* (36)	Luminescence in presence of *N*-decanal vapour, detected: (i) visually or photographically using image intensifiers; (ii) by direct exposure of X-ray films to colonies; (iii) photometrically.	Allows real-time non-lethal detection and spatial localization of activity in colonies. Sensitivity depends on sensitivity of light-detection equipment—can be very high, and quantitative over huge range. Very widely applicable; usually no interfering host activities.

Table I
Continued

Reporter gene	Detection	Comments on utility
xylE from TOL plasmid of *Pseudomonas* (37)	Catechol oxygenase, detected by conversion of catechol to a yellow pigment.	Catechol is cheap and non-lethal. Assay sensitive and simple. Widely applicable. Usually no complicating host activities.
cat-86 from *Bacillus pumilis* (38)	Chloramphenicol acetyltransferase, detected by: (i) chloramphenicol (Cm) resistance; (ii) chromatographic shift of ^{14}C Cm after acetylation; (iii) spectrophotometric assay (Co-ASH formation).	Selection on plates for *cat* expression. Use largely confined to *B. subtilis* though probably more widely applicable.
ampC from *E. coli* chromosome (39)	β-lactamase, detected by: (i) resistance to β-lactams; (ii) cleavage of nitrocefin; (iii) reduction of β-lactam activity.	Selection on plates for *ampC* expression; visual screening (using nitrocefin) for increased or decreased expression. Widely applicable.
neo from Tn*5* (40)	Aminoglycoside phosphotransferase, detected by: (i) resistance to neomycin or kanamycin; (ii) aminoglycoside-dependent incorporation of ^{32}P into material that binds to phosphocellulose; (iii) phosphorylation of amino-glycoside coupled to NADH oxidation.	Selection for aminoglycoside resistance. Probably widely applicable for vegetatively expressed promoters.

In some cases a modified transposon contains a promoterless "reporter" gene, whose activation, by readthrough from adjacent host promoters, can readily be detected and assayed. By this means, additional information about the location, direction and regulation of transcription signals can be obtained. In this book, examples include the use of Tn*5*::*lac* in *M. xanthus* (Chapter 12), of Tn*phoA* on toxin genes in *Vibrio cholerae* (Chapter 15), and of Mu::*lac* derivatives on *lux* DNA of *Vibrio harveyi* (Chapter 4) and ColV DNA of *E. coli* (Chapter 16); we return to this technique in Section II.E. Of course, fusion of cloned genes can also be carried out *in vitro*, and many plasmids and "cassettes" have been constructed to allow this to be done efficiently. A list of reporter genes specifying conveniently assayed enzymes is given in Table 1.

D. Cloning and Analysis of DNA in Gram-positive Bacteria

Although plasmids naturally capable of replicating in both Gram-positive and Gram-negative bacteria have now been discovered (Chapter 1), current cloning vectors designed for use in Gram-negative bacteria do not replicate in Gram-positive organisms. Moreover, the selective markers of these vectors are seldom well-expressed by low (G + C) Gram-positive hosts, though the high (G + C) streptomycetes appear to be more permissive for expression (Chapter 2). For these reasons plasmid or phage replicons of Gram-positive origin have been developed into vectors, usually by the addition of suitable resistance determinants also from Gram-positive bacteria.

As a rule, plasmid vectors for low (G + C) Gram-positive species have a fairly wide host-range within this group; small *Staphylococcus* plasmids were crucial in the development of cloning vectors for *Bacillus* (10), and pAMβ1 and pIP501 are conjugative between a wide variety of genera (11); but *Bacillus* plasmid vectors are not usable in lactic streptococci (12). Several plasmids developed for cloning in the high (G + C) *Streptomyces* have at least some ability to replicate in other actinomycetes. However, various unpublished experiments indicate that there is a significant barrier between the low and high (G + C) groups. Phage vectors usually infect only members of a single genus.

In several cases, plasmid vectors for Gram-positive bacteria have been further modified by the addition of *E. coli* replicons and selective markers to produce, for example, bifunctional cosmid vectors for *Streptomyces* (13) and pBR322/pAMβ1-based vectors capable of transfer by conjugation from *E. coli* into various low (G + C) Gram-positive species using mobilization by the IncP group plasmid RK2 (14).

Transformation of Gram-positive bacteria by plasmids or transfection by phage DNA usually involves the exposure of artificially produced protoplasts (Fig. 1), obtained by lysozyme treatment, to DNA in the presence of polyethylene glycol: a technique first developed for *Streptomyces* (15). Natural transformation may also be used when the host is competent (*Bacillus subtilis* being a prime example), but the vector systems have to be specially developed to allow for the DNA processing involved in natural transformation since monomeric plasmids cannot usually reestablish their replication when introduced by this means into the competent recipient (10).

Transposon mutagenesis in Gram-positive species has been limited by the failure of transposons from Gram-negative bacteria to work well in Gram-positive species, and by the difficulty of demonstrating and developing suitable

---

Figure 1. *Streptomyces* protoplasts. When cell walls are removed by lysozyme treatment from the *Streptomyces* filaments (top), protoplasts (bottom) are released, which can be induced to take up DNA in the presence of polyethylene glycol (15).

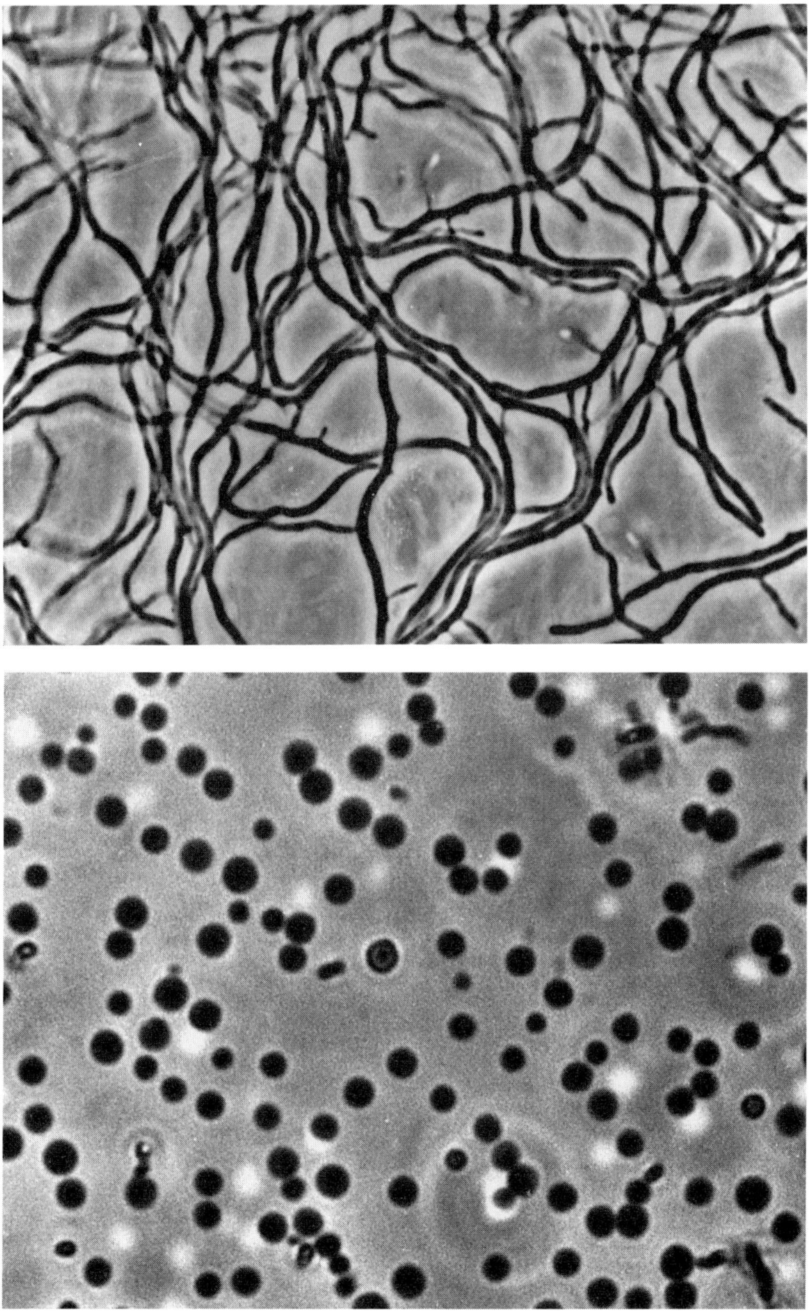

Figure 1.

endogenous systems in organisms where genetic tools and knowledge are more limited than in *E. coli*. The discovery of Tn*3*-like transposons in both low and high (G + C) species (Chapter 2) has changed this situation. Thus, libraries of *B. subtilis* carrying derivatives of Tn*917* have been generated (Chapter 11), and the organization of the primary Tn*917* inserts can readily be changed, to generate gene fusions or to insert *E. coli* replicons for rapid cloning of adjacent DNA into *E. coli*, by natural transformation with linear fragments of differently constructed Tn*917* DNA. The *Streptomyces* transposon Tn*4556* has yet to find extensive use, though early indications are promising (16).

E. Isolation of Sets of Coregulated but Scattered Genes

With the discovery of increasing numbers of sigma factors and other regulatory elements that govern the activity of sets of scattered genes such as sporulation genes, or stress response genes (Chapter 2), attention has recently focused on the enumeration, isolation and detailed analysis of unknown genes that fall into such regulons. This has been achieved in several ways.

Some procedures depend on the presence of at least moderately abundant mRNA species in cells in appropriate physiological or developmental conditions. Thus, in one early influential set of experiments, the occurrence of certain relatively stable mRNA species during *B. subtilis* sporulation allowed the isolation of an *in vivo* radiolabelled probe for a sporulation-specific gene (17), eventually leading on to the discovery that cascades of sigma factors regulate sporulation (Chapter 11). The general use of prokaryotic mRNA as a probe or as a source of cDNA has, however, been prevented by its great instability, and by the difficulty of separating mRNA from ribosomal RNA in the general absence of polyadenylylation of mRNA in prokaryotes (contrasted with its presence in eukaryotes). Despite these difficulties, C. Mathiopoulos and A. L. Sonenshein (personal communication) have used radioactive cDNA probes made with mRNA from suitable *B. subtilis* cells to identify genes expressed during the initiation of sporulation. In an *in vitro* equivalent of these experiments, rare genes that depend on a minor sigma factor (σ^{28} of *B. subtilis*: Chapter 2) were isolated using a radioactive probe generated during *in vitro* transcription of total *B. subtilis* DNA by RNA polymerase containing σ^{28} (18).

Ingenious strategies that bypass the technical demands of these procedures by exploiting gene fusions have also been devised. Among a large collection of Tn*917*::*lac* insertions into the *B. subtilis* chromosome isolated in the laboratory of R. Yasbin, there were many cases where a transcriptional fusion of *lacZ* to a chromosomal promoter had occurred. Promoters responsive to a particular signal could be identified because they gave rise to a positive colour test for β-galactosidase activity only in the relevant circumstances. This has been exploited by A. D. Grossman and co-workers (personal communication) in the identification of promoters that depend on the *spo*OH sigma factor involved in sporulation; by D. Dubnau and co-workers (personal communication) to analyse gene interactions in the development of competence for transformation; and by R. Yasbin and co-workers (personal communication) to isolate mitomycin C-inducible promoters in the study of the SOB regulon (equivalent to the *E. coli* SOS regulon).

III. MUTAGENESIS WITH CLONED DNA

We have already referred to the use of transposons in mutagenizing cloned DNA. Other widely used mutagenic procedures (2) include: the *in vitro* insertion of marker genes as cassettes into restriction sites; deletion of DNA using endo- or exonucleolytic treatments; localized chemical mutagenesis; and site-directed mutagenesis involving oligonucleotide-directed base changes or the insertion or deletion of a few bases at restriction sites.

It is often important to use cloned DNA to replace the wild-type allele, in its natural genomic location, with a defined mutant allele. Three kinds of technique have been used for this purpose. In the procedure of gene replacement, a segment of mutated DNA is used to replace a homologous sequence in the chromosome by double crossing-over. If the mutant segment is selectable, as it is when a marker gene has been inserted (perhaps on a transposon) into or adjacent to it, procedures such as natural transformation and transduction, which lead to the uptake of linear DNA and its efficient recombination with homologous DNA in the chromosome, can often be readily applied (Chapter 18). In *Rhodobacter capsulatus*, with capsduction (Chapter 2) as the delivery system, this procedure is called interposon mutagenesis (Chapter 5).

In a second procedure, a host gene (or operon) is disrupted by a single homologous crossover with an internal fragment of the gene (or operon) cloned in a vector that cannot be maintained independently in the host, but which contains a selectable marker. Such a recombination event between two circular DNA molecules is often referred to as "Campbell-type" recombination, in recognition of A. M. Campbell's deduction that a single crossover between a circular molecule of phage λ DNA and the *E. coli* chromosome gives the integrated λ prophage (19). Such gene disruption has been used in *B. subtilis* with an *E. coli* plasmid as the vector (Chapter 11), in *V. cholerae* using a λ vector in a host containing a cloned *lamB* gene to allow phage adsorption (Chapter 15), and in *Streptomyces coelicolor* using phage ϕC31 vectors lacking the *attP* site by which ϕC31 DNA is usually integrated into the chromosome. In the last case the procedure has also been used for primary cloning (mutational cloning) when the generation of a mutant indicates the isolation of the desired DNA (Chapter 7).

The third procedure, which uses antisense RNA (Chapter 2), is particularly attractive, but its wide applicability is yet to be proven. The principle is to clone part of a gene—preferably including the ribosome binding site—next to a conveniently regulated strong promoter, in an orientation such that activation of the promoter results in production of abundant antisense RNA, which may hybridize with the sense transcript and so prevent its translation. Model systems have confirmed that this procedure can prevent gene expression (20).

IV. BIOCHEMICAL PROCEDURES THAT EXPLOIT CLONED DNA

A. Maxicells, Minicells and *in vitro* Protein Synthesis

If a cloned DNA segment can direct protein synthesis in *E. coli*, the protein product may be detected and its molecular weight and location (cytoplasm/membrane/periplasm) determined using maxicells or minicells. In the maxicell system (1, 21) the gene is expressed in the presence of a radiolabelled amino acid, in a culture that has been heavily irradiated with UV to introduce so many lesions into the host genome that it becomes completely degraded by the repair system. The cloned gene is introduced either after irradiation as part of a λ phage genome or before irradiation as part of a replicon that is a very small target for UV inactivation. The labelled proteins that are synthesized from the cloned gene and its vector are analysed by polyacrylamide gel electrophoresis and fluorography. In the minicell system (1, 22), the gene is cloned in a multicopy plasmid in an *E. coli* mutant strain that produces small daughter cells lacking a chromosome. Such minicells can be purified away from the normal cells by centrifugation, and in them only plasmid-encoded proteins are made. These can again be analysed following radiolabelling.

Unfortunately, maxicell and minicell analysis are largely confined to *E. coli*, though an effective minicell system has been developed in *B. subtilis* (Fig. 2) (23). However, even during the normal growth of bacteria, strongly expressed genes in multicopy vectors may give perceptible high-abundance bands in gels of cell extracts. As an alternative, cell-free systems capable of *in vitro* transcription and translation of cloned DNA are available for some bacteria besides *E. coli* (e.g. *Streptomyces lividans*: 24).

B. Run-off Transcription and S_1 Mapping

Cloned DNA is often used as a source of restriction fragments to direct *in vitro* RNA synthesis by purified RNA polymerase, particularly in the technique of run-off transcription, when transcripts initiated from a particular start-site extend to the end of the template fragment to give products whose size,

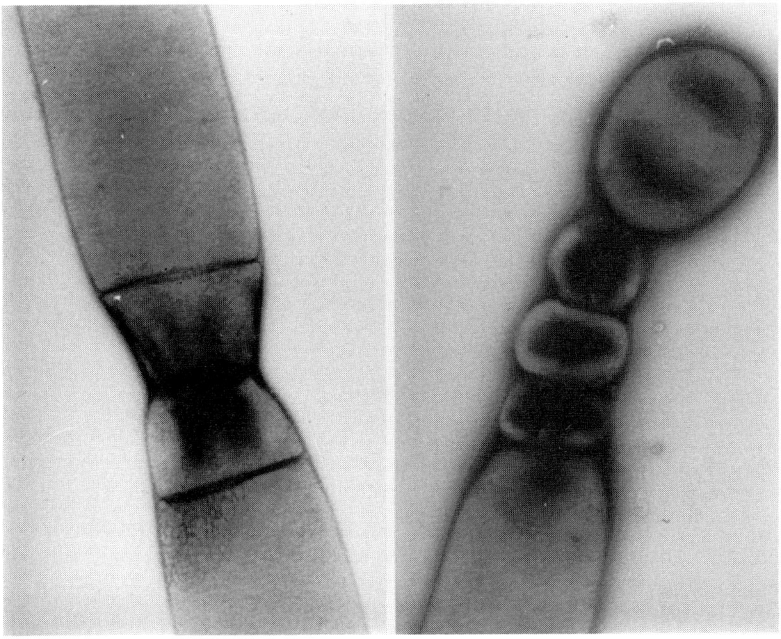

Figure 2. Cells of a mutant strain of *B. subtilis* in two different stages of the production of minicells. Electron micrographs, courtesy of Dr. Margaret Manson.

estimated by gel electrophoresis, allows the position of the start-site and hence of the promoter to be determined. *In vitro* transcription is particularly informative when allied with studies of *in vivo* mRNA. The latter usually depend on protection from digestion by S_1 nuclease (or other single-strand nucleic acid-specific nuclease) of a denaturated probe DNA by mRNA isolated from a suitable host: the size of the protected fragment indicates the length of the overlap between the probe and the mRNA (25). These techniques may also indicate the positions of terminators. Combined study of *in vivo* and *in vitro* transcripts often allows the important distinction to be made between primary transcripts and RNA species that result from post-transcriptional processing (26).

C. Footprinting and Gel Retardation

The interactions of DNA with RNA polymerase and regulatory proteins can be studied in several ways using cloned DNA. Footprinting serves to localize the regions of DNA with which a protein makes contact because such proteins protect the contacted regions from nuclease action or from chemical modifications, or render adjacent regions more sensitive to these treatments (Chapters 9 and 16). Recently, particularly powerful footprinting techniques have been developed which allow the detection and localization of DNA-protein interactions in live bacteria (27). Many proteins that bind to DNA can also be detected by gel retardation: such proteins, even in crude extracts, can complex with the DNA and reduce the migration rate of a specific fragment during electrophoresis (Chapter 19).

D. Blotting

Many chapters refer to various kinds of blots (or transfers), in which restriction digests of DNA (Southern blots), intact RNA (Northern blots) or dissociated proteins (Western blots) are separated by electrophoresis and transferred in denatured form to nitrocellulose or nylon membranes. The presence of a particular nucleic acid sequence or protein is detectable on the membrane by the use of DNA, RNA or immunological probes, which are themselves visualized either by autoradiography or by colour reactions.

Southern blots are used: to find DNA sequences related to a probe, when different hybridization conditions allow detection of more or less perfectly matched sequences; to obtain information about nearby restriction sites; and to quantify the number of copies of a sequence. They are the principal means by which changes in DNA structure in intact genomes are detected.

Northern blots are used to determine the length of transcripts. They are not strictly quantitative, because bacterial mRNA is generally degraded with a half life of a very few minutes, so only part of any given transcript population is present in full-length form. Moreover, excessive amounts of rRNA are always present in mRNA preparations from bacteria, giving heavy bands at ca. 1.7 kb and ca. 3 kb which can cause artifacts in these regions of the gel (28). For these reasons, other related procedures are often used. In dot blots (or slot blots), RNA preparations are adsorbed directly onto membranes without electrophoretic separation so that a rapid estimate of the total amount of hybridizing material, irrespective of whether or not it is full-length, can be made. In a low resolution version of S_1 mapping, RNA is used to protect relatively large unlabelled DNA fragments (in the kilobase size range) against S_1 nuclease (see above) and the protected DNA is analysed by Southern blotting. This allows the positions of RNA endpoints to be approximately localized, and transcript sizes can be estimated (28).

Western blots are used to detect the size and amount of a given protein produced either *in vivo* or during *in vitro* transcription-translation. They depend on the availability of a suitable antiserum. Although it may be prepared against the purified protein, this antiserum is often—in the era of molecular genetics—raised against a fusion protein which has itself been purified by procedures devised for the reporter protein. Alternatively, an antibody raised against a synthetic oligopeptide designed from the DNA sequence of a cloned gene may be used.

V. CURRENT LIMITATIONS AND POSSIBILITIES

Outside *E. coli*, molecular genetic analysis of most bacteria is largely confined to their special features of interest, and this book is principally concerned with describing a selection of these studies. Other aspects of these organisms, such as chromosomal genetics and plasmid and phage biology, are the province of relatively few enthusiasts. Not surprisingly, therefore, there is not such a wealth of well-understood genetic phenomena available for experimental exploitation in these organisms as has accumulated for *E. coli*. There are several aspects of *E. coli* genetic technology that would be particularly helpful in most other bacteria. Examples include: well-developed *in vitro* packaging systems on the λ pattern to allow the direct, highly efficient cloning of very large DNA fragments; efficient regulated expression vectors that would allow the over-production of endogenous proteins for analysis of their properties; and single-stranded DNA vectors that could be used for direct DNA sequencing and localized mutagenesis. Two valuable resources recently established for *E. coli*, which will surely follow for other genetically studied organisms, are a "macro" restriction map of the entire chromosome—obtained by generating very large restriction fragments using enzymes with rare recognition sites, and separating them by pulsed-field gel electrophoresis (29)—and sets of ordered overlapping clones that cover the entire genome (30).

References

(1) Old, R. W. and Primrose, S. B. (1985). "Principles of Gene Manipulation" (3rd Edition). Blackwell Scientific Publications, Oxford.

(2) Winnaker, E-L. (1987). "From Genes to Clones". VCH, Weinheim.

(3) Maniatis, T., Fritsch, E. F. and Sambrook, J. (1982). "Molecular Cloning: A Laboratory Manual". Cold Spring Harbor Laboratory, Cold Spring Harbor, New York.

(4) Young, R. A., Bloom, B. R., Grosskinsky, C. M., Ivanyi, J., Thomas, D. and Davis, R. W. (1985). Dissection of *Mycobacterium tuberculosis* antigens using recombinant DNA. *Proc. Natl. Acad. Sci. USA* **82**, 2583–2587.

(5) Bibb, M. J., Findlay, P. R. and Johnson, M. W. (1984). The relationship between base composition and codon usage in bacterial genes and its use in the simple and reliable identification of protein-coding sequences. *Gene* **30**, 157–166.

(6) Fishman, S. E., Cox, K., Larson, J. L., Reynolds, P. A., Seno, E. T., Yeh, W-K., VanFrank, R. and Hershberger, C. L. (1987). Cloning genes for the biosynthesis of a macrolide antibiotic. *Proc. Natl. Acad. Sci. USA* **84**, 8248–8252.

(7) Koch, C., Vandekerckhove, J. and Kahmann, R. (1988). *Escherichia coli* host factor for site-specific DNA inversion: cloning and characterization of the *fis* gene. *Proc. Natl. Acad. Sci. U.S.A.* **85**, 4237–4241.
(8) Friedman, A. M., Long, S. R., Brown, S. E., Buikema, W. J. and Ausubel, F. M. (1982). Construction of a broad-host-range cosmid cloning vector and its use in the genetic analysis of *Rhizobium* mutants. *Gene* **18**, 289–296.
(9) Kleckner, N., Roth, J. and Botstein, D. (1977). Genetic engineering *in vivo* using translocatable drug-resistance elements. *J. Mol. Biol.* **116**, 125–159.
(10) Ehrlich, S. D., Niaudet, B. and Michel, B. (1982). Use of plasmids from *Staphylococcus aureus* for cloning of DNA in *Bacillus subtilis*. *Curr. Top. Microbiol. Immunol.* **96**, 19–29.
(11) Evans, R. P., Winter, R. B. and Macrina, F. L. (1985). Molecular cloning of a pIP501 derivative yields a model replicon for the study of Streptococcal conjugation. *J. Gen. Microbiol.* **131**, 145–153.
(12) Gasson, M. J. (1987). Genetics of *Streptococcus lactis* 712. *In* "Genetics of Industrial Microorganisms, Volume A" (M. Alačević, D. Hranueli and Z. Toman, eds.), pp. 425–430. Pliva, Zagreb.
(13) Rao, R. N., Richardson, M. A. and Kuhstoss, S. (1987). Cosmid shuttle vectors for cloning and analysis of *Streptomyces* DNA. *Methods in Enzymol.* **153**, 166–198.
(14) Trieu-Cuot, P., Cavlier, C., Martin, P. and Courvalin, P. (1987). Plasmid transfer by conjugation from *Escherichia coli* to Gram-positive bacteria. *FEMS Microbiol. Lett.* **48**, 289–294.
(15) Bibb, M. J., Ward, J. M. and Hopwood, D. A. (1978). Transformation of plasmid DNA into *Streptomyces* at high frequency. *Nature* **274**, 398–400.
(16) Chung, S. T. (1987). Tn*4556*, a 6.8 kb transposable element of *Streptomyces fradiae*. *J. Bacteriol.* **169**, 4436–4441.
(17) Segall, J. and Losick, R. (1977). Cloned *Bacillus subtilis* DNA containing a gene that is activated early during sporulation. *Cell* **11**, 751–761.
(18) Gilman, M. Z., Glenn, J. S., Singer, V. S. and Chamberlin, M. J. (1984). Isolation of sigma-28-specific promoters from *Bacillus subtilis* DNA. *Gene* **32**, 11–20.
(19) Campbell, A. M. (1962). Episomes. *Adv. in Genet.* **11**, 101–145.
(20) Green, P. J., Pines, O. and Inouye, M. (1986). The role of antisense RNA in gene regulation. *Ann. Rev. Biochem.* **55**, 569–597.
(21) Sancar, A., Hack, A. M. and Rupp, W. D. (1979). Simple method for identification of plasmid-coded proteins. *J. Bacteriol.* **137**, 692–693.
(22) Meagher, R. B., Tait, R. C. and Boyer, H. W. (1977). Protein expression in *E. coli* minicells by recombinant plasmids. *Cell* **10**, 521–536.
(23) Reeve, J. (1979). Use of minicells for bacteriophage-directed polypeptide synthesis. *Methods in Enzymol.* **68**, 493–503.
(24) Thompson, J., Rae, S. and Cundliffe, E. (1984). Coupled transcription-translation in extracts of *Streptomyces lividans*. *Mol. Gen. Genet.* **195**, 39–43.
(25) Favoloro, I., Treisman, R. and Kamen, R. (1980). Transcription maps of polyoma virus-specific RNA: analysis by two-dimensional nuclease S_1 mapping. *Methods in Enzymol.* **65**, 718–749.
(26) Baylis, H. A. and Bibb, M. J. (1988). Transcriptional analysis of the 16s rRNA gene of the *rrnD* gene set of *Streptomyces coelicolor* A3(2). *Molec. Microbiol.* (In press).
(27) Borowiec, J. A. and Gralla, J. D. (1986). High-resolution analysis of *lac* transcription complexes inside cells. *Biochemistry* **25**, 5051–5057.
(28) Smith, C. P. and Chater, K. F. (1988). Cloning and transcription analysis of the entire glycerol utilization (*gylABX*) operon of *Streptomyces coelicolor* A3(2) and identification of a closely associated transcription unit. *Mol. Gen. Genet.* **211**, 129–137.

(29) Smith, C. L., Econome, J. G., Schutt, A., Klco, S. and Cantor, C. R. (1987). A physical map of the *Escherichia coli* K12 genome. *Science* **236**, 1448–1453.
(30) Kohara, Y., Akiyama, K. and Isono, K. (1987). The physical map of the whole *E. coli* chromosome: application of a new strategy for rapid analysis and sorting of a large genome library. *Cell* **50**, 495–508.
(31) Schwartz, M. (1985). Gene fusions in bacteria. *In* "Genetics of Bacteria" (J. Scaife, D. Leach and A. Galizzi, eds.), pp. 65–84. Academic Press, London.
(32) Beckwith, J. (1987). The lactose operon. *In* "*Escherichia coli* and *Salmonella typhimurium*. Cellular and Molecular Biology" (F. C. Neidhardt, J. L. Ingraham, K. B. Low, B. Magasanik, M. Schaechter and H. E. Umbarger, eds.), pp. 1444–1452. American Society for Microbiology, Washington, D.C.
(33) King, A. A. and Chater, K. F. (1986). The expression of the *Escherichia coli lacZ* gene in *Streptomyces*. *J. Gen. Microbiol.* **132**, 1739–1752.
(34) Rosenberg, M., Brawner, M., Gorman, J. and Reff, M. (1986). Galactokinase gene fusion in the study of gene regulation in *E. coli*, *Streptomyces*, yeast, and higher cell systems. *In* "Genetic Engineering, Volume 8" (J. K. Setlow and A. Hollaender, eds.), pp. 151–180. Plenum Press, New York.
(35) Hoffman, C. S. and Wright, A. (1985). Fusions of secreted proteins to alkaline phosphatase: an approach for studying protein secretion. *Proc. Natl. Acad. Sci. USA* **82**, 5107–5111.
(36) Schauer, A. L. (1988). Visualizing gene expression with luciferase fusions. *Trends Biotechnol.* **6**, 23–27.
(37) Zukowski, M. M., Gaffney, D. F., Speck, D., Kauffman, M., Findeli, A., Wisecup, A. and Lecocq, J-P. (1983). Chromogenic identification of genetic regulatory signals in *Bacillus subtilis* based on expression of a cloned *Pseudomonas* gene. *Proc. Natl. Acad. Sci. USA* **80**, 1101–1105.
(38) Mongkolsuk, S. and Lovett, P. (1984). Selective expression of a plasmid *cat* gene at a late stage of *Bacillus subtilis* sporulation. *Proc. Natl. Acad. Sci. USA* **81**, 3457–3460.
(39) Jaurin, B. and Cohen, S. N. (1985). *Streptomyces* contain *Escherichia coli*-type (A + T)-rich promoters having novel structural features. *Gene* **39**, 191–201.
(40) Ward, J. M., Janssen, G. R., Kieser, T., Buttner, M. J. and Bibb, M. J. (1986). Construction and characterisation of a series of multicopy promoter-probe plasmid vectors for *Streptomyces* using the aminoglycoside phosphotransferase gene from Tn*5* as indicator. *Mol. Gen. Genet.* **203**, 468–478.

Section II

Specialized Metabolic Capabilities of Bacteria

The small size and simple structure of bacteria imply a greater degree of contact with their environment than for larger, eukaryotic cells. Bacteria have relatively high surface area:volume ratios and there is more direct communication between events at the cell surface and all the structures inside the cell—most particularly, with the genetic material. Thus, bacterial genes for metabolism are usually highly regulated. Moreover, the brevity of bacterial generation times, and hence the intense selection pressures for efficient occupation of the whole range of potential ecological niches, has resulted in the development of many diverse metabolic capabilities with highly refined physiological and genetic control mechanisms. For example, the deployment of metabolic energy to generate light by certain marine bacteria (Chapter 4) is exquisitely dependent on bacterial cell density (luminescence of isolated individual cells would generally be biologically imperceptible, but the combined light emission of massed populations can readily be perceived by animal eyes): the activation of the luminescence-determining genes depends on accumulation of a sufficient concentration of an extracellular hormone-like substance secreted by all the individual cells. (Interestingly, the structure of one of these hormones, from *Vibrio fischeri*, resembles that of A-factor, which fulfils an analogous function in regulating antibiotic production and sporulation in some *Streptomyces* species: Chapter 7.)

The analytical power of gene disruption, modification and deletion to elucidate very complex biological phenomena, a recurrent theme in this book, is well illustrated in the dissection of the *Rhodobacter capsulatus* photosynthetic process (Chapter 5). The process by which light is employed to energize cellular biosynthesis has attracted attention because of its value as a model system for photosynthesis in plants, but it is of course also of great interest and ecological significance in its own right.

A number of bacteria can carry out the unlikely, but globally crucial, conversion of dinitrogen to chemically reactive ammonia. Analysis of nitrogen fixation in several organisms (Chapter 6) has revealed that an amazing complexity of genetic determination and regulation ensures that nitrogenase enzyme is correctly assembled, is synthesized only in conditions in which it can usefully function, and is properly served by associated physiological and metabolic processes. Comparable complexity is now emerging in the genetic determination of antibiotic production (Chapter 7), a process which contrasts with nitrogen fixation in being a pathway leading out of, instead of into, central metabolism, and which is intimately intertwined with morphological differentiation in the producing streptomycetes (giving its regulation an additional layer of interest).

The production of diverse antibiotics and other secondary metabolites provides an example of the remarkable ability of bacteria to synthesize complex and unusual molecules. The inverse situation—ability of bacteria to *degrade* such molecules—is illustrated by the fascinating biochemical genetics of catabolism of aromatic hydrocarbons by *Pseudomonas* species (Chapter 8). This property is very often determined by plasmids. Plasmids and/or transposons frequently carry determinants that allow colonization of sporadically-occurring environments that are not generally suitable for bacterial growth. These include not only "pools" of refractory compounds like aromatic compounds, but also the presence of toxic chemicals such as antibiotics and heavy metals. An example of adaptations to environments of the latter kind is transposon-encoded mercury resistance (Chapter 9), which depends, surprisingly, on highly efficient *uptake* of mercury ions, followed by their intracellular reduction to metabolic mercury, which is then lost by diffusion and evaporation. The regulation of transcription of the mercury resistance operon is controlled by an immediately adjacent regulatory gene which is itself transcribed divergently from the operon, providing the means for additional interplay between the regulatory and structural genes: a situation also observed, but less closely analysed, in the *lux* operon of *Vibrio fischeri* (Chapter 4).

Chapter 4

Regulation of Luminescence in Marine Bacteria

MICHAEL SILVERMAN, MARK MARTIN *and* JOANNE ENGEBRECHT

I. INTRODUCTION

Biological light excited the curiosity of the ancients. Aristotle, in *De Anima*, marvelled at the light produced in darkness. Bacterial bioluminescence, probably the source of the light observed by Aristotle on decaying fish, was first described more than 100 years ago, so this unique property of bacteria was recognized long before the development of most methods for exploration at microscopic and molecular dimensions. That this phenomenon has often been characterized as unique or unusual or fascinating probably derives in part from its appealing aesthetic quality (see self-photograph of luminescent bacteria in Fig. 1). But, appreciation of the peculiar nature of this biological system goes far beyond subjective impressions. It is evident from extensive research, with particularly substantial contributions in recent times in the areas of biochemistry, molecular biology and genetics, that bacterial bioluminescence represents the evolution of a novel design and that further study will enrich understanding of the diversity of biological form and function.

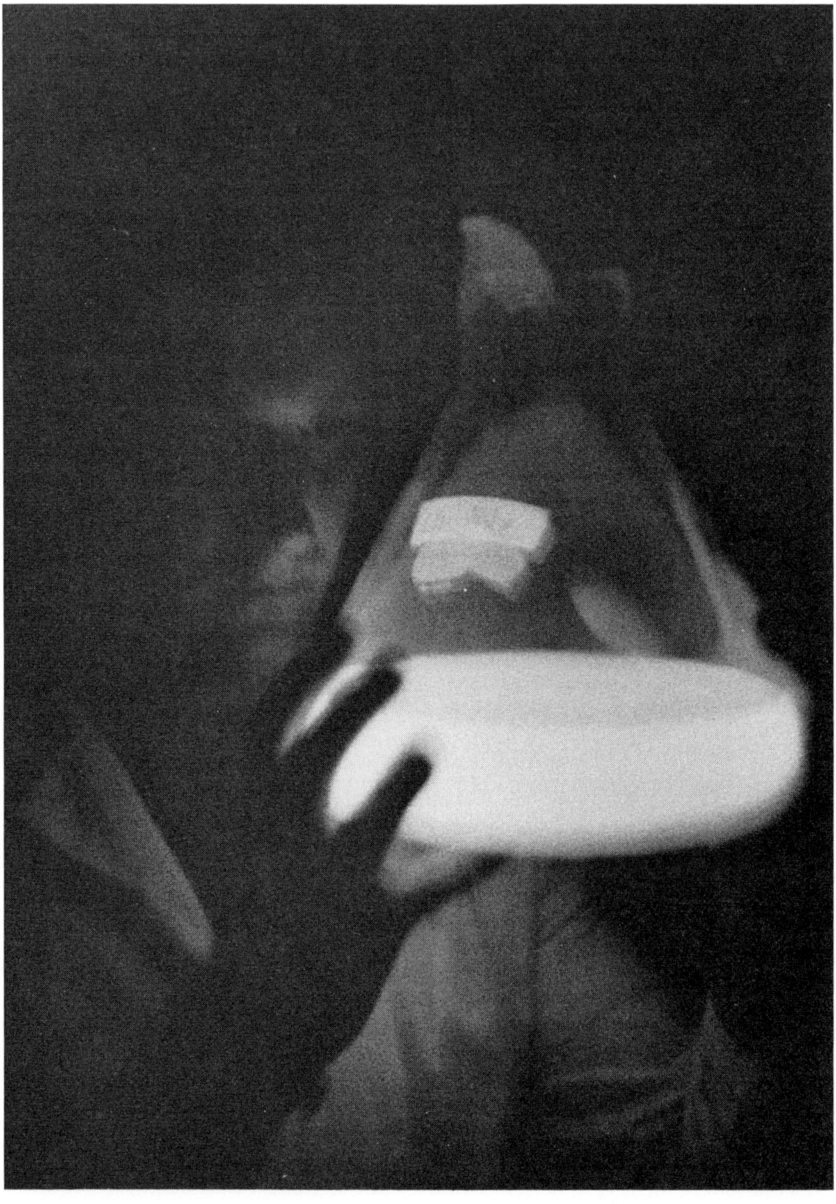

Figure 1. Self-photograph of recombinant *E. coli* containing cloned *lux* genes.

Luminescent bacteria are rare in the terrestrial environment but are widespread in the marine environment where they exist planktonically, as digestive tract symbionts, parasites or saprophytes, or in specialized light organs of certain fish and squids (1). Light produced by bacterial symbionts can be used by the host to attract prey, for intraspecies communication, or to escape from predators, but it is not certain what immediate benefit the bacteria derive from luminescence. One hypothesis is that bioluminescence is an alternative pathway for the transfer of electrons to oxygen. Light emission, at 490 nm, is catalysed by the enzyme luciferase which oxidizes a reduced flavin and a long-chain fatty aldehyde to produce oxidized flavin and the corresponding long-chain fatty acid (2). This reaction is shown in Fig. 2. In contrast to respiratory electron transport, reductant is consumed without the generation of ATP. However, luciferase has a relatively high affinity for oxygen, so electron flow can be sustained at low oxygen tension. Luminescence may thus confer a selective advantage to the bacterium by functioning as a terminal oxidase when alternative pathways for transferring reducing power cannot operate.

Before the application of recombinant DNA technology, very little was known about the organization and regulation of the genes (*lux*) required for bioluminescence. Genes specifying bacterial luciferase have now been cloned from both *Vibrio fischeri* and *Vibrio harveyi*, and light production has been achieved in *Escherichia coli* (3–6). One fragment of DNA cloned from *V. fischeri* was found to encode all the enzymatic and regulatory functions necessary for expression of luminescence in *E. coli*. Analysis of recombinants containing the *V. fischeri lux* system has been extensive; genes, gene functions, gene products, operons, and regulatory circuits have been identified. So, we shall focus primarily on the *V. fischeri* system and the novel regulatory mechanism,

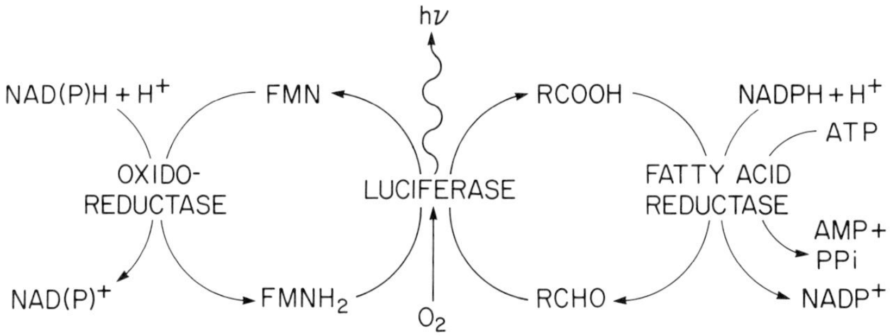

$$RCHO + FMNH_2 + O_2 \xrightarrow{\text{LUCIFERASE}} RCOOH + FMN + H_2O + h\nu$$

Figure 2. Substrates, products and pathways involved in the bacterial bioluminescence reaction.

autoinduction, which controls its expression. Luminescence in many strains of marine vibrio is a variable phenotype. Clones originating from a single cell are not identical with respect to luminescence, but show phenotypes ranging from bright to dim to dark. Each variant form can change or switch to another at a low but detectable frequency. We think that this phenomenon could have a bearing on genetic processes used by microorganisms to generate diversity, and experiments to determine the cause of luminescence variation will be discussed.

II. ORGANIZATION AND FUNCTION OF *lux* GENES

The recombinant plasmid pJE202 carries a segment of DNA that includes the *lux* genes of *V. fischeri*. Production of light by *E. coli* containing pJE202, as with the donor bacterium, does not require the exogenous addition of the fatty aldehyde substrate or the provision of an exogenous promoter. Figure 1 shows a self-exposure of this luminescent *E. coli*. The intensity of light from the recombinant is comparable to that from the parent vibrio, and the unusual regulatory control observed in the vibrio is reproduced in the recombinant host. Production of light by many marine vibrios, including *V. fischeri* and *V. harveyi*, is influenced by the density of the culture. Luminescence (quanta/s/unit of cell mass) is plotted as a function of cell density in Fig. 3A for both *V. fischeri* (strain MJ-1) and *E. coli* with the *lux* recombinant plasmid (pJE202). Light production per unit of cell mass is relatively high initially because the suspension has been inoculated from a dense overnight culture, but the specific luminescence decreases rapidly as the cells in dilute suspension grow. At a particular cell density, light emission increases exponentially until a plateau value is reached. This regulatory phenomenon will be discussed in detail later, but it is clear that the cloned *V. fischeri* DNA encodes the functions necessary for regulation as well as the enzymes for the light reaction.

A 16 kb *Bam*H-I fragment was initially cloned from *V. fischeri*, but the enzymatic and regulatory functions necessary for luminescence could be subcloned to a *Sal* I fragment of 9 kb (4). A summary of the results of extensive genetic analysis of this subclone is shown in Fig. 4. Complementation tests using pairs of polar transposon insertion mutations *in trans* were used to define operons, and two were identified. The direction of transcription of these two *lux* operons was deduced from the properties of *lux* : : *lac* fusions generated by the specialized transposon miniMu*lac*. Only one of the two possible orientations of transposon insertion can align transcription of the target *lux* operon with that of the *lacZ* indicator gene at one end of the transposon. The locations and orientations of many miniMu*lac* insertions in the 9 kb *lux* region were determined by restriction mapping, and those fusion strains that produced β-galactosidase had the *lacZ* gene orientated with the direction of transcription

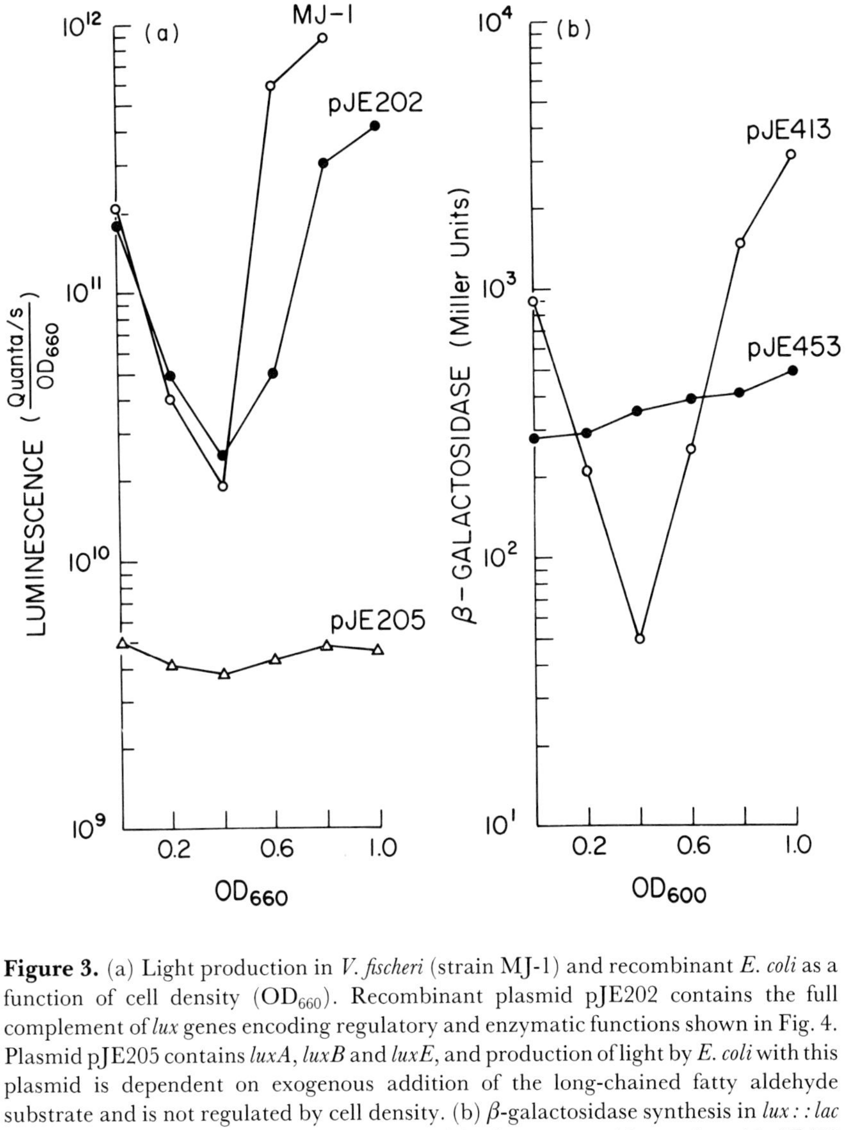

Figure 3. (a) Light production in *V. fischeri* (strain MJ-1) and recombinant *E. coli* as a function of cell density (OD_{660}). Recombinant plasmid pJE202 contains the full complement of *lux* genes encoding regulatory and enzymatic functions shown in Fig. 4. Plasmid pJE205 contains *luxA*, *luxB* and *luxE*, and production of light by *E. coli* with this plasmid is dependent on exogenous addition of the long-chained fatty aldehyde substrate and is not regulated by cell density. (b) β-galactosidase synthesis in *lux : : lac* fusion strains as a function of cell density. *E. coli* harbours recombinant plasmid pJE413 or pJE453 which were derived from pJE201 (similar to pJE202 above). Plasmid pJE413 contains a *luxC : : lacZ* fusion (transcriptional fusion of the indicator gene to operon R), and plasmid pJE453 contains a *luxR : : lacZ* fusion (transcriptional fusion to operon L).

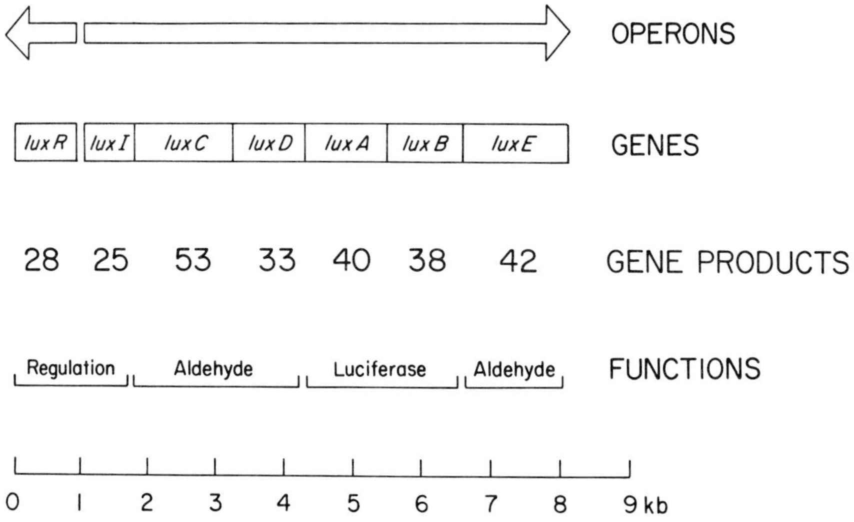

Figure 4. Organization and functions of *lux* genes cloned from *V. fischeri*. Arrows denote operons containing *lux* genes; the leftward arrow marks operon L (containing *luxR*), and the rightward arrow marks operon R (containing *luxI*, *luxC*, *luxD*, *luxA*, *luxB* and *luxE*). The molecular weights ($\times 10^{-3}$) of the *lux* gene products and the functions of the gene products are shown below the gene designations.

of the particular *lux* operon shown in Fig. 4. Definition of *lux* genes required analysis of the complementation of pairs of nonpolar point mutations held *in trans* on separate, compatible replicons. Seven complementation groups corresponding to genes *luxR*, *luxI*, *luxC*, *luxD*, *luxA*, *luxB* and *luxE* were identified (7).

The *luxA* and *luxB* genes encode the α and β subunits of luciferase. The *luxC*, *luxD* and *luxE* genes are required for provision of the aldehyde substrate. Mutants with defects in these latter genes produce light when a long-chain aldehyde, tetradecanal, is provided exogenously. The *luxR* and *luxI* genes have regulatory functions and will be discussed in the next section. No additional gene which could direct the synthesis of an oxidoreductase was found. Apparently this function is supplied by the *E. coli* host, and it is not unique to the luminescence system. The products of the *lux* genes were identified by programming protein synthesis with various *lux* plasmids in *E. coli* minicells (7). The molecular weights ($\times 10^{-3}$) of the particular *lux* gene products are shown in Fig. 4. The molecular weights determined for the *luxA* and *luxB* gene products are in agreement with those reported for the α and β subunits of luciferase, and the products of *luxC*, *luxD* and *luxE* correspond to the molecular weights reported for the components of the fatty acid reductase (8). DNA homologous to *luxA* and *luxB* of *V. fischeri* has been cloned from *V. harveyi*, and the DNA sequence for the *V. harveyi lux* genes corresponds to the amino acid sequence of peptides derived from the α and β subunits of luciferase (5). Although it has not been possible to reproduce the regulatory control characteristic of the native organism in recombinant *E. coli* containing *V. harveyi* DNA, the organization of genes encoding the enzymatic functions, luciferase and fatty acid reductase, appears to be similar in both *V. harveyi* and *V. fischeri* (6).

III. REGULATION OF *lux* EXPRESSION

Luminescence appears to be a social phenomenon which occurs when cells are confined at high intensity. Density-dependent expression of luminescence, which was briefly discussed earlier, is illustrated in Fig. 3A. Light produced by a cell in a high density culture can be two or three orders of magnitude greater than that produced at low density. The bacteria are actually responding to the presence of other cells, or more precisely, to the presence of a substance produced by other cells in the culture. Luminous bacteria such as *V. fischeri* and *V. harveyi* synthesize a signalling molecule, called autoinducer, which accumulates in the environment of the cells (9). When the concentration of this molecule reaches a critical level, induction of luminescence results. Autoinducer has been isolated from culture supernatants of *V. fischeri* and has been determined to be *N*-(β-ketocaproyl) homoserine lactone (10). This compound induces light production in *V. fischeri* but not in *V. harveyi*. Autoinducer-like activity can be detected in culture supernatants of *V. harveyi*, but the activity is labile, and the chemical structure has not been determined. Although *V. fischeri* autoinducer is highly species-specific with regard to production and response, *V. harveyi* luminescence can be induced by a variety of marine species (other than *V. fischeri*), some of which are not luminous (11). *V. fischeri* and *V. harveyi* also occupy different habitats. *V. fischeri* is a symbiont and is found in pure culture within the light organ of the fish *Monocentris japonicus*, but *V. harveyi* occupies the intestinal tracts of fish in mixed culture with other bacteria. Apparently, different communication channel characteristics are required in different ecological situations. The autoinducer of *V. fischeri* is remarkably similar in structure to A-factor, a regulatory molecule which is produced by *Streptomyces griseus* and which causes self-induction of sporulation and streptomycin synthesis (Chapter 7). Perhaps this chemical relationship is an indication that mechanisms used by bacteria to sense their environment have a common origin and that there is a large class of signalling molecules or "bacterial hormones" similar in structure and mode of action.

As shown above, the production of light by *E. coli* carrying pJE202 is also density-dependent (compare pJE202 with MJ-1 in Fig. 3). Furthermore, deletion of the *luxR*, *luxI* region from the recombinant clone abolishes density-dependent control (see pJE205 in Fig. 3). So, regulatory gene(s) are present on the cloned DNA, function in *E. coli*, and map to a specific region of the recombinant DNA. It is thus possible to apply the refined methods developed for *E. coli* to the genetic dissection of the regulatory circuit controlling expression of *V. fischeri lux* genes. Regulation of luminescence by cell density could operate at a variety of control points: at the level of transcription, translation, enzyme function or energy supply. *Lux::lac* gene fusions, generated by transposon miniMu*lac* insertions, were used to prove that density-dependent regulation, or more specifically autoinduction, operates primarily at the level of transcription (4). These fusions coupled transcription of the target *lux* gene to that of the *lacZ* indicator gene of the transposon. In strains with fusions of *lacZ* to any of the *lux* genes encoding enzymes for bioluminescence (*luxC,D,A,B,E* in operon R), synthesis of β-galactosidase is density-dependent. This is illustrated in Fig. 3B which shows the density-dependent expression of β-galactosidase from a strain containing a plasmid with a *luxC::lacZ* fusion (pJE413). The synthesis of β-galactosidase directed by a plasmid with an operon L fusion (*luxR::lacZ* in plasmid pJE453) is shown for comparison. Since *lacZ* expression is regulated by density, and the *lacZ* fusions are transcriptionally coupled to *lux*, density-dependent control of *lux* must operate at the level of transcription. Furthermore, the response of *lacZ* expression to density is similar in rate and extent to that observed with light production, so control by cell density operates primarily at the level of transcription.

The *luxR* and *luxI* genes (Fig. 4) are essential for induction of transcription of operon R, which contains the genes encoding enzymes for the bioluminescence reaction. Recombinant strains with mutations in *luxR* or *luxI* are dark or extremely dim. However, $LuxI^-$ strains can be stimulated to produce the wild-type intensity of light by the exogenous addition of autoinducer, either purified from *V. fischeri* or synthesized chemically (4). This implies that *luxI* encodes a protein required for the synthesis of autoinducer. Autoinducer can be quantitated with a variety of biological assays. For example, a recombinant strain with a *luxI::lacZ* fusion produces β-galactosidase in response to the presence of autoinducer. By using this and other assays, $LuxI^-$ mutants were found to be incapable of synthesizing autoinducer. It has subsequently been shown, using a variety of recombinant constructions, that the *luxI* gene alone is sufficient for autoinducer production. Autoinducer is necessary for transcription of operon R; the most direct demonstration of this relationship is the response of the *luxI::lacZ* fusion strain mentioned above in which autoinducer addition activates transcription of the *lacZ* indicator gene coupled to operon R.

The product of the *luxR* gene (which is in operon L) is also required for expression of operon R; the *luxR* protein is thought to act as a positive effector of operon R transcription (4, 7, 12). A model for *lux* gene control incorporating the results of genetic and physiological studies is shown in Fig. 5. On binding

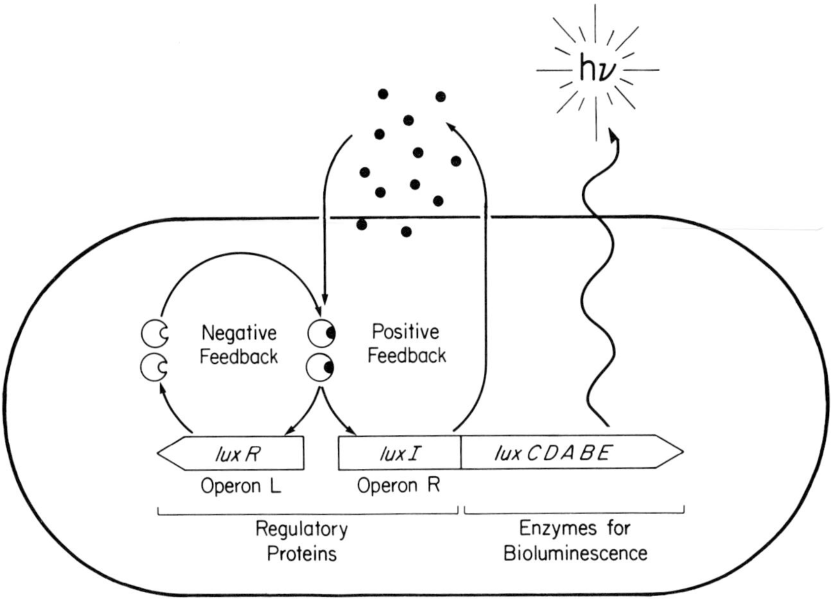

Figure 5. Model for genetic regulation of bioluminescence. Expression of genes encoding enzymes for bioluminescence is controlled by the interaction of a signal molecule, autoinducer (closed circles), synthesized by the *luxI* gene product and a regulatory protein (notched open circle) encoded by the *luxR* gene. A complex of autoinducer and regulatory protein (closed circle in open circle) influences the expression of both operon L and operon R. Consequences of these regulatory interactions are discussed in detail in the text.

autoinducer, the LuxR regulatory protein activates transcription of operon R. Movement of autoinducer out of the cell and sensing autoinducer in the external environment does not require a specific transport system or a receptor for transmembrane communication because autoinducer rapidly diffuses across the bacterial membrane (13). The effective concentration of autoinducer is strongly influenced by cell density because compartmentalization of autoinducer does not occur. However, it is apparent that the autoinducer concentration is not influenced by cell density alone. The *luxI* gene is a part of the operon that is controlled by autoinducer. So, autoinducer synthesis is autoregulated (originally, "auto" referred to the self-directed, species-specific activity of the inducer substance). Put another way, the concentration of autoinducer controls the concentration of autoinducer; as more autoinducer signal accumulates, more transcription of operon R occurs, resulting in the synthesis of more *luxI* product which further accelerates production of autoinducer. One consequence of this positive feedback loop is that expression of genes encoding enzymes for bioluminescence should increase exponentially once the induction process has started, and this is what in fact occurs. One requirement for positive feedback control is a low constitutive level of transcription of operon R to "prime the pump", or start the autoinduction process. Using gene fusions, a low but detectable amount of transcription of operon R has been shown to occur in the preinduction phase of growth.

Another feature of expression of bioluminescence is that light production reaches a plateau of intensity as the culture approaches saturation. Some mechanism may exist to counteract the runaway increase inherent in the positive feedback regulatory circuit. Evidence for negative modulation of operon R induction has been obtained (7). In minicell programming experiments to identify the products of *lux* genes, it was possible to detect labelled polypeptides corresponding to all of the *lux* genes except *luxR*. It appeared that *luxR* expression was being held in check in the dense suspension of minicells used in the experiment. This was indeed the case, because the *luxR* polypeptide could be produced if either *luxI* or *luxR* function were negated by the introduction of missense mutations. So the two genetic functions, *luxI* and *luxR*, which positively regulate operon R also negatively regulate operon L expression (Fig. 5). In a culture of constant density, such as a chemostat or the light organ of a fish, the influence of the opposing regulatory circuits would come into balance, and a constant intensity of light would be produced. In a culture of increasing density, such as that in Fig. 3, the positive and negative feedback effects would be temporarily separated so that light would first increase exponentially and then plateau at a constant intensity. The nature of negative feedback regulation of *luxR* has been investigated (12). In contrast to the transcriptional control of operon R, expression of *luxR* in operon L appears to be regulated at the level of translation of mRNA.

Expression of luminescence requires cAMP and the cAMP receptor protein (CRP). Specifically, the CRP-cAMP complex is necessary for expression of operon L in recombinant *E. coli* (14), and a sequence identical to the consensus DNA site for cAMP-CRP binding has been located adjacent to the operon L promoter (J. Engebrecht and M. Silverman, manuscript in preparation). Expression of luminescence is also influenced by the concentration of iron. Induction of luciferase synthesis in *V. fischeri* grown in an iron-limited medium occurs at a much lower cell density than in a control medium. Iron depletion does not affect induction of *lux* genes in recombinant *E. coli*, and the mechanism by which iron concentration affects *lux* gene regulation in the parent vibrio is not well understood (15). However, iron acquisition is very important to the survival of bacteria in the ocean. It should be interesting to examine the connection between iron and *lux* expression. Furthermore, light emission is stimulated when luminous vibrios or recombinant *E. coli* are exposed to stress conditions such as heat shock, oxygen or nutrient limitation, and transcription of operon R is greatly reduced in *E. coli* defective in the *htpR* gene, indicating participation of sigma factor 32 in the regulation of luminescence (S. Ulitzur, Israel Institute of Technology, personal communication). So, the *lux* regulatory circuit is complex and integrates signals conveying information about cell density, catabolite utilization and possibly a variety of stress conditions including nutrient starvation.

IV. LUMINESCENCE VARIATION

Luminescence among many species of marine bacteria, including *V. harveyi*, *Photobacterium phosphorium* and *Photobacterium leiognathi*, is an unstable or variable phenotype. In 1889, Beijerinck (16) described some of the salient features of luminescence variation. Cultures of luminous bacteria gave rise to colonies which emitted little or no light. The newly-acquired trait was heritable since it persisted through repeated subculturing, but colonies with the original luminescence phenotype occasionally arose. Instability in luminescence was particularly pronounced in old cultures. More recently, it was found that dark variants of *V. harveyi* arose most frequently in non-shaken stationary-phase cultures grown at elevated temperature (17). The dark variants were found to produce detectable levels of light, but far below the visual threshold. Strains freshly isolated from the environment were especially unstable with regard to light production (18), and conversion from bright to dim was accompanied by changes in other characteristics such as flagellation, colony morphology and bacteriophage sensitivity.

Our observations of luminescence variation in *V. harveyi* support and extend the findings of others. Four broad types of spontaneous variant have been identified: extra bright, bright, dim and dark. The difference in light emission among these variants spans a range of approximately 10^6. As others have observed, variants arise frequently in old cultures such as those in stab vials or old petri dishes. Figure 6 shows the types of variants which can be found by streaking out colonies from year-old storage vials originally inoculated with a bright variant. In one case (left) a bright variant and two dim variant types of similar luminosity were recovered, and in another instance (right) bright, dim and dark variant types were recovered. Dark variants are not visible in the self-photograph but leave a void in the confluent area of the streak. Variation is apparently pluripotent and reversible; one variant type can change or switch to a type with the luminescence of its progenitor or to another type. However, variants are distinguished by their relative light production, and until the genetic and molecular basis of this phenomenon is understood the number of distinct luminous types and the switching possibilities for each type cannot be precisely catalogued. It is clear that the propensity for variation differs greatly. For example, some types (such as the extra bright variants) are very unstable while others (such as the darkest variants) seldom give rise to another type.

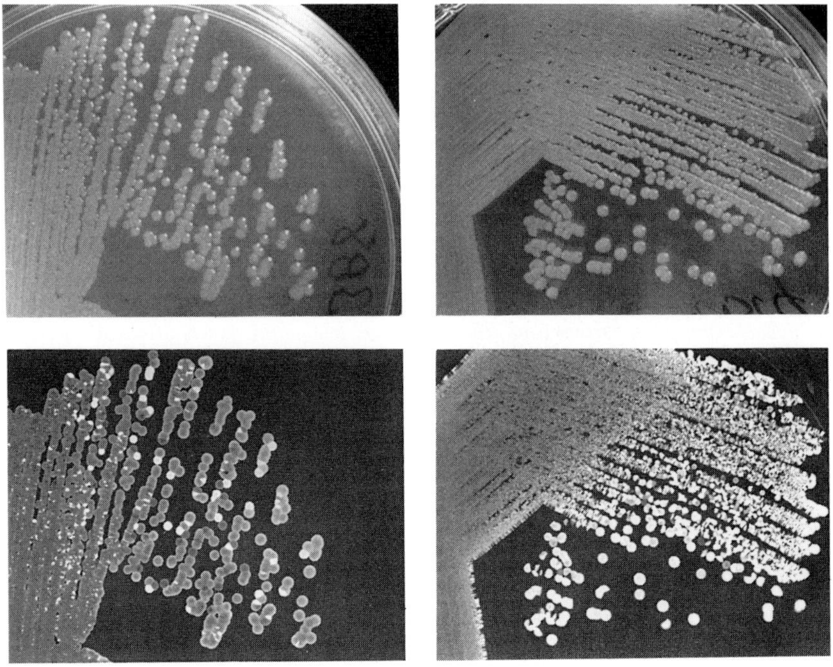

Figure 6. Luminescence variants of *V. harveyi*; streak plate photographed with incident light (top) and self-photograph with bioluminescence (bottom).

Figure 7 shows the extreme instability of luminescence of an extra bright variant. Switching is occurring so rapidly (estimated at 10^{-1} to 10^{-2} events per cell per generation) that every bright colony is heterogeneous and contains sectors of the dimmer variant.

We have also noted the pleiotropy of changes associated with luminescence variation. In addition to those mentioned earlier, changes in cell adhesiveness and in outer membrane protein composition accompany changes in luminescence. Furthermore, differences in luminescence in *V. harveyi* are not due to variability in autoinducer production; all variant types were found to produce essentially identical quantities of autoinducer activity. Dark variants, particularly those of *P. phosphorium* and *P. leiognathi*, can be induced by a variety of substances, including DNA-intercalating agents and genotoxic compounds, to emit light at intensities approaching that of the bright strains (19). The change in light production is not genetically stable, and luminescence decreases rapidly on removal of the compound. The mechanism for this unusual effect is not known, but since expression of the *lux* genes encoding biochemical functions in dark variants can be activated, the deficiency in these variants is

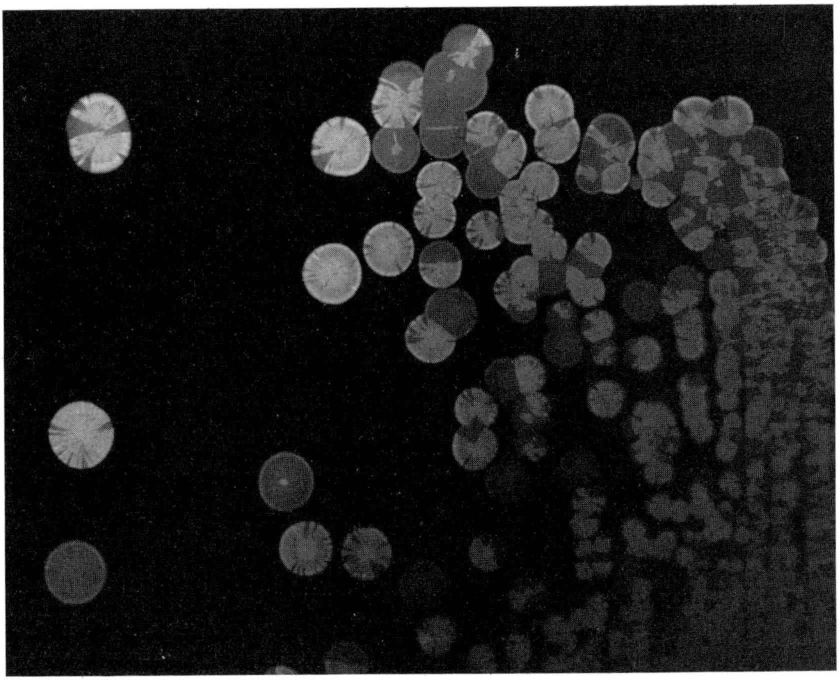

Figure 7. Instability of luminescence of bright variant of *V. harveyi* (self-photograph).

probably not due to a defect in those genes. Rather, it is more likely that luminescence variants result from alterations in a mechanism which controls expression of genes for light production; one possibility is an impairment in the ability to respond to autoinducers.

In the past, variable phenotypes were frequently observed but infrequently studied. Their inherent instability made detailed analysis very difficult, but they nonetheless remained subjects of great interest, in part because their transitory nature was inconsistent with prevailing dogma that genetic material was static and was transmitted unchanged generation after generation. Beijerinck (16) may have believed that phenotypic variations were controlled by a common mechanism, because he noted the similarity of luminescence instability to other phenomena such as pigment variation in what probably was *Serratia marcescens* and loss of virulence of pathogens cultivated in the laboratory. These and other systems of variable gene expression are now being examined by molecular genetic methods. It is evident that such systems are mechanistically related. Examples of well-studied prokaryotic systems include antigenic variation in *Salmonella*, *Neisseria* and *Borellia* (20, 21). With *Salmonella* phase variation, site-specific inversion of DNA containing a promoter element results in oscillatory expression of flagellar antigens. In *Neisseria* (Chapter 13) a large repertoire of different pilin types is produced by homologous recombination between silent *pil* "storage" loci and an active *pil* expression site. Variation in the primary surface protein of *Borellia* is accomplished by recombinational transfer of information from silent genes to expression loci. All of these variable systems are controlled by complex genomic rearrangements, and the function of these genetic devices is to produce antigenic diversity in surface components to allow the pathogens to evade the host immune response.

We speculate that luminescence variation is controlled by genome rearrangement and expect that studing this phenomenon will yield insight into a novel mechanism of gene regulation. Exploration of the molecular mechanism will require isolation of the DNA locus which controls variation. With other systems, the controlling locus has been isolated by cloning the DNA which encodes the protein that varies (i.e. flagellin, pilin, outer membrane protein). We have cloned the *lux* genes which encode the luciferase subunits from *V. harveyi*. Using this DNA as a probe, no rearrangement of genomic sequences among luminescence variants was observed. Luminescence variation could be regulated by another mechanism or the controlling locus could be unlinked to *lux* and could regulate *lux* expression indirectly. We favour the latter explanation since it is clear that many properties in addition to luminescence change as a consequence of variation. One approach for cloning the controlling locus is to screen recombinant molecules for the capacity to restore luminescence function to dark variants. Complementation of a dark variant with *V. harveyi* genes carried on a mobilizable, broad-host-range cosmid vector is now being tested. At least two events could restore luminescence: switching of the dark variant to Lux^+ or complementation to Lux^+. By using a dark variant with a very low frequency of spontaneous switching, it should be possible to detect the latter event. Once the controlling locus has been isolated, exploration of the molecular mechanism could begin. Furthermore, the system could be manipulated to "freeze" variation so that stable phenotypes could be examined to define precisely the function or functions controlled by variation.

V. CONCLUSIONS

It is perhaps unusual that much is known and being learned about the genetic regulation of a system for which the function is poorly understood. However, a clearer understanding of the function of bioluminescence might be inferred from the conditions that induce its expression. Induction of *lux* is maximal in conditions of high cell density, low values of catabolite levels, nutrient concentration, iron concentration and O_2 tension and probably other stressful circumstances, so luminescence may benefit the cell in a variety of hostile environments, or in a situation which combines these conditions such as a stationary phase culture. Genetic resources developed for studying *lux* regulation could be important for ecological analysis. For example, to evaluate the contribution of luminescence to survival in different environments, it would be useful to have isogenic pairs of Lux^+ and Lux^- strains; the latter mutants should contain precisely-defined defects, possibly constructed in *E. coli* and introduced into the marine vibrios by gene replacement methods. Bright and dark variants have been used for this purpose, but since luminescence variation is accompanied by many other cellular changes, differences in growth or survival of variants cannot be unequivocally attributed to the presence or absence of luminescence.

The microbial world certainly contains a great diversity of organisms. Yet, diversity can also exist in a population derived from a single bacterium. Luminescence is often a variable trait, and a diversity of phenotypes can arise from a single cell. We have suggested that variation of expression of *lux* may be an indirect consequence of variation in another cell function. Even if this is so, luminescence will be a useful marker for exploring the mechanism of variation in marine vibrios. The controlling mechanism may be a recombinational switch, like those which generate antigenic variability, and it will be interesting to see how these devices have evolved in marine bacteria. But what is the function of such a mechanism in the ocean? Generation of diversity could be as important for a marine bacterium as it is for a pathogen. Bacteria in the ocean encounter extreme and rapidly changing circumstances, and some variants in a population might be better able to survive in a particular situation. Thus, by generating diversity some descendents would be preadapted to environmental change, would better survive, and would thus ensure the continuation of the species.

Acknowledgements

Work by the authors described in this chapter was supported by a grant from the National Science Foundation. We thank Leticia Wolpert for her expert assistance in preparing this chapter.

References

(1) Nealson, K. H. and Hastings, J. W. (1979). Bacterial bioluminescence: its control and ecological significance. *Microbiol. Rev.* **43**, 496–518.

(2) Ziegler, M. M. and Baldwin, T. O. (1981). Biochemistry of bacterial bioluminescence. *Curr. Topics Bioenergetics* **12**, 65–113.

(3) Belas, R., Mileham, A., Cohn, D., Hilmen, M., Simon, M. and Silverman, M. (1982). Bacterial bioluminescence: isolation and expression of the luciferase genes from *Vibrio harveyi*. *Science* **218**, 791–793.

(4) Engebrecht, J. and Silverman, M. (1983). Bacterial bioluminescence: isolation and genetic analysis of function from *Vibrio fischeri*. *Cell*. **32**, 773–781.
(5) Cohn, D. H., Ogden, R. C., Abelson, J. N., Baldwin, T. O., Nealson, K. H., Simon, M. I. and Mileham, A. J. (1983). Cloning of the *Vibrio harveyi* luciferase genes: use of a synthetic oligonucleotide probe. *Proc. Natl. Acad. Sci. USA* **80**, 120–123.
(6) Miyamoto, C., Byers, P., Graham, A. F. and Meighen, E. A. (1987). Expression of bioluminescence by *Escherichia coli* containing recombinant *Vibrio harveyi* DNA. *J. Bacteriol.* **169**, 247–253.
(7) Engebrecht, J. and Silverman, M. (1984). Identification of genes and gene products necessary for bacterial bioluminescence. *Proc. Natl. Acad. Sci. USA* **81**, 4154–4158.
(8) Boylan, M., Graham, A. F. and Meighen, E. A. (1985). Functional identification of the fatty acid reductase components encoded in the luminescence operon of *Vibrio fischeri*. *J. Bacteriol.* **163**, 1186–1190.
(9) Nealson, K. H. (1977). Autoinduction of bacterial luciferase: occurrence, mechanism, and significance. *Arch. Microbiol.* **112**, 73–79.
(10) Eberhard, A., Burlingame, A. L., Eberhard, C., Kenyon, G. L., Nealson, K. H. and Oppenheimer, N. J. (1981). Structural identification of autoinducer of *Photobacterium fischeri* luciferase. *Biochemistry* **20**, 2444–2449.
(11) Greenberg, E. P., Hastings, J. W. and Ulitzur, S. (1979). Induction of luciferase synthesis in *Beneckea harveyi* by other marine bacteria. *Arch. Microbiol.* **120**, 87–91.
(12) Engebrecht, J. and Silverman, M. (1986). Regulation of expression of bacterial genes for bioluminescence. *In* "Genetic Engineering" (J. K. Setlow and A. Hollaender, eds.) **8**, pp. 31–44. Plenum Publishing Corp., New York.
(13) Kaplan, H. B. and Greenberg, E. P. (1985). Diffusion of autoinducer is involved in regulation of the *Vibrio fischeri* luminescence system. *J. Bacteriol.* **163**, 1210–1214.
(14) Dunlap, P. V. and Greenberg, E. P. (1985). Control of *Vibrio fischeri* luminescence gene expression in *Escherichia coli* by cyclic AMP and cyclic AMP receptor protein. *J. Bacteriol.* **164**, 45–50.
(15) Haygood, M. G. and Nealson, K. H. (1985). Mechanisms of iron regulation of luminescence in *Vibrio fischeri*. *J. Bacteriol.* **162**, 209–216.
(16) Beijerinck, M. W. (1889). Le Photobacterium luminosum, bactérie lumineuse de la Mer du Nord. *Arch. Neerl. Sci. Exactes Nat. Haarlem.* **23**, 401–405.
(17) Keynan, A. and Hastings, J. W. (1961). The isolation and characterization of dark mutants of luminescent bacteria. *Biol. Bull.* (Woods Hole, Mass.) **121**, 375.
(18) Nealson, K. (1972). Factors controlling the appearance of spontaneous dark mutants of luminous bacteria. *Biol. Bull.* (Woods Hole, Mass.) **143**, 471–472.
(19) Ulitzur, S. and Weiser, I. (1981). Acridine dyes and other DNA-intercalating agents induce the luminescence system of luminous bacteria and their dark variants. *Proc. Natl. Acad. Sci. USA* **78**, 3338–3342.
(20) Simon, M. and Silverman, M. (1983). Recombinational regulation of gene expression in bacteria. *In* "Gene Function in Prokaryotes" (J. Beckwith, J. Davies, J. A. Gallant, eds.), pp. 211–227. Cold Spring Harbor Laboratory, Cold Spring Harbor, New York.
(21) Borst, P. and Greaves, D. R. (1987). Programmed gene rearrangements altering gene expression. *Science* **235**, 658–667.

Chapter **5**

Photosynthesis in Rhodospirillaceae

DOUGLAS C. YOUVAN *and* EDWARD J. BYLINA

I. INTRODUCTION

An understanding of many prokaryotic processes can be extrapolated to eukaryotes. This is demonstrated by examples such as the genetic code and ribosome-mediated protein synthesis, but what about bioenergetic processes? Is there a prokaryotic genus we can look to as a model for energy transduction?

Of the well-studied bacterial systems amenable to genetic manipulation, *Escherichia coli* is certainly not the model of choice. In addition to lacking cytochrome *c*, this bacterium is also missing a nearly universal proton pump which is used in other bacteria, chloroplasts, and mitochondria (see below). Furthermore, *E. coli* does not utilize light to drive energy conversion processes, so proposing it as a model for photosynthesis (chloroplast bioenergetics) would be a non sequitur. Is there a better prokaryotic model than *E. coli* for bioenergetics in higher organisms?

Purple nonsulphur bacteria (Rhodospirillaceae) possess energy transducing complexes homologous to those used by mitochondria and chloroplasts. In contrast to the absence from *E. coli* of the above mentioned cytochromes and proton pumps, typical purple nonsulphur bacteria (genus *Rhodobacter*) use these complexes much like higher organisms do. As for photosynthesis, nucleotide sequence data and structural studies (spectroscopy and crystallography) have revealed remarkable similarities between the bacterial photosynthetic reaction centre (the site of light energy conversion in purple nonsulphur bacteria) and photosystem II from higher plants (1, 2).

Rhodospirillaceae are not only good models for bioenergetic studies, but these bacteria may be the progenitors of mitochondria and chloroplasts. Molecular cladistics supports the idea of an evolutionary lineage from ancient Rhodospirillaceae through cyanobacteria and eukaryotic algae to higher plants (the endosymbiont theory). It is indeed possible that both of the plant photosystems are the result of a gene duplication and divergence of the ancestral reaction centre gene from ancient bacteria much like the *Rhodobacter* of today (1). Considering its position in the evolutionary dendrogram and the conditions thought to exist four billion years ago on earth, a proponent of panspermia might conclude that an ancestral *Rhodobacter* was the ultimate seed!

Bacteria of the genus *Rhodobacter* have until very recently been classified as part of a diverse genus known as *Rhodopseudomonas*, one of several genera of Rhodospirillaceae. The new nomenclature (3) divides the most actively studied species into two genera, which include *Rhodobacter capsulatus*, *Rhodobacter sphaeroides*, *Rhodopseudomonas viridis* and *Rhodopseudomonas blastica*. Imhoff's nomenclature follows the work of Woese and his colleagues who have used ribosomal RNA sequences to deduce phylogenetic relationships (4: see also Chapter 1).

Genetic and biophysical studies pertinent to the light reactions of photosynthesis are most advanced for the Rhodospirillaceae (5). In this chapter, we focus on six proteins found in the photosynthetic membrane (Fig. 1) that function in concert to transduce light energy into chemical energy. This process is referred to as photophosphorylation: ATP is regenerated from ADP and inorganic phosphate using the energy of light. When at all possible, we use *Rb. capsulatus* as our model, since genetic studies are most advanced for this species. The last sections of the chapter deal with these genetic studies; in the next two sections we describe the physiology and biochemistry of the photosynthetic apparatus.

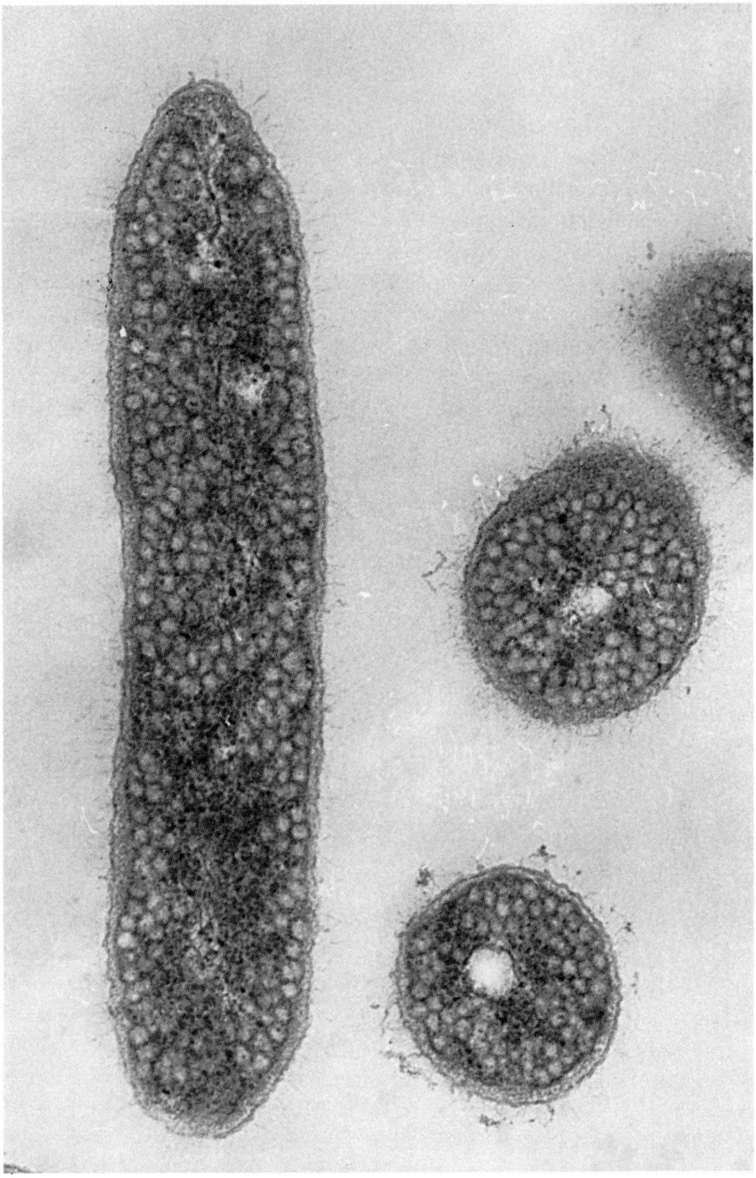

Figure 1. Electron micrograph of a thin section through several *Rhodobacter capsulatus* cells showing invaginations of the bacterial inner membrane—the location of the photosynthetic apparatus. Ruptured bacteria release closed vesicles called chromatophores which carry all of the structural peptides required for photophosphorylation. Magnification ×61 650. Micrograph courtesy of Dr. Barry L. Marrs.

II. STRUCTURE-FUNCTION OF THE PHOTOSYNTHETIC APPARATUS

A. Overall Metabolism

Rhodobacter spp. are unique among contemporary organisms in that they have retained the genetic information for using most if not all of the modes of energy metabolism that their ancestors possessed. These various growth modes function in response to different environmental conditions, in particular the presence or absence of both oxygen and light (Fig. 2). When solar energy is available, there are two metabolic modes, both of which occur in the absence of oxygen. In one case the bacteria obtain their carbon from already reduced compounds (photoheterotrophy); and in the other mode they use carbon dioxide as their sole source of carbon (photoautotrophy). A third anaerobic mode includes fermentative pathways, probably the most primitive means of carbon and energy metabolism. There are two aerobic growth modes: typical respiration in which organic compounds are oxidized for energy and metabolized for carbon (chemoheterotrophy); and chemoautotrophy, in which cellular carbon comes from carbon dioxide and energy is obtained by reacting hydrogen with oxygen, as in a fuel cell. We now know that there is a sixth growth mode which we may refer to as anaerobic respiration, in which compounds other than oxygen (e.g. nitrate, dimethylsulphoxide) are used as

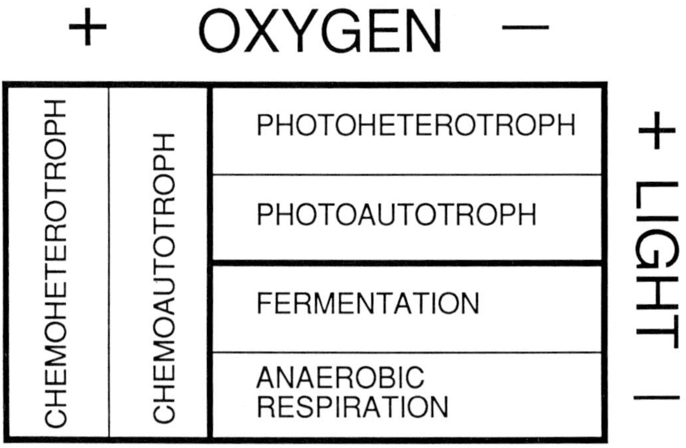

Figure 2. Growth modes of *Rhodobacter capsulatus* in response to oxygen tension and light. The photosynthetic apparatus is derepressed in response to low oxygen tension, regardless of whether light is present. For bioenergetic investigations, it is important that alternative growth modes can be used to support mutants that are impaired or defective in a particular pathway.

terminal electron acceptors for the oxidation of organic compounds. Excess reductant is consumed in the reduction of organic compounds rather than in the fixation of carbon dioxide.

The photosynthetic apparatus functions in the presence of light and in the absence of oxygen (upper right quadrant in Fig. 2). Under these conditions, photophosphorylation is carried out by the combined activities of five membrane proteins and a sixth water soluble protein:

(1) Photosynthetic reaction centre
(2) Cytochrome bc_1 oxidoreductase
(3) Proton ATPase
(4) Light harvesting I
(5) Light harvesting II
(6) Cytochrome *c*

Under intense light, only the first three proteins are essential for photophosphorylation in *Rb. capsulatus*; as we shall see later, the light harvesting and cytochrome *c* genes may be deleted from the bacterial chromosome without abolishing photosynthetic growth. Photosynthetic growth is impaired in some deletion strains since the accessory proteins are important in allowing the maximum efficiency of the photosynthetic light energy conversion processes.

B. Membrane Topography and Chemiosmosis

The photosynthetic membrane, a highly ramified invagination of the inner bacterial membrane, is the *in vivo* site of photophosphorylation. In an electron micrograph of a thin section through *Rb. capsulatus* (Fig. 1), the photosynthetic membrane is visible as what appear to be sections through 300 Å (where $1 \text{ Å} = 10^{-1}$ nm) vesicular spheres. In fact, the interiors of the membrane vesicles are (functionally) continuous with the periplasmic space through connections outside the plane of sectioning. After disruption of the bacteria with a French press, the membrane invaginations pinch off to form closed vesicles. These vesicles are referred to as chromatophores, since they are intensely coloured by bacteriochlorophyll and carotenoid pigments. The five membrane proteins of the photosynthetic apparatus can be isolated from the lipid bilayer of the chromatophores. Cytochrome *c* may be found trapped within the vesicle, while the water soluble portion of the proton ATPase can be found facing outward from the surface of the chromatophore. One of the most convincing arguments for the connectivity of the intravesicular spaces with the periplasm is the observation that cytochrome *c* can be extracted from sphaeroplasts.

A schematic diagram of the bacterial photosynthetic membrane is shown in Fig. 3. We can use it to trace the time course of the light-driven regeneration of

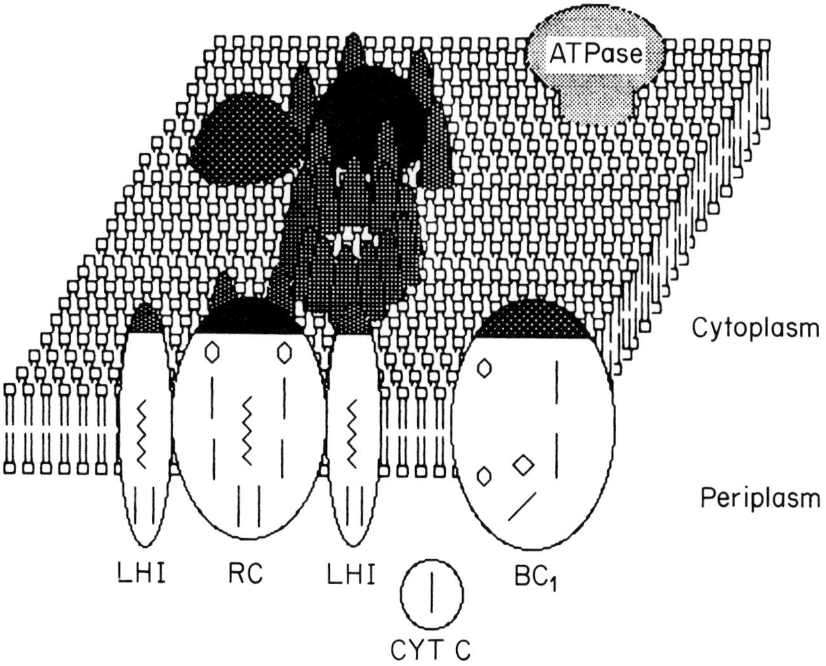

Figure 3. A schematic diagram of the photosynthetic membrane showing the principal structural components that mediate the conversion of light energy into chemical energy. Six membrane-associated complexes function to convert light to chemical energy: light-harvesting antennae (LH II and LH I), photochemical reaction centre (RC), cytochrome c_2 (CYT C), the cytochrome bc_1 oxidoreductase (BC_1), and the proton ATPase (ATPase). See Section II.B for a discussion of chemiosmosis and photophosphorylation.

ATP from inorganic phosphate and ADP. This process is essentially a description of chemiosmosis, which was first proposed by Mitchell (6), as it applies to photosynthetic bacteria such as *Rb. capsulatus* or *Rb. sphaeroides*.

Photophosphorylation begins with the absorption of light by an array of bacteriochlorophylls and carotenoids bound to light harvesting (LH) peptides (7). The peripheral part of this antenna is known as LH II, and its absorption properties are spectroscopically distinct from a more central antenna called LH I. (A discussion of the exact composition of all complexes is deferred to the next section.) Energy absorbed by individual bacteriochlorophylls is shared with others in the array—a process which is referred to as exciton coupling. Reaction centre complexes embedded in the light harvesting array can trap this energy. The light-to-chemical energy conversion process begins when the energy is absorbed by a special pair of bacteriochlorophyll molecules in the reaction centre protein. Prosthetic groups within the reaction centre direct reactions which ultimately result in a positive charge being left on the periplasmic face (inner side of vesicle) of the reaction centre and an excess negative charge (bound to a quinone molecule) on the other side. This process has been studied in great detail by protein chemists and crystallographers and is the subject of a recent article (5).

The light reactions yield a reduced quinone and a positively-charged pair of bacteriochlorophylls (hole) within the reaction centre protein. This charge-separated state represents a form of stored energy since bringing the quinone and hole together would result in a reaction that would release much of the energy present in the original quantum of light. Another membrane protein complex, the cytochrome bc_1 oxidoreductase, uses the stored energy in a productive fashion. The exergonic return of the electron from the quinone to the positively-charged special pair is used to drive the endergonic transport of a proton across the membrane. Most of the electrons pass through cytochrome *c* while some appear to be conducted directly from the bc_1 complex to the reaction centre (see the discussion on deletion strains below). Proton pumping generates another form of stored energy: an electrochemical proton gradient. Continued light absorption, charge separation, and proton pumping drive the accumulation of excess hydrogen ions on the inside of the chromatophore vesicle.

The proton ATPase couples the exergonic flow of protons down their electrochemical gradient to the endergonic regeneration of ATP from ADP and inorganic phosphate. Thus, in this final step, this universal mechanism of chemiosmosis transduces the energy of the proton gradient to another form of chemical energy by shifting the ADP-ATP equilibrium to the ATP side (8).

III. PROTEIN COMPONENTS OF THE PHOTOSYNTHETIC APPARATUS

A. Light-Harvesting Antennae

The LH II antenna in a typical purple nonsulphur bacterium is composed of two small bacteriochlorophyll-binding polypeptides designated α and β. Although crystals of LH II complexes have been obtained, none diffract well enough for structural studies (9). Modelling suggests that the α and β chains form a heterodimer with three bacteriochlorophylls and one or two carotenoid molecules. Although all of the bacteriochlorophylls bound to the antenna are of the same type, absorption occurs at two different wavelengths in the near infrared (800 and 858 nm) because of differences in the binding sites.

The LH I antenna is very similar to the LH II antenna except that only two bacteriochlorophylls are bound per α-β dimer. These bacteriochlorophylls absorb at 878 nm, a considerable red shift from the absorption of bacteriochlorophyll in solution at 770 nm. Protein sequence data show that the LH I and LH II α subunits are homologous but not identical. The same is true for β subunits (10).

B. Photosynthetic Reaction Centre

Reaction centres from typical purple nonsulphur bacteria have three peptide subunits which have been designated: L, M, and H. Together, these subunits total some 900 residues and yield a 100 kd complex (11). L, M, and H are extremely hydrophobic and cross the membrane a total of 11 times (through transmembrane α-helices). Prosthetic groups bound to the reaction centre include four bacteriochlorophyll molecules, two bacteriopheophytins (bacteriochlorophyll without the magnesium), two quinones (typically, ubiquinone), one iron ion, and a carotenoid molecule. Diagnostic features of the near infrared absorption spectra of reaction centres include the two special pair bacteriochlorophylls at 860 nm, the remaining "voyeur" bacteriochlorophylls at 800 nm, and the bacteriopheophytins at 760 nm.

The structure of the photosynthetic reaction centre from *R. viridis* and *Rb. sphaeroides* is known to less than 3 Å resolution, making this the first integral membrane protein with a structure known to atomic resolution (12, 13). Recently it has become clear through protein homology studies and spectroscopy that the bacterial reaction centre is homologous to the photosystem II reaction centre from cyanobacteria and higher plants. The L and M subunits of the bacterial reaction centre are homologous and analogous to two chloroplast-encoded proteins, the so-called D1 and D2 polypeptides.

C. Cytochrome bc_1 Oxidoreductase

Also known as the ubiquinol-cytochrome *c* oxidoreductase, this membrane-bound proton pump is found in many bacteria and in all mitochondria (Complex III). A homologous complex, the cytochrome b_6f oxidoreductase, functions as a proton pump in chloroplasts where it oxidizes plastoquinones (that have been reduced by photosystem II) and reduces plastocyanin (the donor to PS I). In contrast to this tandem arrangement of photosystems and linear electron flow in the chloroplast, the bacterial proton pump participates in a cyclic electron flow in concert with the reaction centre. In a typical *Rhodobacter*, the complex consists of a *b*-type cytochrome, cytochrome c_1 and a Rieske iron-sulphur protein (14, 15). The molecular weight of the bc_1 complex is very similar to that of the reaction centre (ca. 100 kd).

The bc_1 complex uses six redox active sites (three haems, one iron-sulphur centre, and two quinones). An integral membrane protein, cytochrome *b*, is thought to bind two of the haem groups through attachments to transmembrane α-helices. The *b*-type cytochromes are so named because of the non-covalent binding of the haems to the apoprotein. These haems are spectrally distinguishable and show different redox properties: b_H and b_L (high and low oxidation-reduction potentials). Cytochrome c_1 bears a covalently-attached haem, and the Rieske iron-sulphur protein carries one 2Fe-2S centre attached to the apoprotein via the irons to cysteine residues. All of the haems and the iron-sulphur centre participate in single electron transfer reactions whereas the quinone undergoes two-electron (plus two-proton) reduction via semiquinone intermediates.

The bc_1 complex directs a so-called "Q-cycle", or quinone cycle, in which the endergonic transport of protons across the membrane is coupled to the exergonic transport of electrons. In this scheme, quinol generated on the cytoplasmic side of the membrane in reaction centres diffuses to the Q_Z site of the bc_1 complex on the periplasmic side of the membrane. One electron from this quinol eventually reduces cytochrome c_2 after passing through the Rieske cluster and the c_1 haem, while the other electron eventually reduces a quinone at the Q_C site on the cytoplasmic side of the membrane after passing through the two *b*-type haems. The oxidation of the quinol at the Q_Z site is accompanied by the release of two protons into the periplasmic space, effectively shuttling protons across the membrane. Quinol generated at the Q_C site can also diffuse across the membrane to be used at the Q_Z site. The vectorial nature of these reactions serves to complete the electron circuit with the reaction centre and to generate a transmembrane proton gradient.

D. Cytochrome *c*

Cytochrome *c* is a small water-soluble protein bearing a single haem group which mediates one-electron oxidation-reduction reactions. The cytochrome *c* apoprotein has a characteristic sequence cys-X-Y-cys-his . . . met which facilitates covalent haem attachment from the cysteine residues to the ring edges and from the histidine and methionine to the iron atom.

Cytochrome functions as the mobile electron carrier between the oxidizing side of the reaction centre and the reducing side of the proton pump. In a chromatophore preparation, one can envisage cytochrome molecules visiting the oxidized reaction centre where an electron would be dropped off (haem iron changing from ferrous to ferric). This cytochrome-reaction centre docking takes place on the internal surface of the membrane vesicle and the donated electron would be conducted to and neutralize the positively charged special pair.

All prokaryotic and eukaryotic cytochrome *c* molecules have very conserved three-dimensional structures. The conservation and role of cytochromes in almost all electron transfer pathways (primitive bacteria to higher organisms) led Dickerson (16) to conclude: "The history of this ancient family of electron-transferring proteins (cytochrome *c*) suggests that our metabolic ancestors may have been photosynthetic bacteria for which respiration was only a stand-by energy mechanism."

E. Proton ATPase

The proton ATPase is the only photosynthetic apparatus component that we shall discuss which is also found in *E. coli* (17). It is also the only component for which the genetic determinants are not currently cloned or sequenced in *Rb. capsulatus*. This complex is highly conserved between bacteria, chloroplasts and mitochondria. The bacterial complex may be somewhat simpler than the mitochondrial complex which has several additional peptides. Five unique polypeptides form the extrinsic F_1 subunit ($\alpha_3\beta_3\gamma\delta\varepsilon$) which bears up to three nucleotide binding sites. Three peptides form the intrinsic F_0 subunit which provides the membrane pore for proton translocation.

As is the case with the light-harvesting complex, the X-ray crystallographic analysis of the proton ATPase has not attained sufficiently high resolution to reveal aspects of the internal structure of the protein. Biochemical studies localize the proton translocating activity to the membrane-spanning F_0 component and the ATPase activity to the F_1 subunit. The F_1 subunit is found on the outside of chromatophores or submitochondrial particles. Detergent-solubilized proton pumps (such as bacteriorhodopsin) and the proton ATPase can be reconstituted in artificial vesicles to reconstitute photophosphorylation.

Conformational models for the regeneration of ATP from ADP and inorganic phosphate couple the translocation of two protons through the F_0 subunit (from the high pH to the low pH side of the membrane) with the synthesis of one phosphodiester bond. Uncouplers of this process, include agents which can collapse the proton gradient such as dinitrophenol and compounds like DCCD (dicyclohexylcarbodiimide) which may react with an amino acid residue in F_0 which is along the proton conduction pathway.

IV. PHOTOSYNTHETIC APPARATUS GENES

A. Reaction Centre and Light Harvesting Antennae

Probably the biggest single breakthrough in the molecular biology of bacterial photosynthesis was the discovery of a photosynthetic gene cluster by Marrs (18) and its subsequent characterization. Marrs used "classical" genetics and crosses between photosynthetic bacteria to isolate an R-prime plasmid carrying most of the genes necessary for the differentiation of the photosynthetic apparatus from the constitutively expressed respiratory membrane. This plasmid, pRPS404, complemented all the known (in 1980) chromosomal point mutations that affected the reaction centre, light-harvesting apparatus, carotenoid, and bacteriochlorophyll biosynthesis. Subclones of the 46 kb photosynthetic gene cluster were mapped which complemented various reaction centre mutations. After sequencing these fragments, we found all of the reaction centre and light harvesting I structural genes listed in Table 1.

With the increased availability of oligonucleotides, we decided to use DNA probes based on LH II polypeptide sequences to search for the structural

Table I

Structural genes and polypeptides that are required to convert light into chemical energy in the bacterial photosynthetic apparatus. Genetic-physical maps of the relevant DNA can be found in the five references cited.

Complex/Peptide	Gene	Size (amino acid residues)	References
Reaction centre			
H subunit	*puhA*	254	11
L subunit	*pufL*	282	11
M subunit	*pufM*	307	11
Light-harvesting I			
α polypeptide	*pufA*	58	11
β polypeptide	*pufB*	49	11
Light-harvesting II			
α polypeptide	*pucA*	60	10
β polypeptide	*pucB*	49	10
Cytochrome bc_1 oxidoreductase			
Rieske iron-sulphur protein	*petA*	191	15
Cytochrome b	*petB*	437	15
Cytochrome c_1	*petC*	279	15
Cytochrome c_2	*cycA*	116	21
Proton ATPase (F_1 Subunit)			
α subunit	*uncA*	272	24
β subunit	*uncD*	238	24
γ subunit	*uncG*	146	24
δ subunit	*uncH*	172	24
ε subunit	*uncC*	70	24
Proton ATPase (F_0 Subunit)			
subunit a	*uncB*	n.d.[a]	—
subunit b	*uncF*	n.d.[a]	—
subunit c	*uncE*	n.d.[a]	—

[a] n.d.: not determined.

genes for the peripheral antenna. The LH II β and α-encoding genes were localized on a single restriction fragment which was subsequently shown to complement chromosomal point mutations in LH II. Similarities in gene organization, and nucleotide and polypeptide sequences, led our group to suggest that both sets of light-harvesting genes (LH I + LH II) arose through a duplication of an ancestral light harvesting gene (10).

The restriction fragment bearing LH II does not map within or adjacent to the 46 kb photosynthetic gene cluster. From the work described below, it is now known that none of the other genes encoding components of the photosynthetic apparatus (other than reaction centre and light-harvesting I) are found in the 46 kb photosynthetic gene cluster.

B. Cytochrome bc_1 Oxidoreductase and Cytochrome *c*

The first cloning and nucleotide sequencing of the genes encoding a *Rhodospirillaceae* proton pump were carried out in *Rb. capulatus* although the authors thought at the time that they were working with *Rb. sphaeroides* (19). This problem was later resolved by the redetermination of the nucleotide sequence in genes cloned from authentic *Rb. capsulatus* and *Rb. sphaeroides* (20). In all cases, the order of genes was found to be the same: Rieske iron-sulphur protein (*petA*), *b*-type cytochrome (*petB*), and cytochrome c_1 (*petC*).

Cytochrome *c* (more properly known as cytochrome c_2) was cloned from *Rb. capsulatus* using oligonucleotide probes (21). The structural gene was found, by comparison with restriction digests, to lie outside the other photosynthetic apparatus genetic loci. Hence, of the six proteins constituting the photosynthetic apparatus, only LH I and reaction centres are closely linked. A notable exception is that *R. viridis* uses a membrane-bound cytochrome as the reaction centre electron donor in place of cytochrome c_2. The gene for this cytochrome follows *pufM* in the *R. viridis* reaction centre operon (22, 23).

C. Proton ATPase

The proton ATPase is the only photosynthetic apparatus component that we shall discuss in this chapter that is not cloned or sequenced in *Rb. sphaeroides* or *Rb. capsulatus*. Walker's group was the first to sequence the proton ATPase from a purple nonsulphur bacterium (24). This work was done with *R. blastica*—a species not developed for genetic analysis. Restriction fragment probes from the *E. coli* ATP operon (*unc*, for uncoupler) were used to probe a library of the *R. blastica* genome. The water-soluble F_1-encoding region was restriction mapped and sequenced. Unlike the *E. coli* genes, the *R. blastica* F_0

genes were not found upstream from the F_1 genes, and an extra gene was inserted between the γ- and β-encoding genes.

Characterizing the *unc* genes in a *Rhodobacter* amenable to molecular genetics (such as *Rb. capsulatus*) has not yet been attempted. One reason for this may be that work on photosynthetic bacteria offers no particular advantage over *E. coli*. Deletion of the proton ATPase genes in *Rhodobacter* is likely to yield a photosynthetically defective strain. To be viable, the deletion strain would have to be grown under conditions where ATP could be replenished by other processes, perhaps substrate-level phosphorylation. This strain would still be interesting since a light-driven proton gradient might be generated.

V. *IN VITRO* MUTAGENESIS STUDIES

Five of the six photosynthetic apparatus genes have been deleted from the chromosome of *Rb. capsulatus*. Many of the deletion strains are photosynthetically defective or impaired, so it is important that other growth modes are available (Fig. 2). This is clearly an advantage over studying photosynthesis in higher plants since a photosynthetically defective plant will die shortly after germination.

A. Deletion of Cytochrome Genes

Before Daldal's genetic experiments on cytochrome *c* in *Rb. capsulatus*, it was widely accepted that this mobile electron carrier was essential to complete the electron circuit between the cytochrome bc_1 oxidoreductase and the reaction centre. The prediction was that a cytochrome *c* deletion strain would be photosynthetically defective. There was even some possibility that the deletion would be lethal since cytochrome *c* is also involved in shuttling electrons from the bc_1-complex to cytochrome oxidase under oxidative respiration (Fig. 4). This role for cytochrome *c* is similar to its role in mitochondria: electrons from succinate or NADH dehydrogenase are passed to the bc_1 complex (via the quinone pool) and then to cytochrome *c*, cytochrome oxidase and molecular oxygen. Thus a deletion of the cytochrome *c* gene was expected to eliminate both photosynthesis and respiration. There was an additional question as to whether any of the other metabolic modes (Fig. 2) would be fully independent of these two primary modes.

Daldal and colleagues deleted the cytochrome *c* gene by interposon mutagenesis; a technique developed by Scolnik (25) which involves replacing the structural gene in the chromosome with a restriction fragment from a transposon carrying an antibiotic resistance gene. A bacteriophage-like particle known as the gene transfer agent (which is specific for *Rb. capsulatus*) was used

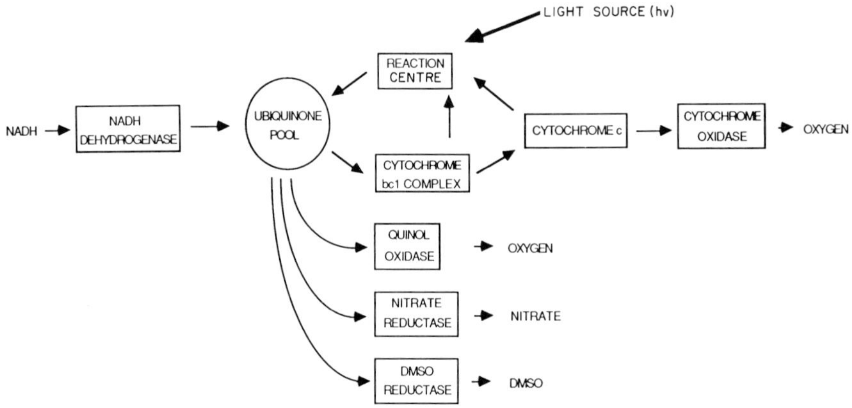

Figure 4. Electron transfer pathways for respiration, photosynthesis, and anaerobic respiration in a typical *Rhodobacter*. Arrows indicate the direction of electron flow in these pathways: linear flow to molecular oxygen in respiration or cyclic flow in photosynthesis. Alternative pathways branch from the ubiquinone pool and are independent of both photosynthesis and aerobic respiration.

to transduce the insertion-deletion mutation into the chromosome. (These techniques have been recently described in a review article (26).) The results of cytochrome *c* deletion were very unexpected. The strain was not only viable but grew at almost wild-type photosynthetic rates. For all practical purposes, there was no mutant phenotype—certainly not one that could be selected or counterselected. The dogma was proven wrong (at least in this species). The mutant was later shown to be somewhat impaired in reaction centre re-reduction kinetics, but it was clear that cytochrome *c* was not essential for photosynthetic growth (27). This study is a perfect example of how a well accepted mechanism can be disproven by a single mutant.

Daldal's group has recently reported deletion of the *b*-type cytochrome component of the bc_1 oxidoreductase. These strains are photosynthetically defective since the reaction centre alone cannot function as a proton pump. The viability of the *b*-deletion strain demonstrates the utility of the diverse metabolism of *Rhodobacter* in bioenergetic studies. A branched respiratory pathway (beginning at the quinone pool) enables this deletion strain to grow by oxidative respiration or anaerobic respiration (using electron acceptors such as dimethylsulphoxide) even in the absence of the bc_1 complex (Fig. 4).

B. Reaction Centre and Light-Harvesting Mutagenesis

Through molecular genetics, various mutations and deletions in genes for the reaction centre and light-harvesting antennae have been combined such

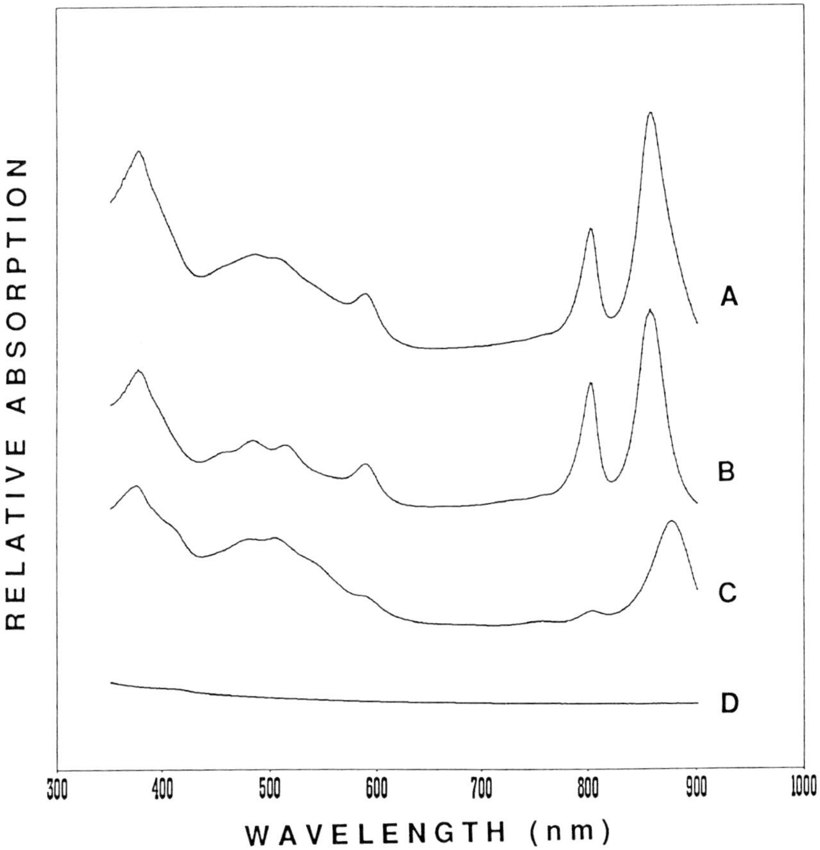

Figure 5. Absorption spectra of chromatophore membranes from some commonly used strains of *Rhodobacter capsulatus*. A: wild-type SB1003; B: Y142; C: MW442; D: U43 (see Section V.B. for genotypes). Ground-state absorption spectra of chromatophore membranes reveal carotenoid absorption near 500 nm, and bacteriochlorophyll absorption at 360, 590, and 800–900 nm. The near-infrared bacteriochlorophyll absorption bands are indicative of the nature of the binding site. Free bacteriochlorophyll in solution absorbs at 770 nm, so the observed red-shifts in chromatophores are useful as phenotypic indicators.

that all possible combinations of LH II, LH I, and RC complexes are observed (28, 29). Figure 5 compares the absorption spectra of chromatophores of several mutant strains of *Rb. capsulatus* in order to highlight the absorption properties of various components. Strain U43 is deleted for RC and LH I genes and also carries a point mutation in LH II which results in the loss of all three components. This serves as a "blank" background against which we can view the expression of each component. Strain Y142 expresses only LH II with characteristic absorption at 800 and 850 nm. Strain MW442 has no LH II, so we can see LH I absorption at 878 nm and small peaks due to reaction centre absorption at 760 and 800 nm. A third reaction centre component that is due to the special pair absorption is buried under the much larger 878 nm peak. Expression of plasmid-borne *puc* genes (encoding LH II) in the deletion background yields spectra identical to Y142, while expression of the *puf* genes (encoding reaction centre and LH I) yields a strain with chromatophore spectra identical to MW442.

VI. GENETIC ENGINEERING IN REACTION CENTRES

Through *in vitro* mutagenesis studies, it has recently become possible to make precise and predetermined changes in the sequence of reaction centre and light-harvesting genes (30, 31, 32). Using the same deletion strains as discussed in the previous section, it is possible to assay the photosynthetic phenotypes of the mutations. This is of particular interest when researchers wish to test the role of a single amino acid in the light reactions. Since the atomic resolution crystallographic structure of the reaction centre is known, there are many questions regarding the role of amino acids in modifying the chemistry of the prosthetic groups. We shall discuss the impact that these ongoing mutagenesis studies have made on our understanding of reaction centre function in two areas: bacteriochlorophyll binding and herbicide resistance.

A. Requirements for Bacteriochlorophyll Binding

Bacteriochlorophyll molecules consist of a tetrapyrrole ring with a magnesium ion bound in the centre as a replacement for two protons. As with the iron in haem, the magnesium in bacteriochlorophyll may be bound to a protein through a coordinate-covalent bond to the lone pair electrons of the N-3 nitrogen on the histidine imidazole ring. One goal of *in vitro* mutagenesis experiments is to determine whether there are functional replacements for this type of protein-metal interaction in membrane-spanning proteins.

Although there are no X-ray data for LH I, Zuber and co-workers have used interspecies protein homology data to support the hypothesis that the 32nd residue (histidine-H) in the α-chain of LH I is the axial ligand to bacteriochlorophyll (7). Arginine (R) and all other residues that we have tested at this position result in the loss of LH I from chromatophore membranes. Expressing a plasmid which has a mutation blocking LH I assembly in the U43 background (strain U43(Hα32R)) reveals the *in situ* light absorption properties of reaction centres. Presumably, the energy of the histidine-

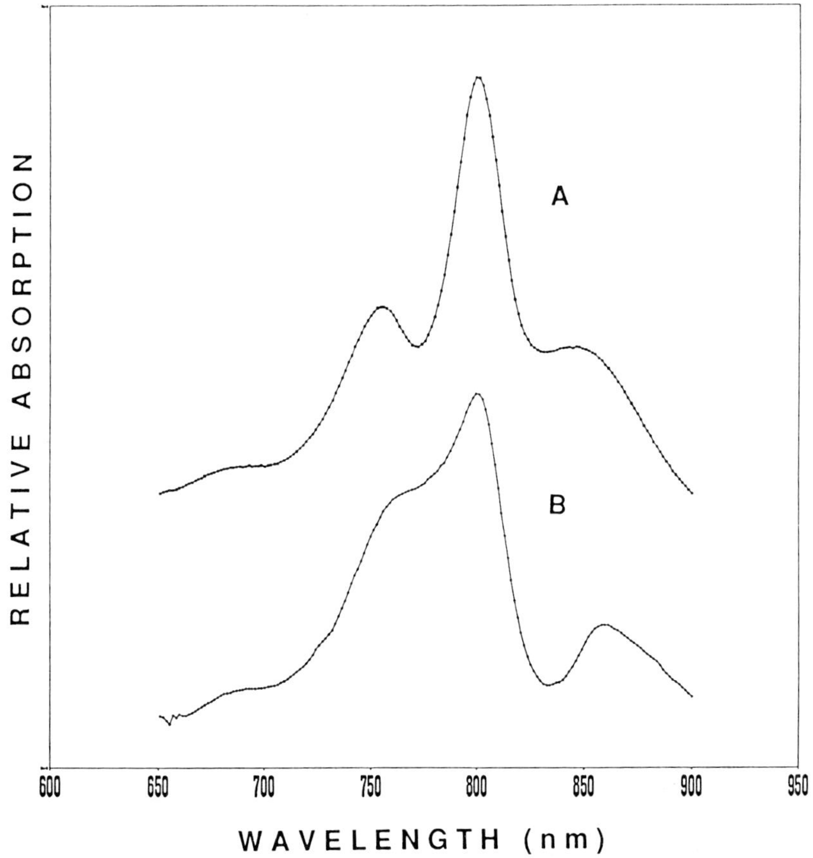

Figure 6. Absorption spectra of purified and membrane-bound reaction centres. A: purified reaction centres in an aqueous-detergent solution; B: chromatophore membranes from U43 (Hα32R). Direct observation of the reaction centre is possible in this light-harvesting mutant since all absorption at 878 nm has been lost by a site-directed mutation. Additional absorption at 770 nm in the mutant membrane may be due to free bacteriochlorophyll absorption, while the apparent blue-shift of the long wavelength absorption band in purified reaction centres may be due to detergent solubilization artifacts.

magnesium interaction is essential in stabilizing the complex, or the bacteriochlorophyll interaction is an essential intermediate in the LH folding pathway (32).

In contrast to LH I, recent studies demonstrate that residues other than histidine may be functional at some of the bacteriochlorophyll binding sites within the reaction centre. Serine and glutamine have been substituted for histidine may be functional at some of the bacteriochlorophyll binding sites within the reaction centre. Serine and glutamine have been substituted for histidine at the accessory chlorophyll and special pair chlorophyll binding sites (respectively), and these modified reaction centres appear to be pseudo-wild-type. The side chains of serine and glutamine contain atoms (oxygen and nitrogen) which could donate lone-pair electrons in the coordinate-covalent bond with magnesium.

In at least one site, substitution of a chlorophyll-binding histidine with hydrophobic residues (lacking lone-pair electrons) results in the replacement of the bacteriochlorophyll molecule with a bacteriopheophytin. Replacement of an M-side histidine (residue number 200) with leucine or phenylalanine yields reaction centres containing a "heterodimer", where one of the two bacteriochlorophyll molecules in the special pair is replaced with a bacteriopheophytin (31). Such strains are photosynthetically defective, but the isolated reaction centre proteins are proficient for *in vitro* charge separation reactions.

Genetically modified reaction centres are important to spectroscopists and theorists interested in the mechanism of electron transfer within the reaction centre. Altered complexes may be blocked or impaired at certain steps, facilitating the observation of a transient state which is not observable in wild-type. Other modified reaction centres, which appear normal, are valuable in evaluating models which depend on a critical amino acid residue. Directed mutations which result in pseudo-wild-type phenotypes are especially powerful in disproving models and circumvent the problem of not being able to predict structure from sequence.

What is the nature of the bacterial processes which synthesize the components of the photosynthetic apparatus? Specifically, how are reaction centres assembled? Answering these questions is a remaining goal for molecular biologists studying complex assembly pathways involving interactions among structural proteins, assembly proteins, biosynthetic enzymes, and pigments. In turn, these assembly principles are essential factors for geneticists to consider before constructing modified reaction centres for biophysical analyses. The experiments described above, where bacteriopheophytin has been substituted for bacteriochlorophyll, are such an example.

B. Molecular Mechanism of Herbicide Resistance

Herbicides such as atrazine are competitive inhibitors of reaction centre quinone reduction. Presumably, a herbicide-resistant reaction centre would bind quinones more tightly than herbicides. The X-ray structure of the quinone binding site is now known to atomic resolution (13). Crystallographers have also been successful in diffusing herbicides into these reaction centre crystals and in determining the atomic resolution structure of the bound inhibitor (33). Using these data, we were able to construct several resistant derivatives of *Rb. capsulatus* by making single amino acid changes in the quinone binding pocket (30, 34). Since atrazine inhibits PS II in a similar manner, our studies may be important for future genetic engineering programmes in which herbicide-resistant mutations could be introduced into the genomes of agriculturally important crops.

References

(1) Youvan, D. C. and Marrs, B. L. (1984). Molecular genetics and the light reactions of photosynthesis. *Cell* **39**, 1–3.

(2) Trebst, A. (1987). The three-dimensional structure of the herbicide binding niche on the reaction centre polypeptides of photosystem II. *Zeitschrift für Naturforschung* **42c**, 742–750.

(3) Imhoff, J. F., Truper, H. G. and Pfennig, N. (1984). Rearrangement of the species and genera of the phototrophic "Purple Nonsulphur Bacteria". *Int. J. System. Bacteriol.* **34**, 340–343.

(4) Woese, C. R., Gibson, J. and Fox, G. E. (1980). Do genealogical patterns in purple photosynthetic bacteria reflect interspecific gene transfer? *Nature* **283**, 212–214.

(5) Youvan, D. C. and Marrs, B. L. (1987). Molecular mechanisms of photosynthesis. *Sci. Am.* **256** (6), 42–48.

(6) Mitchell, P. (1961). Coupling of phosphorylation to electron and hydrogen transfer by a chemiosmotic type of mechanism. *Nature* **191**, 144–148.

(7) Zuber, H. (1986). Structure of light-harvesting antenna complexes of photosynthetic bacteria, cyanobacteria and red algae. *TIBS* **11**, 414–419.

(8) Prince, R. C. (1985). Redox-driven proton gradients. *BioSci.* **35**, 22–26.

(9) Cogdell, R. J., Wooley, K. J., Mackenzie, J. G., Lindsay, H., Michel, H., Dobler, J. and Zinth, W. (1985). Crystallization of the B800-850-complex from *Rhodopseudomonas acidophila* strain 7750. *Springer Ser. Chem. Phys.* **42**, 85.

(10) Youvan, D. C. and Ismail, S. (1985). Light-harvesting II (B800-850-complex) structural genes from *Rhodopseudomonas capsulata*. *PNAS* **82**, 58–62.

(11) Youvan, D. C., Bylina, E. J., Alberti, M., Begusch, H. and Hearst, J. E. (1984). Nucleotide and deduced polypeptide sequences of the photosynthetic reaction centre, B870 antenna, and flanking polypeptides from *Rhodopseudomonas capsulata*. *Cell* **37**, 949–957.

(12) Deisenhofer, J., Epp, O., Miki, K., Huber, R. and Michel, H. (1985). Structure of the protein subunits in the photosynthetic reaction centre of *Rhodopseudomonas viridis* at 3 Å resolution. *Nature* **318**, 618–624.

(13) Allen, J. P., Feher, G., Yeates, T., Rees, D. C., Eisenberg, D. S., Deisenhofer, J., Michel, H. and Huber, R. (1986). Structural homology of reaction centres from *Rhodopseudomonas sphaeroides* and *Rhodopseudomonas viridis* as determined by X-ray diffraction. *Proc. Natl. Acad. Sci. U.S.A.* **83**, 8589–8593.

(14) Daldal, F., Davidson, E. and Cheng, S. (1987). Isolation of the structural genes for the Rieske Fe-S protein, cytochrome *b* and cytochrome c_1. *J. Mol. Biol.* **195**, 1–12.

(15) Davidson, E. and Daldal, F. (1987). Primary structure of the bc_1 complex of *Rhodopseudomonas capsulata*. *J. Mol. Biol.* **195**, 13–24.

(16) Dickerson, Richard E. (1980). Cytochrome *c* and the evolution of energy metabolism. *Sci. Am.* **242**, 137–153.

(17) Nielsen, J., Hansen, F. G., Hoppe, J., Friedl, P. and von Meyenburg, K. (1981). The nucleotide sequence of the *atp* genes coding for the F_0 subunits a, b, c and the F_1 subunit g of the membrane bound ATP synthase of *Escherichia coli*. *Mol. Gen. Genet.* **184**, 33–39.

(18) Marrs, B. L. (1981). Mobilization of the genes for photosynthesis from *Rhodopseudomonas capsulata* by a promiscuous plasmid. *J. Bacteriol.* **146**, 1003–1012.

(19) Gabellini, N. and Sebald, W. (1986). Nucleotide sequence and transcription of the *fbc* operon from *Rhodopseudomonas sphaeroides*. *Eur. J. Biochem.* **154**, 569–579.

(20) Davidson, E. and Daldal, F. (1987) *fbc* operon, encoding the Rieske Fe-S protein, cytochrome *b*, and cytochrome c_1 apoproteins previously described from *Rhodopseudomonas sphaeroides*, is from *Rhodopseudomonas capsulata*. *J. Mol. Biol.* **195**, 25–29.

(21) Daldal, F., Cheng, S., Applebaum, J., Davidson, E. and Prince, R. (1986). Cytochrome c_2 is not essential for photosynthetic growth of *Rhodopseudomonas capsulata*. *PNAS* **83**, 2012–2016.

(22) Michel, H., Weyer, K. A., Gruenberg, H., Dunger, I., Oesterhelt, D. and Lottspeich, F. (1986). The "light" and "medium" subunits of the photosynthetic reaction centre from *Rhodopseudomonas viridis*: isolation of the genes, nucleotide and amino acid sequence. *EMBO J.* **5**, 1149–1158.

(23) Deisenhofer, J., Michel, H. and Huber, R. (1985). The structural basis of photosynthetic light reactions in bacteria. *TIBS* **10**, 243–248.

(24) Tybulewicz, V. L. J., Falk, G. and Walker, J. E. (1984). *Rhodopseudomonas blastica atp* operon. *J. Mol. Biol.* **179**, 185–214.

(25) Scolnik, P. A. and Haselkorn, R. (1984). Activation of extra copies of genes coding for nitrogenase in *Rhodopseudomonas capsulata*. *Nature* **307**, 289–292.

(26) Scolnik, P. A. and Marrs, B. L. (1987). Genetic research with photosynthetic bacteria. *Annu. Rev. Microbiol.* **41**, 703–726.

(27) Prince, R. C., Davidson, E., Haith, C. E. and Daldal, F. (1986). Photosynthetic electron transfer in the absence of cytochrome c_2 in *Rhodopseudomonas capsulata*: cytochrome c_2 is not essential for electron flow from the cytochrome bc_1 complex to the photochemical reaction centre. *Biochem.* **25**, 5208–5214.

(28) Youvan, D. C., Ismail, S. and Bylina, E. J. (1985). Chromosomal deletion and plasmid complementation of the photosynthetic reaction centre and light harvesting genes from *Rhodopseudomonas capsulata*. *Gene* **38**, 19–30.

(29) Bylina, E. J., Ismail, S. and Youvan, D. C. (1986). Plasmid pU29, a vehicle for mutagenesis of the photosynthetic *puf* operon in *Rhodopseudomonas capsulata*. *Plasmid* **16**, 175–181.

(30) Bylina, E. J. and Youvan, D. C. (1987). Genetic engineering of herbicide resistance: saturation mutagenesis of isoleucine 229 of the reaction centre L subunit. *Z. Naturforsch.* **42c**, 769–774.

(31) Bylina, E. J. and Youvan, D. C. (1988). Directed mutations affecting spectroscopic and electron transfer properties of the primary donor in the photosynthetic reaction centre. *Proc. Natl. Acad. Sci. USA* (In press).

(32) Bylina, E. J., Robles, S. and Youvan, D. C. (1988). Directed mutations affecting the putative bacteriochlorophyll binding sites in the light harvesting I antenna of *Rhodobacter capsulatus*. *Israel J. Chem.* (Submitted).

(33) Michel, H., Epp, O. and Deisenhofer, J. (1986). Pigment-protein interactions in the photosynthetic reaction centre from *Rhodopseudomonas viridis*. *EMBO J.* **5**, 2445–2451.

(34) Bylina, E. J., Jouine, R. V. M. and Youvan, D. C. (1988). A genetic system for rapidly assessing herbicides that compete for the quinone binding site of photosynthetic reaction centers. Bio/Technology (In press).

Chapter **6**

The Genetics of Nitrogen Fixation

CHRISTINA KENNEDY

I. THE DIVERSITY OF NITROGEN-FIXING BACTERIA

All organisms must take in nitrogen compounds so that they can synthesize proteins, nucleic acids and other nitrogen-containing molecules. However, only certain bacteria can utilize dinitrogen gas, N_2, from the atmosphere. The enzymatic conversion of N_2 to NH_3 is called nitrogen fixation. This reaction is rather a special one because the N≡N molecule is relatively unreactive yet can be reduced to NH_3 under the relatively mild conditions of pH and temperature found inside cells. A perspective is gained from contrasting biological nitrogen fixation with the chemical Haber process for N_2 reduction where high temperatures and pressures are necessary for making N≡N reactive.

Bacteria capable of nitrogen fixation are called diazotrophs (N_2-eating) (1). They can be found in the soils of most climates (including extremes of temperature and moisture), in fresh or salt waters, and sometimes in exotic places such as sludge deposits of metal ore mines or in the hindgut of termites. Many species have symbiotic associations with plants. Some diazotrophs can convert CO_2 to sugars using energy from light (phototrophs such as *Anabaena* or *Rhodobacter*) or from oxidation of inorganic molecules (autotrophs such as *Thiobacillus* or *Bradyrhizobium*). These organisms are on the very bottom rung of the food chain. Diazotrophs can be found in many different groups of bacteria (Chapter 1). An important feature for bacterial evolution is that nitrogen fixation is not limited to the large group of Eubacteria, which includes the familiar Gram-positive and Gram-negative organisms. Certain methane-producing species of the Archaebacteria, considered to be a separate kingdom of prokaryotes (2), can also fix nitrogen.

A. Nitrogenase is Sensitive to Oxygen

The common aspect of all these diazotrophs is that they contain nitrogenase, the trivial name for a group of complex enzymes that catalyse the conversion of N≡N to NH_3. All known nitrogenase enzymes are very sensitive to inactivation by molecular oxygen; this sensitivity to O_2 makes the wide range of diazotrophs even more remarkable because in many of them, such as *Azotobacter*, *Rhizobium*, and *Azospirillum*, aerobic metabolism is essential for life. The latter two groups of bacteria adopt a semiaerobic lifestyle in which they fix nitrogen only when O_2 concentrations are low. *Azotobacter* species have evolved complex mechanisms for maintaining low intracellular O_2 concentrations when normal aerobic conditions (or, in the laboratory, even higher O_2 concentrations) prevail. In organisms like *Klebsiella* or *Rhodobacter*, anaerobic growth is an option (they are facultative anaerobes), and nitrogen fixation occurs only under conditions of very low or no O_2. Others, such as *Clostridium* or *Desulfovibrio* species, are strict anaerobes where O_2 is no more of a problem for nitrogen fixation than for their life in general.

B. Nitrogen Fixation is Often Directly Beneficial to Other Organisms

Another way of looking at types of nitrogen-fixing bacteria is in terms of value to other organisms. The so-called free-living diazotrophs—and this includes most of them—fix nitrogen for themselves. On the other hand, the associative or symbiotic nitrogen fixers live in close contact with plants which, in nitrogen-deficient conditions, depend on their bacterial partner for optimal growth. Examples include the well-known symbiosis between rhizobia and legume plants. This partnership is the one most significant to world agriculture. Root nodules develop in legumes in response to infection by a suitable species of *Rhizobium* or *Bradyrhizobium* (Chapter 19). Other examples include the infection of *Azolla*, a water fern, by the cyanobacterium *Anabaena*, which provides fixed nitrogen to rice crops; the symbiosis of fungi, green algae and cyanobacteria to create the sturdy lichens; and the root nodule association of the actinomycete, *Frankia* species, which allows trees such as *Alnus* or *Casuarina* to grow in poor soils. There are many other plant-bacteria associations and also at least one documented example of a significant nitrogen-fixing association between a bacterium and an animal. This occurs between a species of termite and the bacterium *Citrobacter freundii*. While the route of transfer of fixed NH_3 to termite proteins has not been elucidated, the association probably provides a little nutrition to the insect. Common to all of these associations is that the bacterial partner gives the means for both organisms to use atmospheric N_2 for growth. The eukaryotic partner generally supplies carbon-containing substrates for bacterial energy metabolism.

II. THE *nif* GENES OF *KLEBSIELLA PNEUMONIAE*

A. *K. pneumoniae nif* Genes are Arranged in a Single Chromosomal Cluster

The bacterial genes necessary for synthesis of nitrogenase, called *nif* genes, have been studied for nearly two decades. The first diazotroph examined in detail was the free-living *K. pneumoniae* (3). The main reason for its choice was

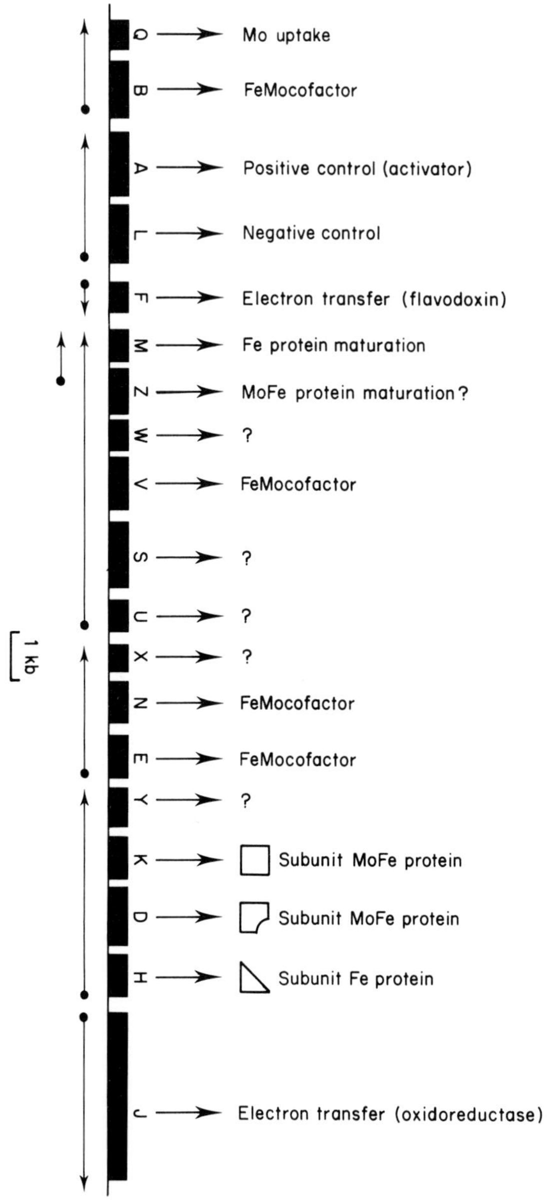

Figure 1. Arrangement of *Klebsiella pneumoniae nif* genes. Arrows show the location of promoters and extent of each transcriptional unit. Functions of the various gene products are shown.

that it is related to *Escherichia coli*, in which the techniques of bacterial genetics were first developed. Methods of bacterial gene transfer (conjugation and transduction), which came before the molecular era of recombinant DNA technology, when applied to *K. pneumoniae* in the early 1970s showed that *nif* genes were located next to the operon for histidine biosynthesis. These genes were studied in subsequent years by the genetic and biochemical analysis of Nif^- mutants and by cloning and physical analysis of individual *nif* genes. The result of these studies was a 24 kb map of 17 distinct and contiguous *nif* genes (organized in eight transcriptional units), a large number considering that functional nitrogenase contains only three polypeptides. Two other *nif* genes, *nifZ* and *nifW*, were added to the map in 1987 (4). The 19 *nif* genes of *K. pneumoniae* are shown in Fig. 1.

B. Nitrogenase Biochemistry and *nif* Gene Regulation are Complex

Parallel biochemical analysis of *K. pneumoniae* Nif^- mutants has led to a greater understanding of when and how nitrogenase is synthesized and functions to catalyse the conversion of N_2 to NH_3. The current model was developed by a combination of genetic and biochemical techniques. "When" is during conditions of low fixed nitrogen and very low O_2 availability; regulatory proteins then interact with *nif* gene promoters to activate transcription. "How" is by a complex interaction of nitrogenase subunits with other proteins used for nitrogenase maturation, cofactor synthesis and electron transfer. The combined result is a paradigm for *nif* gene structure, function and regulation that serves as a basis for studying all these aspects in other nitrogen-fixing organisms. The rest of this chapter will outline the *Klebsiella* model and describe similarities and differences among several other diazotrophs.

III. THE ASSEMBLY OF ACTIVE NITROGENASE

Nitrogenase consists of two interactive proteins: the MoFe protein (originally called component I) and the Fe protein (component II). By convention, components I and II from different organisms are prefaced by the initials of the species of origin; e.g. components I and II from *K. pneumoniae* are referred to as Kp1 and Kp2, from *A. chroococcum* as Ac1 and Ac2, and from *Clostridium pasteurianum* as Cp1 and Cp2. Other names for these proteins are used more frequently in the USA than elsewhere: dinitrogenase for MoFe protein and dinitrogenase reductase for Fe protein.

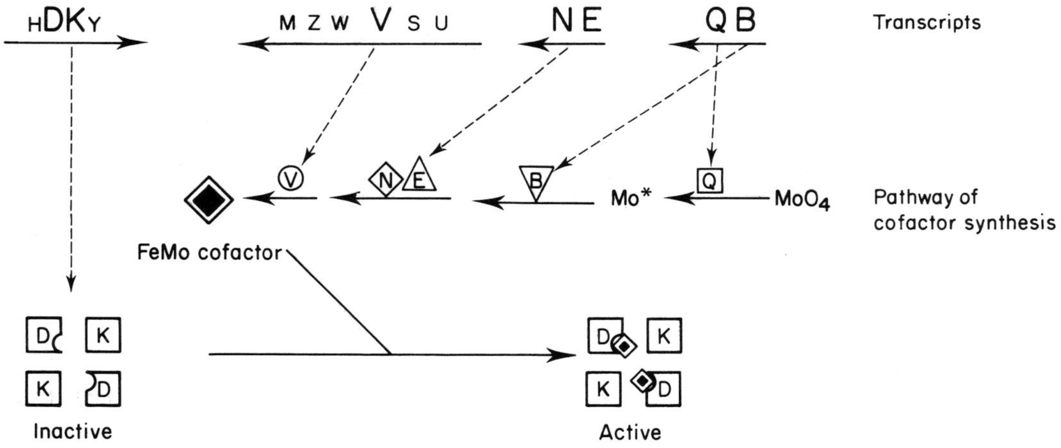

Figure 2. Interaction of *nif* gene products required for *Klebsiella pneumoniae* nitrogenase MoFe protein (Kp1)[5].

A. MoFe Protein Needs Eight Gene Products

The MoFe protein carries the site of N_2 binding and reduction. The *nifD* and *nifK* genes encode the 56 000 (α) and 60 000 (β) M_r (Molecular ratio) polypeptide subunits. There are two of each subunit in the MoFe protein ($\alpha_2\beta_2$ structure). The MoFe cofactor of the protein, a cluster of one Mo: six–eight Fe: eight S- (two per MoFe protein tetramer) is essential for N_2 binding and reduction (as for other substrates such as acetylene). Also necessary for activity are two types of FeS clusters. Cofactor synthesis and insertion into the nascent polypeptides requires five *nif* gene products, those of *nifQBVEN* (5). A plausible scheme for the interaction of gene products is shown in Fig. 2.

B. Active Fe Protein Requires Two Gene Products

The Fe protein is a dimer of two identical subunits of about 30 000 M_r, encoded by *nifH*. However one other *nif* gene product, that of *nifM*, is required to make active Fe protein (4, 6). How the *nifM* gene product modifies the nascent *nifH* product subunits is not known. One possibility is that it is used

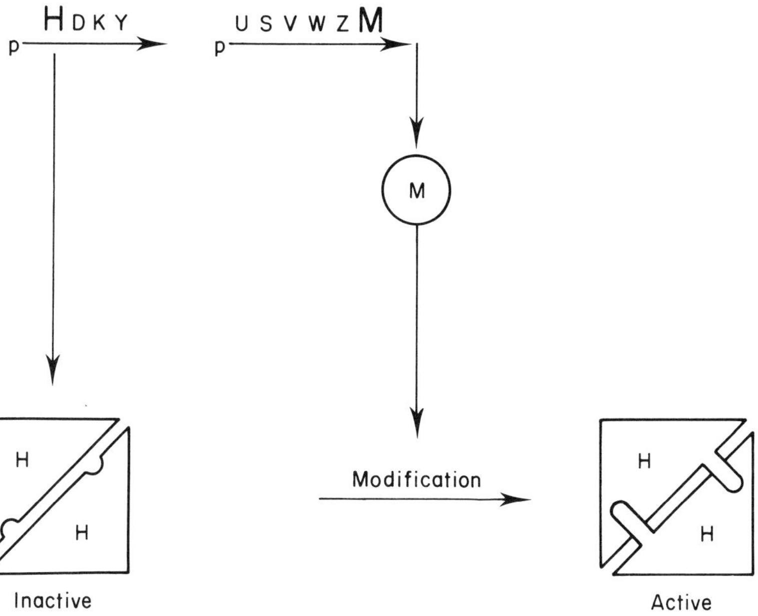

Figure 3. Interaction of *nif* gene products required for *Klebsiella pneumoniae* nitrogenase Fe protein (Kp2).

to make or insert the iron-sulphur cluster (4Fe–4S) that is part of the Fe protein.

The structure of the operons containing the *nifH* and *nifM* genes, and the interaction of their products, are shown in Fig. 3.

The functions of seven other *nif* genes are not yet known. These are *nifW, S, U, X, Y,* and *Z.* It will not be surprising if most of their gene products are involved in nitrogenase activation or maturation.

IV. THE BIOCHEMISTRY AND PHYSIOLOGY OF NITROGENASE

Six electrons must be transferred to N≡N along with six H^+ from the aqueous environment for the formation of two NH_3. Other N≡N analogues, such as acetylene (CH≡CH) or cyanide (C≡N), can be reduced by nitrogenase. The two-electron reduction of acetylene to ethylene (CH_2=CH_2) is important in the laboratory because this product is easier to measure than NH_3. Purified nitrogenase studied *in vitro* can be supplied with electrons from a strong inorganic reducing agent such as the dithionite ion. Electron transfer to nitrogenase *in vivo* in *K. pneumoniae* is from pyruvate to a pyruvate-flavodoxin oxidoreductase encoded by *nifJ* and then to a flavodoxin encoded by *nifF* (Fig. 4). Flavodoxins are small (M_r about 20 000) flavin mononucleotide (FMN)-containing monomeric electron-transfer proteins. Reduced *nifF* flavodoxin then transfers electrons to nitrogenase Fe protein. Electrons then transfer from reduced Kp2 to Kp1 of nitrogenase in probably eight separate and distinct

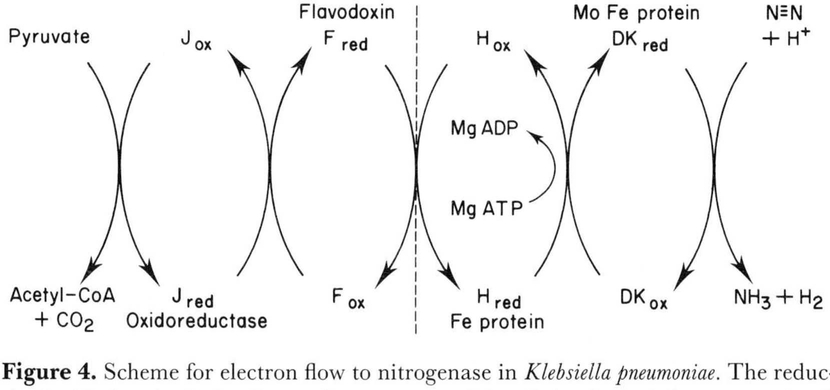

Figure 4. Scheme for electron flow to nitrogenase in *Klebsiella pneumoniae*. The reduction/oxidation (redox) scheme shown to the right of the dashed line occurs in all nitrogen fixing bacteria. That to the left may be specific for *K. pneumoniae* and other facultative anaerobes; aerobes and anaerobes probably have other proteins for coupling electron flow from reduced metabolites to nitrogenase.

steps and each of these is accompanied by hydrolysis of two ATP molecules. Also at each electron transfer step, components I and II separate and then reassociate before the next electron transfer (7). Surprisingly, the dissociation of the two proteins is the rate-limited step in the catalytic cycle and causes nitrogenase to be an extremely slow enzyme.

Of importance to physiology and the mechanism of the enzyme is that H^+ is always reduced to H_2 gas by nitrogenase. Therefore H_2 evolution can be used as a measure of nitrogenase activity. The scheme shown in Fig. 4 shows the flow of electrons from pyruvate to NH_3 for nitrogen fixation in *K. pneumoniae*. In other organisms, in particular strict aerobes, the protein(s) participating in early stages of electron transfer may well be different from the *nifJ*-encoded oxidoreductase since redox potentials of the coupled reactants may vary. However, the interaction of nitrogenase components is the same in all diazotrophs that have been studied so far.

A few words should be said about the large demand for ATP by nitrogenase. In addition to that needed for activity (16 molecules per molecule of NH_3 formed) *nif* gene transcription and translation require much energy during derepression. Nitrogenase can account for up to 40% of the total protein being made. Some energy is also needed for assimilation of NH_3 by glutamine synthetase, an ATP-dependent reaction.

V. THE THREE NITROGENASES OF *AZOTOBACTER*

One of the most interesting recent developments in nitrogen fixation research is the discovery of nitrogenases without molybdenum. While this came as a surprise, one can find earlier evidence for such nitrogenases from physiological and biochemical experiments with *Azotobacter* reported in the 1930s (8). This work was largely forgotten when data for Mo-nitrogenase accumulated to underpin the idea that nitrogen fixation always depended on molybdenum.

A. Deletion of *Azotobacter nifHDK* does not Eliminate Nitrogen Fixation

The suggestion for alternative nitrogenase systems was based on the fact that Nif^- point mutants of *A. vinelandii* carrying defective nitrogenase structural genes grew and fixed N_2 when the culture medium contained no added molybdenum (9). However, confirmation of this idea depended on applying recombinant DNA techniques to *Azotobacter*. To prove the existence of nitrogenase enzymes that did not depend on the conventional *nifHDK* genes, it was necessary to delete these genes and study the ability of resulting mutants to fix nitrogen under various conditions. This was accomplished in several steps.

First, genomic DNA restriction fragments from *A. vinelandii* were cloned and tested for hybridization to *nifH*, *D* or *K* gene probes from *K. pneumoniae*. Analysis of hybridizing plasmids showed that *nifHDK* genes were contiguous, as in *K. pneumoniae*. Then deletions within the cloned *nifHDK* region of *A. vinelandii* were created by removing specific restriction fragments, and the mutated DNA was introduced back into the *A. vinelandii* genome by transformation. Natural recombination mechanisms *in vivo* resulted in replacement of wild-type genes with the *nif*-DNA carrying the deletion mutations. Such *nifHDK* deletion mutants could still fix nitrogen, but only in the absence of molybdenum. Later it was shown that molybdenum represses transcription of genes for the alternative nitrogenase (10).

B. An Alternative Nitrogenase Contains Vanadium

The nitrogenase made in *A. vinelandii* (or *A. chroococcum*) mutants in the absence of molybdenum contains vanadium and growth of the mutants is stimulated by addition of vanadium. Vanadium nitrogenase of both *Azotobacter* species has a structure similar to the two-component, molybdenum enzyme but the VFe protein has three, not two, polypeptides (11) encoded by *KGD* genes in *A. chroococcum*. It contains a Fe-V-cofactor which can substitute for FeMo cofactor in *K. pneumoniae* nitrogenase. V nitrogenase reduces acetylene less efficiently than Mo nitrogenase and produces both ethylene and ethane from this substrate, although ethane is produced in small amounts (typically 2–4% of the amount of ethylene made) (15). Since Mo nitrogenase fails to produce a significant amount of ethane from acetylene, ethane production is diagnostic for non-Mo nitrogenase.

C. Organization of *Azotobacter nif* Genes

Genes encoding the two nitrogenase systems of *Azotobacter* are similar in the two species (Fig. 5). These include a major *nif* cluster with *nifFMZWVSUXNEYTKDH*, similar to that of *K. pneumoniae*, but not containing the flanking genes *nifQBA* or *nifJ* (see 10 for review, also 6, 9, 13, 14). The *nifB* and *nifA* genes are adjacent to each other and located at least 20 kb distant from the major cluster (14, 15, 16). Genes encoding the V-nitrogenase include *vnfH* (*vnf* for V nitrogen fixation) in *A. chroococcum* which is next to a gene encoding a ferredoxin-like protein, with *vnfDGK* located 2.5 kb away from *vnfH* (11). A similar *nifH2* and ferredoxin-like gene were recently found in *A. vinelandii* (17). Other genes specific for V-nitrogenase are not yet identified but probably include a second *nifEN* region (14).

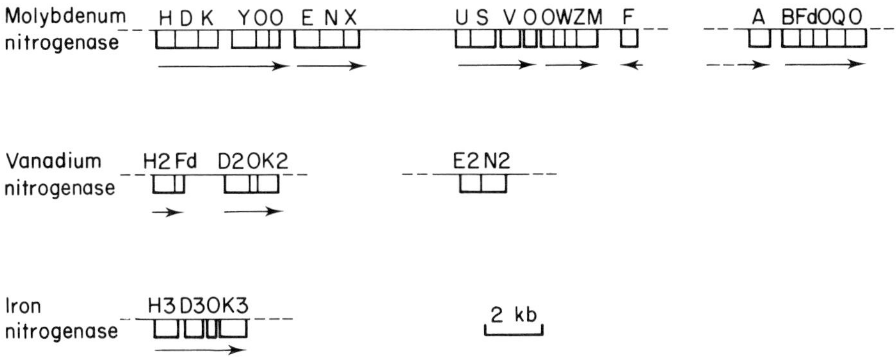

Figure 5. Organization of *nif* genes in Azotobacters. Arrows indicate probable transcription units and directions of transcription. (O = open reading frame of unknown function.)

D. A Third Mo- and V-Independent Nitrogenase is Present in *A. vinelandii*

A third nitrogenase has now been discovered in *A. vinelandii* but is apparently absent from *A. chroococcum* (18). Confirmatory evidence for the third system was provided by deleting DNA encoding both the Mo- and V-nitrogenase structural components (R. Pau, personal communication). While this enzyme has substantial Fe, it contains neither molybdenum nor vanadium in significant amounts. Its synthesis is apparently repressed by addition of either of these metals. Genes encoding its structure are found within a third *nifH*-hybridizing region of the *A. vinelandii* genome (P. Bishop, personal communication).

E. Some Genes are Common to all Three Nitrogenase Systems

While the *nifHDK* genes encoding the three different systems are distinct, at least two *nif* genes are required for activity of all three enzymes in *A. vinelandii*. These are *nifM* and *nifB* (10, 16). Their requirement was shown by the fact that mutations in these genes abolish nitrogen fixation under any condition (in the presence of Mo, V or neither metal). Therefore, the *nifM* gene product is needed for modifying all three nitrogenase Fe proteins and the *nifB* gene product for making all three cofactors. (While a cofactor for the third system has not been identified, it is presumed to be present by generalizing from the structure of the other two enzymes.)

VI. *nif* GENES IN OTHER ORGANISMS

A little is known about *nif* genes and their regulation in some of the organisms mentioned at the beginning of this chapter. They all contain one or more genes homologous to *nifH*, the *nif* gene that is most highly conserved in DNA structure among all nitrogen fixing organisms. *nifDK* genes have also been identified in most organisms examined, while homologues to some of the other *nif* genes of *Klebsiella* and *Azotobacter* occur in some of them (e.g. in *Azospirillum*; 19). Considerable variations in *nif* gene arrangement are found in different organisms, e.g. *Rhizobium* and *Bradyrhizobium* (Fig. 6) and *nif* genes can be plasmid-borne in many organisms, e.g. *Enterobacter* (20), *Desulfovibrio* (21) and *Frankia* (22). *Klebsiella* now seems to be exceptional since all its *nif* genes are contiguous.

A. *Rhizobium nif* Genes are Often Linked to Genes for Symbiosis

In fast-growing *Rhizobium* species such as *R. leguminosarum* (which nodulates peas) and *R. meliloti* (which nodulates lucerne or alfalfa), groups of *nif* genes are interspersed with those involved in nodulation, called *nod* genes, located on Sym plasmids (Chapter 19). The *nifHDK* genes are contiguous and cotranscribed in these organisms. Regions hybridizing to *K. pneumoniae nifS*, *nifE* and *nifN* have been identified in *R. meliloti* (Fig. 6). In contrast to the arrangement in *Klebsiella* and *Azotobacter*, *nifN* and *nifE* are separated by about 15 kb (23).

In the slow-growing *Bradyrhizobium japonicum* (which nodulates soybeans), *nifH* and *nifKD* are separated by 17 kb of DNA with *nifEN*, *nifS* and *nifB* in between (24). These genes have been identified by a combination of mutational analysis, hybridization with *K. pneumoniae nif* gene probes and by sequencing (Fig. 6). There is no evidence that *nif* genes are plasmid-borne in *Bradyrhizobium*.

Other genes involved in nitrogen fixation in *Rhizobium* and *Bradyrhizobium* species, but not known to be homologous to *nif* genes of *Klebsiella*, are called *fix* (e.g. *fixABC* which have been found in all rhizobia examined—see Fig. 6). Mutants with defective *fix* genes form ineffective (non-nitrogen-fixing) nodules and, in the case of slow-growing species, have lost the ability of their parent wild-type strain to fix nitrogen in plant-free cultures (25). Some *fix* genes may correspond to known *nif* genes (e.g. *fixF* of *R. meliloti* was recently found to be equivalent to *nifN*; 23) while others could be involved in electron transfer to nitrogenase in aerobic organisms. Likely candidates for this function are the *fixABC* genes which hybridize to DNA from *Azotobacter* species (aerobes) but not to DNA from *Klebsiella* (a facultative anaerobe) (25). Still other *fix* genes may be found only in *Rhizobium* and *Bradyrhizobium* species and may be specifically required for some aspect of nitrogen fixation in root nodules.

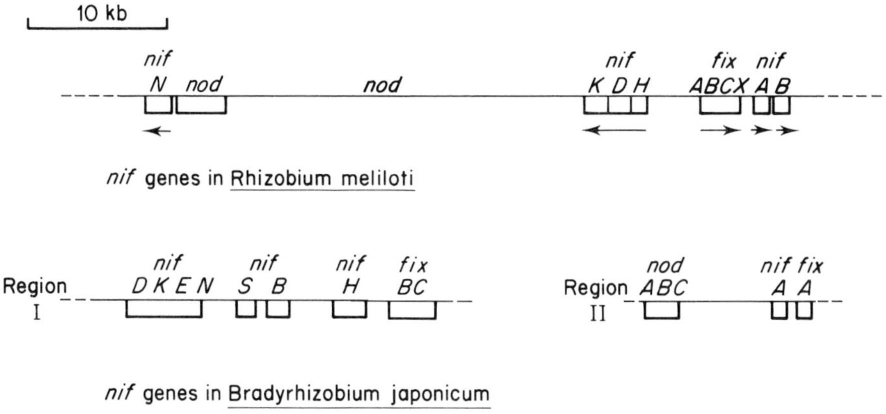

Figure 6. Organization of some *nif* genes in *Rhizobium meliloti* and *Bradyrhizobium japonicum*.

Rhizobium phaseoli, a fast-growing strain which nodulates beans, carries two nearly identical copies of *nifHDK* and a third copy of *nifH* all located on a Sym plasmid (26). These reiterated *nif* genes probably provide extra copies of genes necessary for a single type of nitrogenase enzyme, rather than being associated as in *Azotobacter* with different types of nitrogenase. Two copies of *nifH* have also been identified in *Rhizobium* strain ORS571 (now named *Azorhizobium sesbania*; 27) which forms nodules on the stems as well as roots of the tropical legume *Sesbania*; it is the only legume-nodulating species known that can grow on agar plates in a low O_2 atmosphere with N_2 as a sole nitrogen source. Whether the latter ability is due to a difference in *nif* gene structure or regulation or in some aspect of more general physiology is not known.

B. *Rhodobacter nif* Genes Are Not Clustered

In the photosynthetic *Rhodobacter capsulatus*, mutations affecting nitrogen fixation occur in four different regions of the chromosome (28). One region has contiguous *nifHDK* located next to a *nif* regulatory gene. Another region has three regulatory genes. A third carries genes probably involved in FeMo cofactor synthesis (*nifE*) and nitrogenase maturation (*nifS*). Region four has been poorly characterized but mutants mapping in this region include those unable to grow on several nitrogen sources (an Ntr^- phenotype) and also one Nif^-Ntr^+ mutant that has significant nitrogenase activity in cell extracts supplied with dithionite. Therefore an electron transfer protein may be encoded in region four along with Ntr-related regulatory proteins (see Section VIII).

C. The *nif* Genes of Gram-positive Bacteria and Archaebacteria

Much less is known about *nif* genes in Gram-positive diazotrophs, in the symbiotic actinomycete genus *Frankia*, or in Archaebacteria, mainly because these organisms lack suitable gene transfer methods; also, many of them are difficult to grow in liquid culture or as single colonies on agar. However, all have *nifHDK*-hybridizing regions. In *C. pasteurianum*, multiple copies of *nifH* have been found by hybridization and DNA sequencing (29). At least one of the extra copies of *nifH* probably encodes the Fe-protein of a second V-based nitrogenase, because this organism produces low but significant amounts of ethane from acetylene if molybdenum is absent and vanadium is present in the culture medium (12).

VII. REARRANGEMENT OF *nif* GENES IN *ANABAENA*

A method for regulating *nif* gene expression through excision and rearrangement of *nif*-DNA has been found in the filamentous cyanobacterium *Anabaena*.

All cyanobacteria are photosynthetic and evolve O_2; some of them also fix nitrogen, an oxygen-sensitive process. In many unicellular (e.g. *Gloeothece*) or filamentous types (e.g. *Plectonema*), the cells are undifferentiated and all can carry out both photosynthesis and nitrogen fixation. These organisms have contiguous *nifHDK* genes (34). How they manage to fix nitrogen while evolving oxygen is not understood. In contrast, filamentous organisms such as *Anabaena* have two different arrangements of *nif* genes in two cell types (Fig. 8: 28): in vegetative cells *nifD* is split (into *nifD′* and *D″*—Fig. 8) by 11 kb of intervening DNA not required for nitrogen fixation, while in the nitrogen-fixing heterocysts, *nifD* and *nifK* are contiguous and cotranscribed (Fig. 8). Also in vegetative cells, 55 kb of DNA separate *nifS* from *nifB*, but in heterocysts these genes are contiguous and probably cotranscribed (31). The *nifD-nifK* site-specific rearrangement requires the product of a gene called *xisA* which is within the excised 11 kb segment. This segment, called an excison, is flanked on either side by 11 bp direct repeats (GGATTACTCCG) which are probably recognized by the *xisA* gene product. Expression and function of *xisA* occurs in *E. coli* since the *nifD′*-11 kb excison-*nifD″nifK* region subcloned on pBR322 undergoes the same rearrangement in *E. coli* (32). The *nifS-nifB* rearrangement probably involves a similar site-specific recombination event although its function in *E. coli* has not been reported.

The development of nitrogen-fixing heterocysts is irreversible: they cannot de-differentiate into vegetative photosynthetic cells. The apparently irreversible excision of DNA which separates *nifD* from *nifK* and *nifS* from *nifB* is a part of the multifaceted mechanism for reconciliation of photosynthesis and nitrogen fixation in filamentous cyanobacteria.

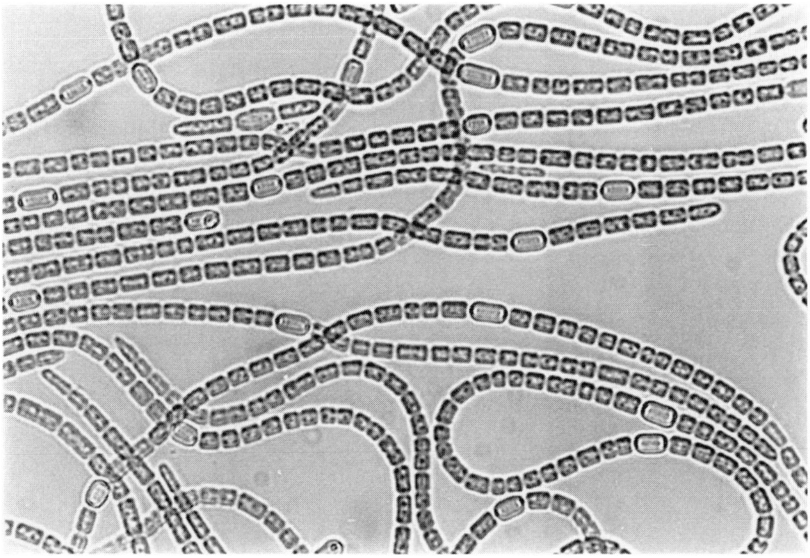

Figure 7. Filaments of *Anabaena* showing two cell types—vegetative cells interspersed with nitrogen-fixing heterocysts.

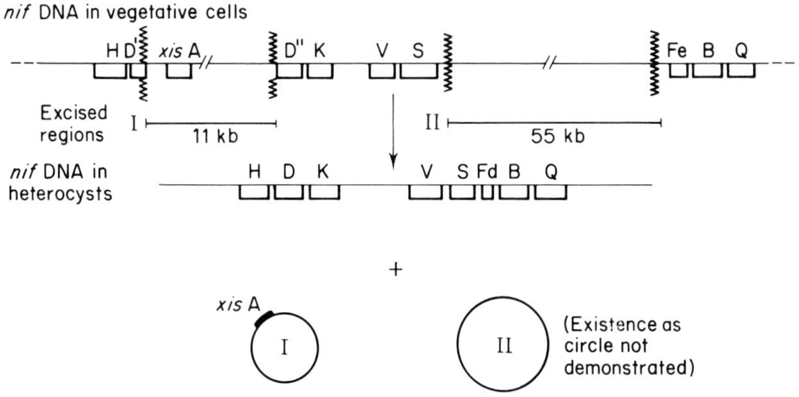

Figure 8. Rearrangement of *Anabaena nif* genes that occurs during differentiation of heterocysts.

VIII. REGULATION OF EXPRESSION OF *nif* GENES

A. Environmental Factors Regulate Nitrogenase Synthesis

The environment of diazotrophs determines whether or not *nif* genes are expressed. In all organisms examined, both excess NH_3, which makes nitrogenase unnecessary, and excess O_2, which destroys nitrogenase activity, prevent nitrogenase synthesis. The genetic regulatory mechanisms through which these environmental effectors control *nif* gene expression are only partly understood. Further complications are present in organisms such as *Azotobacter* where molybdenum and vanadium concentrations affect synthesis of the three different nitrogenase enzymes.

B. Recognition of *nif* Promoters Requires a Modified Form of RNA Polymerase

A good place to begin this discussion is to describe a general structure of *nif* promoters which has been found in *Klebsiella*, *Azotobacter*, *Rhizobium*, *Thiobacillus*, and *Desulfovibrio* (3, 10, 13, 16, 19, 24, 33, 34, 35). In these organisms, *nif* promoters share common nucleotide sequences in two regions upstream from the site at which transcription begins. The first of these, at −12 to −24 bp, has the consensus sequence shown in Fig. 9, in which four nucleotides are most important. These −12 GC and −24 GG nucleotides are necessary for recognition of *nif* promoters by a minor form of RNA polymerase containing a sigma factor, $\sigma 54$, encoded by a gene variously known as *ntrA*, *rpoN*, or *glnF* (36, 37). As expected from this model, *ntrA* mutants of *K. pneumoniae*, *A. vinelandii* and *R. meliloti* are all Nif^- (3, 10, 33). These *nif* promoters are different in structure and recognition factor from most other prokaryotic promoters which are recognized by the major RNA-polymerase holoenzyme containing the most abundant species of sigma factor (Chapter 2). In enteric bacteria, this is σ^{70} encoded by *rpoD*. In *K. pneumoniae*, experiments with *ntrA-lac* fusions indicate that the *ntrA* gene is transcribed at similar levels in high or low levels of fixed nitrogen or O_2. Therefore, levels of transcription

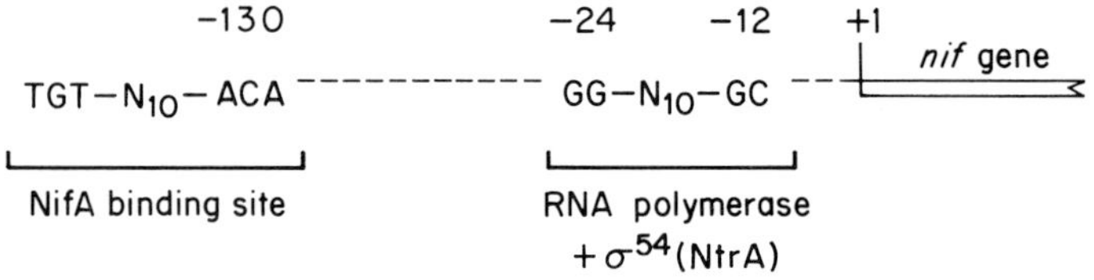

Figure 9. Structure of NifA-dependent *nif* promoters.

of *ntrA* are probably not involved in NH_3 or O_2 control of nitrogenase synthesis. We must look elsewhere for the mechanisms by which these environmental effectors control *nif* gene expression.

C. Upstream Sequences and Activator Proteins Are Needed for *nif* Gene Expression

The other important and common region of most *nif* promoters is found 100–200 bp upstream of the transcription start site. This sequence is TGT-N_{10}-ACA (Fig. 9), called the upstream activator sequence (UAS), and is the site of interaction with the *nifA* gene product. Therefore, in addition to $\sigma54$, *nif* promoters require the *nifA* gene product for their expression and, as expected from this model, *nifA* mutants of *Klebsiella*, *Azotobacter* and *Rhizobium* species are Nif^-. Exactly how the *nifA* gene product bound to UAS acts to initiate transcription by RNA polymerase bound to the −12, −24 sequence is not known, but one model predicts that a direct interaction between the UAS and the −12, −24 sequence occurs if *nifA* product is bound to the UAS (38).

It is interesting that other genes not controlled by NH_3 found in various Gram-negative organisms have the −12, −24 consensus and require a functional *ntrA* gene for expression. These include some genes for cyclic hydrocarbon degradation in *Pseudomonas* (Chapter 8), some involved in anaerobic fermentation in *E. coli*, genes for pilin synthesis in *Pseudomonas*, *Caulobacter* (Chapter 10) and *Neisseria* species (Chapter 13), and genes for dicarboxylic acid transport in *Rhizobium* (36). Some of these genes have been shown to require activators for expression and so a general rule may have emerged: that promoters requiring $\sigma54$ for RNA polymerase recognition also need a partner activator protein which may interact with upstream DNA sequences. No such general requirement has been observed for promoters recognized by RNA polymerase containing other sigma factors. Perhaps this reflects the absence from *ntrA* of DNA sequences common to most other sigma factors (39).

One *K. pneumoniae nif* promoter, that of *nifLA*, requires an upstream activator protein different from the *nifA* gene product. Its upstream activator is the product of *ntrC*, a gene found in a complex operon comprising *glnA*, *ntrB* and *ntrC* (37). *glnA* encodes glutamine synthetase, used in assimilation of NH_4^+. For the *ntrC* product to function as an activator, it must first be phosphorylated by a reaction catalysed by the *ntrB* gene product in the absence of NH_3. Therefore, fixed nitrogen regulates expression of *nif* genes in *K. pneumoniae* by preventing phosphorylation of *ntrC* product, which is required for *nifLA* expression. The other *nif* genes are not expressed without the *nifA* product. A second level of NH_3 control is exerted through the *nifL* product (see following section).

D. O_2 Control and the *nifL* Gene Product

What about O_2? The major mechanism for O_2 control of *nif* gene expression in *K. pneumoniae* involves the product of *nifL*, the other gene in the *nifLA* operon. In the presence of O_2 (and also fixed nitrogen), *nifL* protein interacts with *nifA* protein to prevent its activator function. Whether *nifL* protein directly binds to O_2 or, as seems more probable, interacts with another redox molecule remains unknown.

E. Regulation of *nif* Genes in *Rhizobium* and *Azotobacter*

Both *Rhizobium* (and *Bradyrhizobium*) and *Azotobacter* species have a *nifA* gene as well as *nif* promoter structures with −12, −24 and UAS sequences resembling those of *nifA*-dependent promoters (10, 33). However, in neither genus does *nifA* expression require an *ntrC* gene product. Both do have an *ntrC* gene, and *ntrC*::Tn*5* mutants are unable to grow on nitrate as a nitrogen source. A gene recently identified in *A. vinelandii*, called *nfrX*, may take the place of *ntrC* in controlling *nifA* expression, and in *R. meliloti* the product of the *fixL* gene (Fig. 6) is required for *nifA* expression (40, 41). Whether *nfrX* in *Azotobacter* responds to high NH_3 is not known. Also unknown is how O_2 regulates *nif* gene expression in *Azotobacter*. While a region with some homology to *nifL* has been located upstream of *nifA* in *A. vinelandii* (15), no role for it in nitrogen fixation has yet been assigned. Other differences from the *K. pneumoniae* model are that in *R. meliloti nifA* expression is prevented by O_2 (42), and in *B. japonicum* the *nifA* protein itself is inactivated by aerobic growth conditions (at least when carried on a plasmid in *E. coli*: 43).

Examination of *ntrA*, *ntrC*, *nfrX* and *nifA* mutants of *A. vinelandii* for growth on different metals has revealed that the *ntrA*-encoded $\sigma54$ is required for expression of genes of all three systems. The *nifA* and *nfrX* gene products are necessary for production of Mo-nitrogenase as well as the third uncharacterized enzyme (40) and the *ntrC* product is involved in expression of genes for V-nitrogenase. Thus, in *A. vinelandii*, genes encoding activator proteins have diverged to control expression of three different enzyme systems for nitrogen fixation.

IX. CONCLUDING REMARKS

The diversity of nitrogen-fixing bacteria and their habitats is reflected in variations among them in *nif* gene structure, organization and regulation. These variations have led to important and interesting questions for the future—such as how metals act as regulators of *nif* gene expression in

Azotobacter; the function of the third subunit in component I of the alternative nitrogenases that do not contain molybdenum; the function of many other *nif*, *fix*, and related gene products; the detailed mechanism by which regulatory proteins interact with *nif* promoter DNA sequences; and, more widely, the signals for *nif* gene activation/repression in Gram-positive bacteria and Archaebacteria. In the long term, it may be possible to apply such knowledge to make the associations between nitrogen-fixing bacteria and plants more effective in terms of amounts of fixed nitrogen supplied to the plant, and eventually to engineer plants that can fix nitrogen directly from the atmosphere without a bacterial partner. The complexity of nitrogen fixation genetics and biochemistry portends a tortuous but interesting path for both basic and applied research.

Acknowledgements

The author thanks many colleagues for sharing their results and ideas; John Postgate and Roger Thorneley for criticizing the manuscript; and Beryl Scutt for typing.

References

(1) Postgate, J. R. (1982). "The fundamentals of nitrogen fixation", pp. 1–252. Cambridge University Press, Cambridge (UK).
(2) Jones, W. J., Nagle, D. P. and Whitman, W. B. (1987). Methanogens and the diversity of archaebacteria. *Microbiol. Rev.* **51**, 135–177.
(3) Dixon, R. A. (1984). The genetic complexity of nitrogen fixation. *J. Gen. Microbiol.* **130**, 2745–2755.
(4) Paul, W. and Merrick, M. (1987). The nucleotide sequence of the *nifM* gene of *Klebsiella pneumoniae* and identification of a new *nif* gene: *nifZ*. *Eur. J. Biochem.* **170**, 259–265.
(5) Ugalde, R. A., Imperial, J., Shah, V. K. and Brill, W. J. (1985). Biosynthesis of the iron-molybdenum cofactor and the molybdenum cofactor in *Klebsiella pneumoniae*: effect of the sulphur source. *J. Bacteriol.* **164**, 1081–1087.
(6) Howard, K. S., McLean, P. A., Hansen, F. B., Lemley, P. V., Koblan, K. S. and Orme-Johnson, W. J. (1985). *Klebsiella pneumoniae nifM* gene product is required for stabilization and activation of nitrogenase iron protein in *Escherichia coli*. *J. Biol. Chem.* **261**, 772–778.
(7) Smith, B. E., Campbell, F., Eady, R. R., Eldridge, M., Ford, C. M., Hill, S., Kavanagh, E. P., Lowe, D. J., Miller, R. W., Richardson, T. H., Robson, R. L., Thorneley, R. N. F. and Yates, M. G. (1987). Biochemistry of nitrogenase and the physiology of related metabolism. *Phil. Trans. R. Soc. Lond. B* **317**, 131–146.
(8) Robson, R. L., Eady, R. R., Richardson, T. H., Miller, R. W., Hawkins, M. and Postgate, J. R. (1988). The alternative nitrogenase of *Azotobacter chroococcum* is a vanadium enzyme. *Nature* **32**, 388–390.
(9) Bishop, P. E., Jarlenski, D. M. L. and Hetherington, D. R. (1980). Evidence for an alternative nitrogen-fixation system in *Azotobacter vinelandii*. *Proc. Natl. Acad. Sci. USA* **77**, 7342–7346.

(10) Kennedy, C. and Toukdarian, A. (1987). Genetics of azotobacters: applications to nitrogen fixation and related aspects of metabolism. *Ann. Rev. Microbiol.* **41**, 227–248.
(11) Eady, R. R., Robson, R. L., Paw, R. N., Woodley, P., Lowe, D. J., Miller, R. W., Thorneley, R. N. F., Smith, B. E., Gormal, C., Fisher, K., Eldridge, M. and Bergoträm, J. (1988). The vanadium nitrogenase of *Azotobacter chroococcum*. *In* "Nitrogen Fixation: Hundred Years After" (H. Bothe, F. J. de Bruijn and W. Newton, eds.), pp. 81–86, Gustav Fischer, Stuttgart.
(12) Dilworth, M. J., Eady, R. R., Robson, R. L. and Miller, R. W. (1987). Ethane formation from acetylene as a potential test for vanadium nitrogenase *in vivo*. *Nature* **327**, 167–168.
(13) Bennett, L. T., Jacobson, M. R. and Dean, D. R. (1988). Isolation, sequencing, and mutagenesis of the *nifF* gene encoding flavodoxin from *Azotobacter vinelandii*. *J. Biol. Chem.* **263**, 1364–1369.
(14) Evans, D., Jones, R., Woodley, P. and Robson, R. (1988). Further analysis of nitrogen fixation (*nif*) genes in *Azotobacter chroococcum*: identification and expression in *Klebsiella pneumoniae* of *nifS*, *nifV*, *nifM* and *nifB* genes and localization of *nifE/N-*, *nifU-*, *nifA-* and *fixABC*-like genes. *J. Gen. Microbiol.* **134**, 931–942.
(15) Bennett, L. T., Cannon, F. C. and Dean, D. (1988). Nucleotide sequence and mutagenesis of the *nifA* gene from *Azotobacter vinelandii*. *Mol. Microbiol.* **2**, 315–321.
(16) Joerger, R. D. and Bishop, P. E. (1988). Nucleotide sequence of the *nifB-nifQ* region of *Azotobacter vinelandii*. *J. Bacteriol.* **170**, 1475–1487.
(17) Raina, R., Reddy, M. A., Ghosal, D. and Das, H. K. (1988). Characterization of the gene for the Fe-protein of the vanadium dependent alternative nitrogenase of *Azotobacter vinelandii* and construction of a Tn*5* mutant. *Mol. Gen. Genet.* (In press.)
(18) Chisnell, J. R., Premakumar, R. and Bishop, P. E. (1988). Purification of a second alternative nitrogenase from a *nifHDK* deletion strain of *Azotobacter vinelandii*. *J. Bacteriol.* **170**, 27–33.
(19) Elmerich, C., Bozouklian, J., Vieille, C., Fogher, C., Perroud, B., Perrin, A. and Vanderleyden, J. (1987). *Azospirillum*: genetics of nitrogen fixation and interaction with plants. *Phil. Trans. R. Soc. Lond. B* **317**, 183–192.
(20) Singh, M., Kleeburger, A. and Klingmuller, W. (1983). Location of nitrogen fixation (*nif*) genes on indigenous plasmids of *Enterobacter agglomerans*. *Mol. Gen. Genet.* **190**, 373–378.
(21) Postgate, J., Kent, H. M. and Robson, R. L. (1988). DNA from diazotrophic *Desulfovibrio* strains is homologous to *Klebsiella pneumoniae* structural *nif* DNA and can be chromosomal or plasmid-borne. *FEMS Microbiol. Letts.* **33**, 159–163.
(22) Simonet, P., Haurat, J., Normand, P., Bardin, R. and Moiroud, A. (1986). Localization of *nif* genes on a large plasmid in *Frankia* sp. strain ULQ0132105009. *Mol. Gen. Genet.* **204**, 492–495.
(23) Aguilar, O. M., Reilander, H., Arnold, W. and Pühler, A. (1987). *Rhizobium meliloti nifN (fixF)* gene is part of an operon regulated by a *nifA*-dependent promoter and codes for a polypeptide homologous to the *nifK* gene product. *J. Bacteriol.* **169**, 5393–5400.
(24) Ebeling, S., Hahn, M., Fischer, H. M. and Hennecke, H. (1987). Identification of *nifE-*, *nifN-* and *nifS*-like genes in *Bradyrhizobium japonicum*. *Mol. Gen. Genet.* **207**, 503–508.
(25) Gubler, M. and Hennecke, H. (1986). *fixA*, *B* and *C* genes are essential for symbiotic and free-living, microaerobic nitrogen fixation. *FEBS Lett.* **200**, 186–192.
(26) Quinto, C., de la Vega, H., Flores, M., Fernández, L., Ballado, T., Soberón, G. and Palacios, R. (1982). Reiteration of nitrogen fixation gene sequences in *Rhizobium phaseoli*. *Nature* **299**, 724–726.
(27) Norel, F. and Elmerich, C. (1987). Nucleotide sequence and functional analysis of the two *nifH* copies of *Rhizobium* ORS571. *J. Gen. Microbiol.* **133**, 1563–1576.
(28) Haselkorn, R. (1986). Organization of the genes for nitrogen fixation in photosynthetic bacteria and cyanobacteria. *Ann. Rev. Microbiol.* **40**, 525–547.

(29) Chen, K. C-K., Chen, J-S. and Johnson, J. L. (1986). Structural features of multiple *nifH*-like sequences and very biased codon usage in nitrogenase genes of *Clostridium pasteurianum*. *J. Bacteriol.* **166**, 162–172.
(30) Apte, S. K. and Thomas, J. (1987). Nitrogen fixation genes (*nifKDH*) in the filamentous nonheterocystous cyanobacterium *Plectonema boryanum* do not rearrange. *J. Genet.* **66**, 101–110.
(31) Haselkorn, R., Golden, J. W., Lammers, P. J. and Mulligan, M. E. (1987). Rearrangement of *nif* genes during cyanobacterial heterocyst differentiation. *Phil. Trans. R. Soc. Lond. B.* **317**, 173–181.
(32) Lammers, P. J., Golden, J. W. and Haselkorn, R. (1986). Identification and sequence of a gene required for a developmentally regulated DNA excision in *Anabaena*. *Cell* **44**, 905–911.
(33) Ronson, C. W., Nixon, B. T., Albright, L. M. and Ausubel, F. M. (1987). *Rhizobium meliloti ntrA* (*rpoN*) gene is required for diverse metabolic functions. *J. Bacteriol.* **169**, 2424–2431.
(34) Pretorius, I. M., Rawlings, D. E., O'Neill, E. G., Jones, W. A., Kirby, R. and Woods, D. R. (1987). Nucleotide sequence of the gene encoding the nitrogenase iron protein of *Thiobacillus ferrooxidans*. *J. Bacteriol.* **169**, 367–370.
(35) Postgate, J. R., Kent, H. M. and Robson, R. L. (1988). Nitrogen fixation by *Desulfovibrio*. *In* "The Nitrogen and Sulphur Cycles" (J. A. Cole and S. Ferguson, eds.), Cambridge University Press, Cambridge, pp. 457–471.
(36) Dixon, R. A. (1987). Genetic regulation of nitrogen fixation. *In* "The Nitrogen and Sulphur Cycles" (J. A. Cole and S. Ferguson, eds.), Cambridge University Press, Cambridge, pp. 417–438.
(37) Merrick, M. J. (1988). Regulation of nitrogen assimilation by bacteria. *In* "The Nitrogen and Sulphur Cycles" (J. A. Cole and S. Ferguson, eds.), Cambridge University Press, Cambridge, pp. 331–361.
(38) Buck, M., Cannon, W. and Woodcock, J. (1987). Transcriptional activation of the *Klebsiella pneumoniae* nitrogenase promoter may involve DNA loop formation. *Mol. Microbiol.* **1**, 243–249.
(39) Merrick, M. J. and Gibbins, J. R. (1985). The nucleotide sequence of the nitrogen-regulation gene *ntrA* of *Klebsiella pneumoniae* and comparison with conserved features in bacterial RNA polymerase sigma factors. *Nucleic Acids Res.* **13**, 7607–7620.
(40) Santero, E., Toukdarian, A., Humphrey, R. and Kennedy, C. (1988). Identification and characterization of two nitrogen fixation regulatory regions, *nifA* and *nfrX*, in *Azotobacter vinelandii* and *A. chroococcum*. *Mol. Microbiol.* **2**, 303–314.
(41) Kahn, D., David, M., Batut, J., Daveran, M.- L., Garnerone, A.- M., Hertig, C., Paques, F., Ya, L. R. and Boistard, P. (1988). Cascade activation of *nif* genes in *Rhizobium meliloti*. *In* "Nitrogen Fixation: Hundred Years After". (H. Bothe, F. J. de Bruijn and W. Newton, eds), pp. 357–361, Gustav Fischer, Stuttgart.
(42) Ditta, G., Virts, E., Palomares, A. and Choong-Hyun, K. (1987). The *nifA* gene of *Rhizobium meliloti* is oxygen regulated. *J. Bacteriol.* **169**, 3217–3223.
(43) Fischer, H. M. and Hennecke, H. (1987). Direct response of *Bradyrhizobium japonicum nifA*-mediated *nif* gene regulation to cellular oxygen status. *Mol. Gen. Genet.* **209**, 621–626.

Chapter **7**

Antibiotic Biosynthesis in Streptomyces

K. F. CHATER *and* D. A. HOPWOOD

I. INTRODUCTION TO *STREPTOMYCES* BIOLOGY

Gram-positive, obligately aerobic *Streptomyces* bacteria are abundant in most soils. For a large part of their nutrition, they utilize insoluble organic debris by the production of a variety of extracellular hydrolytic enzymes such as cellulases, hemicellulases, amylases, proteases and nucleases. They are morphologically adapted to this way of life (1) by growing as a mycelium of branching hyphae that penetrate the interstices of the substrate: the coherent mycelial mass allows a high concentration of enzymes to be produced locally and, since the hyphae are connected throughout a colony, with relatively few cross-walls, nutrients obtained at advancing hyphal tips can presumably be transmitted inside the hyphae back to nutrient-limited regions.

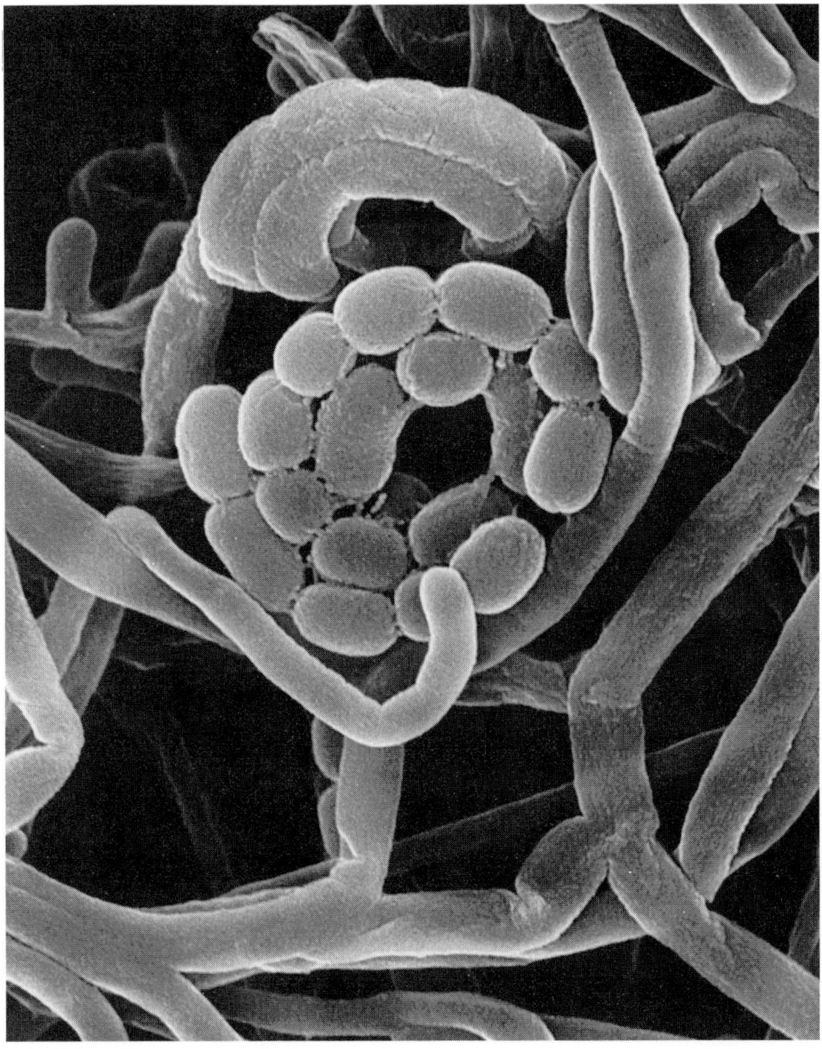

Figure 1. Scanning electron micrograph of aerial mycelium and spores of *Streptomyces lividans*. Photograph courtesy of J. Burgess.

While the mycelial habit makes for efficient exploitation of insoluble substrates, dispersal requires separation of viable subunits from the mycelial mass. Streptomycetes achieve this in a reproductive phase after a period of vegetative colonial growth. In response to nutrient limitation the substrate mycelium gives rise to specialized, spore-bearing aerial hyphae (Fig. 1), which can be thought of as "parasitic" on the lysing substrate mycelium. At about the same time, the remarkable ability of streptomycetes to produce antibiotics is usually manifested. It is this aspect of their biology which concerns us here.

II. ANTIBIOTIC PRODUCTION

A. How are Antibiotics of Adaptive Value?

Antibiotics are compounds that inhibit the growth of other microorganisms by some specific interference in their normal biochemistry. The specificity of action of antibiotics for particular targets, and hence for particular groups of organisms, is the reason why many of them are so valuable in medical, veterinary and agricultural practice, and therefore as industrial products. They are also invaluable as tools in biochemistry: try to imagine the development of the subject without mitomycin C, actinomycin D, rifampicin, chloramphenicol, streptomycin or puromycin as highly specific inhibitors of DNA replication, transcription or translation at its various steps. Of the thousands of naturally-occurring antibiotics discovered so far, about two-thirds are produced by streptomycetes and their close relatives amongst the actinomycetes. Interestingly, the other major producers—*Bacillus* spp. (Chapter 11), myxobacteria (Chapter 12) and mycelial fungi—are also spore-forming, soil-dwelling microbes. The timing of antibiotic production, and its association with sporulating organisms, suggest that an important role of antibiotics may be connected either with the regulation of differentiation or with protection of the producer at stages of its development when it is vulnerable to invasion by competing microbes (for example, during the growth of aerial hyphae on lysing substrate mycelium or, as a component of spores, during germination). In spite of the adaptive significance of antibiotics—whatever it turns out to be—the loss of antibiotic production by pathway-specific mutations does not usually cause observable morphological changes, and certainly is not lethal (in the laboratory). Antibiotic production is therefore not usually an essential part of the differentiation process.

B. Antibiotic Production is Often Genetically Unstable, but is Seldom Plasmid-Specified

The dispensability of antibiotic production and the remarkable differences in the spectra of compounds produced by different *Streptomyces* soil isolates might lead one to suspect that plasmids should generally be involved in their determination. Classically, a sign of plasmid determination of a property is its genetic instability, since plasmids are usually thought of as less stably maintained than chromosomal segments. Antibiotic resistance and production are indeed often unstable phenotypes in *Streptomyces*, but, as we illustrate by a few examples in Table I, this is seldom because they are plasmid-specified; rather it is because they may be contained in remarkably long stretches of chromosomal DNA that can readily be deleted. It is important to emphasize two points: not all sets of antibiotic production genes are subject to high frequency deletion—for example, no actinorhodin nonproducers in *Streptomyces coelicolor* have been found to contain deletions of the biosynthetic genes; and some determinants relevant to antibiotic production are carried on plasmids (methylenomycin production and resistance), or are suspected of being so (A-factor synthesis in *Streptomyces griseus*, but not in *S. coelicolor*, where it is chromosomally determined).

Table I

Examples of genetic instabilities involving resistance to, or production of, secondary metabolites in *Streptomyces*

Species	Mutant phenotype	Genetic changes	Selected references
S. achromogenes ssp. *rubradiris*	Spectinomycin-resistant (or sensitive)	Amplification (or deletion) of resistance gene	50
S. coelicolor A3(2)	Methylenomycin-sensitive Methylenomycin-negative	Loss of large linear SCP1 plasmid carrying production and resistance genes	21
S. fradiae	Tylosin-sensitive Tylosin-negative	Deletion of large segment (of chromosome?), including tylosin production and resistance genes	48
S. glaucescens	Streptomycin-sensitive Melanin-negative	Amplification and rearrangement of chromosomal segments; deletion of long chromosomal segments that include *sph* and *melC* genes	45
S. griseus	Sporulation-defective Streptomycin-sensitive Streptomycin-negative A-factor-negative	Loss of gene for A-factor production, perhaps caused by plasmid loss	4
S. lividans	Chloramphenicol-sensitive Arginine-negative Tetracycline-sensitive	>50 kb deletion from chromosome, including *argG* gene. Massive amplification of adjacent sequences	46, 47
S. reticuli	Sporulation-negative Melanin-negative Leucomycin-negative	Amplifications and deletions of chromosomal DNA, including the *mel* gene	49

C. Antibiotic Production is Highly Regulated

Antibiotics range in molecular weight from just over 100 to about 1200. Chemically they are extremely diverse. Their skeletons normally arise from intermediary metabolites such as acetate, glucose and amino acids by multi-step biosynthetic pathways. Sometimes, separate precursor parts (such as macrolide rings and substituted sugars) are joined together for final modifications, such as hydroxylation or methylation, before acquiring antibiotic activity. It is an intriguing paradox that comparatively simple molecules like antibiotics, made by stepwise biosynthesis in which each step requires a specific enzyme, are much more complex in their genetic determination than giant protein molecules, made by template synthesis and each typically specified by a single structural gene. The genetic complexity of antibiotic biosynthesis colours every aspect of the molecular biology of this process.

In laboratory and industrial fermentations, antibiotic production is usually carried out in vigorously aerated liquid cultures, very different from the natural habitat. Nevertheless, antibiotic synthesis is generally switched on late in growth, just as in a colony growing on a solid substrate. It is often sensitive to particular components of the culture medium, different fermentations being repressed by nutrients that include readily utilized carbon sources (often glucose), abundant available nitrogen (usually ammonia) or high levels of inorganic phosphate (2). Exhaustion of such repressing components of the medium may contribute to the switching on of some antibiotic pathways. Activation of certain pathways may also involve the action of hormone-like substances. The best-known examples are in *S. griseus* and *Streptomyces bikiniensis* where a diffusible effector molecule, A-factor (2*S*-isocapryloyl-3*S*-hydroxymethyl-γ-butyrolactone), produced in minute quantities probably by a single enzymatic step from unidentified intermediary metabolites, is essential for the production of streptomycin, and for sporulation (3, 4). It is interesting to note that the structure of A-factor is rather similar to that of the luminescence autoinducer of *Vibrio fischeri* (Chapter 4).

D. Antibiotic Producers Have Self-Protective Resistance Mechanisms

Many antibiotics could potentially inhibit the growth of, or even kill, their producers. This is averted by a variety of resistance mechanisms (5). In some cases, the potential target of action may be modified and thereby rendered insensitive to inhibition by the antibiotic. Thus, specific methylation of ribosomal RNA protects the producers (*Saccharopolyspora erythraea* and *Streptomyces azureus*) of the antiribosomal antibiotics erythromycin and thiostrepton. In other cases, resistance may be an integral part of biosynthesis. For example, in the biosynthesis of the glutamine synthetase inhibitor bialaphos by *Strep-*

tomyces hygroscopicus, acetylation of an intermediate in the pathway is followed by its intracellular conversion to an inactive, acetylated form of bialaphos (6). A final activation of this compound by deacetylation is presumably associated with secretion of the active antibiotic from the cell. Separate cloning of the acetylation gene into a different streptomycete showed that it can mediate resistance to externally applied bialaphos (7). Sometimes, active export of an antibiotic, effectively the final step in production, might by itself provide resistance, as is perhaps the case for methylenomycin in *S. coelicolor*; here the resistance gene appears to code for a transmembrane protein that may export the antibiotic (produced by epoxidation of the final inactive intermediate in the biosynthetic pathway) from the cells (8). The interesting possibility that there might be a direct interaction between the epoxidase and the resistance protein remains to be explored.

Sometimes there may be more than one gene for resistance in a particular antibiotic-producing organism. For example, cloned *Streptomyces fradiae* genes for phosphorylation or acetylation of neomycin can each separately confer moderate neomycin resistance on *Streptomyces lividans*, but together they give much higher resistance (9); and *Streptomyces kanamyceticus* contains a gene that causes modification of ribosomes to make them kanamycin-resistant, as well as containing a kanamycin acetyltransferase. This is an interesting example because the ribosomal resistance is expressed only in conditions favouring kanamycin production (10).

Why should there often be multiple means of resistance? Such questions are intrinsically difficult to answer, but addressing them may lead to useful experiments. In this case, we speculate that a resistance gene should often be associated with each of the biochemical and genetic units needed to give rise to a particular compound. For example, a macrolide antibiotic such as tylosin would require sets of genes for synthesis of tylactone and each of the three sugars subsequently attached to it: each of these gene sets may at some stage have evolved and become dispersed independently in *Streptomyces* populations and, on introduction into a new host, might have conferred the ability to produce a new antibiotic to which that host could be sensitive. Hence, it would seem likely that each set would gain selective advantage during its dispersal by being linked to a suitable resistance gene. Our tentative prediction, then, is that when two or more genes for antibiotic resistance are present, each will prove to be linked to clusters of biosynthetic genes for the synthesis of the separate component parts of the antibiotic.

III. MOLECULAR GENETICS OF ANTIBIOTIC PRODUCTION

A. Two Model Systems

In the last few years the analysis of antibiotic biosynthetic genes has changed dramatically with the availability of effective systems and strategies for their cloning (11). This follows earlier studies that used the natural genetic systems of *Streptomyces* to identify important features of the production genes. Most recently, the generalized findings that have emerged have led to the cloning and analysis of genes for which there was no previous *in vivo* analysis. In this account, we describe studies of two sets of genes, for actinorhodin and methylenomycin A synthesis in *S. coelicolor*: studies undertaken because of the analytical power of the genetic systems available for this organism. Actinorhodin serves as an example of the polyketide group of compounds which also includes such important antibiotics as tetracyclines, rifamycin, avermectin, macrolides like erythromycin, and anthracyclines such as adriamycin. The main structure of polyketides is made by the iterated condensation of acetyl, propionyl and butyryl residues, in a manner at least superficially resembling the assembly of fatty acids from acetate (12). Methylenomycin is a quite different antibiotic. Interest was stimulated by the discovery that its production is determined by a plasmid (13, 14).

B. Actinorhodin Blocked Mutants are of Several Classes

The identification of the biosynthetic pathway genes by specific mutations was the first crucial step in genetic analysis of the actinorhodin pathway. This was made easier because actinorhodin is pigmented; thus *act* mutants can be identified and classified by inspection, rather than by the more laborious bioassays for antimicrobial activity, or chemical tests, needed for most antibiotics. Mutants that failed to make blue actinorhodin pigment were first grouped into several classes on the basis of colour (15). While some lacked any actinorhodin-related pigments, others accumulated red, yellowish or brown

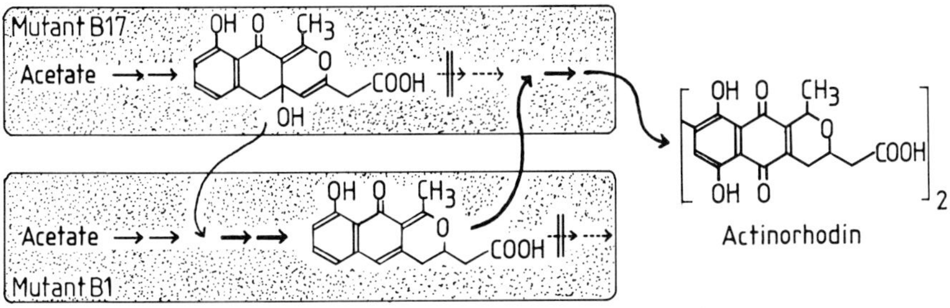

Figure 2. Antibiotic cosynthesis. The drawing illustrates the logic of cosynthesis tests, with specific reference to the cosynthesis of actinorhodin by two blocked mutants (B1 and B17). Mutant B1 accumulates and secretes a late precursor, which B17 is able to take up and convert into actinorhodin. (B17 itself accumulates an earlier precursor, which may itself be used by B1 to augment production of its secreted compound.) The thickness of the arrows indicates relative flux through the different steps.

compounds: presumably precursors of actinorhodin or their shunt products. The next stage in classification used the technique of cosynthesis, which involves the nearby growth of pairs of mutants on an agar plate. A biosynthetic intermediate accumulated by a mutant blocked at a particular biosynthetic step is secreted and is then taken up and converted to the active end product by another mutant blocked at an earlier step (Fig. 2). In this way, the biosynthetic blocks in six classes of mutants (*act*I and III–VII) could be placed in a sequence, while mutants in a seventh class (*act*II) did not cosynthesize actinorhodin with any other mutant (nor did they accumulate any presumptive actinorhodin precursor) and were therefore deduced to be defective in a regulatory gene whose product is required for activation of all the biosynthetic pathway genes.

C. Actinorhodin Biosynthetic Genes are Clustered

From the large number of *act* mutants initially characterized, representatives of each of the seven phenotypic classes were selected for genetic mapping using the plasmid-mediated conjugation system available in *S. coelicolor*. All the *act* genes tested were found to map in the same short interval of the chromosome between two nutritional genes that are separated by less than one-fiftieth of the linkage map (15). Thus, the genes were clustered—an important discovery, which raised the possibility that the whole gene set, and hence the property of actinorhodin synthesis, might be relatively easily transferred between species by either natural or artificial genetic processes.

D. Molecular Cloning and Functional Analysis of the Whole Set of *act* Genes

Many plasmid and phage cloning vectors have been developed for *Streptomyces* (16). DNA that has been subjected to *in vitro* manipulations is efficiently taken up by *Streptomyces* protoplasts in the presence of polyethylene glycol. In order to clone the *act* genes, large pieces (15–35 kb) of *S. coelicolor* DNA were ligated with linearized pIJ922, a low copy-number plasmid vector. The DNA was then introduced into a representative *act* mutant. Two transformant colonies produced blue actinorhodin pigment, and thus carried the cloned act^+ gene needed to restore antibiotic production to the *act*V mutant recipient (17). By this method, two overlapping pieces of cloned DNA were identified. When they were introduced into mutants of the other classes, one clone "complemented" class VI (as well as V) and the other complemented all classes of mutants except VI. The two cloned segments were then joined together in a single plasmid by *in vitro* manipulations, to restore the original genomic sequence and produce a plasmid clone able to complement all the *act* mutants

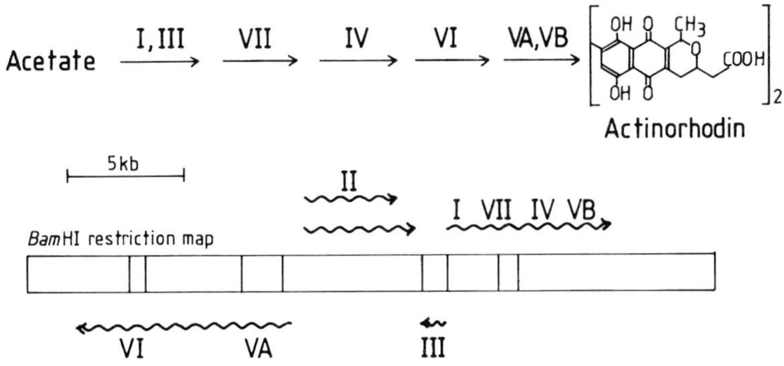

Figure 3. The genetics of actinorhodin synthesis in *Streptomyces coelicolor*. Upper part: gene-pathway relationships. Lower part: map of *Bam*HI sites (vertical bars) in the *act* gene cluster, and location and direction of known transcripts (wavy arrows).

tested; moreover this plasmid specified actinorhodin biosynthesis when introduced into a different host (*Streptomyces parvulus*) not known to make any antibiotic related to actinorhodin, and so it evidently carried the whole set of *act* genes. The regions of the DNA containing particular *act* genes were then identified by subcloning smaller pieces into *act* mutants of each class (18); all the genes fell in a 26 kb DNA segment. One end of the cluster contained genes for late steps in the pathway (genes VI and V) and the other end contained those for early steps (genes, I, III, VII and IV; Fig. 3). (Later, class V mutants were divided on chemical criteria into two subclasses (19), one of which (VB) was found to map with the early genes.) Between these two regions lies *act*II, previously identified as a putative positive control or activator gene. This role was supported by the finding that actinorhodin is greatly overproduced when one or two extra copies of the *act*II region are introduced into the wild-type strain.

E. Cloning the Plasmid-Linked Genes for Methylenomycin A Synthesis

During the early studies of *S. coelicolor* genetics, a genetic determinant responsible for recombination in matings was found to be transferred independently of the chromosome, indicating the presence of a conjugative plasmid which was termed SCP1 (20). SCP1$^+$ strains inhibited SCP1$^-$ strains through production of a diffusible material: thus SCP1 carried production and resistance genes for the inhibitor, later identified as the antibiotic methylenomycin A (13). The modes of biosynthesis and action of this unusual compound (see Fig. 4 for its structure) are still unknown. Proof that SCP1 carried biosynthetic (*mmy*) genes for methylenomycin, rather than genes controlling antibiotic

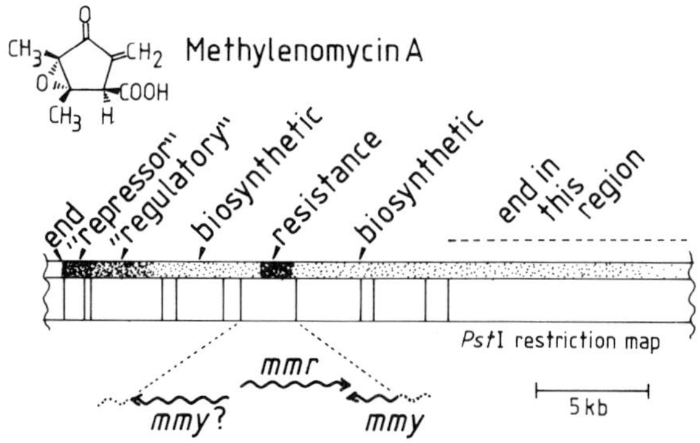

Figure 4. Methylenomycin A and its biosynthetic gene (*mmy*) cluster. The approximate limits of the biosynthetic, resistance and regulatory genes on the SCP1 plasmid are indicated by changes in shading. Transcripts in the *mmr* region, and their directions, are indicated by wavy arrows (see also Figures 6–8).

production indirectly, was obtained when *mmy* mutants, defined as pathway-specific by their ability to cosynthesize methylenomycin in certain pairwise combinations, were shown to donate their characteristic cosynthetic phenotype along with SCP1 during conjugation with SCP1$^-$ recipients (14).

It would have been very convenient if purified SCP1 DNA could have been isolated. Unfortunately, a large range of plasmid isolation procedures failed to realize this objective. Recent evidence from pulsed field gel electrophoresis shows that SCP1 is a large (about 350 kb) linear DNA molecule (21), which would therefore not have been isolated by previously available procedures since they all depended on DNA circularity. A genetic trick was therefore used to obtain a library of cloned SCP1 fragments (22). Total DNA of a *S. parvulus* derivative, into which SCP1 had been transferred by conjugation, was introduced in fragments into a vector based on a temperate phage (ϕC31), the ligated DNA being used to transfect *Streptomyces* protoplasts. Among the phage plaques obtained, some contained cloned pieces of SCP1 DNA. These could be identified, because of the choice of the particular phage derivative used as cloning vector. This was KC400, which lacks the normal *attP* site that otherwise leads to integration of the ϕC31 prophage into the host chromosome by crossing over with the chromosomal *attB* site of the host. Clones carrying SCP1 inserts—unlike KC400 itself—could lysogenize an SCP1-containing *S. coelicolor* strain, by crossing over between the cloned fragment and the corresponding region of the resident SCP1 plasmid, but they were unable to lysogenize an SCP1$^-$ strain. The lysogens were readily identifiable because KC400 carries a viomycin resistance gene, making the lysogens resistant to this antibiotic. (The majority of the clones, carrying *S. parvulus* DNA, could not lysogenize *S. coelicolor* because DNA sequences of the two species are too diverged for crossing over to occur at a detectable frequency.) It was expected that some of the phage clones carrying SCP1 DNA would be directed to integrate into the *mmy* biosynthetic genes, disrupting their expression and giving a Mmy^- phenotype. (The logic of this procedure, called mutational cloning, is explained below.) Indeed, about 3% of the lysogens were methylenomycin nonproducers. The phages released from these lysogens through a reversal of the integration event were analysed to provide information on the organization of the *mmy* genes.

Once part of a cluster of genes is cloned, it is usually straightforward to clone adjacent fragments, most commonly by using the initially cloned fragment as a radioactive probe for homologous sequences in a library of larger cloned fragments (either in *Streptomyces* or in *E. coli*), though other equally convenient strategies exist. Thus, by combining such an approach with Southern blotting, all the SCP1 fragments detected by mutational cloning were found to fall in a 12 kb section of SCP1 (23). About 5 kb of DNA outside this region to the left is involved in the negative regulation of methylenomycin production, as judged by over-production of the antibiotic by strains with deletions or phage insertions in this region. Thus, a segment of at least 17 kb is involved in methylenomycin production (Fig. 4). Further analysis has shown that all the essential *mmy* genes are in a single cluster, since a 28 kb cloned SCP1 fragment that includes the production genes originally identified confers methylenomycin production on *S. lividans* (L. J. Woodburn, personal communication). It is interesting that the methylenomycin resistance gene (*mmr*) lies in the cluster, between biosynthetic genes (23). Increasing numbers of cases are now being found that follow this example, to the extent that direct cloning of a resistance gene from a producing organism has become a useful strategy for the cloning of linked biosynthetic genes (6, 24).

F. Transcription of the *act* and *mmy* Gene Clusters is Complex

Initial indications of transcriptional complexity of the *act* and *mmy* clusters came from further use of the technique of mutational cloning. As shown in Fig. 5, cloned fragments in vectors like KC400 should cause mutations in lysogens only if they are wholly internal to a transcription unit; they are expected to be nonmutagenic if they overlap transcriptional boundaries. Mutational cloning analysis of *mmy* and *act* DNA revealed evidence of transcriptional discontinuities within both clusters (18, 23). This has been confirmed by studies of mRNA isolated from antibiotic-producing cultures (Figs. 3 and 4), using the technique of low resolution S1 mapping. Thus, the actinorhodin cluster seems to be organized as a group of some four operons (one of them monocistronic), two transcribed from one strand of the DNA and two from the other (F.

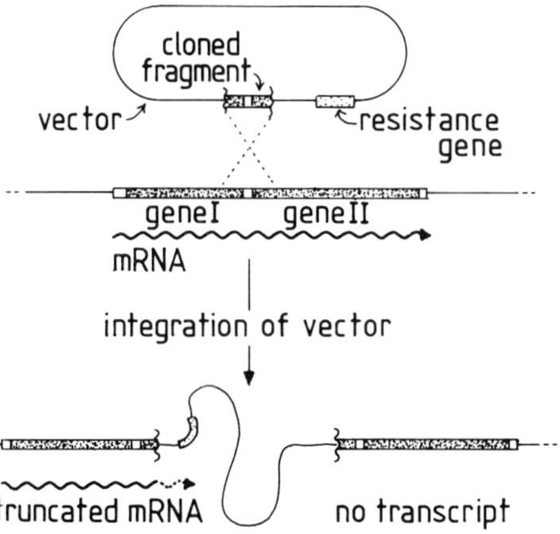

Figure 5. The logic of mutational cloning (gene disruption). A vector unable to replicate can be "rescued" by recombination between a cloned fragment and its chromosomal homologue (these events can be detected by selection for the resistance gene on the vector). Cloned fragments containing the gene I–II promoter or DNA downstream of gene II will not give rise to a mutant phenotype, but when fragments have both ends between these regions, vector integration interrupts the transcription unit and potentially generates a mutant phenotype. (The scheme is more complicated if promoters read out from the vector.)

Malpartida and S. E. Hallam, personal communication: Fig. 3), while the methylenomycin cluster contains at least three transcripts (23, 25: Fig. 4), again utilizing both strands of the DNA.

G. Possible Regulatory Interplay of Resistance and Biosynthetic Genes

In the *mmy* cluster, the resistance gene is read as a monocistronic transcript from a single initiation point (25). The *mmr* promoter is placed "back-to-back" with another promoter which gives a diverging transcript initiating only 81 bp from the *mmr* start site (this puts the two putative −35 regions only about 10 bp apart). It is not yet proven that this transcript extends into the nearby production genes, but this seems very likely. Furthermore, the downstream end of the *mmr* transcript overlaps with a convergent transcript from a biosynthetic gene, with the overlapping sequence (which would give rise to a rather stable stem-loop structure in the RNA) appearing to act as a transcriptional terminator in both directions (25; Fig. 6).

Such intimate transcriptional interplay between resistance and biosynthetic genes would allow for their coordinated regulation, and might ensure that

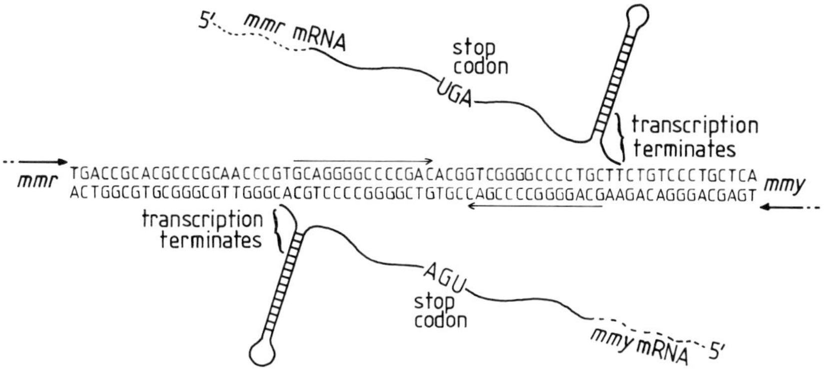

Figure 6. Overlapping terminators of converging *mmr* and *mmy* transcripts. The long inverted repeat indicated by horizontal arrows in the centre of the diagram is thought to lead to hairpin loop formation in the overlapping mRNA species, and recognition of this structure by RNA polymerase then results in termination of transcription. The brackets show the principal 3′-ends of *in vivo* transcription determined by S_1 mapping (25).

biosynthesis of a potentially lethal antibiotic would not take place without the expression of resistance. A strategy in which expression of the resistance gene had "priority" over that of at least one biosynthetic gene, perhaps with the resistance gene product actually playing a role in the activation of biosynthetic genes, could make sense to the organism.

Detailed analysis of the transcription of resistance genes from other antibiotic producers is giving even more complex results than those obtained for *mmr* (Fig. 7). Thus, the genes *ermE* for resistance to erythromycin and *aph* for resistance to neomycin from the producers of these antibiotics (*S. erythraea* and *S. fradiae* respectively), are both transcribed from complex promoter regions in which there are multiple diverging transcriptional start sites (26, 27). Both genes are preceded by two promoters (P1 and the more upstream P2). [Interestingly, the transcripts promoted by P1 start either at the first base of the coding sequence (for *aph*) or only one base before it (for *ermE*), leaving no possibility for a conventional ribosome binding site on these transcripts: Chapter 2.] Transcription in the opposite direction is controlled by as many as three or four promoters and there is overlap between some of the transcripts in the leftward and rightward directions. In the *ermE* region, DNA sequencing has revealed a putative protein-coding region that would be traversed by the leftward transcripts, whereas no such gene is apparent from the limited length of DNA so far available to the left of *aph*. However, the presumption is that in each case leftward transcription leads to expression of antibiotic biosynthetic genes. A further example in Fig. 7 is provided by *Streptomyces glaucescens* where

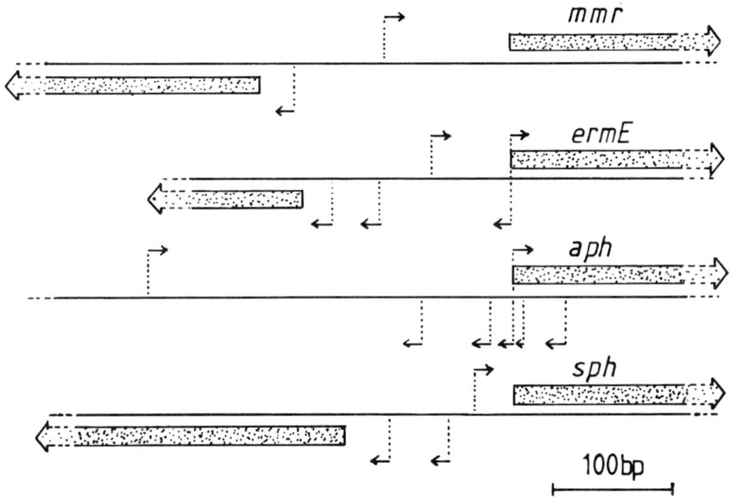

Figure 7. The transcriptional organization upstream of the *mmr* (methylenomycin resistance: 25), *ermE* (erythromycin resistance: 26), *aph* (neomycin resistance: 26) and *sph* (streptomycin resistance: 28) genes. Transcription start sites are shown by small arrows. Shaded large arrows show protein-coding sequences. The leftward protein-coding sequences are deduced from DNA sequence, and their functions have not been determined: however, in each case it is likely that they are involved in production of the relevant antibiotic.

the *sph* gene, for streptomycin resistance, is transcribed in the opposite direction from a putative biosynthetic transcript with two possible start sites (28). As in the case of *mmr*, all these complex transcription patterns give ample scope for regulatory interactions between the antibiotic resistance gene and genes for biosynthesis, perhaps involving competitive or cooperative interaction between promoters, and changes in the relative availability of different holoenzyme forms of RNA polymerase (discussed in Section H below), or other factors that might affect the choice of different promoters; or even the involvement of complementary RNA (Chapter 2). In the case of *ermE* and *aph*, there is experimental evidence that transcription of the resistance gene can vary inversely with divergent transcription from the complementary strand (26).

Much more needs to be done to deduce the precise basis of control in these divergent promoter regions. Meanwhile, it is tempting to compare the features of the intertranscript region upstream of *mmr* with the region containing the P_R and P_{RM} promoters of phage lambda (29, 30). As seen in Fig. 8, the initiation points for the transcripts directed by the two lambda promoters are separated by only 82 bp. This nontranscribed region contains three operators, which are

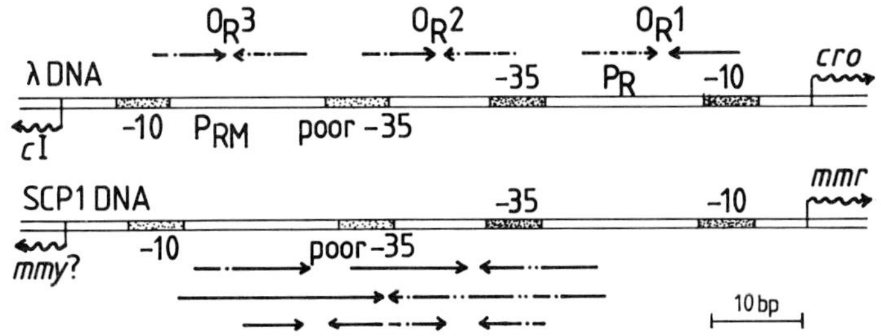

Figure 8. Comparison of the diverging promoter region upstream of the *mmr* gene of SCP1 (28) with that of the P_R-P_{RM} region of bacteriophage lambda (29, 30). Imperfect matches in repeated sequences are shown as dots in the arrows. Wavy arrows show transcripts and their directions. Stippled areas are −10 and −35 regions of promoters. O_R1, O_R2 and O_R3 are related binding sites for the products of the *c*I and *cro* genes (redrawn from ref. 28).

recognized by the products of the *c*I (repressor) and *cro* genes. Depending on the relative concentrations of these two proteins, transcription from one or the other promoter may be repressed, or repression of transcription from one promoter may be associated with *enhanced* transcription from the other. In the *mmr* region (Fig. 8), the two divergent promoters are almost the same distance apart as in the lambda example (there are 81 instead of 82 bp between the two transcription start sites). Although no attempt has yet been made to identify DNA-binding proteins that might recognize this region, or to subject it to site-directed mutagenesis or other forms of analysis, the existence of several regions of direct or inverse symmetry suggests the likelihood that operators will be defined in it. Thus, subtle changes in the differential regulation of the two promoters could ensure transcription of only the resistance gene under some circumstances and allow expression from the diverging promoter under others. Conceivably the *mmr* gene product could itself interact with and regulate the divergent promoter region (in line with the regulatory role of resistance genes suggested earlier). A membrane location would not exclude this: the trans-membrane protein Tox R is a transcriptional activator in *Vibrio cholerae* (Chapter 15).

H. Are Multiple Forms of RNA Polymerase Involved in the Selective Expression of Antibiotic Production Genes?

RNA polymerase preparations from *S. coelicolor* have been shown to contain at least three forms that differ in their possession of different sigma factors, and hence in their promoter selectivity (31, 32). Indeed the *dagA* gene of *S. coelicolor* has four transcriptional start sites, at least three of which are transcribed by a different RNA polymerase form (32). Hence streptomycetes possess a very heterogeneous transcriptional apparatus, and it seems very possible to speculate that this may be important in selective gene expression during developmental processes, including antibiotic production. At the time of writing, we are unaware of any experiments which directly address this question. However, it is possible to make two relevant observations. First, although the sequences of some of the promoters discovered upstream of the resistance genes in Fig. 7 resemble the conventional bacterial consensus sequences in their −10 and −35 regions, others do not (25–28); and the only promoter for a well-characterized antibiotic biosynthetic gene so far studied, i.e. *act*III, also departs from the consensus (S. E. Hallam, personal communication). It will be surprising indeed if the same form of RNA polymerase recognizes all of these promoters as well as the promoters of typical "vegetative" genes. Secondly, there is now circumstantial evidence that *S. coelicolor*, like *Bacillus subtilis* (Chapter 10), does use sigma factor-directed selective transcription in its morphological development: a gene (*whiG*) needed for the development of aerial hyphae into spores has recently been sequenced, and its deduced gene product bears a striking resemblance to sigma factors from other bacteria (C. J. Bruton, C. Mendez and K. F. Chater, unpublished results). The *whiG* gene product is apparently not generally necessary for the expression of antibiotic production genes, since *whiG* mutants produce at least actinorhodin and methylenomycin (the only antibiotics tested). Interestingly, the presence of *whiG* at high-copy number actually reduces actinorhodin production while stimulating sporulation—an observation perhaps indicating either competition between sigma factors for RNA polymerase core, or that sporulation and antibiotic production represent alternative developmental fates (33).

Thus, our current knowledge leads us to expect a positive answer to the question posed in the title of this section. However, it is too early to speculate on whether there are forms of RNA polymerase exclusively dedicated to antibiotic production, or whether the relevant forms of the enzyme may fulfil other roles in the organism.

I. Other Genes Dispensable for Vegetative Growth are Needed for the Production of *Streptomyces coelicolor* Antibiotics

Since there seems to be a rough correlation between the onset of morphological differentiation and of antibiotic production, it is perhaps not surprising to find that some morphological mutants do not make antibiotics. Most *S. coelicolor* mutants unable to produce normal aerial hyphae (designated *bld* for bald), fail to produce any of the four quite different antibiotics made by the wild-type strain (i.e. actinorhodin, methylenomycin, undecylprodigiosin and

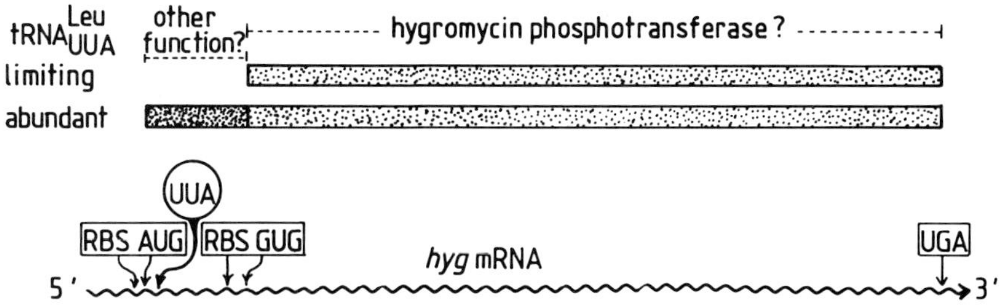

Figure 9. A model in which the presence of a UUA codon in the *hyg* mRNA can allow a regulated change in the nature of the protein product. When charged $tRNA^{Leu}_{UUA}$ (the product of the *bldA* gene) is not available, either early in growth or in a *bldA* mutant, translation from the upstream ribosome binding site (RBS) is arrested at the UUA codon and the downstream RBS is exposed. Translation from this RBS gives a product with hygromycin phosphotransferase activity. When charged $tRNA^{Leu}_{UUA}$ is abundant (later in growth: 36) use of the upstream RBS gives rise to a hygromycin phosphotransferase with an amino-terminal extension of unknown function, which might, for example, serve a role in activating hygromycin biosynthesis.

a "calcium-dependent antibiotic"). These mutations map in one of at least five chromosomal genes, *bldA*, *B*, *D*, *G* and *H*, which lie in different regions of the chromosome, well away from the antibiotic pathway genes (34, and W. Champness, personal communication). One of them, *bldA*, has been cloned and analysed in detail (35, 36). Remarkably, its product appears to be a tRNA (36). It is not yet understood how this putative tRNA, which would recognize the UUA codon for leucine, exerts its pleiotropic influences. Analysis of available *Streptomyces* DNA sequences and other, indirect, evidence suggest that the UUA codon may not be used by genes required for normal growth (E. J. Lawlor and B. L. Daly, personal communication). It has so far been discovered only in a few genes concerned with antibiotic resistance and production. Perhaps the most obvious general possibility is that the tRNA operates through effects on translation. One could envisage that UUA codons play a role in attenuation (37: Chapter 2) of genes for morphological differentiation and antibiotic production.

Alternatively, there could be *bldA*-dependent qualitative differences in the proteins synthesized from UUA-containing transcripts. This possibility was suggested by the finding that the single UUA codons in the *S. glaucescens* streptomycin phosphotransferase gene (28) and the *S. hygroscopicus* hygromycin phosphotransferase gene (38) are both located early in the coding region and upstream of plausible ribosome binding sites (Fig. 9); thus there might be two alternative start points for translation of the mRNAs for these genes.

In addition to the question of how *bldA* works there is a second question: how is *bldA* itself regulated? Dot blot analysis of RNA accumulated during growth on solid medium shows that its gene product is principally accumulated late in vegetative growth, at the time when differentiation and antibiotic production are beginning (36). This result was supported by transcriptional fusion experiments in which the *bldA* promoter region was fused to the promoterless *luxA,B* genes for luciferase (Chapter 4): light emission was strongest after two to three days (39).

A second, morphologically normal, class of pleiotropic antibiotic non-producers was discovered among *S. coelicolor* mutants unable to make A-factor. These mutants, and subsequent DNA sequencing, revealed three genes designated *afsA*, *afsB* and *afsC* (40). *afsA* mutants seem to have a simple block in A-factor production, whereas *afsB* and *afsC* apparently code for components of a positive regulatory system needed for antibiotic production as well as A-factor synthesis (but not for differentiation). Transcripts from the *act* region were strongly reduced in abundance in *afsB* mutants, suggesting a control at the level of transcription (41). Sequencing of *afsB* has suggested that its product is a DNA-binding protein (40). (Since *afsA* and *afsB S. coelicolor* mutants sporulate, and *afsA* mutants still produce antibiotics, A-factor itself is not essential for antibiotic production or differentiation in *S. coelicolor*; this situation contrasts with that in *S. griseus* (see above).)

IV. OVERVIEW, IMPLICATIONS AND PROSPECTS

Antibiotic pathways in *Streptomyces* are usually determined by clusters of genes on the chromosome, although the plasmid-borne genes for methylenomycin A biosynthesis provide an exception. The clusters often contain regulatory genes, which may act positively or negatively by mechanisms which have not yet been studied. The clusters usually (perhaps always) also contain resistance genes, transcription of which is probably intimately coordinated with transcription of the production genes and may help to regulate it. Sometimes, the resistance gene products also serve essentially biosynthetic functions, such as modification of biosynthetic intermediates or perhaps excretion of the end product. In addition to pathway-specific regulation, antibiotic synthesis also depends on more widely-acting regulatory genes, which may in some species be needed for all antibiotic synthesis as well as for normal morphological development.

With the successful cloning and analysis of the genes for a few pathways, new strategies and possibilities have emerged for the study and exploitation of antibiotic production genes. For example, genes for many pathways can be cloned by exploiting their anticipated linkage with resistance genes. Some particularly intriguing further approaches have been suggested by model studies using the cloned *act* genes.

It was found that *act* genes, when introduced into streptomycetes that make

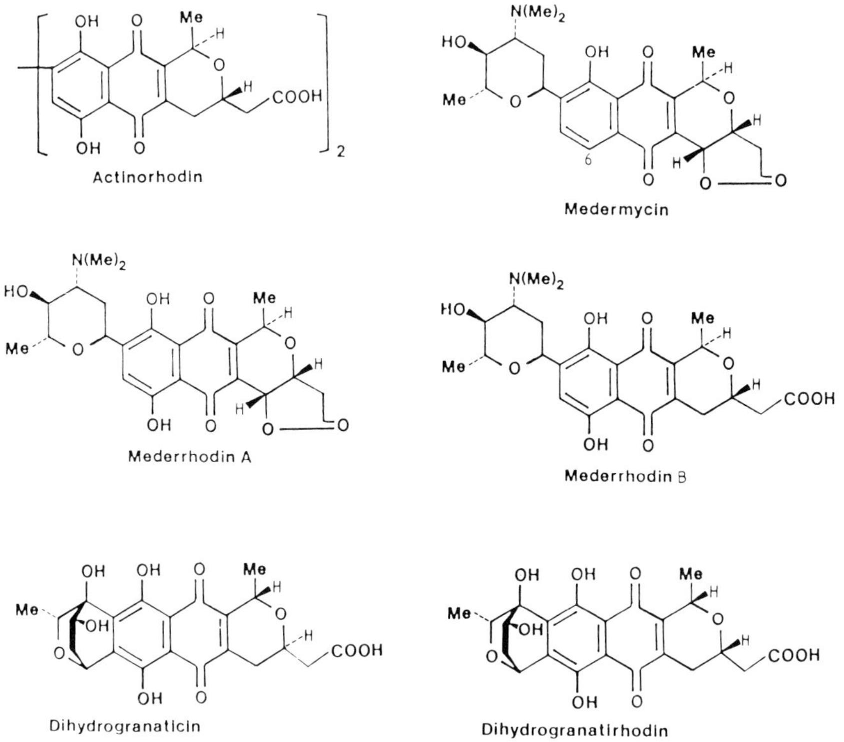

Figure 10. Structures of some isochromanequinone antibiotics. Mederrhodins A and B are chemical "hybrids" between medermycin and actinorhodin. Dihydrogranatirhodin is a "hybrid" between dihydrogranaticin and actinorhodin.

compounds related to actinorhodin, sometimes give rise to hybrid molecules (42, 43) (Fig. 10). This finding suggests eventual practical applications in producing novel antibiotics with potentially advantageous properties. The approach also has analytical value, since the changes in a novel antibiotic give clues about both the function of the products of the gene(s) introduced by cloning, and the specificity (or lack of it) of the biosynthetic enzymes.

In a second study, *act* genes were used as DNA hybridization probes. At least two regions of *act* DNA (*act*I and III) hybridize to parts of gene clusters for other polyketide antibiotics, and can be used as tools in isolating and analysing those genes (44). Moreover, since the hybridizing regions are likely to encode parts of the polyketide assembly machinery, it may be possible in future to engineer hybrid genes (or perhaps less ambitiously to change the specificity of a cloned polyketide synthase by site-directed mutagenesis) so that novel polyketide structures are produced.

The molecular genetics of antibiotic production is currently very exciting, dealing on the one hand with fundamental questions of developmental regulation of gene expression, and on the other with material of considerable medical or agricultural (and therefore also commercial) importance. In the near future, the extensive use of a range of currently-available techniques, including DNA sequencing, *in vitro* transcription studies, binding of putative regulatory proteins to DNA fragments, gene fusions and site-directed mutagenesis, as well as more genetical approaches such as the isolation and analysis of mutants, can be expected to reveal generalizations about the expression of antibiotic production genes, as well as clarifying the regulatory cascades and networks that (we anticipate) interconnect primary with secondary metabolism, and morphological with physiological differentiation.

References

(1) Chater, K. F. (1984). Morphological and physiological differentiation in *Streptomyces*. *In* "Microbial Development" (R. Losick and L. Shapiro, eds.), pp. 89–115. Cold Spring Harbor Laboratory, Cold Spring Harbor, New York.

(2) Martin, J. F. and Demain, A. L. (1980). Control of antibiotic biosynthesis. *Microbiol. Rev.* **44**, 230–251.

(3) Kokhlov, A. S., Anisova, L. N., Tovarova, I. I., Kleiner, F. M., Kovalenko, I. V., Krasilnikova, O. I., Kornitskaya, E. Y. and Pliner, S. A. (1973). Effect of A-factor on the growth of asporogenous mutants of *Streptomyces griseus*, not producing this factor. *Z. Allg. Mikrobiol.* **13**, 647–655.

(4) Hara, O., Horinouchi, S., Uozumi, T. and Beppu, T. (1983). Genetic analysis of A-factor synthesis in *Streptomyces coelicolor* A3(2). *J. Gen. Microbiol.* **129**, 2939–2944.

(5) Davies, J. and Smith, D. I. (1978). Plasmid-determined resistance to antimicrobial agents. *Ann. Rev. Microbiol.* **32**, 469–518.

(6) Murakami, T., Anzai, H., Imai, S., Satoh, A., Nagaoka, K. and Thompson, C. J. (1986). The bialaphos biosynthetic genes of *Streptomyces hygroscopicus*: molecular cloning and characterization of the gene cluster. *Mol. Gen. Genet.* **205**, 42–50.

(7) Thompson, C. J., Movva, N. R., Tizard, R., Crameri, R., Davies, J. E., Lauwereys, M. and Botterman, J. (1987). Characterization of the herbicide-resistance gene *bar* from *Streptomyces hygroscopicus*. *EMBO J.* **6**, 2519–2523.

(8) Neal, R. J. and Chater, K. F. (1987). Nucleotide sequence analysis reveals similarities between proteins determining methylenomycin A resistance in *Streptomyces* and tetracycline resistance in eubacteria. *Gene* **58**, 229–241.

(9) Thompson, C. J., Skinner, R. H., Thompson, J., Ward, J. M., Hopwood, D. A. and Cundliffe, E. (1982). Biochemical characterization of resistance determinants cloned from antibiotic-producing streptomycetes. *J. Bacteriol.* **151**, 678–685.

(10) Nakano, M. M., Mashiko, H. and Ogawara, H. (1984). Cloning of the kanamycin resistance gene from a kanamycin-producing *Streptomyces* species. *J. Bacteriol.* **157**, 79–83.

(11) Hopwood, D. A. (1986). Cloning and analysis of antibiotic biosynthetic genes in *Streptomyces*. *In* "Biological, Biochemical and Biomedical Aspects of Actinomycetes" (G. Szabo, S. Biro and M. Goodfellow, eds.), pp. 3–14. Akademiai Kiado, Budapest.

(12) Dimroth, P., Ringelmann, E. and Lynen, F. (1976). 6-methylsalicylic acid synthetase from *Penicillium patulum*: some catalytic properties of the enzyme and its relation to fatty acid synthetase. *Eur. J. Biochem.* **68**, 591–596.

(13) Wright, L. F. and Hopwood, D. A. (1976). Actinorhodin is a chromosomally-determined antibiotic in *Streptomyces coelicolor* A3(2). *J. Gen. Microbiol.* **96**, 289–297.

(14) Kirby, R. and Hopwood, D. A. (1977). Genetic determination of methylenomycin synthesis by the SCP1 plasmid of *Streptomyces coelicolor* A3(2). *J. Gen. Microbiol.* **98**, 239–252.

(15) Rudd, B. A. M. and Hopwood, D. A. (1979). Genetics of actinorhodin biosynthesis in *Streptomyces coelicolor* A3(2). *J. Gen. Microbiol.* **114**, 35–43.

(16) Hopwood, D. A., Bibb, M. J., Chater, K. F. and Kieser, T. (1987). Plasmid and phage vectors for gene cloning and analysis in *Streptomyces*. *Methods in Enzymol.* **153**, 116–166.

(17) Malpartida, F. and Hopwood, D. A. (1984). Molecular cloning of the whole biosynthetic pathway of a *Streptomyces* antibiotic and its expression in a heterologous host. *Nature* **309**, 462–464.

(18) Malpartida, F. and Hopwood, D. A. (1976). Physical and genetic characterization of the gene cluster for the antibiotic actinorhodin in *Streptomyces coelicolor* A3(2). *Mol. Gen. Genet.* **205**, 66–73.

(19) Cole, S. P., Rudd, B. A. M., Hopwood, D. A., Chang, C-J. and Floss, H. G. (1987). Biosynthesis of the antibiotic actinorhodin: analysis of blocked mutants of *Streptomyces coelicolor*. *J. Antibiot.* **40**, 340–347.

(20) Vivian, A. (1971). Genetic control of fertility in *Streptomyces coelicolor* A3(2): plasmid involvement in the interconversion of UF and IF strains. *J. Gen. Microbiol.* **69**, 353–364.

(21) Kinashi, H., Shimaji, M. and Sakai, A. (1987). Giant linear plasmids in *Streptomyces* which code for antibiotic biosynthesis genes. *Nature* **328**, 454–456.

(22) Chater, K. F. and Bruton, C. J. (1983). Mutational cloning in *Streptomyces* and the isolation of antibiotic production genes. *Gene* **26**, 67–78.

(23) Chater, K. F. and Bruton, C. J. (1985). Resistance, regulatory and production genes for the antibiotic methylenomycin are clustered. *EMBO J.* **4**, 1893–1897.

(24) Stanzak, R., Matsushima, P., Baltz, R. H. and Rao, R. N. (1986). Cloning and expression in *Streptomyces lividans* of clustered erythromycin biosynthesis genes from *Streptomyces erythraeus*. *Biotechnology* **4**, 229–232.

(25) Neil, R. (1987). Structure and regulation of the methylenomycin A resistance gene of *Streptomyces coelicolor* A3(2). Ph.D. Thesis, University of East Anglia, Norwich.

(26) Bibb, M. J. and Janssen, G. R. (1987). Unusual features of transcription and translation of antibiotic resistance genes in antibiotic-producing *Streptomyces*. *In* "Genetics of Industrial Microorganisms" (M. Alačević, D. Hranueli and Z. Toman, eds.), Part B, pp. 309–318. Pliva, Zagreb.

(27) Bibb, M. J., Janssen, G. R. and Ward, J. M. (1985). Cloning and analysis of the promoter region of the erythromycin-resistance gene (*ermE*) of *Streptomyces erythraeus*. *Gene* **38**, E357–E368.

(28) Vögtli, M. and Hütter, R. (1987). Characterization of the hydroxystreptomycin phosphotransferase gene (*sph*) of *Streptomyces glaucescens*: nucleotide sequence and promoter analysis. *Mol. Gen. Genet.* **208**, 195–203.

(29) Gussin, G. N., Johnson, A. D., Pabo, C. O. and Sauer, R. T. (1983). Repressor and Cro protein: structure, function and role in lysogenization. *In* "Lambda II" (R. W. Hendrix, J. W. Roberts, F. W. Stahl and R. A. Weisberg, eds.), pp. 93–121. Cold Spring Harbor Laboratory, Cold Spring Harbor, New York.

(30) Ptashne, M. (1986). A genetic switch: gene control and phage lambda. Cell Press, Cambridge, Mass. and Blackwell, Oxford.

(31) Westpheling, J., Ranes, M. and Losick, R. (1985). RNA polymerase heterogeneity in *Streptomyces coelicolor*. *Nature* **313**, 22–27.

(32) Buttner, M. J., Smith, A. M. and Bibb, M. J. (1988). At least three different RNA polymerase holoenzymes direct transcription of the agarase gene (*dagA*) of *Streptomyces coelicolor* A3(2). *Cell* **52**, 599–607.

(33) Mendez, C . and Chater, K. F. (1987). Cloning of *whiG*, a gene critical for sporulation of *Streptomyces coelicolor* A3(2). *J. Bacteriol.* **169**, 5715–5720.

(34) Merrick, M. J. (1976). A morphological and genetic mapping study of bald colony mutants of *Streptomyces coelicolor*. *J. Gen. Microbiol.* **96**, 299–315.

(35) Piret, J. M. and Chater, K. F. (1985). Phage-mediated cloning of *bldA*, a region involved in *Streptomyces coelicolor* morphological development, and its analysis by genetic complementation. *J. Bacteriol.* **163**, 965–972.

(36) Lawlor, E. J., Baylis, H. A. and Chater, K. F. (1987). Pleiotropic morphological and antibiotic deficiencies result from mutations in a gene encoding a tRNA-like product in *Streptomyces coelicolor* A3(2). *Genes & Devel.* **1**, 1305–1310.

(37) Platt, T. (1985). Control of gene expression in bacteria. *In* "Genetics of Bacteria" (J. Scaife, D. Leach and A. Galizzi, eds.), pp. 255–278. Academic Press, London.

(38) Zalacain, A., Gonzalez, A., Guerrero, M. C., Mattaliano, R. J., Malpartida, F. and Jimenez, A. (1986). Nucleotide sequence of the hygromycin B phosphotransferase gene from *Streptomyces hygroscopicus*. *Nuc. Acids Res.* **14**, 1565–1581.

(39) Schauer, A., Ranes, M., Santamaria, R., Guijarro, J., Lawlor, E., Mendez, C., Chater, K. F. and Losick, R. (1988). Visualizing gene expression in time and space in the morphologically complex, filamentous bacterium *Streptomyces coelicolor*. *Science*. **240**, 768–772.

(40) Horinouchi, S. and Beppu, T. (1987). A-factor and regulatory network that links secondary metabolism with cell differentiation in *Streptomyces*. *In* "Genetics of Industrial Microorganisms" (M. Alačević, D. Hranueli and Z. Toman, eds.), Part B, pp. 41–48. Pliva, Zagreb.

(41) Horinouchi, S., Malpartida, F., Hopwood, D. A. and Beppu, T. (1988). *afsB* stimulates transcription of the actinorhodin biosynthetic pathway in *Streptomyces coelicolor* A3(2) and *Streptomyces lividans*. (In preparation).

(42) Hopwood, D. A., Malpartida, F., Kieser, H. M., Ikeda, H., Duncan, J., Fujii, I., Rudd, B. A. M., Floss, H. G. and Ōmura, S. (1985). Production of "hybrid" antibiotics by genetic engineering. *Nature* **314**, 642–644.

(43) Ōmura, S., Ikeda, H., Malpartida, F., Kieser, H. M. and Hopwood, D. A. (1986). Production of new hybrid antibiotics, mederrhodins A and B, by a genetically engineered strain. *Antimicrob. Agents Chemo.* **29**, 13–19.

(44) Malpartida, F., Hallam, S. E., Kieser, H. M., Motamedi, H., Hutchinson, C. R., Butler, M. J., Sugden, D. A., Warren, M., McKillop, C., Bailey, C. R., Humphreys, G. O. and Hopwood, D. A. (1987). Homology between *Streptomyces* genes coding for synthesis of different polyketides used to clone antibiotic biosynthetic genes. *Nature* **325**, 818–821.

(45) Hasegawa, M., Hintermann, G., Simonet, J-M., Crameri, R., Piret, J. and Hütter, R. (1985). Certain chromosomal regions in *Streptomyces glaucescens* tend to carry amplifications and deletions. *Mol. Gen. Genet.* **200**, 375–384.

(46) Altenbuchner, J. and Cullum, J. (1985). Structure of an amplifiable DNA sequence in *Streptomyces lividans* 66. *Mol. Gen. Genet.* **201**, 192–197.

(47) Dyson, P. and Schrempf, H. (1987). Genetic instability and DNA amplification in *Streptomyces lividans* 66. *J. Bacteriol.* **169**, 4796–4803.

(48) Fishman, S. E., Cox, K., Larson, J. L., Reynolds, P. A., Seno, E. T., Yeh, W-K., van Frank, R. and Hershberger, C. L. (1987). Cloning genes for the biosynthesis of a macrolide antibiotic. *Proc. Natl. Acad. Sci. USA* **84**, 8248–8252.

(49) Schrempf, H. (1983). Deletion and amplification of DNA sequences in melanin-negative variants of *Streptomyces reticuli*. *Mol. Gen. Genet.* **189**, 501–505.

(50) Hornemann, U., Otto, C. J., Hoffman, G. G. and Bertinuson, A. C. (1987). Spectinomycin resistance and associated DNA amplification in *Streptomyces achromogenes* subsp. *rubradiris*. *J. Bacteriol.* **169**, 2360–2366.

Chapter 8

Catabolism of Aromatic Hydrocarbons by Pseudomonas

SHIGEAKI HARAYAMA *and* KENNETH N. TIMMIS

I. INTRODUCTION

Photosynthetic microorganisms and plants are responsible for fixing carbon dioxide and thus channelling inorganic carbon into the complex organic compounds of which living systems consist. Once organisms die, their organic constituents are generally biodegraded, either to simpler organic compounds which serve as precursors for the synthesis of new complex molecules, or to inorganic carbon compounds such as CO_2. Organic carbon may cycle through a number of different living organisms before finally being mineralized to CO_2, thereby completing the carbon cycle. Soil and water microorganisms are important agents of biodecomposition and thus play a crucial role in recycling carbon and in maintaining the biosphere in equilibrium. Organic materials may, however, not necessarily be degraded but instead become converted to new organic compounds under the influence of geophysical and geochemical processes, and thereby exit from the carbon cycle for a period of time. The return to the carbon cycle of material sequestered in coal and oil deposits, etc., occurs following geological processes and human activities that expose it to the biodegradative activities of microorganisms or that result in its combustion.

Whereas the number of different organic molecules resulting from biosynthetic processes is relatively small, and hence the spectrum of distinct pathways for the catabolism of such compounds limited, the variety of organic molecules in the biosphere that have been created by geological processes is enormous (1). It is not surprising, therefore, that the enterprising microbial world, which has evolved in such a way as to be able to profit maximally from the availability of a wide range of organic compounds (many of which are chemically unusual or toxic to some degree), possesses an extremely varied arsenal of catabolic enzymes with which to attack potential growth substrates.

Superimposed on the rich variety of natural organic compounds present in the environment is the increasing number of novel industrial organic chemicals heavily substituted with chemical groups (halides, SO_3, NO_2, etc.) not found, or not extensively present, in organic molecules of biological origin. Although many of these compounds can be processed by existing versatile catabolic pathways, or by readily-occurring adaptations of existing pathways, others are so novel, toxic or stable that they are not easily processed, and thus accumulate in the environment, sometimes causing severe pollution. One of the most important challenges of our time, then, is to find and exploit existing biodegradative routes to eliminate pollutants from the environment or, when this is not possible, to develop experimentally new catabolic pathways that can be effective for this purpose.

Pseudomonas species are known to metabolize a broad range of organic compounds and therefore are an ideal choice as the bacteria to be used for environmental detoxification. Two useful genetic techniques, namely transposon mutagenesis and gene cloning applicable to these species, have been recently developed. The most popular method for transposon mutagenesis involves the use of a mobilizable, narrow-host-range plasmid carrying a drug-resistant transposon: such a plasmid replicates in *Escherichia coli* but not in *Pseudomonas*, and so *Pseudomonas* transconjugants which have acquired the transposon-coded drug-resistance marker through mating with an *E. coli* strain containing this plasmid are usually transposon-insertion mutants (2). Many *Pseudomonas* genes have been cloned in *E. coli* (3); however the latter organism does not necessarily express the biological properties of cloned foreign genes. Therefore, a number of broad-host-range cloning vectors which can replicate in many Gram-negative bacteria have been constructed (4).

Here, we describe the biochemical strategies used by soil microorganisms to degrade aromatic compounds aerobically, the genetics and regulation of the best characterized pathway (that for the degradation of toluene encoded by the TOL plasmid of *Pseudomonas putida*) and current strategies to evolve experimentally new catabolic activities.

II. BIOCHEMICAL STRATEGIES FOR OXIDATIVE CATABOLISM OF AROMATICS

A. Convergent Pathways Lead to a Limited Number of Ring-Cleavage Substrates

A critical feature of catabolic routes is the channelling of structurally diverse substrates into a limited number of central pathways (Fig. 1). This is accomplished through the activity of a large number of enzymes that carry out the initial reactions in the catabolic pathways and that collectively are able to attack a wide range of substrates. Although the spectrum of substrates and initial enzymes is large, the types of reactions carried out are limited. In the aerobic catabolism of aromatic molecules, these reactions generally lead to the formation of dihydroxy aromatic intermediates that have hydroxyl groups either on two adjacent carbons of the aromatic ring (ortho) or on opposing carbons of the ring (para) and are key intermediates for the entry into central pathways (Fig. 2). Depending on whether or not the substrate already carries hydroxyl groups on the aromatic ring, the hydroxylation may be carried out by either mono- or dioxygenase enzymes. The formation of dihydric phenols (aromatic compounds carrying two hydroxyl groups) seems to be an important biochemical strategy of bacteria to destabilize the chemically stable resonant structure of the aromatic ring in order to facilitate its subsequent opening (5, 6). One way to form dihydric phenols is by dioxygenation of the aromatic ring,

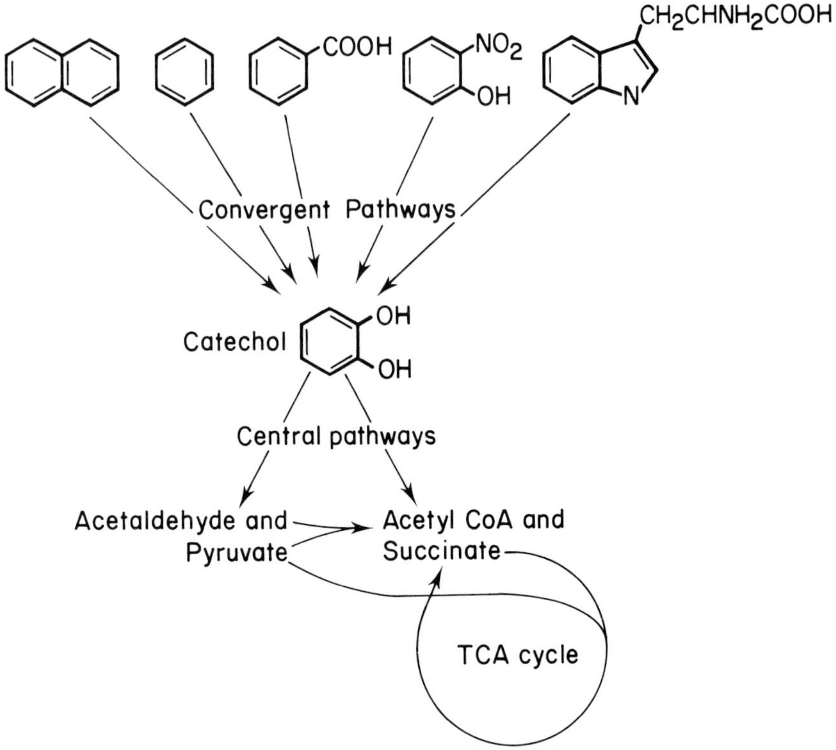

Figure 1. Metabolism of aromatic hydrocarbons through catechol. Diverse aromatic hydrocarbons are transformed to catechol, one of the key intermediates of aromatic degradation, which is further metabolized by central pathways either to pyruvate and acetaldehyde or to succinate and acetyl CoA, which are subsequently decomposed to CO_2 and H_2O through the Krebs cycle.

involving incorporation of two atoms of molecular oxygen to produce *cis*-diols, followed by dehydrogenation (1, 5, 6). The subunit structure and function of aromatic ring dioxygenases is shown in Fig. 3. Note that the single NADH consumed by dioxygenation of the benzene nucleus is regenerated by the subsequent dehydrogenation of the dihydrodiol.

Catabolism of aromatic substrates with more complex substitutions may sometimes involve additional steps that process the substituent group, and the formation of the central dihydric phenols may involve FAD-containing mono-oxygenases. Examples of these pathways are summarized in Fig. 2. Alternative routes to dihydric phenols for aromatic compounds substituted with halo-, nitro- and sulfonate groups are shown in Fig. 4.

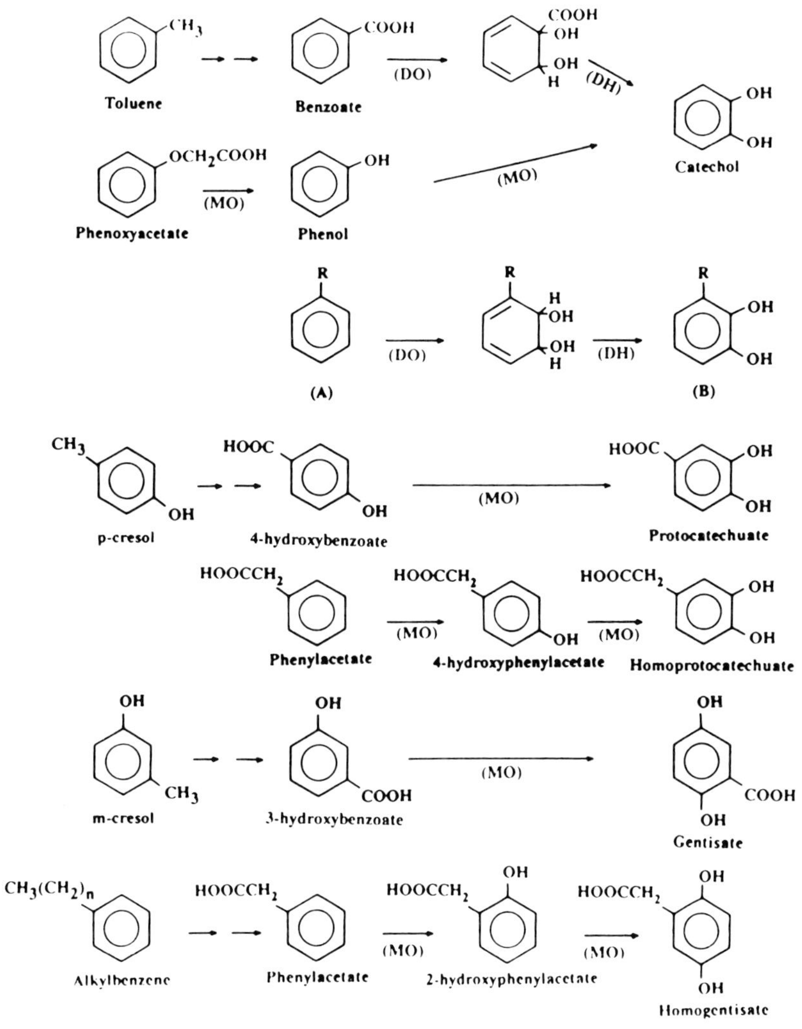

Figure 2. Convergent pathways that lead aromatic hydrocarbons to one of five key pathway intermediates, namely catechol (or substituted catechols), protocatechuate, homoprotocatechuate, gentisate and homogentisate. Addition of hydroxyl group(s) to the benzene nucleus is carried out by either monooxygenase (MO) or dioxygenase (DO) plus dihydrodiol dehydrogenase (DH). (a): toluene ($R = CH_3$); isopropylbenzene ($R = CH_2CH_3$). (b): 3-methylcatechol ($R = CH_3$); 3-isopropylcatechol ($R = CH_2CH_3$). These reaction sequences have been reviewed in references 1, 5, 6, 12, 13 and 37.

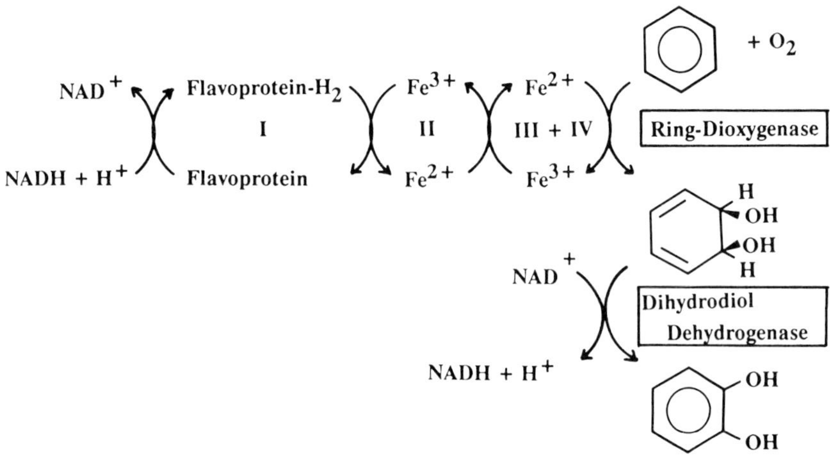

Figure 3. Formation of dihydric phenols catalysed by ring-hydroxylating dioxygenase and dihydrodiol dehydrogenase. Ring-hydroxylating dioxygenases produce dihydrodiols by introducing two atoms of oxygen into the aromatic ring. They consist of three or four polypeptides which together form a short electron transfer chain that transfers electrons from NADH to oxygen. Component I is a flavoprotein and has NADH oxidase activity; II is a ferredoxin and contains one (2Fe-2S*) cluster (S* is the inorganic sulphur in the iron-sulphur cluster). In dioxygenases which oxidize benzoate, components I and II comprise one polypeptide. Components III and IV form the terminal oxygenase and contain (2Fe-2S*) clusters. The molecular weights of polypeptide subunits of different ring dihydroxylating dioxygenases are similar: the terminal oxygenases generally consist of two polypeptides of 20 and 50 kD, and their ferredoxins are usually small polypeptides. Dihydrodiols are oxidized to dihydric phenols by dihydrodiol dehydrogenases which are single component enzymes.

B. The Aromatic Ring is Cleaved by Two Types of Fission

Fission of the dihydroxylated aromatic ring is carried out by dioxygenases which incorporate two atoms of oxygen (Fig. 5). Catechols and protocatechuate can be cleaved by two distinct modes: either between the two hydroxyl groups (ortho or intradiol-cleavage) or proximal to one hydroxyl group (meta or extradiol-cleavage). In the case of gentisate or homogentisate, whose two hydroxyl groups are positioned in para, the ring is opened between carbon-1 and 2 by the action of gentisate 1,2-dioxygenase or homogentisate 1,2-dioxygenase. The ring-cleavage reactions of gentisate and homogentisate mechanistically resemble meta-fission (5, 6). As discussed later, the double hydroxylated rings of polycyclic aromatic compounds are also meta cleaved.

Ring-cleavage enzymes are generally multimeric proteins consisting of a

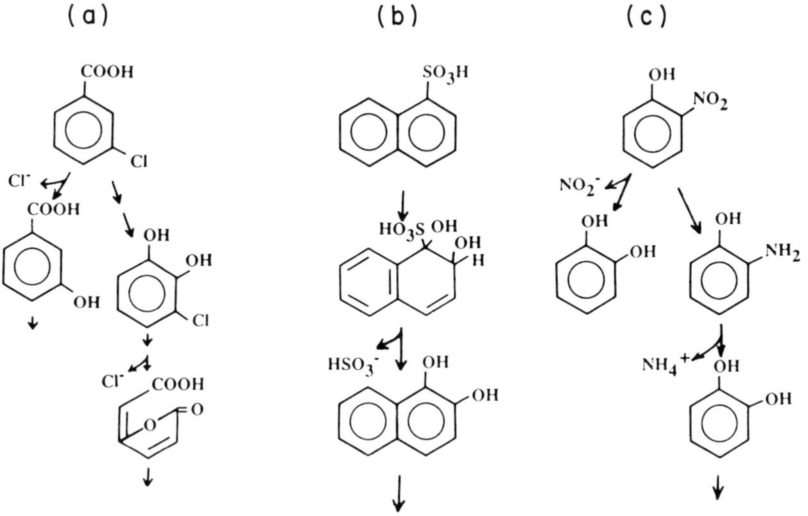

Figure 4. Possible reactions for elimination of substituents of the aromatic ring: (a) 3-chlorobenzoate; (b) naphthalene-1-sulphonic acid; (c) *o*-nitrophenol.

single polypeptide species, although protocatechuate 3,4-dioxygenases from various *Pseudomonas* species are composed of two different polypeptides (7). The meta-fission dioxygenases contain non-haem ferrous irons as prosthetic groups whereas ortho-fission enzymes contain non-haem ferric irons. These irons are essential for the reaction although their precise roles are not yet clear (8).

C. After Ring-Cleavage There are Two Principal Catabolic Routes Towards the Krebs Cycle

Following ring-cleavage, metabolites are directed towards the Krebs cycle by suites of enzymes which comprise the central pathways. Typically, unsubstituted and haloaromatic compounds are metabolized by ortho-cleavage (or β-ketoadipate) pathways (Fig. 5) in which catechol and protocatechuate are cleaved by catechol 1,2-dioxygenase and protocatechuate 3,4-dioxygenase, respectively, and the cleavage products, *cis, cis*-muconate and 3-carboxy-*cis,cis*-muconate are transformed via the common intermediate, β-ketoadipate enol-lactone to succinate and acetyl CoA (5, 6, 9; Fig. 6). Ortho-cleavage pathways do not usually process alkylaromatics because many ortho-cleavage enzymes have very low affinities for alkylcatechols, and even if such compounds are ring-cleaved, deadend intermediates usually accumulate further down the

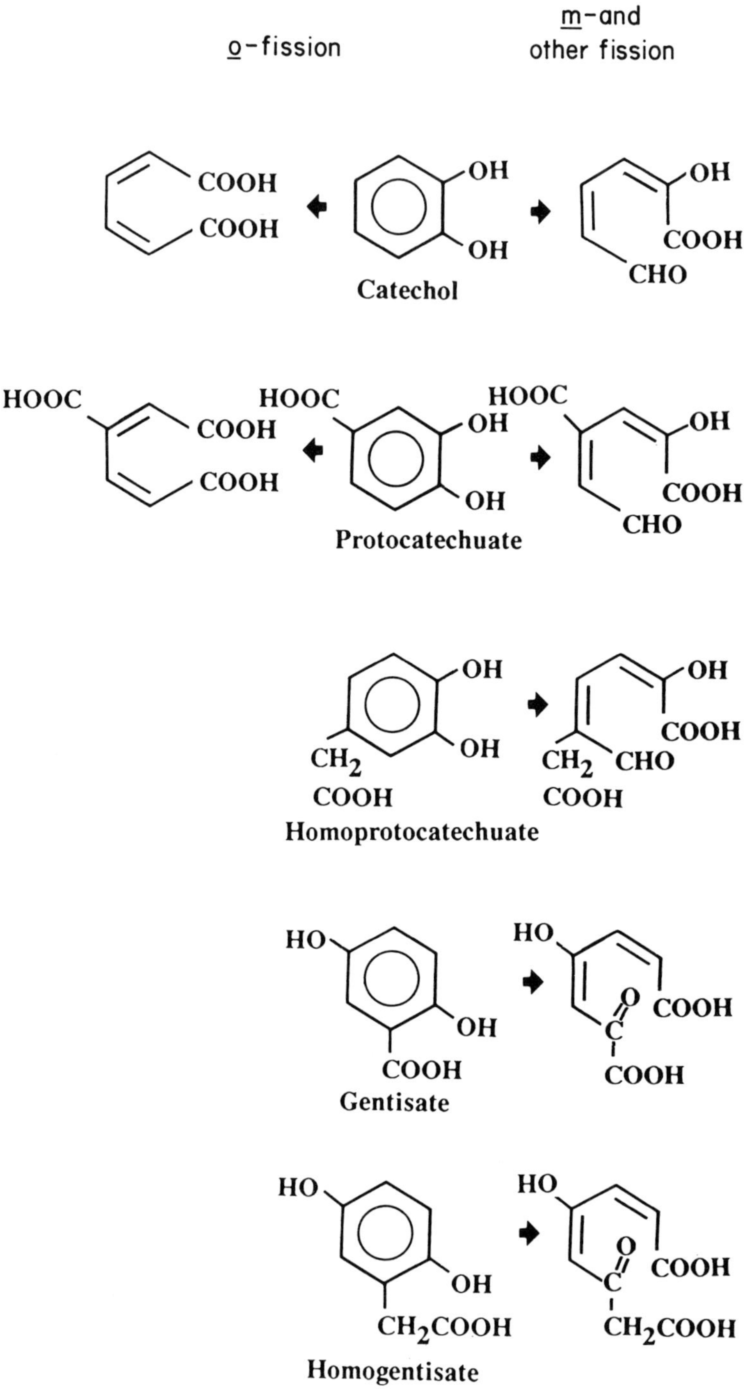
o-fission
m-and other fission
COOH
COOH
OH
OH
Catechol
OH
COOH
CHO
HOOC
COOH
COOH
HOOC
OH
OH
Protocatechuate
HOOC
OH
COOH
CHO
OH
OH
CH2
COOH
Homoprotocatechuate
OH
COOH
CH2
CHO
COOH
HO
OH
COOH
Gentisate
HO
O
C
COOH
COOH
HO
OH
CH2COOH
Homogentisate
HO
O
C
COOH
CH2COOH

Figure 5.

pathway (10). Organisms such as *Pseudomonas* sp. B13 that are able to degrade benzoate and 3-chlorobenzoate have two parallel β-ketoadipate pathways, designated the ortho-cleavage pathway (for benzoate) and the modified ortho-cleavage pathway (for 3-chlorobenzoate), consisting of parallel enzymes with altered substrate specificities (11; Fig. 6).

Alkylaromatics are mostly processed by meta-cleavage which results, in the case of catechols or alkyl catechols, in the formation of hydroxymuconic semialdehydes or their derivatives, which are bright yellow products. These intermediates are further processed by one of two pathway branches: the hydrolytic branch involves the direct formation of 2-hydroxypent-2,4-dienoate or its derivatives, whereas in the dehydrogenative branch this compound is formed by three enzymatic steps (12, 13; Fig. 7). The dehydrogenase has a high affinity for ring-fission products derived from catechol and 4-methylcatechol, but cannot attack the ring-fission product derived from 3-methylcatechol which is therefore catabolized hydrolytically (13, 14). Similar metabolic routes also exist for protocatechuate and homoprotocatechuate. Pathways for degradation of gentisate and homogentisate are less closely related but there is significant similarity in the order of enzymatic reactions to that of the hydrolytic branch of the catechol meta-cleavage pathway (5, 6; Fig. 8). Even though the initial enzymes of such pathways often exhibit relaxed substrate specificities and are able to transform intermediates carrying a variety of substituents, meta-cleavage pathways are not generally suitable for the catabolism of haloaromatics because there are several steps in the pathways which will not occur with the chlorinated pathway intermediates.

D. Metabolism of Polycyclic Hydrocarbons

Degradation of polycyclic aromatic compounds is initiated by dihydroxylation of one of the benzene nuclei by a ring dioxygenase and dihydrodiol dehydrogenase, followed by dioxygenase-catalysed cleavage of the destabilized ring. The carbon skeleton of this ring is then catabolized by a route similar to that for the degradation of catechol via the hydrolytic route of the meta-cleavage pathway. In the degradation of naphthalene, for example, salicylate is a key intermediate prior to the formation of catechol. Biphenyl and

---

Figure 5. Fission of aromatic rings by dioxygenases. Catechol and its substituted derivatives are ring-cleaved either by catechol 1,2-dioxygenase (*o*-fission) or catechol 2,3-dioxygenase (*m*-fission). Substrate specificities and inducer specificities of these enzymes are narrow and usually exclude the misrouting of metabolites. Protocatechuate is *o*-cleaved in fluorescent pseudomonads such as *P. putida* whereas this compound is *m*-cleaved in nonfluorescent pseudomonads.

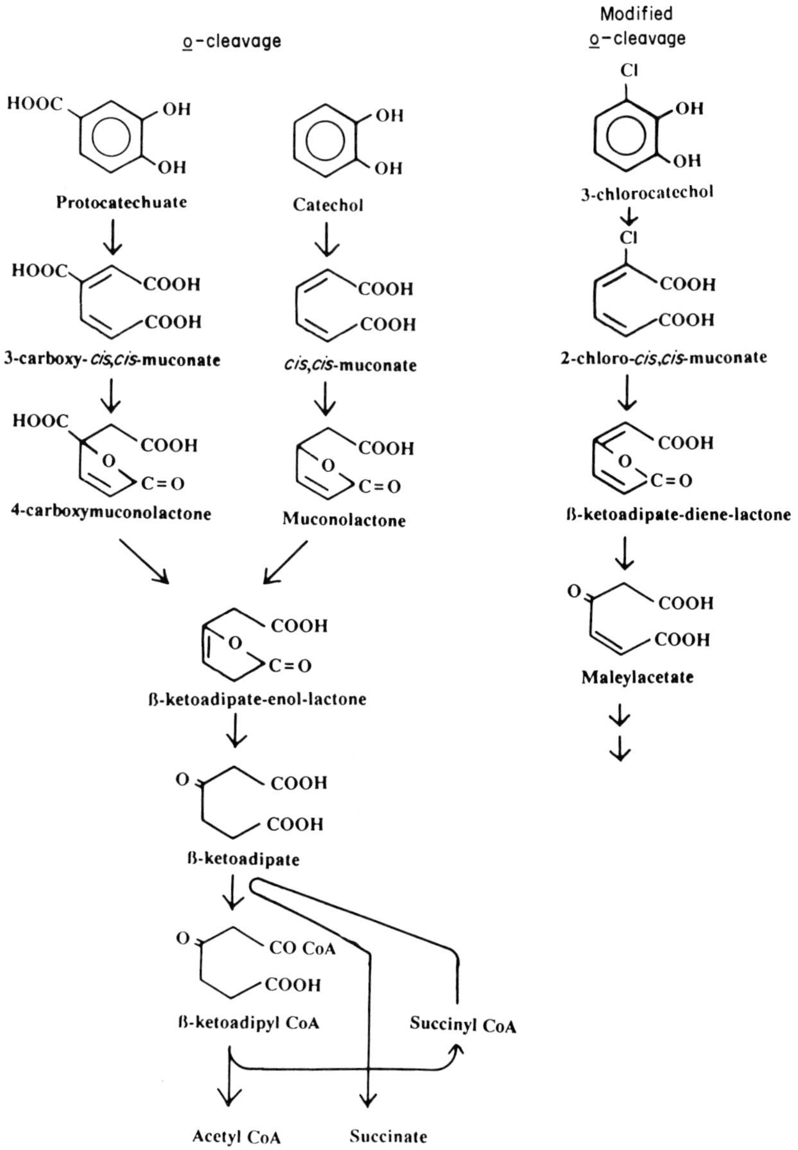

Figure 6. β-ketoadipate pathways for degradation of catechol, protocatechuate and 3-chlorocatechol. The *o*- or β-ketoadipate pathway is a metabolic route which transforms catechol or protocatechuate to succinate and acetyl CoA. After the *o*-fission of the aromatic ring by intradiol ring-cleavage dioxygenases, catechol and protocatechuate are processed to a common intermediate, β-ketoadipate enol-lactone, and further transformed to succinate and acetyl CoA. The pathway for degradation of 3-chlorocatechol is slightly different from other β-ketoadipate pathways and is therefore designated the modified *o*-pathway.

chlorinated biphenyls are degraded via benzoate or chlorobenzoate and 2-hydroxypent-2,4-dienoate, which are either further metabolized or accumulated (6).

III. ORGANIZATION AND REGULATION OF GENES FOR CATABOLISM OF AROMATIC HYDROCARBONS

A. Catabolic Genes are Often, but not Always, Carried on Plasmids

Many bacterial catabolic pathways are specified by conjugative plasmids (15, 16; Table I). These plasmids can readily transfer laterally into new host bacteria and thereby expand the metabolic potential of other members of an ecosystem. Conjugative plasmids are of course important agents of genetic change and evolution in bacteria, and can pick up and bring together from different organisms groups of genes that through mutations can specify new metabolic functions (17). In an environment rich in a particular organic compound, a selective pressure exists to acquire and maintain a plasmid that specifies a corresponding catabolic pathway; many degraders of exotic compounds have been isolated from soil or water contaminated with such compounds. In some cases, the same catabolic genes are located on a plasmid in one organism and on the chromosome in another. In at least one instance, the TOL plasmid, transposition of the catabolic genes from the plasmid to the chromosome has been demonstrated in the laboratory (18).

Table I
Plasmids encoding catabolic functions

Plasmid	Host	Compound(s) catabolized
TOL	*Pseudomonas putida*	Toluene, *p*- and *m*-xylene
NAH	*Pseudomonas putida*	Naphthalene
SAL	*Pseudomonas putida*	Salicylate
pND50	*Pseudomonas putida*	*p*-cresol
pWW31	*Pseudomonas putida*	Phenylacetate
pJP1	*Alcaligenes paradoxus*	2,3-dichlorophenoxyacetate and phenoxyacetate
pJP4	*Alcaligenes eutrophus*	2,4-dichlorophenoxyacetate and 3-chlorobenzoate
pKF1	*Acinetobacter* sp.	4-chlorobiphenyl
pAC21	*Pseudomonas* sp.	4-chlorobiphenyl
pRE1	*Pseudomonas putida*	Isopropyl benzene
pCIT1	*Pseudomonas* sp.	Aniline
pAC25	*Pseudomonas putida*	3-chlorobenzoate
pWR1	*Pseudomonas* sp.	3-chlorobenzoate
pCS1	*Pseudomonas diminuta*	Parathion
pEG	*Pseudomonas fluorescens*	Styrene

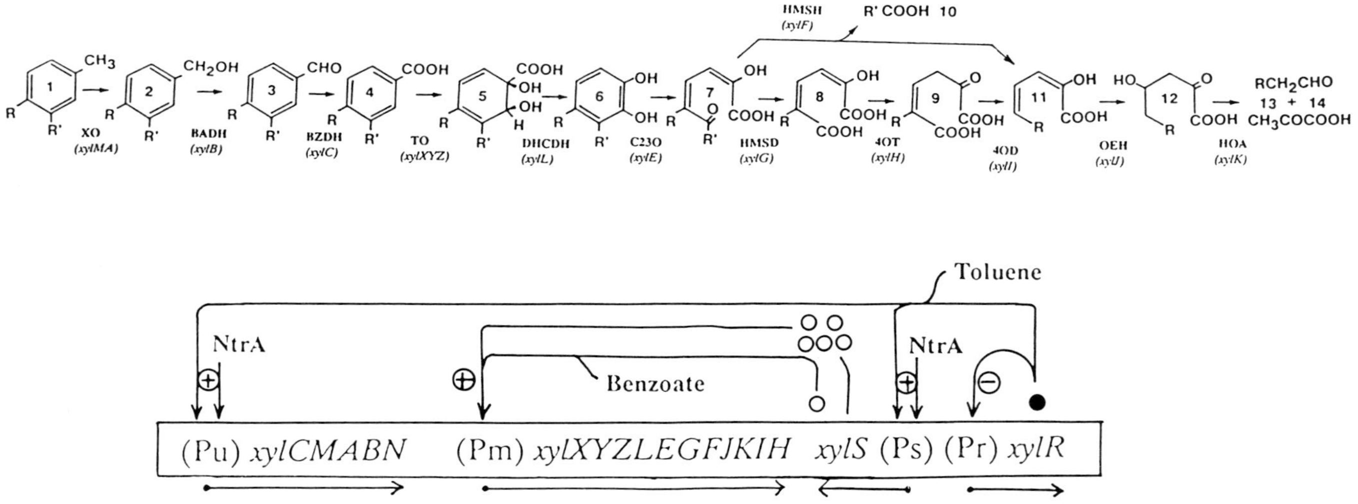
XO (xylMA)
BADH (xylB)
BZDH (xylC)
TO (xylXYZ)
DHCDH (xylL)
C23O (xylE)
HMSD (xylG)
HMSH (xylF)
R' COOH 10
4OT (xylH)
4OD (xylI)
OEH (xylJ)
HOA (xylK)
RCH2CHO
13 + 14
CH3COCOOH
Toluene
NtrA
Benzoate
(Pu) xylCMABN
(Pm) xylXYZLEGFJKIH
xylS (Ps) (Pr) xylR

Figure 7.

Figure 7. Toluene degradative pathway encoded in TOL plasmid pWW0: its genes and regulation. The upper part of the figure shows pathway intermediates, enzymes, and their structural genes. Enzyme abbreviations: XO, xylene oxygenase (multicomponent); BADH, benzylalcohol dehydrogenase; BZDH, benzaldehyde dehydrogenase; TO, toluate 1,2-dioxygenase (multicomponent); DHCDH, 1,2-dihydroxycyclohexa-3,5-diene-carboxylate dehydrogenase; C23O, catechol 2,3-dioxygenase; HMSD, 4-hydroxymuconic semialdehyde dehydrogenase; HMSH, 4-hydroxymuconic semialdehyde hydrolase; 4OT, 4-oxalocrotonate tautomerase; 4OD, 4-oxalocrotonate decarboxylase; OEH, 2-hydroxypent-2,4-dienoate hydratase; HOA, 4-hydroxy-2-oxovalerate aldolase. *xylA* to *xylZ* designate the structural genes for the catabolic enzymes. Compounds: for R = H, R′ = H, (1) toluene, (2) benzylalcohol, (3) benzaldehyde, (4) benzoate, (5) 1,2-dihydroxycyclohexa-3,5-diene-carboxylate, (6) catechol, (7) 2-hydroxymuconic semialdehyde, (8) 2-hydroxyhexa-2,4-diene-1,6-dioate (enol form of 4-oxalocrotonate), (9) 2-oxohex-4-ene-1,6-dioate (keto form of 4-oxalocrotonate), (10) formate, (11) 2-hydroxypent-2,4-dienoate, (12) 4-hydroxy-2-oxovalerate, (13) pyruvate, (14) acetaldehyde; for R = H, R′ = CH_3, (1) *m*-xylene, (2) *m*-methylbenzylalcohol, (3) *m*-methylbenzaldehyde, (4) *m*-toluate, (5) 1,2-dihydroxy-3-methylcyclohexa-3,5-diene-carboxylate, (6) 3-methylcatechol, (7) 2-hydroxy-6-oxohepta-2,4-dienoate, (10) acetate, (11) 2-hydroxypent-2,4-dienoate, (12) 4-hydroxy-2-oxovalerate, (13) pyruvate, (14) acetaldehyde, for R = CH_3, R′ = H, (1) *p*-xylene, (2) *p*-methylbenzylalcohol, (3) *p*-methylbenzaldehyde, (4) *p*-toluate, (5) 1,2-dihydroxy-4-methylcyclohexa-3,5-diene-carboxylate, (6) 4-methylcatechol, (7) 2-hydroxy-5-methyl-6-oxohexa-2,4-dienoate, (8) 2-hydroxy-5-methylhexa-2,4-diene-1,6-dioate, (9) 5-methyl-2-oxohex-4-ene-1,6-dioate, (10) formate, (11) 2-hydroxy-*cis*-hex-2,4-dienate, (12) 4-hydroxy-2-oxohexanoate, (13) pyruvate, (14) propionaldehyde.

The lower part of the figure shows the organization of TOL genes and regulatory circuits controlling their transcription. TOL catabolic genes are organized into two operons. The upper pathway operon encodes catabolic enzymes which transform toluene/xylenes to benzoate/toluates (the function of the last gene of the operon, *xylN*, is not known). The *m*-cleavage pathway operon encodes enzymes which transform benzoate/toluates to pyruvate and acetaldehyde/propionaldehyde. Pu and Pm are promoters of the upper and *m*-cleavage pathway operons, respectively. Transcription from the Pu promoter is positively regulated by the XylR protein which is activated by toluene/xylenes and their alcohol derivatives. This induction requires the NtrA protein, a sigma factor. Transcription from the Pm promoter is positively regulated by the XylS protein when it is activated by benzoate/toluates. This induction is NtrA-independent. The expression of the *xylS* gene from its promoter, Ps, is also positively controlled by XylR when activated by toluene/xylenes. This induction also requires the NtrA sigma factor. When XylS is over-produced, it induces the *m*-cleavage pathway operon even in the absence of its effector, benzoate/toluates.

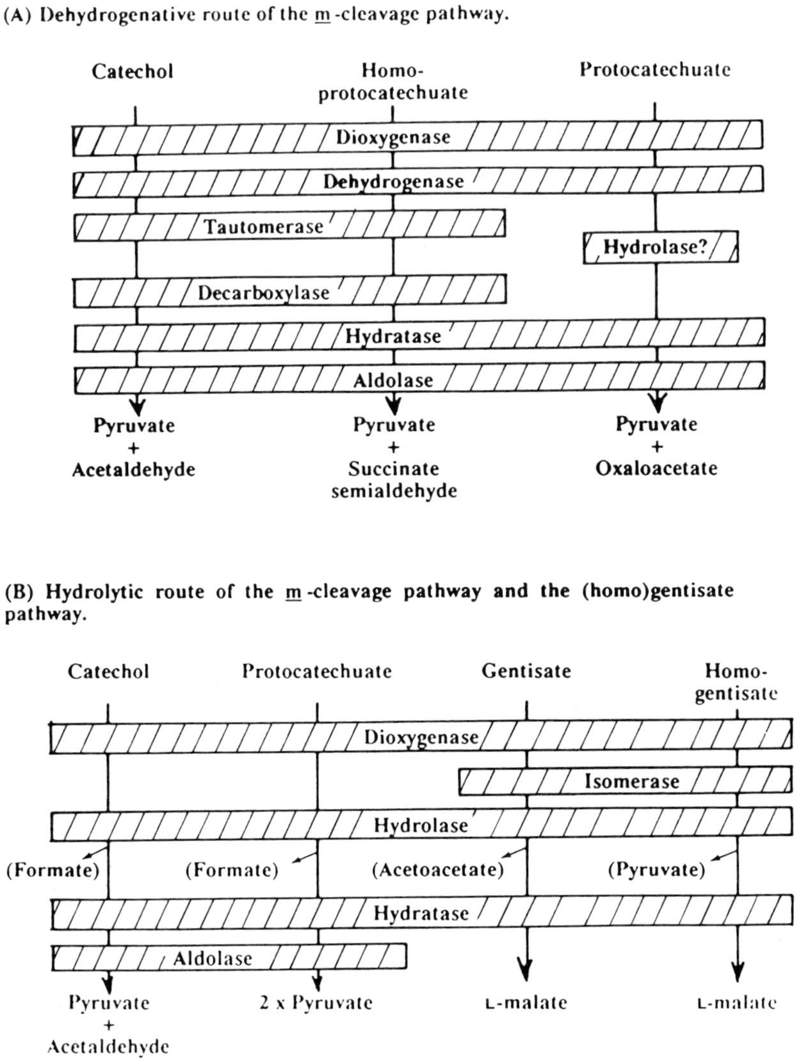

Figure 8. *m*-cleavage pathways and (homo)gentisate pathways. After the *m*-cleavage of catechol and (homo)protocatechuate by dioxygenases, the products are processed by either dehydrogenase (A) or hydrolase (B). The dehydrogenative route for the metabolism of catechol and homoprotocatechuate involves enzymatic steps effected by a dehydrogenase, a tautomerase, a decarboxylase, a hydratase and an aldolase. Protocatechuate is processed by similar enzymatic reactions in the dehydrogenative route, except that steps for tautomerization and decarboxylation are missing (37, 38). The hydrolytic route converts the ring-fission product directly to the substrate of hydratase. Pathways for degradation of gentisate and homogentisate are similar to that of the hydrolytic route of the *m*-cleavage pathway with respect to the order of enzymatic reactions, except that the ring-cleavage products undergo isomerization (37).

The β-ketoadipate pathway for the degradation of benzoate and protocatechuate (Fig. 6) is usually chromosomally encoded (19). This suggests that possession of this pathway may have been beneficial to certain bacteria over a long period of time. The probable primordial importance of the β-ketoadipate pathway is also supported by the fact that many soil bacteria show chemotactic responses either towards benzoate (20) or towards protocatechuate and β-ketoadipate (21).

In contrast to the normal β-ketoadipate pathway, modified β-ketoadipate pathways for mineralization of chlorocatechols (Fig. 6) are plasmid-encoded. Comparison of the sequences of genes of isofunctional chlorocatechol degradative enzymes in different bacteria isolated in Germany, the USA and Australia has shown that the catabolic genes are homologous, if not identical (22). This finding revealed the worldwide spread of the genes encoding this pathway and presumably also of the selective pressures for maintenance of these traits.

B. Catabolic Plasmids May Influence the Expression of Other Catabolic Gene Sets Present in the Same Cell

Many soil organisms possess two or more pathways for the degradation of a substrate, although in any given situation, only one of these pathways is usually induced. In *Acinetobacter calcoaceticus*, two parallel pathways for the transformation of β-ketoadipate to succinate and acetyl CoA exist, only one of which is normally induced because of the different induction specificities of the two pathways: protocatechuate induces one suite of enzymes whereas benzoate or *cis,cis*-muconate induces the other. *Pseudomonas putida* encodes a β-ketoadipate pathway for benzoate that is induced by *cis,cis*-muconate. If *P. putida* bacteria carry the TOL plasmid, which specifies a meta-cleavage pathway, the chromosomally-encoded β-ketoadipate pathway is not induced. This occurs because benzoate induces synthesis of the meta-cleavage pathway enzymes and is metabolized to catechol, which is then preferentially meta-cleaved by catechol 2,3-dioxygenase. Production of the inducer of the β-ketoadipate pathway, *cis,cis*-muconate formed by ortho-cleavage of catechol, is thereby prevented (4, 20).

C. Catabolic Genes are Usually Organized into Operons

Unlike the situation for *E. coli*, clustering is generally not observed for biosynthetic genes of *Pseudomonas*, although it is for catabolic genes (23). This difference in the organization of biosynthetic and catabolic genes in *Pseudomonas* is not yet clearly related to the difference in their function, regulation or evolution. Chromosomal genes encoding degradation of catechol (*cat*) and protocatechuate (*pca*) to succinate and acetyl CoA via β-ketoadipate are located in two separate clusters on the chromosomes of *Pseudomonas aeruginosa*, *P. putida* (19) and *A. calcoaceticus* (24), and each cluster is organized in two or three operons (24). Structural genes for oxidation of benzoate to catechol (*ben*) are also genetically linked to the *cat* operon (19, 25). Biochemical and genetic analyses of the β-ketoadipate pathway enzymes of *P. putida* and *A. calcoaceticus* indicate evolutionary relatedness of the two pathways (9). The nucleotide sequences of *catB*, the gene for the *cis,cis*-muconate lactonizing enzyme, and *clcB*, the gene for the analogous enzyme in a chlorocatechol modified ortho-cleavage pathway, also showed significant homology (26).

Many plasmid-borne catabolic genes are also known to be organized in operons. One operon usually encodes one functional biochemical unit; for example, in the case of the TOL plasmid, genes specifying the oxidation of toluene to benzoate are organized into one operon (the "upper" operon) whereas those encoding the transformation of benzoate to pyruvate and acetaldehyde via the meta-cleavage of catechol are organized into a second operon (the "lower" or meta-cleavage operon: Fig. 8).

In independently isolated plasmids encoding toluene degradation, the order of the catabolic genes within each operon is identical and some restriction sites within the operons are conserved. However, the relative orientation of the two operons and the size of the intervening DNA between the operons vary (27). NAH plasmid catabolic genes specifying the oxidative degradation of naphthalene are also organized into two operons, one specifying the transformation of naphthalene to salicylate (*nah* operon), and a second specifying the transformation of salicylate to catechol and thence to pyruvate and acetaldehyde (*sal* operon; 28). Genes specifying the degradation of catechol to pyruvate and acetaldehyde have the same order in the TOL and NAH plasmids, and there is strong homology between the DNA sequences of the isofunctional enzymes. The structural organization of the TOL and NAH catabolic operons is thus consistent with an evolution of complex pathways through the combination of discrete genetic modules that encode functional biochemical units (29).

D. Regulation of Expression of TOL Plasmid Catabolic Genes involves Specific Positive Regulatory Proteins and a Minor Form of RNA Polymerase Holoenzyme

Regulation of the TOL and NAH plasmid-encoded catabolic pathways is accomplished primarily if not exclusively through modulation of the rates of initiation of transcription of the catabolic operons. Regulation in both instances is positive and the substrates of the pathways are also the inducers of transcription (3).

In the case of the TOL plasmid, the substrates of the upper pathway, toluene/xylenes and (methyl)benzylalcohol, induce transcription of both the upper and the lower operons, whereas substrates of the lower operon, benzoate/toluates, induce transcription of only the lower operon (30, 31; Fig. 8). Two regulatory proteins, XylR and XylS, activate transcription from the operon promoters. The XylS protein, which interacts with benzoate and benzoate analogues to activate transcription from the lower operon promoter, Pm, is present in small amounts in TOL plasmid-carrying bacteria, reflecting low level constitutive expression of its gene from promoter Ps. The XylR protein, which interacts with toluene and benzylalcohol and their analogues to activate transcription from the upper operon promoter, Pu, also autoregulates its own synthesis from promoter Pr. The XylR protein, in the presence of a substrate-effector for the upper pathway, also stimulates expression of the lower operon indirectly, by stimulating a 25 times increase in transcription of *xylS*. Induced XylS protein thereby reaches intracellular levels that cause it to activate transcription of the lower operon in the absence of benzoate or benzoate analogues (Fig. 8).

Transcription from Pm and Pr is mediated by conventional RNA polymerase (Chapter 2), whereas XylR-regulated transcription from Pu and Ps requires the *ntrA* (*rpoN*) product, a sigma factor specific for nitrogen-regulated promoters (Chapters 2 and 6). Inspection of the *rpoN*-dependent promoters Pu and Ps has revealed that their −12 and −24 regions correspond to consensus sequences which are recognized by the *rpoN* sigma factor (30, 32). In fact two other factors involved in the transcription of nitrogen-regulated promoters, NtrC and NifA from *Klebsiella pneumoniae*, (Chapter 6) stimulate transcription from Pu and are able to substitute for XylR (32). These findings indicate that the control of energy metabolism involving aromatic hydrocarbons and that of nitrogen utilization are in some way integrated. Several domains of NifA share homology with those of NtrC (33) and the DNA sequences of C-terminal regions of *ntrC* and *xylR* are homologous (T. Köhler, unpublished). However, there is no significant similarity in DNA sequences between Pu/Ps and *nif* promoters in the putative XylR/NtrC/NifA binding sites upstream of −12 and −24 regions.

IV. UTILITY OF DETERMINANTS OF CATABOLIC PATHWAYS

Several determinants of catabolic plasmids have been found to be useful in the development of recombinant genetic methods. The lower operon promoter plus the *xylS* regulator gene have been incorporated into a broad-host-range vector for the regulated expression of cloned genes in Gram-negative bacteria (4). The TOL plasmid *xylE* gene, whose product transforms catechol to the yellow compound 2-hydroxymuconic semialdehyde, is used as an indicator gene in bacterial and eukaryotic cells. The NAH plasmid *nahA* gene-encoded naphthalene 1,2-dioxygenase and the TOL plasmid *xylAM* gene-encoded xylene oxygenase may also prove useful as indicator genes because of their production of indigo, a water-insoluble blue dye, from indole (4, 34).

V. LABORATORY EVOLUTION OF AROMATIC CATABOLIC PATHWAYS

Recently, there has been considerable interest in the prospect of exploiting the catabolic versatility of pseudomonads for controlling environmental pollutants. This has involved the development of new metabolic capabilities using three basic approaches: chemostat selection for the evolution of organisms able to use a novel substrate; *in vivo* genetic transfers, in which genes for critical enzymes of one organism are recruited into a pathway of another organism through experiments involving natural genetic processes, such as transduction, transformation and especially conjugation; and *in vitro* evolution, in which cloned and well-characterized genes are selectively transferred into a different organism. The *in vitro* approach has been particularly stimulated by the advances in our knowledge of the molecular genetics of catabolic pathways.

A. Expansion of the Substrate Range of a Pathway by Mutation

P. putida containing TOL plasmid pWW0 cannot degrade 4-ethylbenzoate (4EB). Part of the reason for this is that 4EB does not activate the regulatory XylS protein, and therefore does not induce synthesis of the meta catabolic enzymes (Fig. 9(a)). A procedure was therefore developed for selection of XylS regulator protein mutants which are activated by 4EB. One such mutant *xylS* gene was then transferred into *P. putida* bacteria carrying the TOL plasmid. This derivative also failed to grow on 4EB but did degrade the compound to 4-ethylcatechol. Thus, the catabolic enzymes are synthesized in this derivative and 4EB is transformed to 4-ethylcatechol, but the meta-cleavage enzyme does not permit further metabolism (Fig. 9(b)). Characterization of the meta-cleavage enzyme, catechol 2,3-dioxygenase, revealed that 4-ethylcatechol is in fact a "suicide" substrate which causes irreversible inactivation of the enzyme (35; Fig. 9(b)). However, after mutagenesis, it was possible to select, in large populations of *P. putida* carrying the modified *xylS* gene and the TOL plasmid, mutants that could grow on 4EB as sole carbon source, as a result of change in the *xylE* gene for catechol 2,3-dioxygenase (36).

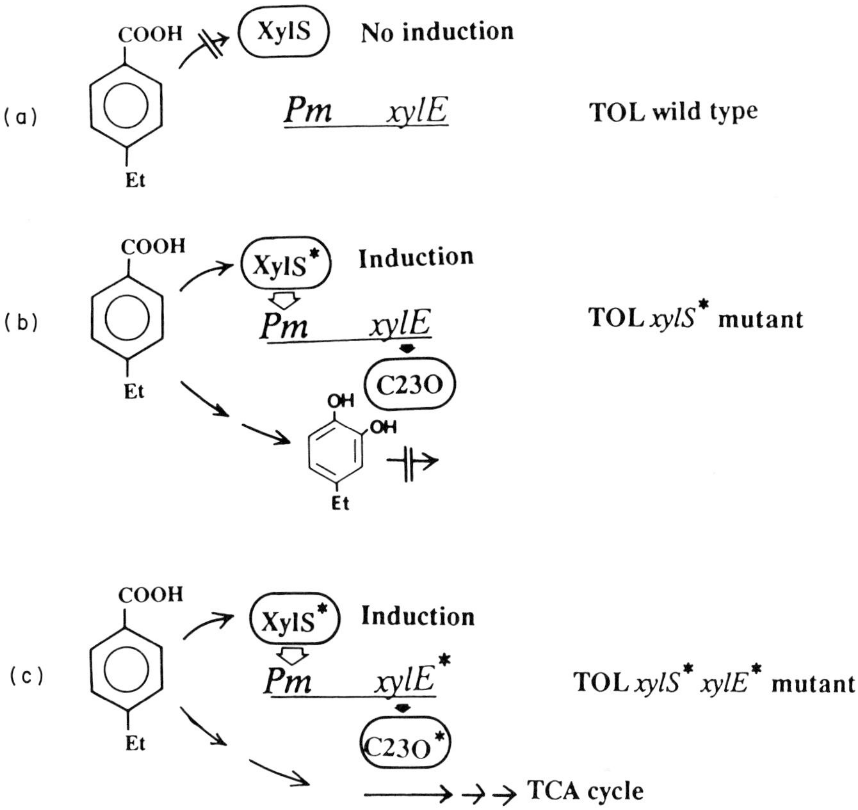

Figure 9. Stepwise improvements of the TOL plasmid-encoded *m*-cleavage pathway to permit mineralization of 4-ethylbenzoate. (a): The TOL-encoded *m*-cleavage pathway cannot metabolize 4-ethylbenzoate (4EB) because this compound does not activate the positive regulatory protein, XylS, and the TOL genes are, therefore, not expressed. (b): A *xylS* mutant, *xylS**, of pWW0, whose product is activated by 4EB, was isolated. All catabolic enzymes are induced in this mutant in response to 4EB, and 4EB is metabolized to 4-ethylcatechol. However, catechol 2,3-dioxygenase (C23O) cannot attack 4-ethylcatechol because this compound is a suicide inhibitor of C23O. (c): A mutant of the structural gene of C23O (*xylE**) was isolated which synthesizes a modified C23O enzyme capable of oxidizing 4-ethylcatechol. A bacterium containing the TOL plasmid carrying both mutations, namely *xylS** and *xylE**, can degrade 4EB completely through the *m*-cleavage pathway.

B. Designing New Catabolic Pathways: Patchwork Assembly of Enzymes and Regulatory Systems

Pseudomonas sp. B13 can degrade 3-chlorobenzoate (3CB) via the modified ortho-cleavage pathway in Fig. 6. However, it cannot degrade 4-chlorobenzoate (4CB) because of the narrow substrate specificity of the first enzyme of the pathway, benzoate 1,2-dioxygenase (Fig. 10(a)). The TOL plasmid genes for toluate 1,2-dioxygenase which can oxidize not only alkylbenzoates but also 3CB and 4CB were cloned into B13. This construct grew on 3CB and 4CB but not on 4-methylbenzoate (4MB) because the metabolism of 4-methylcatechol via the modified ortho-cleavage pathway leads to the formation of the deadend product, 4-methyl-2-enelactone (Fig. 10(b)). Industrial wastes frequently contain mixtures of chloro- and methylbenzoates and in such mixtures this construct cannot grow because of the formation of the deadend product which perturbs the productive metabolism of chlorobenzoates. The gene for the *Alcaligenes* 4-methyl-2-enelactone isomerase, which converts 4-methyl-2-enelactone to 3-methyl-2-enelactone, was then cloned into this strain. This new derivative of B13 could degrade 4MB completely, because 4-methyl-2-enelactone produced from 4MB was isomerized to 3-methyl-2-enelactone which was then utilized by the modified ortho-pathway of B13 (Fig. 10(c)). This derivative grew on mixtures of 3CB + 4MB and 4CB + 4MB and degraded simultaneously the chloro- and methylbenzoates. This pathway was further expanded through mutational activation of the previously cryptic phenol hydroxylase of B13 which allowed metabolism of 4-methylphenol exclusively via ortho-cleavage (10; Fig. 10(d)).

VI. CONCLUDING REMARKS

The metabolic versatility of the pseudomonads may be a reflection of the structural diversity of potentially energy-yielding substrates in the environment which have led to the evolution of an extremely wide range of diverse degradative activities among soil and water microorganisms. Many catabolic pathways seem to have evolved by the assembly on conjugative plasmids of discrete genetic modules that specify functional units of biochemical routes, and by mutational change of specificities of component enzymes and regulators of gene expression. The advantage of the evolution of new genetic determinants on conjugative, broad host-range plasmids is clear: their transmission through many different hosts provides access to an enormous gene pool and allows the acquisition of pathway components in minimal evolutionary time.

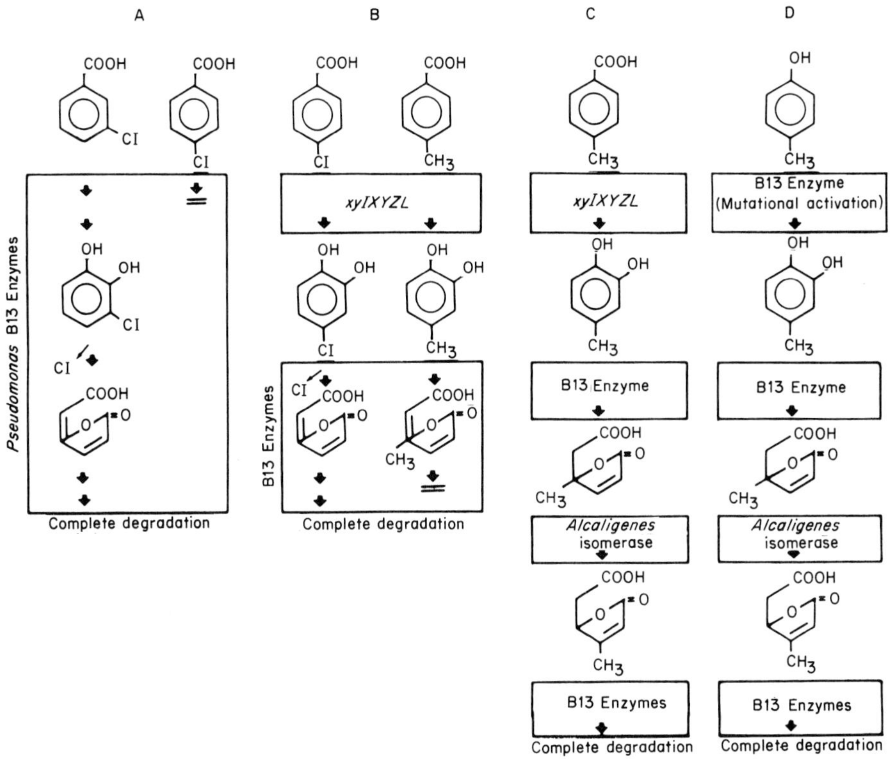

Figure 10. Enzyme recruitment to expand the substrate range of the 3-chlorobenzoate degrader. (A) *Pseudomonas* sp. B13 can degrade 3-chlorobenzoate via the modified *o*-pathway. It cannot degrade 4-chlorobenzoate (4CB) as a result of the narrow substrate-specificity of the first enzyme of the pathway, benzoate 1,2-dioxygenase. (B) Introduction of the TOL genes *xylXYZL*, which encode broader substrate specificity enzymes for transformation of chloro- and methylbenzoates to chloro- and methyl-catechols, together with *xylS*, which encodes the positive regulator of the *xylXYZL* operon, expands the degradation range of B13 to include 4CB. 4-methylbenzoate is metabolized to 4-methyl-2-enelactone which is accumulated as a deadend product. (C) Recruitment of a 4-methyl-2-enelactone isomerase from *Alcaligenes* sp. allows transformation of 4-methyl-2-enelactone to 3-methyl-2-ene-lactone which is degraded by other enzymes of B13. (D) A spontaneous mutation allows expression of the cryptic phenol hydroxylase of B13.

Since many substrates will only rarely appear in any given ecosystem, the expression of catabolic genes is carefully regulated to economize materials and energy. Such control is carried out by specific regulatory systems which are sometimes superimposed on the global regulatory systems.

The evolution of new pathways for novel compounds introduced into the environment through the industrial activities of man is facilitated by another important characteristic of some catabolic enzymes, namely their relaxed substrate specificities. Thus, many enzymes able to attack novel compounds already exist and the evolution of efficient pathways may in some cases require only minor changes in substrate affinities or kinetic properties of existing enzymes. In terms of laboratory evolution of new pathways, a critical part of the strategy will be the construction of well-characterized genetic modules for functional parts of pathways, in particular parts of pathways exhibiting relaxed substrate specificities. The development of a range of such versatile modules will then enable rational mixing and matching to generate rapidly the desired biochemical routes that would otherwise arise naturally by a similar though much slower process.

Acknowledgements

We thank Dr. Barbara Angus for careful reading of the manuscript and Françoise Rey for typing it.

References

(1) Gibson, D. T. and Subramanian, V. (1984). Microbial degradation of aromatic hydrocarbons. *In* "Microbial Degradation of Organic Compounds" (D. T. Gibson, ed.), pp. 181–252. Dekker, New York.

(2) Simon, R., Priefer, U. and Pühler, A. (1983). A broad host range mobilization system for *in vivo* genetic engineering: transposon mutagenesis in Gram-negative bacteria. *Bio/Technology* **1**, 784–790.

(3) Nakazawa, T. and Inouye, S. (1986). Cloning of *Pseudomonas* genes in *Escherichia coli*. *In* "The Bacteria" (J. R. Sokatch and L. N. Ornston, eds.), Vol. 10, pp. 357–382. Academic Press, New York.

(4) Mormod, N., Lehrbach, P. R, Don, R. H. and Timmis, K. N. (1986). Gene cloning and manipulation in *Pseudomonas*. *In* "The Bacteria" (L. N. Ornston and J. R. Sokatch, eds.), Vol. 6, pp. 325–355. Academic Press, New York.

(5) Dagley, S. (1978). Pathway for the utilization of organic growth substrates. *In* "The Bacteria" (L. N. Ornston and J. R. Sokatch, eds.), Vol. 6, pp. 305–388. Academic Press, New York.

(6) Dagley, S. (1986). Biochemistry of aromatic hydrocarbon degradation in pseudomonads. *In* "The Bacteria" (J. R. Sokatch and L. N. Ornston, eds.), Vol. 10, pp. 527–555. Academic Press, New York.

(7) Nozaki, M. (1979). Oxygenases and dioxygenases. *Topics in Curr. Chem.* **78**, 145–186.
(8) Que, L., Lauffer, R. B., Lynch, J. B., Murch, B. P. and Pyrz, J. W. (1987). Elucidation of the coordination chemistry of the enzyme-substrate complex of catechol 1,2-dioxygenase by NMR spectroscopy. *J. Am. Chem. Soc.* **109**, 5381–5385.
(9) Ornston, L. N. and Yeh, W.-K. (1982). Recurring themes and repeated sequences in metabolic evolution. *In* "Biodegradation and Detoxification of Environmental Pollutants" (A. M. Chakrabarty, ed.), pp. 105–126. CRC Press Inc., Boca Raton (Florida).
(10) Rojo, F., Pieper, D. H., Engesser, K.-H., Knackmuss, H.-J. and Timmis, K. N. (1987). Assemblage of ortho-cleavage route for simultaneous degradation of chloro- and methylaromatics. *Science* **238**, 1395–1398.
(11) Reineke, W. (1984). Microbial degradation of halogenated aromatic compounds. *In* "Microbial Degradation of Organic Compounds" (D. T. Gibson, ed.), pp. 319–360. Dekker, New York.
(12) Ribbons, D. W. and Eaton, R. W. (1982). Chemical transformation of aromatic hydrocarbons that support the growth of microorganisms. *In* "Biodegradation and Detoxification of Environmental Pollutants" (A. M. Chakrabarty, ed.), CRC Press Inc., Boca Raton (Florida).
(13) Bayly, R. C. and Barbour, G. (1984). The degradation of aromatic compounds by the meta- and gentisate pathways: biochemistry and regulation. *In* "Microbial Degradation of Organic Compounds" (D. T. Gibson, ed.), pp. 253–294. Dekker, New York.
(14) Harayama, S., Mermod, N., Rekik, M., Lehrbach, P. R. and Timmis, K. N. (1987). Roles of the divergent branches of the meta-cleavage pathway in the degradation of benzoate and substituted benzoates. *J. Bacteriol.* **169**, 558–564.
(15) Frantz, B. and Chakrabarty, A. M. (1986). Degradative plasmids in *Pseudomonas*. *In* "The Bacteria" (J. R. Sokatch and L. N. Ornston, eds.), Vol. 10, pp. 295–323. Academic Press, New York.
(16) Haas, D. (1983). Genetic aspects of biodegradation by pseudomonads. *Experientia* **39**, 1199–1213.
(17) Lessie, T. G. and Gaffney, T. (1986). Catabolic potential of *Pseudomonas cepacia*. *In* "The Bacteria" (J. R. Sokatch and L. N. Ornston, eds.), Vol. 10, pp. 439–481. Academic Press, New York.
(18) Tsuda, M. and Iino, T. (1987). Genetic analysis of a transposon carrying toluene degrading genes on a TOL plasmid p WWϕ. *Mol. Gen. Genet.* **210**, 270–276.
(19) Holloway, B. W. and Morgan, A. F. (1986). Genome organization in *Pseudomonas*. *Ann. Rev. Microbiol.* **40**, 79–105.
(20) Harwood, C. S. and Ornston, L. N. (1984). TOL plasmid can prevent induction of chemotactic responses to aromatic acids. *J. Bacteriol.* **160**, 251–255.
(21) Parke, D., Ornston, L. N. and Nester, E. W. (1987). Chemotaxis to plant phenolic inducers of virulence genes is constitutively expressed in the absence of the Ti plasmid in *Agrobacterium tumefaciens*. *J. Bacteriol.* **169**, 5336–5338.
(22) Ghosal, D., You, I.-S., Chatterjee, D. K. and Chakrabarty, A. M. (1985). Genes specifying degradation of 3-chlorobenzoic acid in plasmids pAC27 and pJP4. *Proc. Natl. Acad. Sci. USA* **82**, 1638–1642.
(23) Holloway, B. W. (1986). Chromosome mobilization and genomic organization in *Pseudomonas*. *In* "The Bacteria" (J. R. Sokatch and L. N. Ornston, eds.), Vol. 10, pp. 217–249. Academic Press, New York.
(24) Doten, R. C., Michell, D. T., Ngai, K.-L. and Ornston, L. N. (1987). Cloning and genetic organization of the *pca* gene cluster from *Acinetobacter calcoaceticus*. *J. Bacteriol.* **169**, 3168–3174.
(25) Neidle, E. L., Shapiro, M. K. and Ornston, L. N. (1987). Cloning and expression in *Escherichia coli* of *Acinetobacter calcoaceticus* genes for benzoate degradation. *J. Bacteriol.* **169**, 5496–5503.

(26) Frantz, B. and Chakrabarty, A. M. (1987). Organization and nucleotide sequence determination of a gene cluster involved in 3-chlorocatechol degradation. *Proc. Natl. Acad. Sci. USA* **84**, 4460–4464.
(27) Keil, H., Saint, C. M. and Williams, P. A. (1987). Gene organization of the first catabolic operon of TOL plasmid pWW53: production of indigo by the *xylA* gene product. *J. Bacteriol.* **169**, 764–770.
(28) Yen, K.-M. and Gunsalus, I. C. (1982). Plasmid gene organization: naphthalene/salicylate oxidation. *Proc. Natl. Acad. Sci. USA* **79**, 874–878.
(29) Harayama, S., Rekik, M., Wasserfallen, A. and Bairoch, A. (1987). Evolutionary relationships between catabolic pathways for aromatics: conservation of gene order and nucleotide sequences of catechol oxidation genes of pWW0 and NAH7 plasmids. *Mol. Gen. Genet.* **210**, 241–247.
(30) Inouye, S., Nakazawa, A. and Nakazawa, T. (1987). Expression of the regulatory gene *xylS* on the TOL plasmid is positively controlled by the *xylR* gene product. *Proc. Natl. Acad. Sci. USA* **84**, 5182–5186.
(31) Ramos, J. L., Mermod, N. and Timmis, K. N. (1987). Regulatory circuits controlling transcription of TOL plasmid operon encoding meta-cleavage pathway for degradation of alkylbenzoates by *Pseudomonas*. *Mol. Microbiol.* **1**, 293–300.
(32) Dixon, R. (1986). The *xylABC* promoter from the *Pseudomonas putida* TOL plasmid is activated by nitrogen regulatory genes in *Escherichia coli*. *Mol. Gen. Genet.* **203**, 129–136.
(33) Drummond, M., Whitty, P. and Wootton, J. (1986). Sequence and domain relationships of *ntrC* and *nifA* from *Klebsiella pneumoniae*: homologies to other regulatory proteins. *EMBO J.* **5**, 441–447.
(34) Harayama, S. and Don, R. H. (1985). Catabolic plasmids: their analysis and utilization in the manipulation of bacterial metabolic activities. *In* "Genetic Engineering" (J. K. Setlow and A. Hollander, eds.), Vol. 8, pp. 283–307. Plenum, New York.
(35) Ramos, J. L., Stolz, A., Reineke, W. and Timmis, K. N. (1986). Altered effector specificities in regulators of gene expression: TOL plasmid *xylS* mutants and their use to engineer expansion of the range of aromatics degraded by bacteria. *Proc. Natl. Acad. Sci. USA* **83**, 8467–8471.
(36) Ramos, J. L., Wasserfallen, A., Rose, K. and Timmis, K. N. (1987). Redesigning metabolic routes: manipulation of TOL plasmid pathway for catabolism of alkylbenzoates. *Science* **235**, 593–596.
(37) Ribbons, D. W., Keyser, P., Eaton, R. W., Anderson, B. N., Kunz, D. A. and Taylor, B. F. (1984). Microbial degradation of phthalates. *In* "Microbial Degradation of Organic Compounds" (D. T. Gibson, ed.), pp. 371–397. Dekker, New York.
(38) Maruyama, K. (1983). Enzymes responsible for degradation of 4-oxalmesaconic acid in *Pseudomonas ochraceae*. *J. Biochem.* **93**, 567–574.

Chapter **9**

Mercury Resistance in Bacteria

N. L. BROWN, P. A. LUND *and* N. NI BHRIAIN

I. INTRODUCTION

Mercury is highly toxic to all living organisms and, unlike many toxic metals, such as copper and nickel, which are actually required in small amounts, mercury has no beneficial functions. At high concentrations mercury is biocidal, while lower concentrations are mutagenic.

Mercury is a widely distributed and mobile element in the environment. Its distribution results from both natural processes, such as the degassing of the earth's crust at volcanic vents, and human activities, such as the mining of cinnabar (mercuric sulphide) ores and the burning of fossil fuels. Together these processes release about 80 000 tonnes of mercury to the biosphere every year. Once released, the mercury becomes part of a biological cycle in which microorganisms transform it to and from organomercurial derivatives, and to elemental mercury. The reservoir of available mercury in the biosphere is estimated at about 2×10^8 tonnes.

Much of the interest in the mechanisms that distribute mercury in the biosphere is caused by the local appearance of large amounts of mercury and its derivatives in the food chain through industrial pollution, such as that at Minemata Bay, Japan, which resulted in 111 cases of serious neurological disorder or death in the 1950s. Many of the transformations which cause mercury to enter the food chain are undertaken by bacteria as part of their mechanisms of resistance. The biochemistry of mercury resistance is now beginning to be understood, largely through the application of molecular genetics.

II. BACTERIAL TRANSFORMATIONS OF MERCURY

A. Mercury has Peculiar Chemical Properties

The toxicity and environmental mobility of mercury are functions of its peculiar chemical properties. The most obvious special attribute of the element mercury is that it is a metal that is liquid at standard temperatures and pressures. It is also volatile. Mercury can exist in three valence states: elemental mercury, Hg(0); or the two oxidized forms (Hg(I) and Hg(II), as ionic salts (Hg_2^{2+} and Hg^{2+}), partially covalent salts or as organometallic derivatives with alkyl and aryl groups. Mercuric (i.e. Hg(II)) compounds are the most common form of mercury.

The oxidized forms are highly reactive with biological systems, having a very high affinity for thiol groups in proteins (1). This is the major chemical basis of mercury toxicity, although mercuric salts also react with nucleotides and lipids. Elemental mercury is much less reactive, but is readily oxidized to the higher valence states. Organomercurials (such as dimethylmercury) and partially covalent mercuric salts (such as mercuric chloride) can readily diffuse through biological membranes, thus increasing their effective toxicity.

B. Mercury Resistance is Very Common and is of Two Types

Mercury-resistant bacteria are common in environments where concentrations of mercury are high, because of the presence of cinnabar ores or industrial pollution (2, 3). Resistance is also common in clinical isolates, because of the widespread use of mercury compounds as topical disinfectants. Bacteria in which mercury resistance has been documented include *Alcaligenes* spp, *Bacillus* spp, *Clostridium* spp, *Escherichia coli*, *Klebsiella pneumoniae*, *Mycobacterium scrofulaceum*, *Pseudomonas* spp, *Serratia marcescens*, *Staphylococcus aureus*, *Streptococcus* spp, *Streptomyces* spp, *Thiobacillus ferrooxidans* and *Yersinia enterocolitica*, as well as many isolates that have not been fully characterized. Thus, resistance is seen in Gram-positive and Gram-negative genera, including obligate aerobes, anaerobes and facultative organisms (2, 3). Such widespread resistance is not surprising because this bactericidal agent has been present in high local concentrations since primordial times.

All mercury-resistant bacteria tolerate inorganic mercuric salts; but they differ in their resistance to organomercurials. Those showing "narrow-spectrum" resistance are resistant to a limited number of these, usually merbromin and fluorescein mercuric acetate; whereas bacteria showing "broad-spectrum" resistance can tolerate a wide range of organomercurials.

C. Bacterial Mercury Resistance is Usually Due to Inducible Enzymatic Detoxification

The best characterized mechanisms of resistance to Hg^{2+} and to organomercurial compounds are enzymatic. Enzymatic detoxification of mercuric ions is due to the cytoplasmic enzyme mercuric reductase (MR), which catalyses the reaction:

$$Hg(II) \rightarrow Hg(0)$$

The elemental mercury produced is essentially non-toxic and is volatile; thus it is easily lost from the immediate environment of the bacterial cell.

In broad-spectrum resistance, another enzyme, organomercury lyase, is responsible for cleaving the C—Hg bond in most organomercurial compounds, to liberate Hg^{2+}, which is then reduced by mercuric reductase. In contrast, the biochemical basis of organomercurial resistance conferred by narrow-spectrum determinants is not understood. In most cases the compounds are not degraded, and resistance may be due to permeability barriers to the organomercurials. There is a host-specific contribution to the patterns of narrow-spectrum resistance to organomercurials. Thus, some plasmids that confer resistance to *p*-hydroxymercurobenzoate in *Pseudomonas aeruginosa* fail to do so in *E. coli* (4).

The expression of mercury or organomercurial resistance can be induced in most cases by preincubation of the bacterial strain with subtoxic levels of mercury. Uninduced strains show a long lag phase of arrested growth when challenged with mercury, after which growth resumes normally, whereas preinduced cultures show no such lag. In a minority of cases, resistance is constitutive (for example, the constitutive MR of *T. ferrooxidans*, and the constitutive production of volatile organomercurials in *Bacteroides rumenicola* and *Clostridium perfringens*).

D. Other Resistance Mechanisms also Exist

In addition to the enzymatic reduction of Hg(II) to Hg(0), several other mechanisms of resistance to mercury have evolved (2, 3, 4). These are less widespread and less well studied than the reductive detoxification system. One such example, found in many bacteria, is the methylation of Hg(II). Methylation requires methylcobalamin as the methyl donor. Volatilization of the methylated ion from the environment of the cell is probably an important factor in the resistance mechanism in methylating strains. Another mechanism is the production of mercuric sulphide. Many bacteria produce hydrogen sulphide, in the presence of which mercury is converted to a precipitate of insoluble and biologically-inactive HgS. This is thought to function as a resistance mechanism in one strain of *Clostridium cochlearium* (5), and may be an important mechanism in sediments and other anaerobic environments. There has been speculation that permeability barriers to Hg^{2+} may also exist, limiting the access of the toxic ion to sensitive intracellular targets. The data supporting this are inconclusive; non-volatilizing resistance in *Enterobacter aerogenes* is associated with changes in the membrane protein composition and decreased Hg^{2+} binding (6).

Mercury resistance in eukaryotes, including microorganisms, is not enzymatic, but is due to the production of a class of inducible proteins, the metallothioneins (7), which sequester mercury by stoichiometric binding. No similar stoichiometric mechanism has been described for mercury resistance in prokaryotes, although metallothionein-like proteins conferring cadmium, zinc and copper resistance have been detected in *Pseudomonas putida* (8).

III. MERCURY RESISTANCE GENES

A. Bacterial Mercury Resistance is Usually Plasmid-Encoded and can be Transposable

In most cases studied, the genes encoding volatilizing mercuric resistance are extrachromosomal (3). The mercury resistance genes in Gram-negative bacteria frequently occur on transposable genetic elements. Some transposons, such as Tn*501* and Tn*2613*, contain only mercury resistance and transposition functions, whereas others, such as Tn*21* or Tn*2603*, carry resistances to several antimicrobial agents (9). Only one broad-spectrum resistance has so far been shown to be carried on a transposon (Tn*3401* from *Pseudomonas fluorescens*; 10).

Transposition of mercury resistance genes has not been shown in Gram-positive organisms.

B. Gram-Negative Narrow-Spectrum Resistance Determinants are the Best Studied

Two Gram-negative narrow-spectrum mercury resistance determinants have provided most of the information on which current understanding of the biochemistry, regulation and expression of mercury resistance are based (11, 12, 13). These are the determinants of the multiple drug resistance plasmid R100, which was originally isolated in *Shigella* (14), and the determinant of transposon Tn*501*, originally isolated on plasmid pVS1 from *P. aeruginosa* (15). Both determinants have been studied in *E. coli*.

Genetic analysis and nucleotide sequence determination have allowed the mercury resistance genes of Tn*501* and R100 to be identified. The genetic organization of Tn*501* is shown in Fig. 1. The *merR* gene encodes a *trans*-acting regulator responsible for the induction of expression of the *mer* operon. The

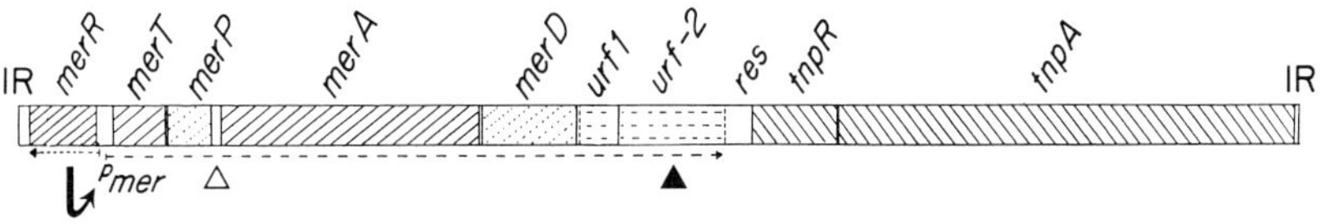

Figure 1. Physico-genetic map of transposon Tn*501*. This is a genetic map established from the DNA sequence, rather than from recombination data. The triangles show differences from the R100 mercury resistance determinant: the open triangle marks the position of the *merC* gene in R100, and the closed triangle marks the 11.2 kb insert in R100 which carries sulphonamide and spectinomycin resistance. The *res*, *tnpR* and *tnpA* genes are transposition functions. The remaining genes are discussed in the text.

operon contains the structural genes for mercuric reductase (*merA*) and mercury transport (*merT* and *merP*), as well as a gene (*merD*) of unknown function. In the R100 determinant, there is an additional gene, *merC*, which may be involved in mercury uptake.

There is remarkable homology between the predicted structures of the gene products of Tn*501* and R100. The properties of these products have allowed a model for the mechanism of mercury resistance to be formulated (Section V). The open reading frames *urf-1* and *urf-2* (Fig. 1) are present in both Tn*501* and R100, and show good homology, but they have not been shown to be involved in mercury resistance. *urf-2* is interrupted by an 11.2 kb sequence containing the sulphonamide and spectinomycin resistance genes in R100.

Mutants in *merA* are more sensitive to Hg^{2+} than are plasmid-free cells. Typically such hypersensitive mutants in *E. coli* are killed by concentrations of mercury as low as 5 μM, whereas plasmid-free strains are resistant to 15 μM Hg^{2+}, and induced resistance strains can tolerate 70 μM Hg^{2+}. Hypersensitivity is presumably due to the cells taking up mercury, but being unable to detoxify it. Hypersensitive mutants bind more mercury following induction with subtoxic amounts of Hg^{2+} than do control strains.

IV. THE GRAM-NEGATIVE STRUCTURAL GENES AND THEIR PRODUCTS

A. The Proteins Encoded by the Mercuric Resistance Determinants

Experiments to identify the mercury-inducible gene products of the R100 and Tn*501* systems in normal resistant cells have proved difficult. Initially, the only mercury-inducible protein that could be identified unequivocally was MR (the MerA protein). In *P. aeruginosa* carrying Tn*501*, MR can constitute 6% of the soluble protein of the cell. The properties of the other gene products were initially predicted from the sequences of their genes. The proteins are very hydrophobic (MerT and MerC), small and processed (MerP) or probably produced in very small amounts (MerD). The MerT, MerP and MerC proteins have been shown to be membrane-associated (13, 16, 17).

Because of the very high affinity of mercury for thiol groups, cysteine residues are likely to be important in the function of the mercury resistance proteins. Pairs of cysteine residues are indeed found in those proteins essential to mercury detoxification.

B. Mercuric Reductase (MR)

The best studied protein of the narrow-spectrum mercury resistance determinants is MR. This is a member of the same class of enzymes as the pyridine nucleotide-disulphide oxidoreductases, lipoamide dehydrogenase (LD) and glutathione reductase (GR). The Tn*501* enzyme has been studied by enzyme kinetics and by protein chemistry (18, 19), and kinetic analysis of the R100 enzyme has also been undertaken (20). DNA sequence analysis of both these *merA* genes (21, 22) and the kinetic data on the enzymes show that they are very similar in structure and mechanism. The MR enzymes are dimers of identical subunits, each of about M_r 59 000, containing one bound FAD per subunit, and they catalyse NADPH-dependent reduction of mercuric salts to elemental mercury in the presence of exogenous thiols. Detailed spectral analysis and protein chemistry have shown that the mechanism and the active site of MR are similar to those of GR. Exogenous thiols are required, and the likely reaction catalysed *in vitro* by MR is:

$$RS{-}Hg(II){-}SR + NADPH + H^+ \Leftrightarrow Hg(0) + NADP^+ + 2RSH$$

MR is highly specific for mercuric and mercurous ions, and no significant reaction with other metal ions is found.

The primary sequences of the MRs of Tn*501* and R100, predicted from the DNA sequences, show 86% identity of amino acid residues and a large number of conservative substitutions. There are also strong homologies with the FAD and NADPH binding sites of GR and LD; and a redox-active cysteine pair (Cys-135, Cys-140) in a highly conserved sequence present in all the disulphide oxidoreductases is also present in MR. The crystallographic structure of human erythrocyte glutathione reductase has been determined at 0.2 nm resolution (23). The primary sequence homology between MR and GR suggests that MR will have a similar three-dimensional structure to GR. Differences between the Tn*501* and R100 MRs are exclusively in regions predicted to be on the surface of the enzyme. This structural and mechanistic homology has been particularly important in developing the model for mercuric ion resistance in Gram-negative bacteria. If MR and GR are aligned for homology (Fig. 2), there are two regions in which the homology is poor. One is the C-terminal region, which in GR is involved in substrate binding, and in MR contains a pair of cysteine residues (Cys-557 and Cys-558 in the Tn*501* enzyme) accessible to solvent. It was suggested (21) that these cysteines may be involved in binding Hg(II), and site-directed mutagenesis experiments indicate that they are indeed essential for MR activity (24).

The second difference between MR and GR is that there is a potential extra domain at the N-terminus of MR (Fig. 2). The first 85 amino acids of MR can be removed by proteolysis, with little effect on the catalytic activity of the enzyme *in vitro* (19). This region also contains a pair of cysteine residues (Cys-10 and Cys-13). This N-terminal region, and particularly the cysteine residues, is conserved in all MRs so far sequenced, and it presumably has a specific function *in vivo*, if not *in vitro*. However, since site-directed mutagenesis of each of these cysteines to alanine on a high copy-number plasmid does not alter the mercury resistance phenotype (24), the *in vivo* function is not clear.

```
          10        20         30        40        50
MKKLFASLALAAVVAPVWAATQTVT-LSVPGMTCSACPITVKKAISEVEGVSKVDVTFE-   MerP
                        : :   :::: :   :: :   : ::    :
                       MTHLKITGMTCDSCAAHVKEALEKVPGVQSALVSYPK   MR
                                  10        20        30

 60         70         80         90
-TRQ-AVV--TFDDAKTS-VQKLT-KAT-ADAGYPSS-VKQ                      MerP
 : : : :  :  :: :  :  :  ::: :::       :
GTAQLAIVPGTSPDALTAAVAGLGYKATLADAPLADNRVGLLDKVRGWMAAAEKHSGNEP   MR
 40        50        60        70
                                        ACRQEPQPQGPPPAAGAVAS   GR
                                                  10

          110       120       130       140       150
VQVAVIGSGGAAMAAALKAVEQGAQVTLIERGTIGGTCVNVGCVPSKIMIRAAHIAHLRR   MR
    ::: :    : :  : : ::      :   ::::::::::: : :   :
YDYLVIGGGSGGLASARRAAELGARAAVVESHKLGGTCVNVGCVPKKVMWNTAVH-----   GR
          30        40        50        60        70

          170       180       190       200       210
ESPFDGGIAATVPTIDRSKLLAQQQARVDELRHAKYEGILGGNPAITVVHGEARFKDDQS   MR
 : :    :         :                  :  :                   :
-SEFMHDHADYGFPSCEGKFNWRVIKEKRDAYVSRLNAIYQNNLTKSHIEIIRGHAAFTS   GR
     80        90        100       110       120       130

          230       240       250       260       270
LTVRLNEGGERVVMFDRCLVATGASPAVPPIPGLKE-SPYWTSTEALASDTIPERLAVIG   MR
     :                : :::  :  :       :  ::        : :   :
DPKPTIEVSGKKYTAPHILIATGGMPSTPHESQIPGASLGITSDGFFQLEELPGRSVIVG   GR
     140       150       160       170       180       190
```

```
 280       290       300       310       320       330
SSVVALELAQAFARLGSKVTVLARNTLFFRE-DPAIGEAVTAAFRAEGIEVLEHTQASQV     MR
    : : :     ::::      :     :  :  :    :      : :::   :  :
AGYIAVEMAGILSALGSKTSLMIRHDKVLRSFDSMISTNCTEELENAGVEVLKFSQVKEV     GR
     200       210       220       230       240       250

 340                 350       360       370       380
AHMDGEFV----------LTTTHGELRADKLLVATGRTPNTRSLALDAAGVTVNAQGAIV     MR
                  :         : :: : :: :::  : :   :      : :
KKTLSGLEVSMVTAVPGRLPVMIMIPDVDCLLWAIGRVPNTKDLSLNKLGIQTDDKGHII     GR
     260       270       280       290       300       310

 390       400       410       420       430       440
IDQGMRTSNPNIYAAGDCTDQPQFVYVAAAAGTRAAINMTGG--DAALDLTAMPAVVFTD     MR
 :    :    ::: ::            :: :::   :        :  ::   : :::
VDEFQNTNVKGIYAVGDVCGKALLTPVAIAAGRKLAHRLFEYKEDSKLDYNNIPTVVFSH     GR
     320       330       340       350       360       370

  450       460       470       480       500       500
PQVATVGYSEAEAHHDGIETDSRTLTLDNVPRALANFD--TRGFIKLVIEEGSHRLIGVQ     MR
:   :::  :  : :         :      :   :     :    : :         :
PPIGTVGLTEDGAIHKYGIENVKTYSTSFTPMYHAVTKRKTKCVMKMVCANKEEKVVGIH     GR
     380       390       400       410       420       430

      510       520       530       540       550       560
AVAPEAGELIQTAALAIRNRMTVQELADQLFPYLTMVEGLKLAAQTFNKDVKQLSCCAG      MR
      :   :  : :      :            :  : :
MQGLGCDEMLQGFAVAVKMGATKADFDNTVAIHPTSSEELVTLR                     GR
     440       450       460       470
```

Figure 2. Alignment of Tn*501* mercuric reductase with human erythrocyte glutathione reductase and the Tn*501* MerP protein to show the regions of homologous amino acid sequence. Amino acid identities are marked with a colon; overlines and underlines mark the main regions of homology.

Interestingly, the N-terminal region of MR shows homology with the MerP gene product (Fig. 2), indicating that the N-terminal region of MR may have arisen by duplication and fusion of a *merP* gene with the ancestral reductase-encoding gene. Recently, part of the *merA* gene of pDU1358 has been sequenced (25; S. Silver, personal communication), and shown to have close homology with the R100 and Tn*501* genes. Data are now available from *S. aureus* (26), which show that the regions originally identified as functionally important in Tn*501* MR are conserved in this Gram-positive MR, which shows great divergence elsewhere in the protein.

C. The Transport Proteins

The MerT protein is membrane-associated. It is predicted to be a polypeptide of 116 amino acids, containing three runs of hydrophobic residues of sufficient length to form membrane-spanning helices. It is predicted to be an integral protein of the inner membrane. It is not known whether the MerT protein is processed, and the predicted folding of the primary translation product across the membrane is shown in Fig. 3. Two cysteine residues (Cys-76, Cys-82) are located in a charged region of the protein and are likely to be on the cytoplasmic face of the membrane. The N-terminus of the protein and a counterion pair (Glu-48, Arg-51) are the only other charged regions of the protein. There is a pair of cysteine residues (Cys-24, Cys-25) in the first transmembrane helix.

The MerP proteins from Tn*501* and R100 are predicted to be primary translation products of 91 amino acids, with N-terminal polypeptide sequences typical of signal peptides. Thus, MerP was suggested to be a periplasmic protein, which would be processed by signal peptidase. Indeed, the R100 MerP has been shown to be a dimeric periplasmic polypeptide of 72 amino acids per subunit. There are two cysteine residues in the predicted sequence, Cys-14 and Cys-17 of the processed product. One of the cysteine residues in each subunit is reported to be involved in subunit interactions. The protein has been crystallized (13), and its three-dimensional structure should soon be elucidated.

The MerC protein is found in the R100 determinant but not in that of Tn*501*, yet the Tn*501* determinant functions well without it. Part of the intracistronic sequence between *merP* and *merA* in Tn*501* shows homology with the *merC* gene of R100, suggesting that Tn*501* may have lost its *merC* gene by deletion. MerC is hydrophobic and membrane-associated, but nothing is known about its specific role in mercury resistance. Attempts to predict its function by comparison of its amino acid sequence with protein data bases have revealed little.

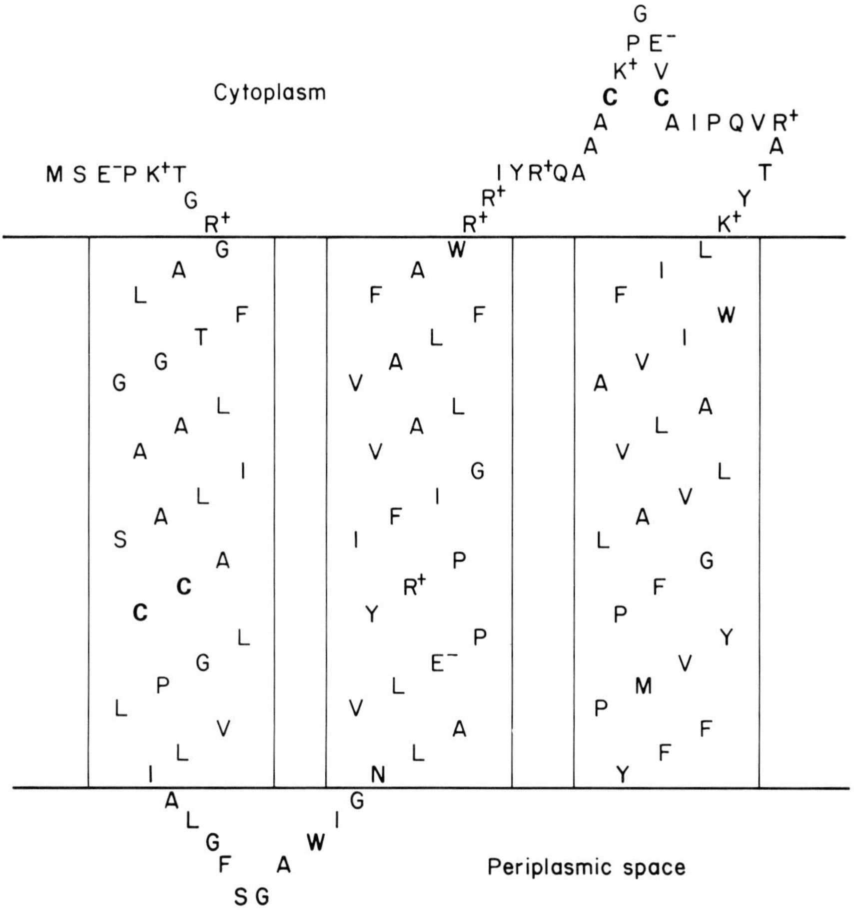

Figure 3. The primary structure of the Tn*501* MerT protein showing the predicted folding through the bacterial inner membrane. The predicted primary translation product is shown, and no account is taken of N-terminal processing. The cysteine residues are indicated in bold and charged residues are marked.

The *merD* gene has no clearly-identified role in mercury resistance. It can be deleted without destroying resistance. Tn*5* insertion mutants of *merD* are normally resistant on low copy-number vectors, but sensitive at high copy-number (27). Mutants deleted for *merD* and promoter-distal genes have altered responses to mercuric ions (28). The MerD protein is highly conserved between Tn*501*, R100 and pDU1358. The predicted amino acid sequence of the MerD protein shows homology to the MerR protein, raising the interesting possibility that the MerD protein is a regulatory protein. As Summers comments (13), speculation on its role must await evidence for expression of the gene. Even more does this comment apply to the potential reading frames *urf-1* and *urf-2* beyond *merD*. They too are conserved in R100, Tn*501* and pDU1358, although *urf-2* is disrupted by the sulphonamide and spectinomycin resistance genes in R100 (28).

V. A MODEL FOR MERCURY RESISTANCE IN BACTERIA

A. The Model

The properties, predicted or known, of the products of the mercury resistance genes of Tn*501* and R100 were originally used to formulate a model for resistance to mercuric ions in Gram-negative bacteria (11). This model is minimal, in that only those gene products known to be essential for the expression of resistance are included. The MerC and MerD proteins are not included, but must somehow fit into the mechanism. The model is illustrated in Fig. 4 and is described below. It is tentative, and provides a hypothesis to be tested experimentally; it may also be modified to include mechanisms for broad-spectrum resistance and for mercury resistance in Gram-positive organisms.

The main features of the model are as follows, and the letters A to E refer to the pairs of cysteine residues shown in Fig. 4.

(1) The *merP* gene product acts as a scavenger protein in the bacterial periplasm, and binds mercuric ions, thus protecting essential thiol groups which are exposed to the periplasmic space. The functional protein may be a dimer (13); it is not known whether one or two mercuric ions are bound per dimer. To satisfy the charge equation, we have assumed that the binding of a mercuric ion to the cysteine pair (A) in MerP releases two protons.

(2) In the first of a series of ligand exchange reactions, the Hg(II) ion is transferred from the MerP protein to the cysteine pair (B) near the outer face of the MerT protein. These cysteine residues are predicted to be partially buried inside the membrane, as is common with the substrate binding sites of other transport proteins (29). A pair of protons would be exchanged in the opposite direction. The Hg(II) would then be transferred to the cysteine pair (C) on the inner face of the MerT protein. There are no charged residues in the protein between transmembrane helices 1 and 2, and this region of the MerT protein may be able to move across the membrane and carry Hg(II) to its inner face.

(3) MR is a cytoplasmic enzyme. The model suggests that MR may become transiently associated with the MerT protein in the membrane, and Hg(II) be transferred initially to the cysteine pair (D) (Cys-10, Cys-13) in the N-terminal

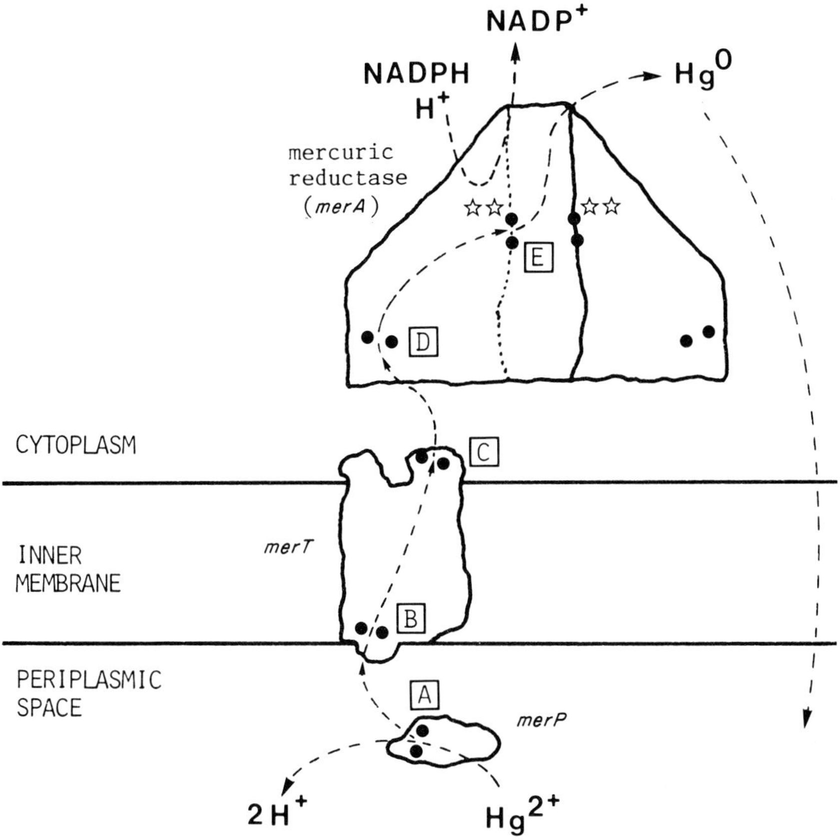

Figure 4. Diagram of the model for mercuric ion detoxification in Gram-negative bacteria (from 11). Each pair of cysteine residues is represented by two filled circles; the letters A to E indentifying these pairs are referred to in the text. The stars represent the redox-active cysteine residues, Cys-135 and Cys-140.

domain of MR. From there the mercury would be passed to the C-terminal cysteine pair (E) (Cys-557, Cys-558) on the same or the other subunit, which constitutes the substrate-binding site. The mechanism of MR is quite well understood, both in its own right and by comparison with the mechanisms of other pyridine nucleotide-disulphide oxidoreductases. The mercuric ion forms a covalent complex with the redox-active thiol group of Cys-135 or Cys-140 (starred in Fig. 4), and is then reduced. After reduction of the divalent Hg(II) to elemental mercury, the non-toxic Hg(0) is released to the cytoplasm.

(4) Elemental mercury can pass out through the cell envelope by diffusion. There is no biochemical requirement for a carrier, but it is possible that one or

more of the additional genes of the *mer* operon, for which no function has been ascribed, may facilitate the loss of Hg(0) from the cell. Hg(0) is volatile and thus mercury will rapidly be lost from the immediate environment of the cell.

B. Why Such a Complex Mechanism?

The mechanism described in the model is biologically advantageous in that a highly toxic and reactive compound is specifically sequestered in the periplasmic space, before it can react with essential thiol groups, and it is not released until it has been detoxified. Thus, the mercuric ion is prevented from interacting with cytoplasmic constituents. Whereas the mechanisms of resistance to cadmium and arsenate are efflux mechanisms (30), such a mechanism would probably be insufficient to maintain resistance to mercuric salts, because of the extreme reactivity of mercury with several chemical groups found in biological macromolecules. Mutants in which the transport genes of Tn*501* have been deleted, but in which MR is still expressed at high level, are sensitive to mercuric salts (31), showing that the uptake of Hg(II) and its delivery to the reductase are an important part of detoxification. Other deletion and site-directed mutagenesis studies of the *mer* genes, which will test the model, are underway in several laboratories.

The transport of mercury into the cell appears to be a dangerous practice. Transport must be closely linked with detoxification if the resistance is to be effective. This may be achieved in part by having excess MR in the cell. Broken cells have a higher capacity for mercury reduction than do whole cells, showing that transport is rate-limiting. Thus the relative levels of expression of the genes for the transport proteins and for MR are likely to be important.

C. The Model can be Applied to Other Systems

The mechanism of mercuric ion resistance is probably very similar in other systems. MRs from Gram-positive and Gram-negative organisms have been studied in varying degrees of purity. With the exception of MR from *T. ferrooxidans*, all Gram-negative MRs show a close immunological relationship (4). However, the MRs are not all identical, and antiserum raised against the Tn*501* enzyme distinguishes three classes of MR: Tn*501*-like, R100-like and those of an intermediate class. Given the close homology between the MRs from Tn*501* and R100, the Gram-negative enzymes must be very closely related. MRs from Gram-positive organisms show no cross reaction with antisera to Gram-negative enzymes and differ from them in heat stability. Yet, even here, the amino acid sequences in the reductase which are postulated to be essential or important to the mechanism of mercury detoxification are conserved.

The genetic organization of the determinants of broad-spectrum mercury resistance is less well understood than those of narrow-spectrum resistance. In the *S. aureus* plasmid pl258, the gene for the organomercury lyase (*merB*) is closely linked to that for the reductase. The mercury resistance region has now been sequenced (26) and the MR gene can be identified by its homology with the Gram-negative gene. The protein is smaller than the Gram-negative protein, and the homology is largely confined to the FAD and NADPH binding domains, the redox-active site and the N-terminal and C-terminal cysteine pairs (Fig. 2); these are precisely the regions predicted to be important in the mechanism of MR *in vivo*. Moreover, there are another seven possible open-reading frames, none of which are closely homologous to MerT, MerP or MerC. Thus, the mechanism of transport of mercury in at least one Gram-positive species is different from that in Gram-negatives, which is not surprising in view of the differences in the cell envelope between the two types. Yet the mechanism of mercury reduction by MR is presumably very similar in both.

Part of the Gram-negative broad-spectrum resistance determinant from plasmid pDU1358 has also been sequenced. This determinant is more directly comparable to those of Tn*501* and R100. The plasmid has two mercuric resistance determinants, one narrow-spectrum and one broad-spectrum, that can operate independently. The broad-spectrum determinant is very similar to the *mer* determinant of Tn*501*, but contains the *merB* gene, encoding the organolyase, between the *merA* and *merD* genes. The pDU1358 determinant does not contain a *merC* gene (S. Silver, personal communication).

It is not known whether the organomercury lyase is cytoplasmic or periplasmic, but it lacks any obvious signal sequences that might be required for export. Mutants of pDU1358 which are *merA-merB* deletions are hypersensitive to mercury, indicating that organomercurials are transported through the membrane on a specific transport system; possibly the MerP and MerT proteins. Thus, organomercurials in Gram-negative bacteria may be transported into the cytoplasm, and cleaved by the organomercury lyase prior to reduction of the mercuric salt to Hg(0). More information on the mechanism of detoxification of organomercurials should be forthcoming now that the organomercury lyase from plasmid R381 has been overexpressed and purified (32).

VI. REGULATION OF EXPRESSION OF THE MERCURY RESISTANCE GENES

A. The Regulation of Mercury Resistance is Similar in Different Determinants

Mercury resistance is usually inducible. The first detailed and systematic study of the mechanism of induction was by the creation of *lacZ* transcriptional fusions in the R100 *mer* determinant. Two classes of mutants were isolated: those in which mercuric salts caused an increase in *lacZ* expression, and those in which they caused little change in expression. Further analysis showed that the first class were transcriptional fusions with the *mer* operon, and Tn*5* insertion mutagenesis of these identified a gene, *merR*, which acted as a repressor in the absence of mercury and as an inducer in its presence. The second class of mutants were transcriptional fusions in the *merR* gene. Complementation experiments were performed on a mutant of the R100 determinant which contains a Tn*5* insertion in *merR* and a *lacZ* transcriptional fusion in *merA* (33). Eleven mercury-resistance plasmids of incompatibility groups C, B, S, L and P, as well as the transposons Tn*501* and Tn*3401*, and specifying broad- or narrow-spectrum resistances, were singly introduced *in trans*, and all but one were shown to contain a *merR* analogue, as shown by the expression of the *lacZ* gene.

B. The *merR* Gene and its Product

The *merR* genes of Tn*501* and R100 have been shown by sequence analysis to be very similar, and are transcribed divergently from the rest of the operon. They encode a protein of about M_r 16 000. Those DNA-binding regulatory proteins for which a crystal structure is available contain a helix-turn-helix secondary structural motif, and other DNA-binding proteins are predicted to contain the same motif (34). Potential helix-turn-helix motifs can be identified in the MerR protein, as can sequences which may interact with mercuric ions (35, 36), but proof of their role in binding DNA is lacking.

The MerR proteins from Tn*501* and R100 have been overproduced and used in DNA binding studies (36, 37). The binding site for the Tn*501* MerR protein has been shown by DNase I footprinting to be between the −35 and −10 sequences of the *mer* promoter, and across the start of the divergent *merR* transcript (Fig. 5).

The *merR* gene of the Gram-negative broad-spectrum resistance determinant of pDU1358 does not cross-hybridize with that of R100 by Southern blotting, yet it can cross-complement a *merR*$^-$ mutant of R100 (25). This broad spectrum *merR* gene responds not only to Hg^{2+}, but also to phenylmercuriacetate as an inducer. Determination of the structure of this gene may help to locate the DNA-binding and mercury-binding regions of other MerR proteins.

C. Studies of Mutants Reveal Unusual Properties of the *mer* Promoter

DNA sequence analysis and high-resolution S1 mapping of Tn*501* transcripts has produced the picture of the promoter shown in Fig. 5. The −35 and

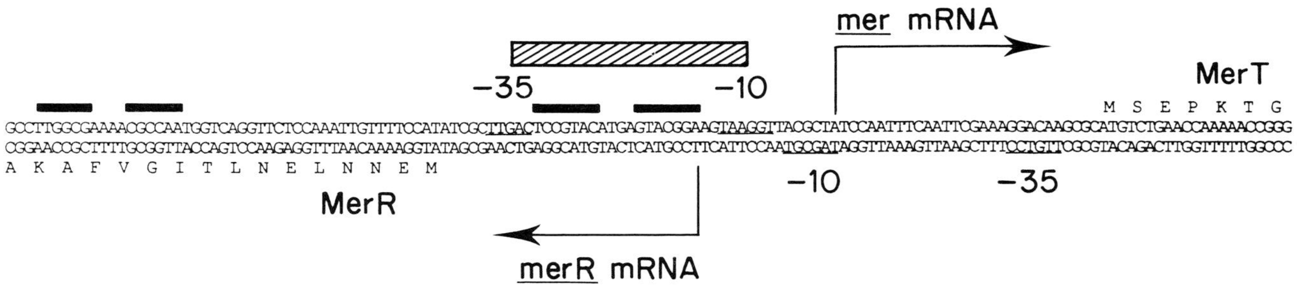

Figure 5. Diagram of the Tn*501* regulatory region, showing the features discussed in the text. The box marks the region protected by the MerR protein; the dyad symmetries centred on −79 and −24 with respect to the *mer* operon are marked with overlines.

−10 sequences for the *merR* promoter are only provisionally identified, because there are no mutants of this promoter. The −35 and −10 sequences of the major *mer* transcript show good homology to those of highly expressed eubacterial promoters (TTGACA-N(16-18)-TATAAT; 38), but they are separated by a large spacer of 19 bp.

Two dyad symmetrical sequences have been identified in the promoter region. One of these is centred at position −79 relative to the *mer* promoter, and the other is centred at position −24, between the −10 and −35 sequences. Studies of purified MerR protein show that it binds to the dyad at position −24(36). This is ideally situated to repress both transcription of the *mer* operon, thus regulating expression of the *mer* structural genes, and to autoregulate its synthesis by repressing transcription of the *merR* gene.

Analysis of the *mer* promoter by deleting upstream sequences has revealed some surprising things (39). The constitutive low level activity of the *mer* promoter in the absence of *merR* is unchanged even when the −35 and −10 sequences are deleted. The deletion of a −35 sequence with little effect on promoter activity has been observed before (e.g. the *E. coli gal* promoter; 40). The lack of effect of deleting the −10 sequence can be explained by there being a second potential −10 sequence (TACGCT) just downstream from the first, and only when this is deleted is promoter activity abolished. Promoter mutants have confirmed that there are two potential −10 sequences in the *mer* promoter; the upstream one is normally used (39). The importance of the downstream −10 sequence *in vivo* is not known.

Deletion analysis of the *mer* promoter in the presence of *merR* has shown that both induction and repression are effected by the binding of the MerR protein to the dyad symmetrical sequence at position −24. The difference between the repressor and the inducer forms of the MerR protein is due to the presence of mercury, the coinducer, which presumably binds to the MerR protein. Repression is not totally a function of the downstream dyad, since repression is partially alleviated when the upstream dyad at position −79 is deleted, suggesting that interactions with both dyads are required for full repression. The upstream dyad is not required for induction.

D. How Does the MerR Protein Regulate Expression of the *mer* Promoter?

It is easy to see how the MerR protein might repress transcription from the *mer* promoter by binding to the DNA between the −35 and −10 sequences. Because of the overlap of the *mer* and *merR* promoters, this binding should also autoregulate synthesis of MerR. However, the mechanism of induction is less obvious. Most inducible promoters have poor −35 sequences and contain an inducer binding site upstream of this, whereas the *mer* promoter has a good −35 sequence, with the MerR-binding site in the 19 bp spacer region.

Promoter up-mutations, in which the spacer has been reduced to 18 bp, have been isolated. These are repressible but not inducible. A hypothesis for the mechanism of induction is that the binding of the MerR—Hg(II) complex to the MerR-binding site on DNA (or the binding of Hg(II) to a pre-existing MerR—DNA complex) causes twisting or bending of the DNA, such that favourable contacts can be made between the promoter and RNA polymerase. This requires that the MerR protein—Hg(II) complex and RNA polymerase recognize overlapping DNA sequences, presumably from opposite sides of the double helix.

VII. OVERVIEW AND PROSPECTS

The genetic study of mercury resistance, particularly the molecular approaches such as DNA sequence analysis, has allowed a number of hypotheses to be proposed. Some of these have been described in detail, such as the biochemical mechanism of mercuric ion resistance, and the regulation of expression of the mercuric resistance genes. Others, such as the mechanisms of evolution of transposons (41) and of antibiotic resistance determinants (28), have not been considered here. Much of the future work will be concerned with the direct testing and refining of these hypotheses.

One of the most important properties of the mercuric resistance determinants, which is often overlooked, is their ubiquity. The genes for the majority of the determinants are inducible. This provides the opportunity to study mechanisms of genetic regulation in a wide variety of different bacterial species.

Acknowledgements

We are grateful to colleagues for access to data before publication, particularly to Tim Foster, Tom O'Halloran, Simon Silver, Anne Summers, Chris Walsh and members of their laboratories. Work from the authors' laboratory was supported by the Medical Research Council, the Science and Engineering Research Council and the Royal Society, of which NLB is an E.P.A. Cephalosporin Fund Senior Research Fellow.

References

(1) Leach, S. J. (1960). The reaction of thiol and disulphide groups with mercuric chloride and methylmercuric iodide. *J. Aust. Chem. Soc.* **13**, 520–546.

(2) Summers, A. O. and Silver, S. (1978). Microbial transformations of metals. *Ann. Rev. Microbiol.* **32**, 637–672.

(3) Robinson, J. B. and Tuovinen, O. H. (1984). Mechanisms of microbial resistance and

detoxification of mercury and organomercury compounds: physiological, biochemical and genetic analyses. *Microbiol. Rev.* **48**, 95–124.

(4) Silver, S. and Kinscherf, T. G. (1982). Genetic and biochemical basis for microbial transformations and detoxification of mercury and mercurial compounds. *In* "Biodegradation and Detoxification of Environmental Pollutants" (A. M. Chakrabarty, ed.), pp. 85–103. CRC Press Inc., Boca Raton, Fla.

(5) Pan-Hou, H. S. and Imura, N. (1982). Involvement of mercury methylation in microbial mercury detoxification. *Arch. Microbiol.* **131**, 176–177.

(6) Pan-Hou, H. S., Nishimoto, M. and Imura, N. (1981). Possible role of membrane proteins in mercury resistance of *Enterobacter aerogenes. Arch. Microbiol.* **130**, 93–95.

(7) Hamer, D. H. (1986). Metallothioneins. *Ann. Rev. Biochem.* **55**, 913–951.

(8) Higham, D. P., Sadler, P. J. and Scawen, M. D. (1984). Cadmium-resistant *Pseudomonas putida* synthesizes novel cadmium proteins. *Science* **225**, 1043–1046.

(9) Tanaka, M., Yamamoto, T. and Sawai, T. (1983). Evolution of complex resistance transposons from an ancestral mercury transposon. *J. Bacteriol.* **153**, 1432–1438.

(10) Foster, T. J. (1987). Genetics and biochemistry of mercury resistance. *CRC Crit. Rev. Microbiol.* (In press.)

(11) Brown, N. L. (1985). Bacterial resistance to mercury: *reductio ad absurdum? Trends Biochem. Sci.* **10**, 400–403.

(12) Silver, S., Rosen, B. P. and Misra, T. K. (1986). DNA sequencing analysis of mercuric and arsenic resistance operons of plasmids from Gram-negative and Gram-positive bacteria. *In* Fifth International Symposium on the Genetics of Industrial Microorganisms" (M. Alačević, D. Hranueli and Z. Toman, eds.), pp. 357–371. Pliva, Zagreb.

(13) Summers, A. O. (1986). Organization, expression and evolution of genes for mercury resistance. *Ann. Rev. Microbiol.* **40**, 607–634.

(14) Nakaya, R., Nakamura, A. and Murata, Y. (1960). Resistance transfer agents in *Shigella. Biochem. Biophys. Res. Commun.* **3**, 654–659.

(15) Stanisich, V. A., Bennett, P. M. and Richmond, M. H. (1977). Characterization of a translocation unit encoding resistance to mercuric ions that occurs on a non-conjugative plasmid in *Pseudomonas aeruginosa. J. Bacteriol.* **129**, 1227–1233.

(16) Ni Bhriain, N. and Foster, T. J. (1986). Polypeptides specified by the mercuric resistance (*mer*) operon of plasmid R100. *Gene* **42**, 323–330.

(17) Jackson, W. J. and Summers, A. O. (1982). Polypeptides encoded by the *mer* operon. *J. Bacteriol.* **149**, 479–487.

(18) Fox, B. S. and Walsh, C. T. (1982). Mercuric reductase: purification and characterization of a transposon-encoded flavoprotein containing an oxidation-reduction-active disulfide. *J. Biol. Chem.* **257**, 2498–2503.

(19) Fox, B. S. and Walsh, C. T. (1983). Mercuric reductase: homology to glutathione reductase and lipoamide dehydrogenase. Iodoacetamide alkylation and sequence of the active site peptide. *Biochemistry* **22**, 4082–4088.

(20) Rinderle, S. J., Booth, J. E. and Williams, J. W. (1983). Mercuric reductase from R-plasmid NR1: characterization and mechanistic study. *Biochemistry* **22**, 869–876.

(21) Brown, N. L., Ford, S. J., Pridmore, R. D. and Fritzinger, D. C. (1983). Nucleotide sequence of a gene from *Pseudomonas* transposon Tn*501* encoding mercuric reductase. *Biochemistry* **22**, 4089–4095.

(22) Misra, T. K., Brown, N. L., Haberstroh, L., Schmidt, A., Goddette, D. and Silver, S. (1985). Mercuric reductase structural genes from plasmid R100 and transposon Tn*501*: functional domains of the enzyme. *Gene* **34**, 253–262.

(23) Thieme, R., Pai, E. F., Schirmer, R. H. and Schulz, G. E. (1981). Three-dimensional structure of glutathione reductase at 2 Å resolution. *J. Mol. Biol.* **152**, 763–782.

(24) Walsh, C. T., Moore, M. J. and DiStefano, M. D. (1987). Conserved cysteines of mercuric

ion reductase: an investigation of function via site-directed mutagenesis. *In* "Flavins and Flavoproteins: Proceedings of the Ninth International Symposium". Walter de Gruyter, Berlin and New York.

(25) Griffin, H. G., Foster, T. J., Silver, S. and Misra, T. K. (1987). Cloning and DNA sequence of the mercuric- and organomercurial-resistance determinants of plasmid pDU1358. *Proc. Natl. Acad. Sci. USA* **84**, 3112–3116.

(26) Laddaga, R. A., Chu, L., Misra, T. K. and Silver, S. (1987). Nucleotide sequence and expression of the mercurial resistance operon from *Staphylococcus aureus* plasmid pl258. *Proc. Natl. Acad. Sci. USA* **84**, 5106–5110.

(27) Ni Bhriain, N., Silver, S. and Foster, T. J. (1983). Tn*5* insertion mutations in the mercuric ion resistance genes derived from plasmid R100. *J. Bacteriol.* **155**, 690–703.

(28) Brown, N. L., Misra, T. K., Winnie, J. N., Schmidt, A. and Silver S. (1986). The nucleotide sequence of the mercuric resistance operons of plasmid R100 and transposon Tn*501*: further evidence for *mer* genes which enhance the activity of the mercuric ion detoxification system. *Mol. Gen. Genet.* **202**, 143–151.

(29) Ames, G. F. (1986). Bacterial periplasmic transport systems: structure, mechanism and evolution. *Ann. Rev. Biochem.* **55**, 397–425.

(30) Trevors, J. T., Oddie, K. M. and Belliveau, B. H. (1985). Metal resistance in bacteria. *FEMS Micro. Rev.* **32**, 39–54.

(31) Lund, P. A. and Brown, N. L. (1987). Role of the *merT* and *merP* gene products of transposon Tn*501* in the induction and expression of resistance to mercuric ions. *Gene* **52**, 207–214.

(32) Bergley, T. P., Walts, A. E. and Walsh, C. T. (1986). Bacterial organomercurial lyase: overproduction, isolation and characterization. *Biochemistry* **25**, 7186–7192.

(33) Foster, T. J. and Ginnity, F. (1985). Some mercurial resistance plasmids from different incompatibility groups specify *merR* regulatory functions that both repress and induce the *mer* operon of plasmid R100. *J. Bacteriol.* **162**, 773–776.

(34) Pabo, C. O. and Sauer, R. T. (1984). Protein-DNA recognition. *Ann. Rev. Biochem.* **53**, 293–321.

(35) Lund, P. A., Ford, S. J. and Brown, N. L. (1986). Transcriptional regulation of the mercury-resistance genes of transposon Tn*501*. *J. Gen. Microbiol.* **132**, 465–480.

(36) O'Halloran, T. and Walsh, C. (1987). Metalloregulatory DNA-binding protein encoded by the *merR* gene: isolation and characterization. *Science* **235**, 211–214.

(37) Heltzel, A., Gambill, D., Jackson, W. J., Totis, P. A. and Summers, A. O. (1987). Overexpression and DNA-binding properties of the *mer* regulatory protein from plasmid NR1 (Tn*21*). *J. Bacteriol.* **169**, 3379–3384.

(38) Harley, C. B. and Reynolds, R. P. (1987). Analysis of *Escherichia coli* promoter sequences. *Nucl. Acids Res.* **15**, 2343–2361.

(39) Lund, P. A. and Brown, N. L. (1988). Regulation of transcription from the *mer* and *merR* promoters of the transposon Tn*501*. *J. Mol. Biol.* (Submitted).

(40) Ponnambalam, S., Webster, C., Bingham, A. and Busby, S. (1986). Transcription initiation at the *Escherichia coli* galactose operon promoters in the absence of the normal −35 region sequences. *J. Biol. Chem.* **261**, 16043–16048.

(41) Grinsted, J. and Brown, N. L. (1984). A Tn*21* terminal sequence within Tn*501*: complementation of *tnpA* gene function and transposon evolution. *Mol. Gen. Genet.* **197**, 497–502.

Section **III**

Morphological Differentiation— Flagella, Spores and Multicellular Development

Morphological differentiation has often been thought of as a property of higher organisms. However, some bacteria also show fascinating morphological complexity, and the development of molecular genetic methods for some of these bacteria has led to a flowering of studies about various aspects of bacterial differentiation which are both intrinsically of great interest and potentially informative about the developmental biology of eukaryotes.

The questions of developmental biology start from basic aspects of cell biology such as DNA replication, cell division and protein secretion that are well-studied in *E. coli* and fall outside the scope of this book. The next level of complexity has to do with subcellular structures, represented in eukaryotes by organelles such as mitochondria and chloroplasts that are absent from prokaryotes. In bacteria, the specific targeting of proteins to photosynthetic membranes in *Rhodobacter*, or from the mother cell into the forespore compartment in sporulating *Bacillus* cells, can be expected to attract considerable attention in the future, even though it is apparently little-studied at present. More attention has been paid to another subcellular structure, the polar flagellum of *Caulobacter*. Chapter 10 addresses several questions about this remarkable machine. How is it assembled? How is assembly directed to a particular position on the *Caulobacter* cell? How are the genes regulated to produce the correct amounts of the various components, and to allow flagellum synthesis only at a particular time in the dimorphic cell cycle of *Caulobacter*? How does *Caulobacter* use its flagellum to swim in the most advantageous direction? And, how is the biogenesis of this chemotactic ability synchronized with the cell cycle and flagellum formation? Not surprisingly, a complex picture is emerging, involving more than 40 genes in a variety of chromosomal locations. Many of these are controlled by a complicated cascade of regulatory genes. Remarkably, transcription of many of the genes appears to involve an alternative sigma factor at least analogous—perhaps homologous— with σ^{54}, which in other bacteria is involved in transcription of genes for nitrogen assimilation (Chapter 6) and aromatic hydrocarbon degradation (Chapter 8).

The theme of complex cascades of regulatory genes, involving several alternative sigma factors, is taken a step further when the development of a whole *B. subtilis* cell into an endospore is analysed (Chapter 11). Nearly all of more than 60 known genes essential for sporulation appear to play integral roles in this regulatory cascade. In contrast, it seems that many of the structural components of spores are encoded by further families of genes, individual members of which are virtually dispensable for sporulation, and which have been amenable to study only through gene fusion experiments and reverse genetic procedures that exploit cloned DNA in the generation of mutants.

Multicellular developmental features are exhibited by a few bacteria (perhaps by all, according to a recent article entitled "Bacteria as Multicellular Organisms", by J. A. Shapiro in June 1988, *Scientific American*, pp. 62–69). Examples already referred to in Section II of this book, but not dealt with in detail, include actinomycetes such as *Streptomyces*, with their spatially, functionally and morphologically distinct substrate and sporulating aerial mycelium (Chapter 7), and those cyanobacteria, such as *Anabaena*, in which the filaments of vegetative cells contain regularly spaced terminally differentiated nitrogen-fixing heterocysts (Chapter 6). Another example, as yet even less closely studied, is provided by swarming bacteria such as *Proteus* (Shapiro, op. cit.). Perhaps the most extensive analysis is of cell-to-cell interactions involved in the movement and development into fruiting bodies of swarms of *Myxococcus xanthus* (Chapter 12) in which transposon-mediated gene fusions have been exploited to reveal dependent, sequential expression of genes that both require and mediate cascades of signals between cells. The genetic complexity of myxococcal differentiation is considerable: the enumeration of transposon-mediated gene fusions that show developmental regulation suggests that several hundred genes may be involved.

Chapter **10**

Differentiation in **Caulobacter:** *Flagellum Development, Motility and Chemotaxis*

AUSTIN NEWTON

I. INTRODUCTION

Caulobacter species are Gram-negative bacteria with a wide distribution in soil, fresh water and sea water (1). The characteristic property of these dimorphic bacteria is their asymmetric pattern of cell division in which two structurally different progeny cells are produced at each division. One of these

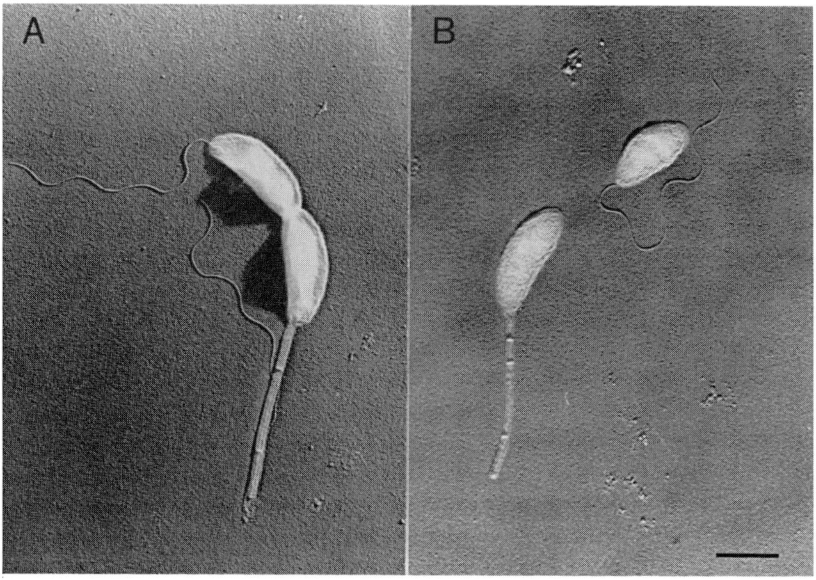

Figure 1. Electron micrograph of *C. crescentus* cells. (A) Dividing cell with polar flagellum and stalk showing bands along its length. (B) Newly-divided stalked cell and swarmer cell. Scale bar = 1 μm. Photograph courtesy of Jean Poindexter.

cells is the motile swarmer cell that carries a single polar flagellum and the other is a nonmotile, stalked cell which has a cellular stalk formed by an outgrowth of the cell wall and membranes. The unique structural elements of the two cell types can be seen in the predivisional cell of the fresh water species *Caulobacter crescentus* (Fig. 1), which is the most widely studied of the caulobacters.

The life cycle of *C. crescentus* is similar in many respects to that observed during stem cell development in higher organisms in which the mother cell plus a new cell type are formed at each division. Thus, the stalked cell of *C. crescentus* (a) divides repeatedly to produce the same stalked cell plus a new swarmer cell (b) in a pattern that can be represented as a --> a + b. The stalked cell continues to divide in this fashion, while each new swarmer cell first loses its flagellum and forms a stalk before entering the division cycle. The metabolic versatility provided by the two cell types may account for the ability of caulobacters to survive in a variety of nutritionally poor environments. The swarmer cell is motile and chemotactic (Section IV), and it can be considered the dispersal phase of the life cycle during which the cell moves to the optimum location in the medium for cell growth. Cell growth and proliferation in the stalked cell phase of the cell cycle may be facilitated by the presence of the stalk, or prostheca, which can anchor the cell at one location by attachment via the adhesive holdfast at its tip and provide additional cell surface for the absorption of nutrients (1).

The pattern of cell division in *C. crescentus* and the precisely defined timing of developmental events in the cell cycle has made this organism a powerful prokaryotic model for the study of cell differentiation (2, 3). Studies of development in these cells are now facilitated by the availability of generalized transduction, conjugation, transposon mutagenesis and a detailed genetic linkage map (4), as well as an extensive set of molecular techniques. In this chapter I consider the regulation of one set of related developmental events that include flagellum formation, motility and chemotaxis.

II. DEVELOPMENTAL PROGRAMMES AND CELL DIFFERENTIATION

The swarmer and stalked cells of *C. crescentus* inherit different developmental programmes, as shown in Fig. 2, this is reflected in the patterns of DNA

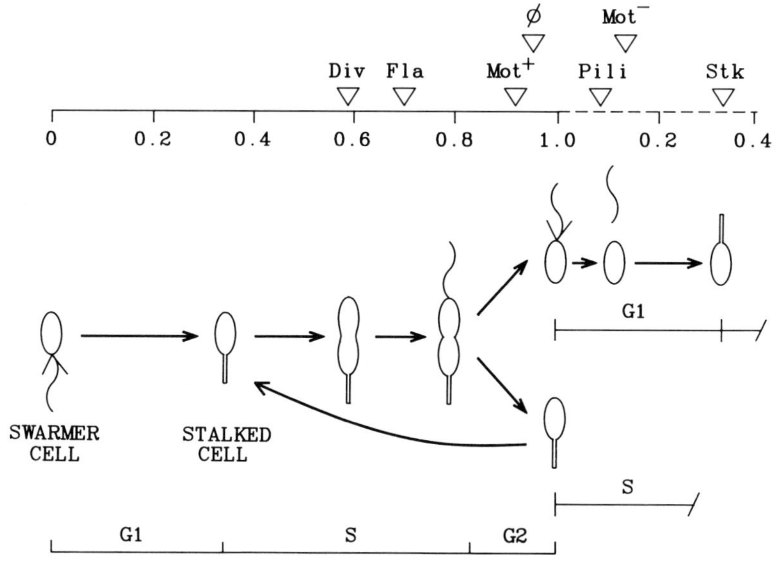

Figure 2. *C. crescentus* cell cycle and timing of developmental events. Times in the cell cycles are indicated as fractions of a cell cycle where 1.0 unit is the time for division of the swarmer cell. Developmental events indicated are initiation of cell division (Div), onset of flagellin and hook protein synthesis (Fla), gain of motility (Mot^+), activation of polar bacteriophage OLC72 receptors (ϕ), pili assembly (Pili), loss of motility (Mot^-), and stalk formation (Stk). Cell cycle periods G1, S and G2 are defined in the text.

synthesis in the two progeny cells. The stalked cell initiates chromosome replication immediately, and after a 90 min period of DNA synthesis (S-period) in a salts-glucose medium, the cell enters a 30 min post-synthetic gap (G2-period). The swarmer cell, on the other hand, undergoes a pre-synthetic gap (G1-period) of approximately 60 min before initiating DNA synthesis. During this period, the cell loses its flagellum, forms a stalk, and prepares to initiate chromosome replication (5). This new stalked cell then follows the same cell cycle just described for the "mother" stalked cell.

Another expression of developmental programming in *C. crescentus* is the ordered sequence of stage-specific, structural events leading to polar morphogenesis. The first event to define the incipient swarmer cell is the appearance in mid-S-period of a constriction or division site on the growing stalked cell and the subsequent initiation of flagellum formation at the cell pole opposite to the stalk. If the time required for the separation of the swarmer cell is fixed as 1.0, then we can estimate that flagellum formation occurs at 0.7 to 0.8 and that flagellum activation occurs just before cell separation at 0.9 to 0.95 of the swarmer cell cycle (2). The formation of DNA bacteriophage ϕLC72 receptors can also be detected at approximately 0.95 by the adsorption of the bacteriophage particles to the flagellated cell pole (6). Thus, at cell division, a motile swarmer cell is produced with a constellation of surface structures, including the flagellum, bacteriophage receptors, and an adhesive holdfast, all localized to the one cell pole.

The swarmer cell differentiates into the stalked cell by a sequence of developmental events that is a continuation of the one just described. Pili are assembled at the base of the flagellum shortly after cell division at 0.1 (7), the cell loses motility, sheds the flagellum, and, finally, forms the cellular stalk at the point of flagellum attachment at the end of G1 (Fig. 2). The central conclusion from these studies is that formation of the swarmer cell and its differentiation into a stalked cell result from an ordered sequence of events that are restricted to the new cell pole (see Fig. 2; 6, 7). Genetic results considered in Section V of this chapter suggest that information required for the timing of these events and for the localization of structures to the one cell pole in *C. crescentus* is specified by underlying cell cycle steps.

III. REGULATION OF FLAGELLUM BIOSYNTHESIS

A. Flagellar Gene Expression is Periodic in the Cell Cycle

The flagellum is the most accessible structure for studies of gene expression and spatial localization in *C. crescentus*. It is composed of a basal body-rod assembly, a hook, and a filament (Fig. 3). The *C. crescentus* flagellum is similar to flagella of the enteric bacteria, except that the basal body has five instead of

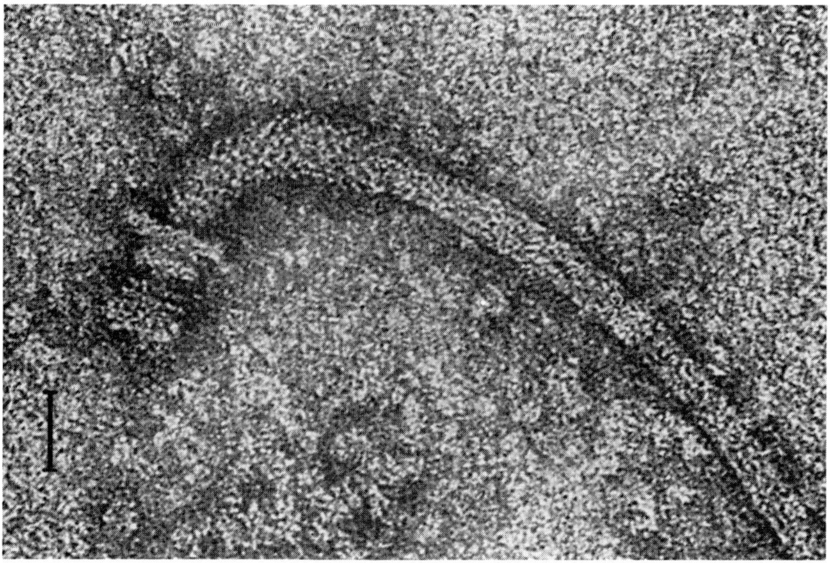

a

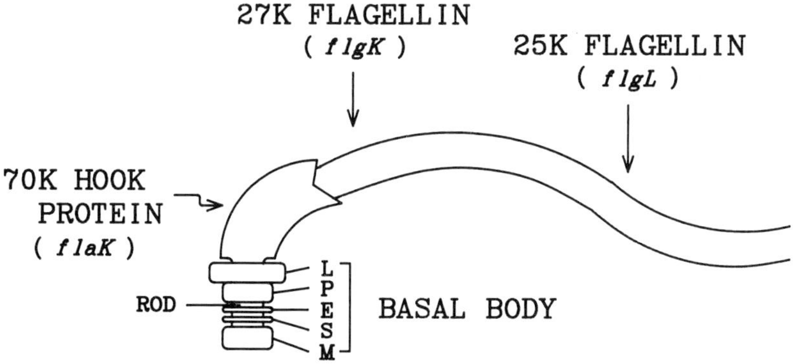

b

Figure 3. Structure of *C. crescentus* flagellum. (a) Electron micrograph of flagellum taken from Johnson *et al.* (8). Scale bar = 30 nm. (b) Schematic drawing of flagellum showing the M, S, E, P and L rings of basal body.

four rings which are embedded in the inner membrane, peptidoglycan, and outer membrane of the cell envelope (8). The hook contains a single 70 kD hook protein subunit, while the flagellar filament contains two (Fig. 3) and possibly three distinct, but related, flagellin subunits, the 29, 27 and 25 kD flagellin. These proteins cross-react immunologically and they have similar amino acid sequences. The 27 kD flagellin is assembled adjacent to the hook and the 25 kD flagellin is assembled at the end of the flagellar filament distal to the hook (Fig. 3; 2, 3). There are now immunological data to suggest that a small amount of the 29 kD flagellin is assembled between the hook and the 27 kD flagellin (L. Shapiro, personal communication).

Highly purified swarmer cells can be readily obtained from exponential cultures of *C. crescentus* by either a plate release technique or density gradient centrifugation, and the availability of synchronous cultures allows the regulation of developmental events during the cell cycle to be determined by biochemical techniques (2). Specific radioimmunoassay for flagellar proteins has shown that the hook protein, the 27 kD flagellin and the 25 kD flagellin are all synthesized periodically at the time of flagellum formation (Fig. 4; 9). Synthesis of the 27 kD flagellin ends before cell division, but synthesis of the 25 kD flagellin continues after cell division in the swarmer cell where it is

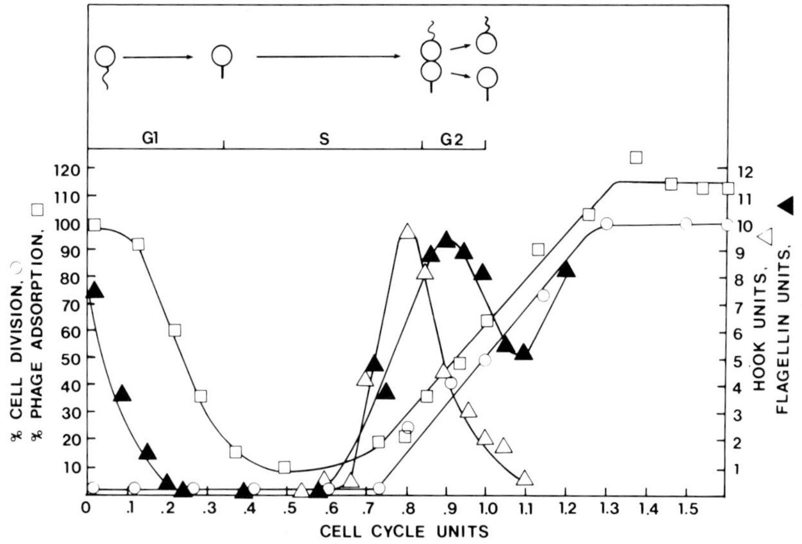

Figure 4. Periodic expression of flagellar genes in the cell cycle. Rates of hook protein (△) and 25 kD flagellin synthesis (▲) were measured by radioimmune assays on samples of cells from a synchronous culture that had been pulse labelled with ^{35}S-methionine. Appearance and loss of the polar OLC72 receptor sites (□) were measured by adsorption of bacteriophage particles. From Huguenel and Newton (6).

assembled at the tip of the growing filament. The use of rifampicin in synchronous cultures to block transcription has shown that initiation of hook protein and flagellin synthesis requires *de novo* RNA synthesis and that expression of these genes is restricted to a brief period before cell division. These experiments also suggested that the 25 kD flagellin is translated after cell division from a relatively stable messenger RNA that is transcribed in the predivisional cell and then segregated to the new swarmer cell (9).

An important conclusion from these results and those to be discussed below is that flagellum biosynthesis in *C. crescentus* is controlled in part at the level of transcription. The results also raise a number of more general questions about the expression and regulation of the *fla* gene family. Are periodic expression and transcriptional control general features of *fla* gene regulation? Are the *fla* genes expressed in a temporal sequence that corresponds to the order of gene product assembly into the flagellum? What mechanism(s) accounts for the activation and shut-off of *fla* genes at the beginning and end of the synthetic periods? And perhaps most importantly, what "clock" ultimately determines when these genes are activated in the cell cycle? Several of these questions have been answered and good progress has been made on others using a variety of molecular and genetic approaches.

B. Organization of Motility Genes on the Chromosome

C. crescentus mutants defective in motility, arising spontaneously or after Tn*5* insertion, can be isolated at frequencies of up to 1% by screening for colonies that form small or no swarms when stabbed into semisolid nutrient agar (10, 11). The mutants can be classified as flagellar (*fla*, or alternatively *flb* or *flg*) mutants, which assemble a defective flagellum, motility (*mot*) mutants, which form an apparently normal flagellum but are paralyzed, or chemotactic (*che*) mutants, which are motile, but defective in a component of the chemosensory apparatus responsible for cellular responses to attractants or repellents in the medium. Pleiotropic (*ple*) mutants that are defective for both motility and formation of active polar, bacteriophage receptors (Fig. 2) have also been identified. *pleA* mutants do not form a flagellum or the polar bacteriophage receptors, while *pleC* mutants form a flagellum, but they cannot execute subsequent developmental events, including activation of the flagellum and normal pattern of stalk formation (10).

Over 30 *fla* genes required for flagellum formation have been identified and mapped on the *C. crescentus* chromosome. Approximately 24 of them are located in one of three clusters: the basal body cluster (*flbON* cluster), the hook gene cluster (*flaK* cluster) and the flagellin gene cluster (*flaEY* cluster); the remainder of the *fla* genes are scattered on the chromosome (Fig. 5). *fla* genes were initially mapped by *in vivo* genetic techniques using conjugation and

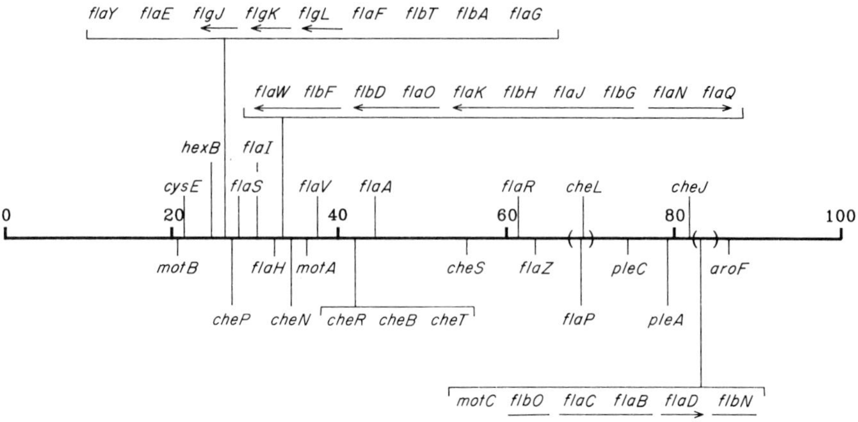

Figure 5. Genetic map of *C. crescentus* chromosome showing positions of *fla*, *che* and *mot* genes. (Data are taken from references 10, 11, 16, 17, 19, 24.) The map is divided into 100 map units with the *fla*EY cluster, hook cluster and basal body cluster located at approximately 28, 30 and 82 units, respectively. Transcription units defined by complementation of Tn*5* insertion mutants are indicated by lines below the genes; arrows indicate transcription units where the transcription start sites and direction of transcription have been determined by S1 nuclease mapping (see text).

transduction. More recently, many of them have been physically mapped using cloned *fla* genes as DNA probes, as described below, and the protein products of several *fla* genes have now been identified (Table I).

Since a method for efficient transformation is not yet available in *C. crescentus*, cloning has been carried out in *Escherichia coli* using lambda or plasmid vectors. The procedures used to identify these *C. crescentus* genes may also be applicable to other non-enteric, Gram-negative bacteria. Genes in the *flaEY* cluster were cloned by two techniques. In the first, a mutant *flaE* gene containing a Tn*5* insertion was cloned by selecting for a hybrid plasmid containing the kanamycin resistance marker encoded by the transposon. The cloned insert was then used as a radiolabelled probe to isolate the wild-type gene and flanking sequences from a *C. crescentus* plasmid library (12). In the second method, flagellin genes in the *flaEY* cluster were identified by screening a *C. crescentus* library using a cDNA probe that had been prepared from RNA enriched for flagellin transcripts by immunoprecipitating polysomes from predivisional cells with anti-flagellin immunoglobulin (13).

One *C. crescentus fla* gene has also been isolated by its ability to produce the specific flagellar protein in *E. coli*. Radioimmune blotting was used to identify *flaK* in a bacteriophage lambda library by screening individual plaques for the production of the *C. crescentus* hook protein (14). More recently, it has been feasible to screen for relatively large *C. crescentus* inserts by complementation

Table I
Products or functions of *fla* and *che* genes in *C. crescentus*

Gene[a]	Cluster[a]	Product or function[b]
flgJ	*flaEY*	29 kD flagellin
flgK	*flaEY*	25 kD flagellin
flgL	*flaEY*	27 kD flagellin
flbG	Hook gene cluster	Initiation of hook assembly
flaJ	Hook gene cluster	Termination of hook assembly
flaK	Hook gene cluster	70 kD hook protein
flbO	Basal body cluster	Basal body assembly
flaC	Basal body cluster	Basal body assembly
flaB	Basal body cluster	Basal body assembly
flbN	Basal body cluster	Basal body assembly
flaD	Basal body cluster	P ring of basal body
cheR	Basal body cluster	Methyltransferase
cheB	Basal body cluster	Methylesterase

[a] See Fig. 5.
[b] See text for references to assignments of functions.

using a derivative of the specialized vector pLAFR1, a derivative of plasmid pRK290 containing the bacteriophage lambda *cos* site (15). Since this cosmid vector, like plasmid pRK290, replicates stably in *E. coli* and *C. crescentus* cells, a pLAFR1 library packaged and cloned in *E. coli* can be transferred by conjugation to *C. crescentus* cells and tested directly for complementation of a defective gene.

After a *fla* gene has been cloned, it can be used as a radiolabelled probe to identify other *fla* mutations located in the same gene cluster. Insertion or deletion mutations located adjacent to the cloned sequence disrupt the chromosome and their presence can be detected by the altered pattern of DNA restriction fragments when probed on Southern blots with the cloned sequence. In one set of experiments (16), labelled DNA fragments containing the hook protein gene *flaK* were used to screen the genomic DNA from a collection of 50 *fla* mutants that had been isolated by Tn*5* mutagenesis; 19 of the insertions were found to map in a 17 kb gene cluster (Fig. 6). The functional organization of the *fla* genes, which make up the hook gene cluster, was determined by complementing each of the insertion mutants for motility with DNA fragments of different lengths that had been subcloned in vector pRK290 (16). Tn*5* insertions are polar for expression of downstream genes and, consequently, complementation by this test demands that the subcloned fragment should carry an intact transcriptional unit, including the promoter and all structural genes. This study (Fig. 6) showed that there are at least four transcription units (I–IV) in the hook gene cluster, including transcription

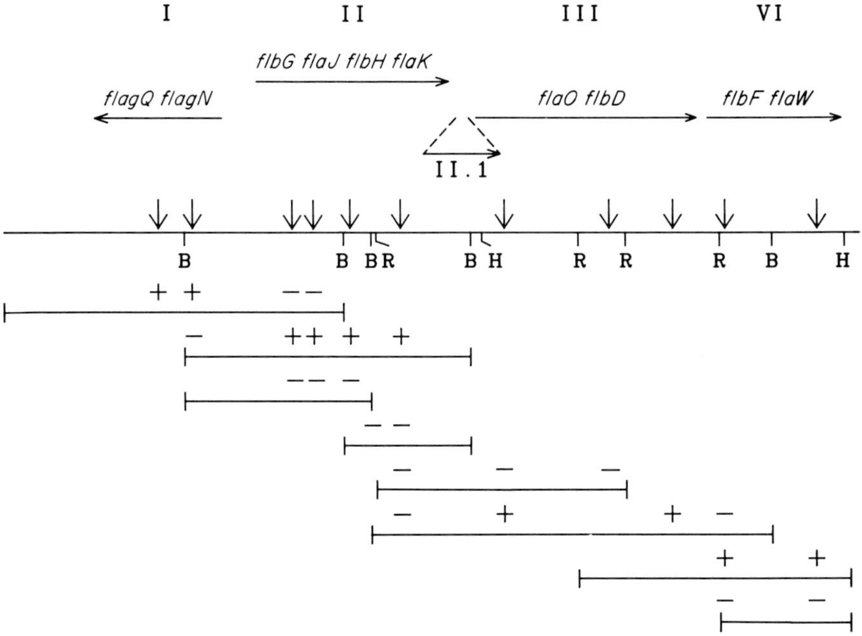

Figure 6. Complementation analysis of insertion mutations in the hook gene cluster. The locations of 11 spontaneous mutations and Tn*5* insertions are indicated by vertical arrows above the DNA restriction map. These mutations were examined for complementation on swarm plates by the DNA fragments indicated below the restriction map (+, Mot^+; −, Mot^-). These results showed that the *fla* genes are organized in transcription units I, II, III and IV (16; unpublished). S1 nuclease mapping later confirmed the extent and direction of transcription and located the transcription start sites; see horizontal arrows (18, 20, G. Ramakrishnan and A. Newton, unpublished).

unit II which contains *flaK*. Complementation has also been employed to define several transcription units in the basal body cluster (Fig. 5; 17).

The results of S1 nuclease protection assays in the hook gene cluster have confirmed the organization of transcription units deduced from the complementation analysis, allowed the direction of transcription to be determined unambiguously, and located the transcription start sites on the genetic map (18–20). These studies also identified a fifth transcription unit (Section II.1; Fig. 6) whose function has not been determined. S1 nuclease mapping and DNA sequence analysis were used almost exclusively to determine the transcriptional organization of flagellin genes in the *flaEY* cluster (19). The 29, 25, and 27 kD flagellin genes (*flgJ, flgK, flgL*) are organized in three separate transcription units that are arranged in a tandem array (Fig. 5). These genes could not be studied using complementation tests, since mutations in flagellin genes, which were not available until recently, do not give a clear *fla*$^{-}$ phenotype (see next Section). In general, however, it has been possible to use a combination of genetic and molecular techniques to define rapidly the map position and functional organization of *fla* genes in *C. crescentus*.

C. Flagellum Assembly May Follow a Linear Morphogenetic Pathway

A linear pathway for flagellum formation in *Salmonella typhimurium* and *E. coli* has been proposed in which assembly is from the inside of the cell to the outside of the cell, with the basal body formed first, the hook next, and the flagellar filament last (21). Electron micrographs of flagellum preparations from *C. crescentus fla* mutants suggest that an analogous pathway of flagellum assembly is operative in this bacterium (17). No assembled structures can be detected in *flaB, flaC, flbN* or *flbO* mutants of the basal body cluster (Fig. 7, I), while *flaD* mutants form a partially assembled basal body with only the three inner rings (II) (Fig. 7, II).

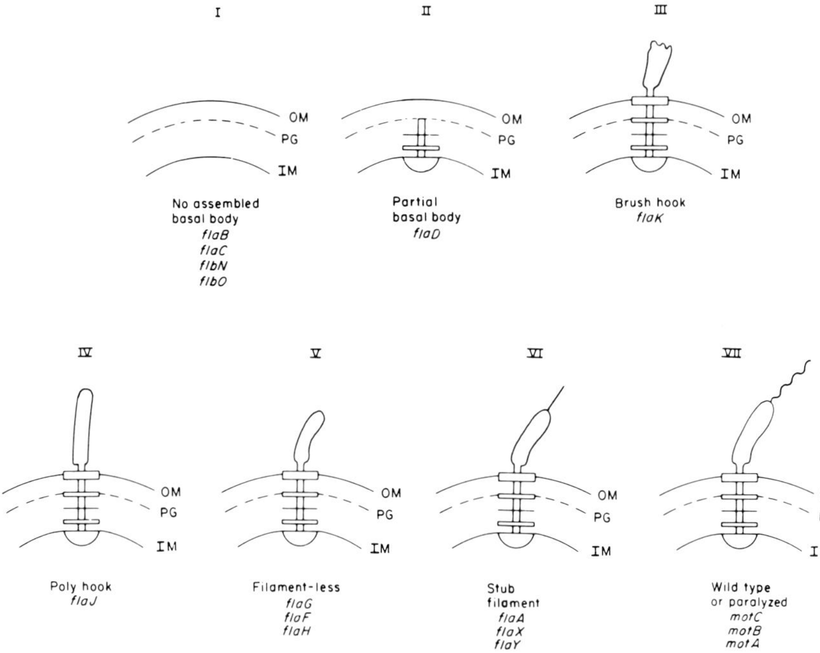

Figure 7. Proposed morphogenetic pathway for flagellum assembly. Structures diagrammed were isolated from the *fla* mutants indicated (see text for descriptions). From Hahnenberger and Shapiro (17).

Hook assembly is directly dependent on at least two genes, *flaK*, the hook protein structural gene (14), and *flaJ*, a gene that controls the length of the hook (9). Null mutants of *flaK* produce no hooks, while temperature-sensitive *flaK* mutants produce a defective hook (Fig. 7, III). By contrast, *flaJ* mutants fail to terminate hook protein assembly (9) and produce abnormally long hooks or "polyhooks" (Fig. 7, IV). A number of other *fla* mutants defective in flagellin gene expression and filament assembly produce a complete basal body and hook assembly with no filament (Fig. 7, V), while *flaA*, *flaX* (*flaE*) and *flaY* mutants produce a defective or "stub" filament (Fig. 7, VI). An apparently intact flagellum is formed in the three *mot* mutants that have been examined (Fig. 7, VII).

Although the pathway of flagellar assembly appears similar in *C. crescentus* and the enteric bacteria, one feature distinguishing *C. crescentus* is the presence of the three flagellin subunits. Interestingly, point mutations in the flagellin genes had not been isolated. Therefore, to determine the function of these genes, it was necessary to develop a gene replacement procedure to mutagenize the 25 (*flgL*), 27 (*flgK*) and 29 kD flagellin (*flgJ*) genes (S. Minnich, N. Ohta, N. Taylor and A. N., manuscript in preparation). Plasmids with the cloned flagellin gene sequences were mutagenized *in vivo* by Tn*5* in *E. coli* or *in vitro* by the replacement of specific DNA fragments by a streptomycin resistance cassette. The mutagenized sequences, which were cloned in a "suicide" plasmid incapable of replication in *C. crescentus*, were then transferred to a fla^+ *C. crescentus* strain by conjugation. A recombinant containing the gene replacement is either kanamycin or streptomycin resistant, depending on the construct, and arises by a double crossover between the chromosome and the plasmid insert containing the mutated flagellin gene. All replacement mutants were motile when examined by light microscopy. This demonstrates that none of the three flagellin genes is absolutely essential for motility and probably explains the previous failure to identify flagellin mutants. Tn*5* insertion mutants of *flgL* displayed wild-type motility on swarm plates and synthesized 25 kD flagellin, a finding that is consistent with the presence of additional copies of the 25 kD flagellin gene outside of the *flaEY* cluster (19).

Tn*5* insertion mutants of the 27 and the 29 kD flagellin genes had impaired motility on swarm plates, however, with the 29 kD flagellin mutant displaying the most severe defect (S. Minnich, *et al.*, manuscript in preparation). The role of the 27 and the 29 kD flagellins has not been determined in further detail, since the swarm assay in soft agar reflects both cell motility and chemotaxis.

D. Temporal Control of *fla* Gene Expression, and Regulatory Cascades

All of the *C. crescentus fla* genes examined to date are under temporal cell cycle regulation, as observed initially for hook protein and the flagellin genes

(Fig. 4). S1 nuclease assays carried out on 14 *fla* genes in eight different transcription units (Fig. 5), including transcription units I–IV of the hook gene cluster (18, 20, and unpublished) and the three flagellin genes of the *flaEY* cluster (19, 22), have shown that messenger RNA from these genes accumulates periodically in the cell cycle. These findings, along with those obtained using transcriptional fusions of *fla* gene promoters to either the neomycin phosphotransferase gene (23) or *lacZ* (N. Ohta and A. Newton, unpublished) and the requirement of *de novo* transcription for flagellar protein synthesis (9), have provided persuasive evidence that the periodic expression of *fla* genes is transcriptionally regulated. This conclusion has now been confirmed and extended by the analysis of *fla* gene promoters (Section III.E).

It is of particular interest that not all of the *fla* genes examined are expressed with identical periodicities. Where the protein products of *fla* genes have been identified, the genes are expressed sequentially and in an order corresponding to the time of product assembly into the flagellum. Thus, synthesis of hook protein and flagellin is initiated relatively late in the cell cycle (0.6 to 0.7 units), and occurs in the same order as subunit assembly (Figs. 3(b) and 7). The flagellin transcripts also appear in a sequence that corresponds to the order of flagellin assembly, with transcripts detected first from the 29, then the 27, and finally from 25 kD flagellin gene (Fig. 3(b); 19, 22). Recent observations also show that the transcript from *flaD*, which may code for the P ring of the basal body, appears relatively early in the cell cycle (K. Hahnenberger and L. Shapiro, personal communication).

How is this complex temporal pattern of gene regulation determined and coordinated in the cell cycle with flagellar morphogenesis? One level of control is the organization of the *C. crescentus fla* genes in a regulatory hierarchy wherein expression of each *fla* gene requires the prior expression of a *fla* gene(s) located above it in the hierarchy. This hierarchy has been elucidated by determining the effect of different *fla* mutations on the expression of other *fla* genes for which an assay is available. S1 nuclease assays have been particularly useful in these experiments, since they allow the activity of a gene to be determined even when the protein product and function is unknown. Results of these assays have shown that mutations in transcription units III or IV prevent expression of transcription units I and II and of the 25 and 27 kD flagellin genes (16, 18, 20), while mutations in transcription units I or II prevent expression of the 25 and 27 kD flagellin genes, but not of transcription unit III (20). This pattern of gene expression, coupled with the observation that all of the mutations can be complemented in *trans* by the wild-type gene (16), has led to the conclusion (24) that this set of *fla* genes is positively regulated by a cascade of *trans*-acting genes: transcriptions units III, IV --> transcription units I, II --> *flgK*, *flgL* (25, 27 kD flagellin genes). Similar studies have been carried out by determining the effects of different *fla* mutations on the levels of hook protein and flagellin synthesis (16, 23) and on *fla* gene promoter function as assayed in fusions to the neomycin phosphotransferase gene (23). A summary of the regulatory interactions between *fla, mot* and *che* genes reported to date is summarized in Fig. 8.

A regulatory hierarchy for *fla* gene expression has been described previously in *E. coli* and *Salmonella* (23). In these cells it appears to coordinate the levels of the *fla* gene products with assembly (21). The regulatory cascade of *trans*-acting *fla* genes in *C. crescentus* may play an analogous role, but it is also sufficient to account for the order of gene expression in the *Caulobacter* cell cycle. Thus, transcription units III and IV, both of which are required for expression of transcription units I and II (Fig. 8), are expressed before the two latter transcription units in the cell cycle (A. N., L.-S. Chen, N. Ohta and D. Mullin, *et al.*, manuscript in preparation; G. Ramakrishnan and A.N., unpublished). Additional evidence that the sequence in which *fla* genes are expressed is determined in part by their positions within the regulatory cascade is provided by the observations that shifting the expression of transcription unit III to a later time in the cell cycle also delays the periodic expression of the hook protein gene and the 25 and 27 kD flagellin genes (N. Ohta and A. Newton, unpublished), all of which are below transcription unit III in the hierarchy (Fig. 8). These experiments were carried out by fusing the genes in transcription unit III to the 5′ regulatory region of transcription unit II.1, a promoter whose expression is very late in the cell cycle and independent of *fla* gene expression (Fig. 6; 18).

The "clock" that initiates the regulatory cascade and ultimately determines the timing of *fla* gene expression in the cell cycle has not been identified. In contrast to *E. coli*, where there is no temporal control of flagellum biosynthesis and the master control genes for *fla* gene expression are under positive regulation by the cyclic AMP/CAP complex (21), it seems reasonable to speculate that in *C. crescentus*, a gene(s) at the top of the hierarchy is expressed in response to a cell cycle signal and thereby initiates the regulatory cascade. One candidate for the cellular clock activating this crucial *fla* gene(s) is chromosome replication. Evidence for this proposal is provided by the observation that DNA synthesis is necessary for *flaK* (hook protein gene) expression (9), but that duplication of the gene is not sufficient for its expression (24). Thus, the DNA synthetic requirement may be for expression of a gene above the hook protein gene in the regulatory cascade.

A poorly understood aspect of *fla* gene regulation is how these genes are shut off at the end of a synthetic period. *flaJ* is particularly interesting in this respect, since previous results have shown that this gene is required directly or indirectly to terminate expression of transcription unit II (9). Thus, in synchronous cultures, *flaJ* mutants continue synthesis of hook protein in swarmer cells instead of terminating synthesis before division (9), as observed in *fla*$^+$ strains (Fig. 4). An autoregulatory role has also been proposed for *flaK* on expression of transcription units I and II (16, 18) and for genes in

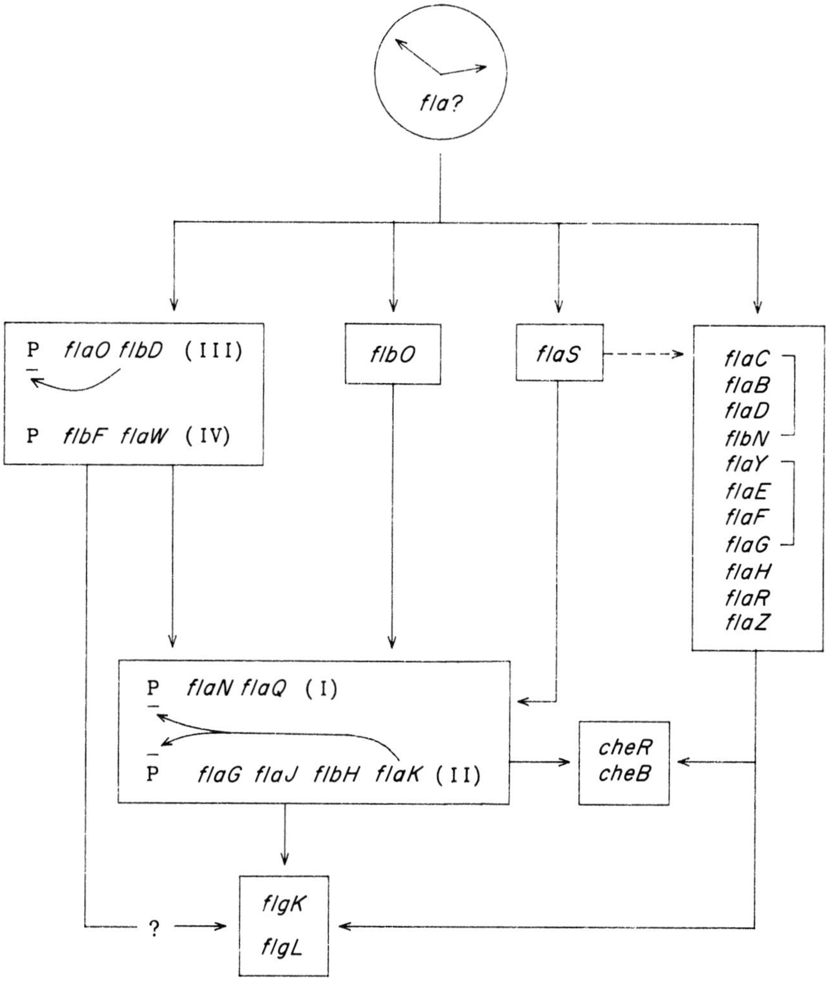

Figure 8. Regulatory hierarchy controlling flagellar and chemotactic gene expression. Arrows outside boxes indicate positive regulation by genes higher in the regulatory cascade and arrows within the boxes indicate negative (−) regulation (see text). Data taken from assays of hook protein and flagellin synthesis (16, 24) and of messenger RNA levels determined by S1 nuclease assays (18, 19, 24) in spontaneous and Tn*5* insertion *fla* mutants. Recent results have shown that *flaS* is required for expression of *flaC*, *flaB*, *flaD*, *flbN*, *flaY* and *flaE* (see dashed arrow), as well as *cheR* and *cheB* (L. Shapiro, personal communication). The dependence of *flaF*, *flaG*, *flaH*, *flaR* and *flaZ* on *flaS* has not been examined. This hierarchy is necessarily incomplete because the activity of many *fla* genes cannot yet be assayed.

transcription unit III (see model in Fig. 8; unpublished). It remains to be seen, however, whether these interactions regulate only the level and not the timing of *fla* gene expression, as seems to be the case for *flaJ*.

E. *fla* Genes are Transcribed from Specialized Promoters

A direct approach to studying regulation of the *fla* genes is to characterize the promoters and transcription factors that are required for their expression. In this laboratory the *flaN, flaK* and *flaO* promoters in the hook cluster (transcription units I, II and III, respectively; 18, 20) and the *flgJ, flgK,* and *flgL* promoters in the *flaEY* cluster (29, 25 and 27 kD flagellin genes, respectively; 19) have been examined for common sequence features. The first conclusion from these studies was that none of the sequences includes either the canonical −10, −35 promoter sequence of *E. coli* (TTGACA-$N_{16\text{–}19}$-TATAAT-$(N)_{6\text{–}9}$) or sequence elements similar to the promoter of the *C. crescentus trpFBA* operon (C. Ross and M. Winkler, personal communication), which may be considered to contain "housekeeping" genes in these bacteria. The *flaN, flaK, flgK* and *flgL* promoters do have strongly conserved nucleotide sequence elements at −12 and −24, however, with the consensus sequence TT*GG*CCC-N5-TT*GC* (Table II; 20). This −12, −24 promoter motif and consensus sequence is strikingly similar to that of the nitrogen fixation (*nif*) gene promoters of *Klebsiella pneumoniae* and *Rhizobium* sp. (25) and the glutamine synthetase (*glnA*) promoters of the enteric bacteria (26; Chapter 6). A third conserved sequence element, termed II-1, with the consensus sequence C-C-CGGC--AAA--GC-G has also been identified at approximately −100 in regions of dyad symmetry midway between divergent

Table II

fla gene promoters of *C. crescentus*

Promoter	−24	−12	+1
CcflaN (transcription unit I)	TTGGCGCC----	TTGC -----------	
CcflaK (transcription unit II)	TTGGCCCG----	TTGC ---------	C
CcflgK (25 kD flagellin gene)	TTGGCCCG----	TTGC ----------	A
CcflgL (27 kD flagellin gene)	CTGGCCCG----	TTGC -----------	C
Cc consensus (σ54-type promoters)	TTGGCCCG N4	TTGC	
Ec glnA	TTGGCACA----	TCGCT	
CcflaO (transcription unit III)	TTTACCTT----	TGGG---------	C
CcflgJ (29 kD flagellin gene)	CTCGGCGT----	TTTG--------	C

C. crescentus (*Cc*; 19, 20) and the *E. coli* (*Ec*; 25) sequences have been published previously. The nucleotide at the +1 position is indicated on the *C. crescentus* sequence where the start sites have been accurately determined. Conserved GG and GC dinucleotides at −24 and −12 are underlined.

transcription units I and II and upstream from the transcriptional start sites of the flagellin genes *flgK* and *flgL* (19).

The function of the three conserved sequence elements in the *flaK* promoter of transcription unit II has now been examined using site-directed mutagenesis to introduce point mutations and small deletions. Activity of the modified promoters was measured using *in vivo* S1 nuclease assays and the results of these experiments showed that the GG and GC dinucleotides at −12 and −24 (underlined in Table II) and the 10 nucleotide spacing between the two conserved dinucleotides is essential for promoter function (D. Mullin and A. Newton, unpublished). Both of these requirements are consistent with the nucleotide sequence features necessary for *nif* and *glnA* promoter activity in other systems. These studies also showed that deletion of any part of the 5′ end of element II-1 at −100 destroys promoter activity (D. Mullin and A. Newton, unpublished). Thus, the conserved sequence elements at −100, −24 and −10 are all required for recognition of the *flaK* promoter and presumably for expression from other *C. crescentus* promoters that contain these sequence elements (Table II).

The *nif* promoters of *K. pneumoniae, Rhizobium* spp. and *glnA* promoters of enteric bacteria (27; Chapter 6) are recognized by a specific sigma-54 which is the product of the *rpoN* (*ntrA*) gene, and it was proposed previously that transcription of *C. crescentus fla* genes with the characteristic promoter sequences at −12 and −24 might require a specialized sigma factor similar to the sigma-54 (19). Recent *in vitro* experiments support this suggestion. Sigma-54 RNA polymerase plus the transcriptional activator NRI (the *ntrC* gene product) purified from *E. coli* (26) recognize the same transcription start sites on the *flaK* and *flaN* genes utilized by *C. crescentus in vivo*. The specificity of the sigma-54 RNA polymerase recognition of the *C. crescentus fla* gene promoters is underlined by the failure of the *E. coli* sigma-70 RNA polymerase to recognize either of the two promoters. The *flaO* promoter of transcription unit III and the *flgJ*, or 29 kD flagellin gene promoter, on the other hand, have little homology to the *ntrA* promoter consensus sequence, and the *flaO* promoter is not recognized *in vitro* by either sigma-54 or the sigma-70 RNA polymerase of *E. coli* (A. Ninfa, D. Mullin, and A.N., unpublished). It is possible then that transcription unit III contains another class of promoters that is recognized by a different RNA polymerase holoenzyme or, alternatively, that there is a requirement for some additional transcriptional factor(s).

In *E. coli* and *Salmonella*, the *ntrA* promoters are regulated by activators (NRI in the case of the *E. coli glnA* gene; 26, Chapter 6) that bind to enhancers approximately 100 bp upstream from the transcription start site, and not by the availability of the sigma-54 factor. The overall organization of the conserved sequence elements at −100, −24 and −12 in the four *ntrA*-like *fla* promoters of *C. crescentus* (Table II) and the requirement of II-1 sequence for transcription from the *flaK* promoter (D. Mullin and A.N., unpublished) suggest that the II-1 element also functions as an enhancer. This element, which shares no homology with the *nif* or *glnA* enhancers, may confer the specificity for *fla* gene regulation in *C. crescentus*. As a working model we have proposed that transcription from the *flaK* promoter requires interaction between the II-1 element and an activator protein encoded by a gene higher in the regulatory cascade, perhaps a gene in transcription unit III (Fig. 8; 20). A II-1 element is also present 49 bp upstream from the start site of transcription unit III. It has been suggested that this element mediates different regulatory effects depending on its location in the promoter. Thus, the same regulator protein, could act in *trans* both to activate transcription unit II and to repress transcription unit III (Fig. 8; 20).

In summary, the promoter analysis and *in vitro* studies support the conclusion that the differential initiation of transcription is an important mechanism controlling the periodic expression of *fla* genes in the *C. crescentus* cell cycle. The results also suggest that the order of gene expression may depend on the sequential synthesis of activator proteins by genes in the regulatory hierarchy and the use of multiple sigma factors specific for the *fla* gene promoters. Proof of this model will require the isolation of *fla* gene products from *C. crescentus* that confer *in vitro* the transcriptional specificities already observed *in vivo*.

IV. CONTROL OF CHEMOTAXIS AND POSITIONING OF DIFFERENTIATED STRUCTURES

Bacterial cells have evolved a number of mechanisms to optimize their growth by taking advantage of changes in nutrient conditions. This is seen in their metabolic responses to carbon or nitrogen sources and in chemotaxis; i.e., their ability to move in response to chemical gradients in the medium. Several hexose sugars and amino acids have been identified as specific chemoattractants for *C. crescentus* (11). Considering the dimorphic life cycle of *C. crescentus* and the limited time that a cell is motile, it is not surprising that the chemotactic response of this organism is a developmental event restricted primarily to the swarmer cell stage of the cell cycle. Much of the work on chemotaxis in this organism has been carried out in the laboratories of L. Shapiro and B. Ely.

The response of enteric bacteria to a chemical determines the relative frequency of smooth swimming to tumbling. This results in their ability to swim toward an attractant or away from a repellent (28). The chemotactic response in *C. crescentus* results from changes in the direction of rotation of the polar flagellum and the direction of swimming (11). Clockwise rotation of the flagellum results in forward swimming (stalk first in the dividing cell), which may be analogous to "smooth" swimming in multiflagellated, or peritrichous, bacteria like *E. coli*, and anticlockwise rotation results in the short, backward swimming, which may be analogous to "tumbling" in the enteric bacteria (28). These responses to gradients of attractants or repellents in enteric bacteria are correlated with the level of methylation of the methyl-accepting-chemotaxis proteins (MCPs). The MCPs are methylated and demethylated by two soluble enzymes, the methyltransferase and methylesterase, respectively (28).

Eight *che* genes have been identified in *C. crescentus* and mapped to six different locations on the chromosome (11). This organization is in contrast to *E. coli* and *Salmonella* where all of the *che* and *mot* genes, along with several *fla* genes, are clustered at 41 min on the chromosome (28). Three of the *C. crescentus* genes, *cheR*, *cheB* and *cheT*, are clustered (Fig. 5), however, and mutations in any of them are characterized by the inability of cells to reverse the direction of flagellar rotation. Rotation of the flagellum in *cheR* mutants is clockwise and it is exclusively anticlockwise in *cheB* and *cheT* mutants. Mutations in the other four *che* genes do not change the frequency of reversal between clockwise and anticlockwise rotation. *In vitro* assays of *che* mutants for methyltransferase activity and methylesterase activity suggest that *cheR* is the structural gene for the transferase and that *cheB* codes for the esterase; the other *che* mutants, including *cheT*, have normal levels of transferase and esterase activities (11).

The developmental regulation of chemotaxis was demonstrated by *in vivo* methylation experiments on intact cells and *in vitro* with combinations of extracts and isolated membrane preparations from different cell types. These experiments showed that the methyltransferase, methylesterase, and MCPs are present just before cell division in the predivisional cell and in the chemotactic swarmer cell, and that the three activities are lost by the swarmer cell at the time of stalked cell differentiation (29). More detailed studies of chemotaxis have been facilitated by the fortuitous observation that antibodies prepared against the Tar protein, a *Salmonella* MCP, specifically immunoprecipitate *C. crescentus* MCPs and, similarly, that antibodies against the *Salmonella* methylesterase cross react with the *C. crescentus* esterase. Radioimmunoassays on synchronous cell cultures using these antibodies have demonstrated that the MCPs and the 36 kD methylesterase of *C. crescentus* are synthesized periodically in the predivisional cell at a time that coincides closely with the periods of hook protein and flagellin synthesis. Pulse-chase experiments also showed that radiolabelled MCPs synthesized in the predivisional cell segregated almost exclusively with the flagellar apparatus to the swarmer cell at division (29). This result suggested that newly-synthesized chemotactic proteins are compartmentalized in the swarmer portion of the predivisional cell.

A direct test of this last conclusion was performed using an immunoaffinity technique that separates membrane vesicles derived from the flagellated cell pole from vesicles derived from the remainder of the cell envelope (30). When this procedure was applied to pulse-labelled predivisional cells under lysis conditions that generate membrane vesicles approaching the size of a swarmer cell, radioimmunoassays showed that the labelled MCPs were located predominantly in the flagellated vesicles; i.e., in vesicles derived from the swarmer portion of the predivisional cell. Results of similar experiments on much smaller membrane vesicles showed that the MCPs were present in both flagellated and nonflagellated vesicles; i.e., the MCPs were not localized exclusively to a membrane compartment at the base of the flagellar rotor. Flagellar proteins are targeted for assembly to a very restricted polar domain after synthesis, and we can speculate from the above results that the chemotactic proteins are targeted to a larger membrane domain corresponding to the incipient swarmer cell portion of the predivisional cell (30).

V. PROSPECTS—THE CELL CYCLE AS A REGULATOR OF TEMPORAL AND SPATIAL PATTERNING

The developmental programme for swarmer cell differentiation in *C. crescentus* contains information for temporal control of expression of both the *fla* and *che* genes. An important element in the execution of this programme is the differential initiation of transcription, which depends in part on *ntrA*-like promoters and specialized transcriptional factors. This novel aspect of *fla* gene regulation in *C. crescentus* is particularly interesting in light of recent evidence that some *fla* genes in *E. coli* and *Salmonella* have promoters with strong sequence homology to the sigma-28 promoters described originally in *Bacillus subtilis* (31). It is tempting to speculate that during evolution, genes or regulons for the same function may have appropriated different, specialized promoter sequences and minor sigma factors.

The developmental programme in *C. crescentus* must also contain, in a form yet to be identified, positional information that directs the localized assembly of the proteins to different subcellular compartments within the predivisional cell. The possibility that the temporal and spatial components of this programme require steps in the DNA synthetic and cell division pathways, respectively, has been discussed previously (6, 9, 24). One goal of future work on developmental regulation in *C. crescentus* is to elucidate the mechanisms that couple the execution of the ordered expression of *fla* genes and the subsequent targeting of their products for localized assembly to the completion of steps in the cell cycle.

The interest in these mechanisms extends beyond the control of flagellum biosynthesis and function. Understanding how a single cell differentiates into new cell types is of fundamental importance in developmental biology. The study of this problem in *C. crescentus* should provide valuable insights into mechanisms that regulate development in more complex systems.

Acknowledgements

Special thanks to N. Ohta for suggestions on the manuscript and drawing several of the figures, to L. Egloff for help in preparing the manuscript and to L. Shapiro and B. Ely for reading the manuscript and communicating results prior to publication. Work from this laboratory was supported by Public Health Service Grants GM22299 and GM25644 from the National Institutes of Health and Grant MV-386 from the American Cancer Society.

References

(1) Poindexter, J. (1981). The Caulobacters: Ubiquitous unusual bacteria. *Microbiol. Rev.* **45**, 123–179.

(2) Newton, A., Ohta, N., Huguenel, E. and Chen, L.-S. (1985). Approaches to the study of cell differentiation in *Caulobacter crescentus. In* "The Molecular Biology of Microbial Differentiation" (P. Setlow and J. Hock, eds.), pp. 267–276. Am. Soc. Microbiol., Washington, D.C.

(3) Shapiro, L. (1985). Generation of polarity during *Caulobacter* cell differentiation. *Annu. Rev. Cell Biol.* **1**, 173–207.

(4) Ely, B. (1987). Genetic map of *Caulobacter crescentus. In* "Genetic Maps", (Stephen J. O'Brien, ed.), pp. 242–244, Cold Spring Harbor, Cold Spring Harbor Laboratory, New York.

(5) Degnen, S. T. and Newton, A. (1972). Chromosome replication during development in *Caulobacter crescentus. J. Mol. Biol.* **64**, 671–680.

(6) Huguenel, E. and Newton, A. (1982). Localization of surface structures during prokaryotic differentiation: role of cell division in *Caulobacter crescentus. Differentiation* **21**, 71–78.

(7) Sommer, J. and Newton, A. (1988). Sequential regulation of developmental events during polar morphogenesis in *Caulobacter crescentus*: Assembly of pili on swarmer cells requires cell separation. *J. Bacteriol.* **170**, 409–415.

(8) Johnson, R. C., Walsh, M. P., Ely, B. and Shapiro, L. (1979). Flagellar hook and basal complex of *Caulobacter crescentus. J. Bacteriol.* **138**, 984–989.

(9) Sheffrey, M. and Newton, A. (1981). Regulation of periodic protein synthesis in the cell cycle: control of initiation and termination of flagellar gene expression. *Cell* **24**, 49–57.

(10) Ely, B., Croft, R. H. and Gerardot, C. J. (1984). Genetic mapping of genes required for motility in *Caulobacter crescentus. Genetics* **108**, 523–532.

(11) Ely, B., Gerardot, C. J., Fleming, D. L., Gomes, S. L., Frederikse, P. and Shapiro, L. (1986). General nonchemotactic mutants of *Caulobacter crescentus. Genetics* **114**, 717–730.

(12) Purucker, M., Bryan, R., Amemiya, K., Ely, B. and Shapiro, L. (1982). Isolation of a *Caulobacter* gene cluster specifying flagellum production by using non-motile Tn*5* insertion mutants. *Proc. Natl. Acad. Sci. USA* **79**, 6797–6801.

(13) Milhausen, H., Gill, P. R., Parker, G. and Agabian, N. (1982). Cloning of developmentally regulated flagellin genes from *Caulobacter crescentus* via immunoprecipitation of polyribosomes. *Proc. Natl. Acad. Sci. USA* **79**, 6847–6851.

(14) Ohta, N., Chen, L.-S. and Newton, A. (1982). Isolation and expression of cloned hook protein gene from *Caulobacter crescentus. Proc. Natl. Acad. Sci. USA* **79**, 4863–4867.

(15) Friedman, A. M., Long, S. R., Broron, S. E., Buikema, J. W. and Ausubel, F. M. (1982). Construction of a broad-host-range cosmid cloning vector and its use in the genetic analysis of *Rhizobium* mutants. *Gene* **18**, 289–296.

(16) Ohta, N., Swanson, E., Ely, B. and Newton, A. (1984). Physical mapping and complementation analysis of transposon Tn*5* mutations in *Caulobacter crescentus*: organization of transcriptional units in the hook gene cluster. *J. Bacteriol.* **158**, 897–904.

(17) Hahnenberger, K. M. and Shapiro, L. (1987). Identification of a gene cluster involved in flagellar basal body biogenesis in *Caulobacter crescentus*. *J. Mol. Biol.* **194**, 91–103.
(18) Chen, L.-S., Mullin, D. and Newton, A. (1986). Identification, nucleotide sequence, and control of developmentally regulated promoters in the hook operon region of *Caulobacter crescentus*. *Proc. Natl. Acad. Sci. USA* **83**, 2860–2864.
(19) Minnich, S. A. and Newton, A. (1987). Promoter mapping and cell cycle regulation of flagellin gene transcription in *Caulobacter crescentus*. *Proc. Natl. Acad. Sci. USA* **84**, 1142–1146.
(20) Mullin, D., Minnich, S., Chen, L.-S. and Newton, A. (1987). A set of positively regulated flagellar gene promoters in *Caulobacter crescentus* with sequence homology to the *nif* gene promoters of *Klebsiella pneumoniae*. *J. Mol. Biol.* **195**, 939–943.
(21) Macnab, R. (1987). Flagella. *In* "*Escherichia coli* and *Salmonella typhimurium* Cellular and Molecular Biology" Vol. 1, (F. C. Neidhardt, *et al.* ed.), pp. 70–83. Am. Soc. Microbiol., Washington, D.C.
(22) Loewy, Z. G., Bryan, R. A., Reuter, S. H. and Shapiro, L. (1987). Control of synthesis and positioning of a *Caulobacter crescentus* flagellar protein. *Genes and Development* **1**, 626–635.
(23) Champer, R., Dingwald, A. and Shapiro, L. (1987). Cascade regulation of *Caulobacter* flagellar and chemotaxis genes. *J. Mol. Biol.* **194**, 71–80.
(24) Ohta, N., Chen, L.-S., Swanson, E. and Newton, A. (1985). Transcriptional regulation of a periodically controlled flagellar gene operon in *Caulobacter crescentus*. *J. Mol. Biol.* **186**, 107–115.
(25) Ausubel, F. M. (1984). Regulation of nitrogen fixation genes. *Cell* **37**, 5–6.
(26) Reitzer, L. J. and Magasanik, B. (1985). Expression of *glnA* in *Escherichia coli* is regulated at tandem promoters. *Proc. Natl. Acad Sci. USA* **82**, 1979–1983.
(27) Hoopes, B. C. and McClure, W. R. (1987). Strategies in Regulation of Transcription Initiation. *In* "*Escherichia coli* and *Salmonella typhimurium* Cellular and Molecular Biology", Vol. 2 (F. C. Neidhardt, *et al.* ed.), pp. 1231–1240. Am. Soc. Microbiol., Washington, D.C.
(28) Macnab, R. (1987). Motility and chemotaxis. *In* "*Escherichia coli* and *Salmonella typhimurium* Cellular and Molecular Biology", Vol. 1 (F. C. Neidhardt, *et al.* ed.), pp. 732–759. Am. Soc. Microbiol., Washington, D.C.
(29) Gomes, L. and Shapiro, L. (1984). Differential expression and positioning of chemotaxis proteins in *Caulobacter*. *J. Mol. Biol.* **178**, 551–568.
(30) Nathan, P., Gomes, S. L., Hahnenberger, K., Newton, A. and Shapiro, L. (1986). Differential localization of membrane receptor chemotaxis proteins in the *Caulobacter* predivisional cell. *J. Mol. Biol.* **191**, 433–440.
(31) Helmann, J. D. and Chamberlin, M. J. (1987). DNA sequence analysis suggests that expression of flagellar and chemotaxis genes in *Escherichia coli* and *Salmonella typhimurium* is controlled by an alternative σ-factor. *Proc. Natl. Acad. Sci. USA* **84**, 6422–6424.

Chapter **11**

Pathways of Developmentally Regulated Gene Expression in Bacillus subtilis

RICHARD LOSICK, LEE KROOS, JEFFERY ERRINGTON *and* PHILIP YOUNGMAN

I. INTRODUCTION

A wide range of microbes have the capacity to respond to conditions of nutrient limitation by forming a so-called "resting cell" that has greater resistance to environmental stress than the corresponding vegetative form. This kind of adaptive response is exhibited in a rather elaborate form by several species of bacteria and has served as a very useful experimental model for understanding the regulation of genes expressed during developmental cell

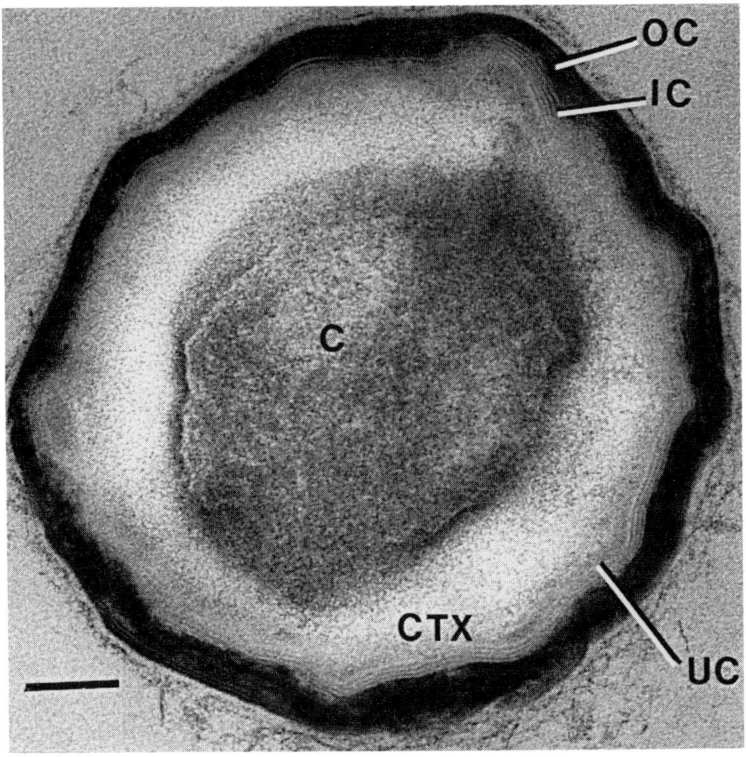

Figure 1. Electron micrograph of a *B. subtilis* endospore showing the distinct outer (OC), inner (IC) and under coat (UC) layers separated from the spore core (C) by the low density cortex (CTX). Scale bar = 0.1μm. The photograph was kindly provided by Philip C. Fitz-James of the University of Western Ontario.

differentiation. Examples include sporulation by filamentous bacteria of the genus *Streptomyces*, the formation of cysts by members of the nitrogen-fixing genus *Azotobacter* and the conversion of cells of the blue-green bacterium *Anabaena cylindrica* to akinetes. Although many kinds of bacterial resting cells are found in nature, none are more elaborate in their morphology and ultrastructure or more striking in their resistance to heat, desiccation, radiation and other adverse environmental conditions than the endospores of *Bacillus* and certain other genera of Gram-positive bacteria (Fig. 1). Formation of the endospore—so-called because it is produced within the parent vegetative cell—and the basis for its remarkable resistance properties have fascinated scientists since sporulation was discovered more than 100 years ago (1, 2). In recent years, the challenge of investigating how a vegetative cell undergoes metamorphosis into a dormant cell has centred on the species *Bacillus subtilis* in which highly developed tools of traditional and molecular genetics have produced significant progress in understanding underlying mechanisms.

One of the principal objectives of research on sporulation in *B. subtilis* is to identify genes involved in endospore formation and to understand how their activation is integrated into a temporally ordered programme of developmentally regulated gene expression. Although much progress has been made in identifying sporulation genes (so far more than 80 are known; 3, 4), an understanding of the regulatory networks that govern their expression is only just beginning to emerge (5). A complete understanding will require:

(a) determining the time at which individual genes are switched on and off during sporulation and

(b) elucidating how the activation of each developmental gene depends on the products of other such genes.

Over the last few years, detailed studies have been carried out in several laboratories on the regulation of a dozen or more genes whose transcription is induced at various times during sporulation. In this chapter, we have attempted to organize the results of these studies into an integrated scheme. Although regulatory studies have been carried out on only a few of the several scores of genes that comprise the overall programme of sporulation gene expression, it is already becoming possible to identify pathways of developmentally regulated gene expression and to propose ways in which morphological development and gene activation may be coordinated during development.

II. SPORULATION AND GERMINATION

The cycle of events leading to the formation, germination and outgrowth of spores is divided into morphologically defined stages (Fig. 2; 3, 6). Vegetative or growing cells are said to be in stage 0. As cells enter sporulation, the chromosome condenses into an axial filament. This event is conventionally called stage I of sporulation, although no mutations are known that block sporulation at this stage. The hallmark of the sporulation process and the first unequivocal morphological manifestation of sporogenesis is the formation of an asymmetrically positioned septum in stage II, which partitions the cell into two unequal compartments. The two compartments contain presumably identical chromosomes, but they have divergent developmental fates. The smaller compartment is called the forespore and is destined to become the endospore. The larger compartment is the mother cell and participates in the formation of the endospore, but lyses to liberate the mature spore when development is complete. During stage III, the mother cell membrane of the sporulation septum migrates toward the forespore pole of the cell and engulfs the forespore in a second membrane layer. This engulfment pinches off the

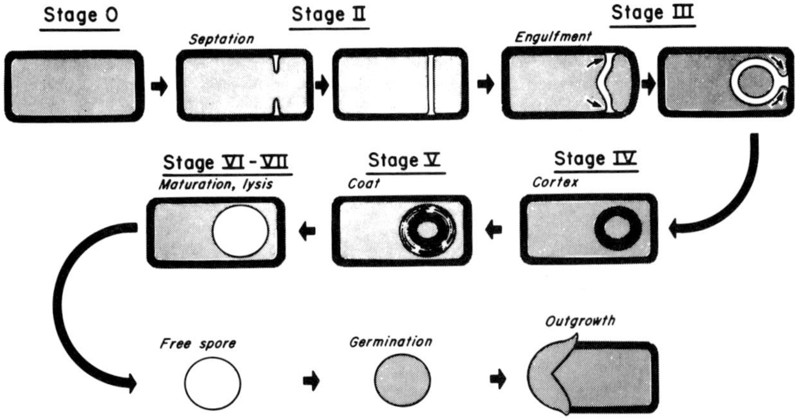

Figure 2. The stages of sporulation and germination. Reproduced with permission from reference 6.

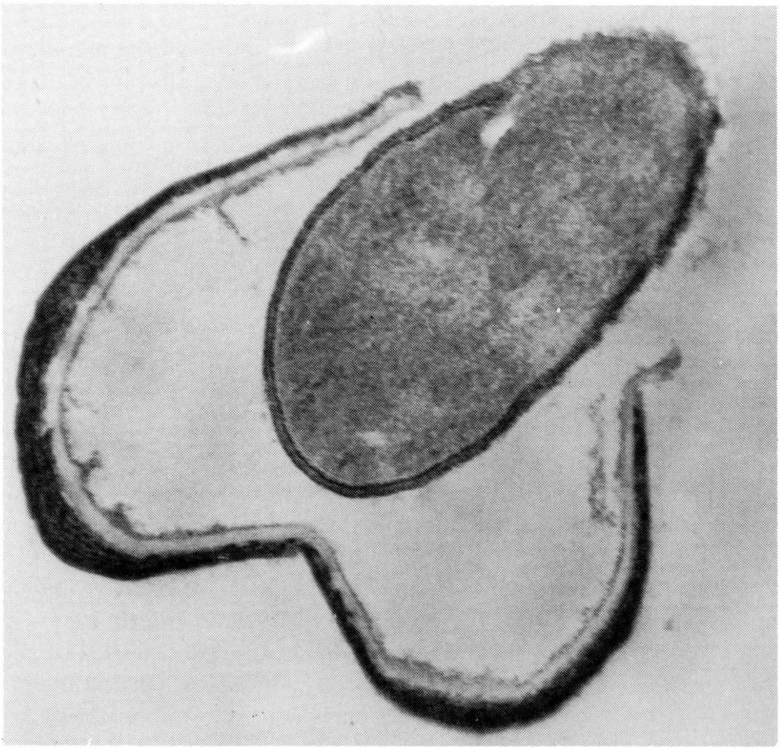

Figure 3. Electron micrograph of a germinating spore. The photograph was kindly provided by C. Robinow of the University of Western Ontario.

forespore as a protoplast within the mother cell. During stage IV, a layer of cell wall-like material called the cortex is deposited between the inner and outer membranes that surround the spore protoplast. In stage V, the coat, a tough protein shell within which the mature spore is encased, is deposited around the exterior of the cortex outer membrane. This and the following stage, VI (maturation), during which the characteristic resistance properties of the spore develop, can occur almost normally in the absence of protein synthesis, presumably by self-assembly of proteins synthesized somewhat earlier in sporulation (7). The completely mature spore (Fig. 1) is released during stage VII by lysis of the mother cell. The entire process lasts for 6 to 8 hours at 37° C.

Conversion of the spore back into a growing cell can occur readily and is believed to involve two principal morphological stages referred to as germination and outgrowth. First, in response to small molecules known as germinants, the optically refractile spore becomes hydrated and undergoes conversion to a heat-sensitive, phase-dark state. Germination is very rapid (occurring in less than a minute) and does not involve macromolecular synthesis. Efficient germination requires that the spores be activated by heat or other treatments that enhance the response to germinants. During outgrowth, the germinated spore emerges from its protein coat and undergoes metamorphosis into a vegetative rod (Fig. 3). From the start of outgrowth to the formation of the first symmetrically positioned cell division septum takes more than an hour.

III. GENES INVOLVED IN SPORULATION AND GERMINATION

Genes involved in the sporulation cycle have been traditionally identified by the isolation of mutants blocked in sporulation, germination and outgrowth (Fig. 4). For example, many (but not all) sporulation (Spo$^-$) mutants can be recognized by their failure to develop the brown colour characteristic of colonies of wild-type sporulating cells. As discussed below, the brown colour is determined by a spore structural protein whose synthesis is switched on during stage V. The white colony phenotype of many Spo$^-$ mutants is due to their failure to turn on the pigment-determining gene.

Sporulation mutations generally block sporulation at a particular stage and are classified accordingly. Thus, stage II mutations prevent sporulation from proceeding beyond the stage of septum formation and stage V mutations arrest development at the stage of coat synthesis. Germination (Ger$^-$) mutants produce colonies with normal pigmentation, but they can be distinguished from the wild-type by their failure to turn tetrazolium dye to a red colour (an indicator of metabolic activity) when it is applied in an agar overlay to chloroform-treated colonies of sporulated bacteria.

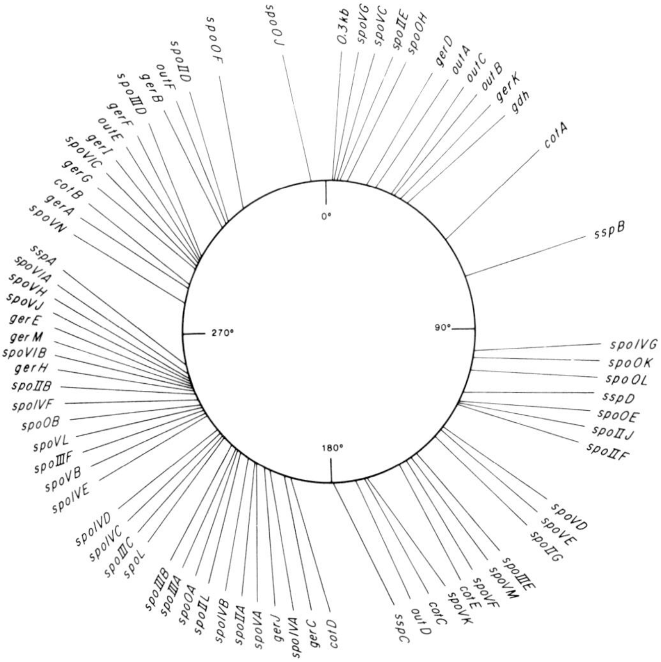

Figure 4. Genetic map of developmental genes. Based on the genetic map of reference 4. See Table I for explanation of symbols.

Mutations that are closely linked genetically and that block the developmental cycle at the same stage are considered to be in the same locus, and an upper case letter is used to indicate each such locus (3, 4). Thus, mutations that block the cycle at the engulfment stage (III) of sporulation are grouped into the six loci *spoIIIA, spoIIIB, spoIIIC, spoIIID, spoIIIE* and *spoIIIF*. Similarly, mutations that block germination are grouped into 12 loci, *gerA–gerK* and *gerM*. In this way, the many hundreds of mutations that have been identified as blocking the developmental cycle at some stage of sporulation, germination or outgrowth have been grouped into more than 60 *spo, ger* and *out* loci. The total number of genes required in the cycle is probably much greater than 60, however. The map has not been saturated for *spo, ger* or *out* mutations, and many existing mutations have not been mapped with sufficient precision to assign them confidently to a known locus. For example, insertional mutagenesis using the transposon Tn*917* (8), which greatly facilitates genetic mapping, has revealed several new sporulation and germination loci. Also, physical and genetic studies with cloned DNAs containing *spoIIA* (9), *spoIIE* (10), *spoIIG* (11), *spoVA* (12) and *gerA* (13) have shown that each is an operon composed of two or more genes required in the differentiation cycle. Finally, the application of cloning vectors based on the temperate *B. subtilis* phage ϕ105 (Fig. 5) has made it possible to clone *spo* genes on a large scale. Indeed, most of the *spo* loci have now been cloned (14 and J. Errington, unpublished results). An important advantage of this procedure is that recombinant phages are selected directly for complementation of *spo* mutations in suitable recipient strains. The phages selected must therefore contain not only intact structural genes but also any regulatory sequences necessary for their normal expression. Physical and genetic analyses of the new clones should provide a much more complete understanding of the genetic complexity of sporulation loci.

A complementary approach to identify genes involved in the sporulation cycle has been the use of "reverse genetics". This approach has been highly successful in the cloning and genetic analysis of genes encoding two classes of structural proteins of the spore. The outside of the spore, a tough protein shell known as the coat, is composed of a dozen or more different polypeptides, which together make up 50–60% of the spore protein. The coat provides a barrier that protects the spore from noxious environmental agents and also influences the responsiveness of the spore to germinants (15). The inside of the spore contains several small, acid-soluble polypeptides (known as SASPs), some of which are believed to be bound to the chromosome (16). These proteins, which together make up 8–10% of the spore protein, play a key role in the resistance of spores to UV light, and, by means of their rapid degradation during germination, provide an important source of amino acids during outgrowth. Five genes encoding spore coat polypeptides (*cot* genes) and five genes encoding the SASPs (*ssp*) have been cloned by the use of synthetic oligonucleotide hybridization probes based on partial NH_2-terminal amino acid sequences or by the use of specific antibodies for immunodetection of clones (17–20; L. Zheng, unpublished results). Null mutations of the *cot* and *ssp* genes have been created by recombinational replacement of chromosomal genes with the corresponding cloned gene that has been inactivated by means of *in vitro* recombinant DNA techniques.

Remarkably, all of the *in vitro*-constructed *cot* and *ssp* mutants produce normal-looking spores. Thus, none of the *cot* and *ssp* genes could have been identified by the traditional genetic approach of isolating mutants defective in sporulation or germination. This result is surprising and interesting considering the fact that these genes encode the major structual proteins of the spore.

Although not required in the formation of optically refractile spores, some of the *cot* and *ssp* gene products contribute significantly to the structure and properties of the mature spore. For example, mutants of *cotD* (encoding a 11 kD protein rich in sulphur-containing amino acids) and *cotE* (encoding, a 24 kD protein) are partially defective in germination (17), and *cotE* spores are highly sensitive to lysozyme, unlike wild-type spores (L. Zheng, unpublished

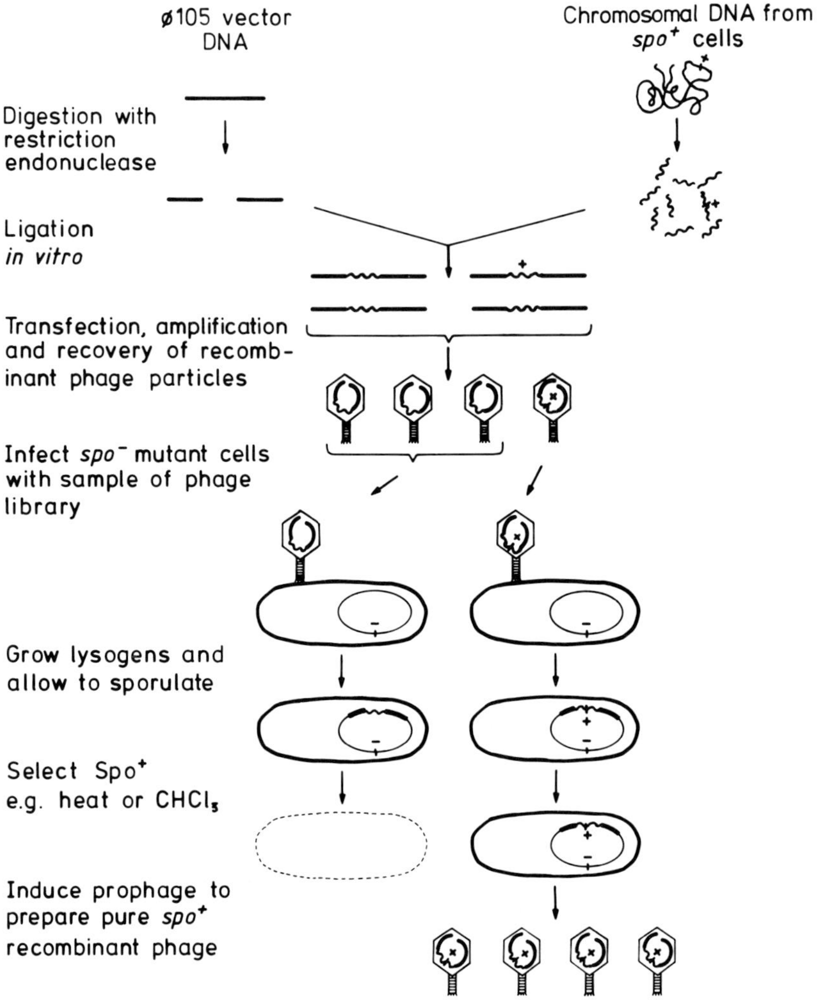

Figure 5. Cloning sporulation genes in bacteriophage Φ105. Vector and chromosomal DNA samples are digested with a suitable restriction endonuclease, usually *Bam*HI or an enzyme that gives the same type of cohesive ends. After ligation *in vitro* and transfection, the recombinant phages are amplified by plaquing in a lawn of sensitive bacteria, and the progeny are recovered to generate a phage suspension that constitutes a genomic library. Samples of the phage suspension may be tested for the ability to transduce Spo^- bacteria to Spo^+ by a simple direct selection for Spo^+ bacteria; i.e. those that can form resistant spores. The survivors of treatment with agents such as heat or chloroform vapour, which kill nonsporulating cells, contain recombinant prophages that carry a fragment of DNA which complements the defective *spo* gene in the recipient. The procedure is very convenient and a single genomic library may be used to clone many different *spo* genes.

Table I
Classes of developmental genes of *Bacillus subtilis*

Gene class	Comments
cot	Genes encoding spore coat proteins.
ger	Genes required for the normal germination properties of the mature spore (i.e. loss of refractility and response to germinants). Some *ger* mutations impair sporulation and hence may also be considered to be *spo* mutations. Many *ger* genes are expressed at intermediate to late stages of sporulation.
out	Genes involved in the outgrowth stage of spore germination. *out* mutations prevent the resumption of macromolecular synthesis which occurs prior to the appearance of the first cell division septum. Some *out* genes may be preferentially expressed during outgrowth.
spo	Genes required for sporulation, but not growth. *spo* mutations usually block sporulation at a characteristic stage and are designated by the stage at which development is arrested (*spo0, spoII, spoIII* etc.).
ssp	Genes encoding a partially homologous family of small, acid-soluble proteins (SASPs) located in the spore core.

results). In addition, the *cotE* gene product seems to play an important morphogenetic role in spore coat assembly, because spores of a *cotE* mutant lack several coat proteins, including those that form the electron-dense outermost layer of the coat apparent in electron micrographs (L. Zheng and P. Fitz-James, unpublished results). Mutation of another spore coat gene (*cotA*) has no significant effect on the resistance or germination properties of the spore, but blocks the appearance of the dark brown pigment characteristic of colonies of sporulating cells (17). *cotA* is identical to a previously known gene called *pig*, mutations in which produce albino colonies but cause no defect in spore formation. Thus, the characteristic brown colour of sporulating colonies is due to a structural component of the spore coat, the 65 kD protein product of *cotA*, which is deposited during stage V.

The properties of *ssp* mutants suggest that spore structural proteins are partially redundant. *sspA* and *sspB* mutants are not significantly defective in spore formation, germination or outgrowth, but spores of the double mutant are substantially impaired in outgrowth (21). Likewise, although spores of *sspA* mutants exhibit some UV-sensitivity (*sspB* mutants do not), spores of the double mutant are significantly more sensitive (21).

Thus, formation of the fully mature and resistant spore and its metamorphosis into a growing cell require the action of many genes of several kinds, including *spo, ger, out, cot* and *ssp* genes (Table I). A complete picture of the range of genes involved in the sporulation cycle has required the use of a variety of experimental approaches, including techniques of both traditional and molecular genetics.

IV. DEVELOPMENTAL GENES ARE SWITCHED ON IN AN ORDERED TEMPORAL SEQUENCE

To understand the programme of gene expression during the course of sporulation and germination, we must describe the expression of each of the many genes involved in the developmental cycle in terms of an underlying temporal sequence of gene activation. An important approach to this problem has been the cloning of developmental genes and the use of the cloned DNAs to monitor expression of individual *spo, ger, out, ssp* and *cot* genes. For example, cloned DNAs can be used as probes in nuclease-protection and Northern hybridization experiments to follow the appearance of specific developmental transcripts. Cloned developmental genes and operons can also be fused to reporter genes, such as *lacZ, xylE, lux* and *cat*, to create transcriptional and translational gene fusions. This permits convenient and highly sensitive monitoring of gene activity by the use of simple and sensitive assays for reporter gene products β-galactosidase, catechol 2,3-dioxygenase, luciferase and chloramphenicol acetyltransferase. An alternative approach to gene

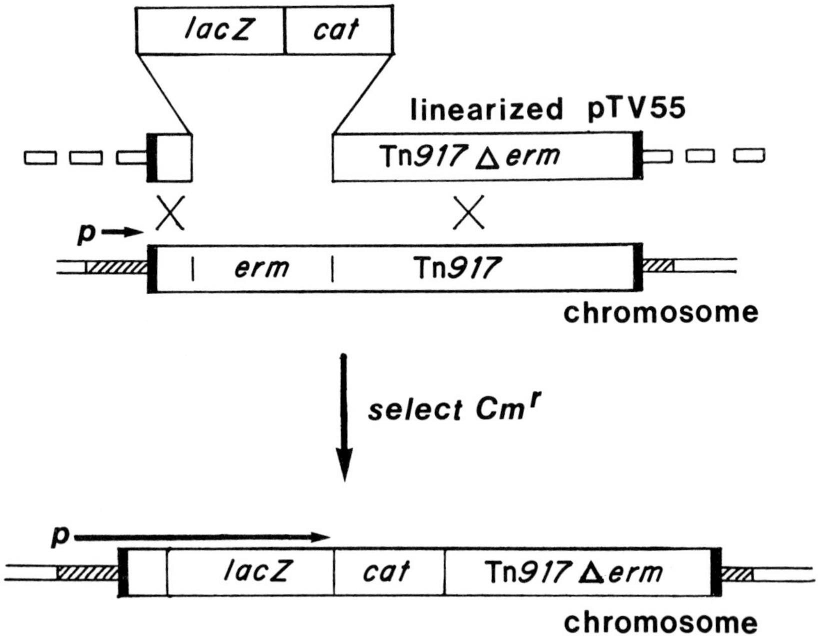

Figure 6. The conversion of a simple insertional mutation of Tn*917* in the *B. subtilis* chromosome into a *lacZ* transcriptional fused gene by transformation with linearized plasmid pTV55 DNA containing a Tn*917* derivative substituted with a promoterless *lacZ* gene and a chloramphenicol resistance gene (Cm^r) (*cat*) that has its own promoter.

cloning involves the use of fusion-generating derivatives of the transposon Tn*917*, such as Tn*917lac* (22, 23). Insertion of this transposon into developmental genes often creates transcriptional fusions as an automatic consequence of the transposition event. Methods are also available for the conversion of existing Tn*917* insertions into transcriptional fusions by recombinational replacement with Tn*917lac* (Fig. 6; 23).

The expression patterns of more than a score of developmental genes have been analysed by such methods (4). Not unexpectedly, several different temporal patterns of induction can be distinguished. Some genes (principally *spo0* genes) are found to be expressed in growing cells. Others, including many (but not all) *spoII* to *spoV* genes and *ger, ssp* and *cot* genes, are induced at various times during the course of sporulation, ranging from the time at which cells enter sporulation to the stage at which the coat is deposited. The *outB* gene is maximally transcribed during the outgrowth period. As an illustration of the diversity of times at which different genes are activated, Fig. 7 shows the

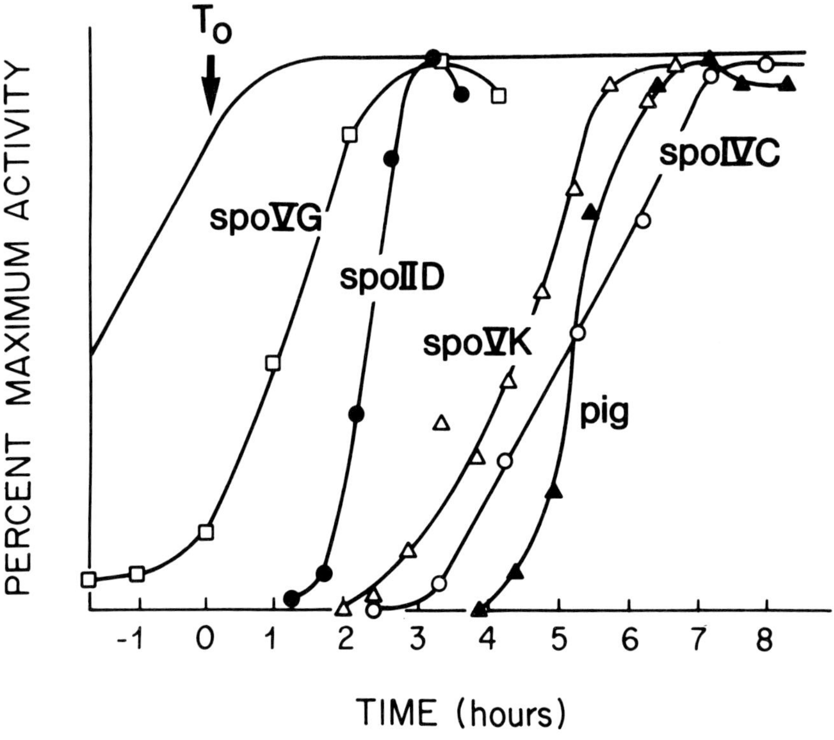

Figure 7. The temporal pattern of induction of developmental genes studied by the use of Tn*917lac*. The figure shows the patterns of transcription-fusion-directed β-galactosidase synthesis for the indicated genes. T_0 indicates the onset of sporulation.

temporal pattern of expression of several genes whose time of induction was studied by use of Tn*917lac*.

V. COMPARTMENTALIZATION OF GENE EXPRESSION

After stage II, the sporangium is partitioned into two compartments, each of which has its own chromosome. Clever but indirect genetic experiments (24) involving the transformation of late-blocked sporulation mutants at or near the time of sporulation initiation with *spo*$^+$ DNA made it possible to assign the expression of certain *spo* genes to the forespore compartment and the expression of others to the mother cell. The rationale for these experiments is that the completion of a round of DNA replication before formation of the septum can lead to one chromosome of the sporangium being mutant and the other being wild-type. These two chromosomes are then randomly segregated between the forespore and mother cell compartments. Mutations in genes whose expression is required only in the mother cell are recognized as giving rise to a significant proportion of spores that are genotypically *spo*$^-$. Conversely, mutations in genes whose expression must take place in the forespore are identified as giving rise to spores that are exclusively *spo*$^+$.

The use of *lacZ* operon fusions and methods for the subcellular localization of β-galactosidase have made it possible to demonstrate compartmentalization of gene expression directly. In these experiments, cells bearing an operon fusion of *lacZ* to a particular developmental gene are harvested at a late stage of sporulation and are then subjected to procedures that allow for differential breakage of the mother cell and forespore compartments, whose contents can be separately assayed for the presence of β-galactosidase (25). Examples of genes induced after stage II expressed only in the forespore are *spoVA* (25), *sspA* and *sspB* (26). Examples of genes whose expression is confined to the mother cell are *spoIIIC* (27), *spoIVC* and *cotA* (S. Panzer, unpublished results).

VI. DEPENDENCE PATTERNS OF DEVELOPMENTAL GENE EXPRESSION: FOUR EXAMPLES

To understand how developmental genes are organized into regulatory networks, it is necessary to determine the pattern of dependence of each individual *spo, ger, out, cot* and *ssp* gene on the expression of large numbers of other developmental genes. This task has become greatly simplified with the availability of methods for the construction of "portable" gene fusions that can be introduced into the chromosome of developmental mutants by specialized transduction or by transformation and Campbell recombination. Portable gene fusions can also be manipulated in ways (e.g. analysis of point mutations and deletions) that permit the identification of their promoters and associated regulatory sequences. When promoter and regulatory sequences have been identified by mutational analysis, it becomes possible to use cloned copies of developmental genes as templates for *in vitro* transcription studies and thereby to identify biochemically factors involved in gene expression and regulation. The four examples discussed below illustrate the kinds of information that can be derived from such studies. The examples deal sequentially with genes transcribed at progressively later times during sporulation: *spoVG, spoIIE, spoIID* and *cotA*.

spoVG is a small sporulation gene (400 bp; 4) whose transcription is induced at the onset of sporulation. However, judging from the phenotype of a *spoVG* mutant, its product is not required until the late stages of development. In experiments in which cells were treated with the sporulation-inducing agent decoyinine (28), stimulation of *spoVG* RNA synthesis could be detected within 15 min after addition of the drug, a finding that indicates that transcriptional activation of *spoVG* is a very early event in sporulation. An unusual feature of *spoVG* is the complexity of its transcription initiation region, which is composed of two overlapping promoters, P1 and P2, and an upstream AT-rich "box", which strongly enhances transcription from both start sites. This box is largely composed of alternating stretches of A's and T's, a structural feature commonly associated with the upstream regions of strongly utilized sporulation and nonsporulation promoters in *B. subtilis*. The AT-box and the overlapping P1 and P2 promoters are known to comprise the entire *cis*-acting *spoVG* regulatory region, because this region, contained in a 157 bp DNA cassette, is sufficient to mimic the normal pattern of *spoVG* transcription when fused to *lacZ* and moved to a different chromosomal location.

Expression of *spoVG* is impaired to some degree by mutations in several *spo0* loci, but particularly by mutations in *spo0A, spo0B* and *spo0H*. Interestingly, *spo0A* and *spo0H* influence the transcription of *spoVG* through separate pathways of negative and positive control. Evidence that the inhibitory effect of *spo0A* mutations on *spoVG* is indirect and is exerted through a pathway of negative control comes from the findings that the dependence of *spoVG* expression on *spo0A* can be by-passed by mutations at *abrB* (mutations that also suppress some other aspects of the phenotype associated with *spo0A* mutants) and by a *cis*-acting mutation within the *spoVG* promoter (29). It is believed that the *abrB* gene product acts, directly or indirectly, to block transcription of *spoVG* and that the *spo0A* gene product somehow causes inactivation of the *abrB* protein. The product of the *spo0H* gene, however, is a positively-acting regulatory protein that interacts directly with the *spoVG* promoter. It is an RNA polymerase sigma factor called σ^{H} (formerly σ^{30}) present in vegetative and early sporulating cells which directs transcription from the upstream *spoVG* promoter P1 both *in vivo* and *in vitro* (30, 31).

The *spoIIE* locus was recently shown to consist of a single polycistronic operon that becomes transcriptionally active between the first and second hours of sporulation (10). Mutations in all stage 0 loci, except *spo0J*, abolish or severely reduce *spoIIE* expression. Mutations in *spo0J* reduce transcription to about 15% of the wild-type level. Despite the fact that *spoIIE* remains inactive for at least 90 min after the initiation of sporulation, its expression is unaffected by mutations in all known stage II loci. This result rules out the possibility that any of the known sporulation-induced sigma factors (i.e. the *spoIIGB* and *spoIIAC* gene products; see below) are responsible for the activation of *spoIIE*. An interesting feature of the *spoIIE* promoter is the presence within what is presumed to be its RNA polymerase-binding region of sequences that conform perfectly to the "−10" and "−35" consensus sequences characteristic of promoters recognized by Eσ^A, the principal form of RNA polymerase present in vegetative *B. subtilis* bacteria. The spacing between these Eσ^A-like consensus sequences is 21 bp, however, which is significantly larger than the 17–19 bp typically observed. This raises the possibility that the *spoIIE* operon may be transcribed by a modified form of Eσ^A. Another striking feature of the *spoIIE* promoter is that sequences extending more than 118 bp upstream from its transcription start site are required for expression at wild-type levels. The *spoIIE* operon may be of considerable interest from the stand-point of its gene products as well: mutations in each of its three open-reading frames prevent the proteolytic processing of σ^E, an event that is believed to control critical changes in gene expression during the second hour of sporulation (see below).

spoIID is an example of a sporulation gene whose transcription is induced at the second to third stage of development (32, 33). The genetic requirements for induction of *spoIID* are more complex than those governing the expression of *spoVG* and *spoIIE*. Transcription of *spoIID* depends on many of the stage 0 genes and on the stage II operons *spoIIA*, *spoIIE* and *spoIIG* (32, 33). Transcription of *spoIID in vitro* can be achieved with RNA polymerase associated with the sigma factor called σ^E (formerly σ^{29}). σ^E is the product of the promoter-distal gene (*sigE* or *spoIIGB*) of the *spoIIG* operon (34, 35). The primary product of *sigE* is an inactive precursor to σ^E of 31 kD called P^{31} or pro-σ^E, which is processed to the mature sigma factor by proteolytic removal of 29 amino acids from the NH_2-terminus (34).

The appearance of Eσ^E holoenzyme is the primary event required for the induction of *spoIID*, and the complex genetic dependences of *spoIID* expression can be largely understood as relating directly or indirectly to events involved in the enzymatic processing of pro-σ^E to its active form. Thus, pro-σ^E processing depends on the product of the promoter-proximal gene of the *spoIIG* operon (called *spoIIGA*), which may be the actual processing enzyme (11; P. Stragier, C. Bonamy and C. Karazyn-Campelli, personal communication), and also on the products of all three genes of the *spoIIE* operon (10; P. Youngman and W. Haldenwang, unpublished results). In addition, recent experiments by P. Stragier, C. Bonamy and C. Karazyn-Campelli (personal communication) and by W. Haldenwang (personal communication) indicate that pro-σ^E processing is prevented by a mutation in the promoter-proximal gene (*spoIIAA*) of the *spoIIA* operon and delayed significantly by a mutation in the promoter-distal gene (*spoIIAC*). Expression of the *spoIIA*, *spoIIE* and *spoIIG* operons depends, in turn, on the products of various *spo0* genes.

P. Stragier, C. Bonamy and C. Karazyn-Campelli (personal communication) speculate that the putative processing enzyme SpoIIGA, whose predicted amino acid sequence suggests that it is a membrane protein, must integrate into the properly assembled sporulation septum in order to catalyse the conversion of pro-σ^E to its active form. Morphogenesis of the sporulation septum would depend, in turn, on the products of the *spoIIE* and *spoIIA* operons. Interestingly, the nucleotide sequence of *spoIIE* suggests that at least some of its products are membrane proteins (P. Guzman, J. Westpheling and P. Youngman, unpublished results) and the sequence of the *spoIIA* operon suggests that it encodes an RNA polymerase sigma factor, the predicted product of the promoter-distal gene *spoIIAC* (36). According to the model of P. Stragier, C. Bonamy and C. Karazyn-Campelli (personal communication), the *spoIIAC*-encoded sigma directs the synthesis during sporulation of ancillary proteins (possibly the products of certain vegetative septation genes) involved in the formation of the sporulation septum.

As a final example, we consider the case of the spore coat protein gene *cotA*. As discussed above, the 65 kD protein product of *cotA* is a component of the spore coat that determines the brown pigment characteristic of colonies of wild-type sporulating cells (17). Transcription of *cotA* is induced at a relatively late time in sporulation (stages IV–V). This RNA synthesis is controlled from a promoter region that is located just upstream of the initiation codon for the *cotA* open-reading frame (37). The upstream boundary of the promoter precedes the transcription start site by no more than 55 bp. Interestingly, expression of this spore coat protein gene is absent or strongly impaired in almost all mutants blocked prior to stage V, the stage during which coat synthesis and assembly take place. This is surprising because it is unlikely that all stage 0–IV genes are regulatory determinants in the usual sense. It suggests instead that transcriptional activation of *cotA* may be coupled in some way to morphogenetic or physiological events occurring at an intermediate stage of sporulation. In other words, the developing sporangium may have to proceed through stage III or IV in order to be competent to activate spore coat gene expression. Thus, expression of genes like *cotA* may be kept in register with the course of sporogenesis by a feedback mechanism in which the transcriptional apparatus somehow senses a landmark event associated with the developing sporangium, in analogy with the proposed dependence (see above) of *spoIID* induction on formation of the sporulation septum.

VII. PATHWAYS OF DEVELOPMENTALLY REGULATED GENE EXPRESSION

A principal long-range goal of studies on the regulation of developmental genes in *B. subtilis* is to organize all *spo, ger, out, ssp* and *cot* genes into a regulatory network that describes the dependence of the expression of each gene on all other developmental genes. The derivation of such a regulatory network will probably require the construction (by *in vitro* or *in vivo* methods as discussed above) of a fusion of each developmental gene to a reporter gene and the introduction of these gene fusions into large numbers of developmental mutants. It can then be determined whether and to what extent the gene fusion is expressed. For genes whose transcription is switched on early in the developmental cycle, it may suffice to determine dependence only in mutants blocked early in development; it can generally be assumed that genes induced early in sporulation will not depend for their expression on genes acting at late times. However, genes switched on at intermediate or late stages may need to be studied in a nearly complete collection of developmental mutants in order to obtain a full picture of their dependences on other genes.

Although many developmental genes have been cloned from *B. subtilis*, only a few of the cloned DNAs have been used to monitor gene expression in a sufficiently large number of developmental mutants for the dependence pattern of the gene to be compared with those of other genes. As a preliminary effort to describe the programme of developmental gene expression in *B. subtilis*, we have attempted to integrate the results of several laboratories on 13 genes that have been studied the most extensively. For simplicity, we have limited ourselves to genes whose transcription is induced during the course of sporulation. (Thus, we are not considering certain *spo0* genes that are expressed in growing cells or *out* genes, which are turned on during the outgrowth stage.) Also, in considering the dependence of the expression of one gene on another, we have arbitrarily ignored cases in which only a partial dependence is observed. Thus, the induction of a gene is considered to be dependent on another gene only if its level of expression is at least five to ten times lower in mutant cells than in the wild-type. We emphasize that our proposed scheme is provisional, because it is based on the results of several different investigators who used partially overlapping sets of developmental mutants representing several different genetic backgrounds in determining dependences.

Fig. 8 maps the dependence patterns of these 13 developmental genes in a stepwise hierarchical scheme. The scheme is adapted from that devised by Kroos and Kaiser (38; see also Chapter 12) to describe the dependence patterns of developmental genes in *Myxococcus xanthus* on extracellular factors governing fruiting body formation. An important feature of the diagram (38) is that it summarizes complicated information in a simple form.

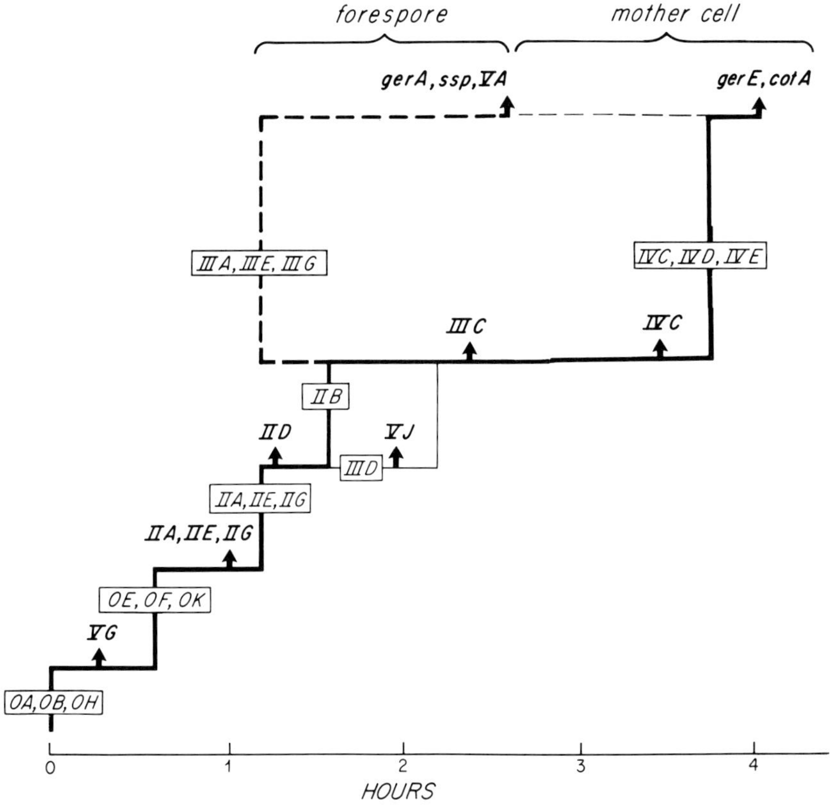

Figure 8. Dependence patterns of developmental gene expression. See the text for an explanation of the figure. The indicated dependences are based on work cited in Sections V and VI of the text, reviewed in reference 4, reported in references 39 and 40 or are based on unpublished work from the laboratories of the authors or communicated to us by colleagues in the *B. subtilis* field, including A. Moir, C. Moran, P. Piggot, P. Setlow and P. Stragier. A few of the dependences shown are inferential and not based on direct measurement. These are the dependence of *spoIIG* expression on *spo0E*, *spo0F* and *spo0K*, the dependence of *gerA*, *spoIIIG*, and the implied lack of strong dependence of *spoIIIC* expression on *spoIIIG*. *spoIIIG* is a newly-discovered gene (P. Stragier, personal communication) that encodes an RNA polymerase sigma factor for *ssp* and certain other forespore specific genes (P. Setlow and P. Stragier, personal communication).

The 13 genes are highlighted in bold above the arrows in Fig. 8. These genes were studied by the use of gene fusions or by the use of a cloned copy of the gene as a hybridization probe. The mutants in which dependences were determined are indicated in the boxes. The time of induction of each gene is indicated by its position along the abscissa. Thus, genes that are switched on early in sporulation are near the left of the figure and genes that are switched on at late times are near the right-hand side of the figure. Position along the ordinate represents the level of dependence on boxed-in genes beneath it in the step-graph. Since dependences are cumulative with increasing numbers of vertical steps, a higher position along the ordinate indicates dependence on a greater number of genes. For example, expression of *spoVG* depends only on *spo0A, spo0B* and *spo0H* (i.e. *spoVG* expression was found to be substantially blocked only in *spo0A, spo0B* and *spo0H* mutants). On the other hand, induction of *spoIIE*, which is indicated at a higher step than *spoVG* induction, depends not only on *spo0A, spo0B* and *spo0H* but also on *spo0E, spo0F* and *spo0K*.

The 13 genes can be thought of as "markers" of an underlying programme of developmental gene expression. The diagram maps the positions of the marker genes in a network reflecting pathways of developmental gene expression. For approximately the first two hours of sporulation, marker genes can be seen to fit generally in a simple linear pathway of cumulative dependences. After the second hour, the remaining genes are accommodated in the diagram by two principal branches. Members of the *gerA, spoVA* and *ssp* branch are genes whose expressions are known to be confined to the forespore, whereas *spoIIIC, spoIVC, gerE* and *cotA* are genes whose expressions are confined to the mother cell. Thus the diagrammatic branch can be correlated with a physical branch: the compartmentalization of gene expression that occurs after stage II.

Interestingly, expression of the latest mother cell-specific genes, *gerE* and *cotA*, depends on the *spoIIIA, spoIIIE* and *spoIIIG* gene products, which are otherwise required only for expression of genes on the forespore branch. This is shown in the diagram by the thin broken line creating a loop that connects the forespore branch with the terminus of the mother cell branch. Thus, production of a forespore may be the morphogenetic event (see above) to which *cotA* transcriptional activation is coupled. Note, also, the loop that denotes the dependence of the mother cell branch on the *spoIIID* gene product. This loop shows that *spoVJ* expression depends on the *spoIIID* gene product but not on the product of *spoIIB*. Undoubtedly, several other branches will emerge as efforts are made to fit other marker genes into the diagram and as the bases for the dependence relationships begin to be understood at the molecular level.

VIII. OVERVIEW, IMPLICATIONS AND PROSPECTS

The sporulation cycle in *B. subtilis* involves the regulated activation of many (probably more than 100) genes located at many chromosomal positions. The expression of these genes is regulated temporally and spatially. Coordinately controlled gene sets are switched on in an ordered sequence that lasts for several hours. At intermediate to late developmental stages, the expression of some genes is compartmentalized between the forespore and mother cell chambers of the sporangium.

Several kinds of genes are involved in the sporulation cycle. These include *spo, ger* and *out* genes, which are principally identified by the traditional approach ("forward" genetics) of isolating mutants blocked in differentiation, and *ssp* and *cot* genes, which are identified by the strategy of cloning the structural genes for spore-specific proteins ("reverse" genetics). Genes identified by mutations that block the sporulation cycle very often turn out upon further analysis (i.e. nucleotide sequencing) to be true transcriptional determinants (e.g. *spo0A, spo0F, spo0H, spoIIA, spoIIG, spoIIIC* and *gerE*) or genes for morphogenetic proteins that are indirectly required for the activation of transcriptional regulatory proteins (e.g. *spoIIE*). Conversely, genes identified as coding for the major structural components of the spore generally have only a subtle and largely dispensable role in the sporulation cycle. This suggests that mutations in only special kinds of genes (those whose impairment has highly pleiotropic consequences) cause blocks at discrete stages.

One explanation for the high proportion of *spo* genes that are transcriptional determinants or are required for the activation of regulatory proteins is that much redundancy is built into the structural genes whose products are directly involved in the metamorphosis of a cell into a spore (as has been demonstrated for *ssp* genes). Thus in many cases, the absence of any individual gene does not by itself cause a noticeable block in sporulation or germination. Another explanation is that many of the morphogenetic events of sporulation may involve the products of genes that are also required in vegetative growth. Mutations in such genes would not be expected to appear in traditional screens of developmental mutants.

Two implications follow from these considerations. First, the systematic cloning (14) and sequencing of genes required in the sporulation cycle will continue to be highly rewarding in the identification of key regulatory genes. Second, a comprehensive identification of the genes that comprise the regulons of sporulation will require schemes that emphasize expression (22) rather than mutant phenotypes. It is noteworthy that a high proportion of developmentally regulated genes of the differentiating bacterium *Myxococcus xanthus* (see Chapter 10), identified on the basis of being expressed during development, were not found to be required for fruiting body or spore formation.

The application of powerful genetic (e.g. the use of gene fusions) and biochemical (e.g. the use of *in vitro* transcription systems) methods have begun to provide a basis for organizing genes of the sporulation cycle into pathways that describe the complex interdependences of developmental gene expression. The elucidation of an underlying regulatory network that accounts for the entire programme of sporulation and germination gene expression still remains a formidable challenge but increasingly a realistic one.

Acknowledgements

We are grateful to S. Cutting for helpful discussions, to P. Stragier for communicating his results prior to publication, and to M. Yudkin for advice on the manuscript. This work was supported by NIH grants GM18568 to R.L. and GM35495 to P.Y.; L.K. is a postdoctoral fellow of the Helen Hay Whitney Foundation.

References

(1) Cohn, F. (1876). Untersuchungen über Bakterien: IV Beiträge zur Biologie der Bacillen. *Beiträge zur Biologie der Pflanzen* **2**, 249–276.

(2) Koch, R. (1876). Die Aetiologie der Milzbrand-Krankheit, begrundet auf die Entwicklungsgeschichte des *Bacillus anthracis*. *Beitrage zur Biologie der Pflanzen* **2**, 277–310.

(3) Piggot, P. and Coote, J. (1976). Genetic aspects of bacterial endospore formation. *Bacteriol. Rev.* **40**, 908–962.

(4) Losick, R., Youngman, P. and Piggot, P. J. (1986). Genetics of endospore formation in *Bacillus subtilis*. *Ann. Rev. Genet.* **20**, 625–669.

(5) Mandelstam, J. and Errington, J. (1987). Dependent sequences of gene expression controlling spore formation in *Bacillus subtilis*. *Microbiol. Sci.* **4**, 238–244.

(6) Losick, R. and Youngman, P. (1984). Endospore formation in *Bacillus subtilis*. *In* "Microbial Development" (R. Losick and L. Shapiro, eds.), pp. 63–88. Cold Spring Harbor, Cold Spring Harbor Laboratory, New York.

(7) Jenkinson, H. F., Kay, D. and Mandelstam, J. (1980). Temporal dissociation of late events in *Bacillus subtilis* sporulation from expression of genes that determine them. *J. Bacteriol.* **141**, 793–805.

(8) Sandman, K., Losick, R. and Youngman, P. (1987). Genetic analysis of *Bacillus subtilis spo* mutations generated by Tn*917*-mediated insertional mutagenesis. *Genetics* **117**, 603–617.

(9) Fort, P. and Piggot, P. (1984). Nucleotide sequence of sporulation locus *spoIIA* in *Bacillus subtilis*. *J. Gen. Microbiol.* **130**, 2147–2153.

(10) Guzman, P., Westpheling, J. and Youngman, P. (1988). Characterization of the promoter region of the *Bacillus subtilis spoIIE* operon. *J. Bacteriol.* (In press).

(11) Jonas, R., Weaver, E. A., Kenney, T. J., Moran, C. P. and Haldenwang, W. G. (1987). The *spoIIG* operon of *Bacillus subtilis* encodes both σ^E and a gene necessary for its activation. *J. Bacteriol.* (In press).

(12) Fort, P. and Errington, J. (1985). Nucleotide sequence and complementation analysis of a polycistronic sporulation operon, *spoVA*, in *Bacillus subtilis*. *J. Gen. Microbiol.* **131**, 1091–1105.

(13) Zuberi, A. R., Fevers, I. M. and Moir, A. (1986). Identification of three complementation units in the *gerA* spore germination locus of *Bacillus subtilis*. *J. Bacteriol.* **162**, 756–762.

(14) Errington, J. and Jones, D. (1987). Cloning in *Bacillus subtilis* by transfection with bacteriophage vector ϕ105J27: Isolation and preliminary characterization of transducing phages for 23 sporulation loci. *J. Gen. Microbiol.* **133**, 493–502.

(15) Aronson, A. I. and Fitz-James, P. (1976). Structure and morphogenesis of the bacterial spore coat. *Bacteriol. Rev.* **40**, 360–402.

(16) Fliss, E. R., Connors, M. J., Loshon, C. A., Curiel-Quesada, E., Setlow, B. and Setlow, P. (1985). Small, acid-soluble spore proteins of *Bacillus*: products of a sporulation-specific, multigene family. *In* "Molecular Biology of Microbial Differentiation" (J. A. Hoch and P. Setlow, eds.), pp. 60–66. American Society for Microbiology, Washington, D.C.

(17) Donovan, W., Zheng, Z., Sandman, K. and Losick, R. (1987). Genes encoding spore coat proteins from *Bacillus subtilis*. *J. Mol. Biol.* **196**, 1–10.
(18) Connors, M. J., Mason, J. M. and Setlow, P. (1986). Cloning and nucleotide sequencing of genes for three small, acid-soluble proteins from *Bacillus subtilis* spores. *J. Bacteriol.* **166**, 417–425.
(19) Connors, M. J. and Setlow, P. (1985). Cloning of a small, acid-soluble spore protein gene from *Bacillus subtilis* and determination of its complete nucleotide sequence. *J. Bacteriol.* **161**, 333–339.
(20) Hackett, R. H. and Setlow, P. (1987). Cloning, nucleotide sequencing, and genetic mapping of the gene for small, acid-soluble spore protein gamma of *Bacillus subtilis*. *J. Bacteriol.* **169**, 1985–1992.
(21) Mason, J. M. and Setlow, P. (1986). Essential role of small, acid-soluble spore proteins in resistance of *Bacillus subtilis* spores to UV light. *J. Bacteriol.* **167**, 174–178.
(22) Youngman, P., Zuber, P., Perkins, J. B., Sandman, K., Igo, M. and Losick, R. (1985). New ways to study developmental genes in spore forming bacteria. *Science* **228**, 285–291.
(23) Perkins, J. and Youngman, P. (1986). Construction and properties of Tn*917lacZ*, a transposon derivative that mediates transcriptional gene fusions in *Bacillus subtilis*. *Proc. Natl. Acad. Sci. USA* **83**, 140–144.
(24) Lencastre, H. and Piggot, P. (1979). Identification and different sites of expression for *spo* loci by transformation of *Bacillus subtilis*. *J. Gen. Microbiol.* **114**, 377–389.
(25) Errington J. and Mandelstam, J. (1986). Use of a *lacZ* fusion to determine the dependence pattern and the spore compartment expression of sporulation operon *spoVA* in *spo* mutants of *Bacillus subtilis*. *J. Gen. Microbiol.* **132**, 2977–2985.
(26) Mason, J., Hackett, R. H. and Setlow, P. (1988). Regulation of expression of genes coding for small, acid-soluble proteins of *Bacillus subtilis* spores: studies using *lacZ* fusions. *J. Bacteriol.* **170**, 239–244.
(27) Turner, S. M., Errington, J. and Mandelstam, J. (1986). Use of *lacZ* gene fusion to determine the dependence pattern of sporulation operon *spoIIIC* in *spo* mutants of *Bacillus subtilis*: a branched pathway of expression of sporulation operons. *J. Gen. Microbiol.* **132**, 2995–3003.
(28) Freese, E., Heinze, J. E. and Galliers, E. M. (1979). Partial purine deprivation causes sporulation of *Bacillus subtilis* in the presence of excess ammonia, glucose and phosphate. *J. Gen. Microbiol.* **115**, 193–205.
(29) Zuber, P. and Losick, R. (1987). Role of AbrB in Spo0A- and Spo0B-dependent utilization of a sporulation promoter in *Bacillus subtilis*. *J. Bacteriol.* **169**, 2223–2230.
(30) Carter, L. and Moran, C. P. (1986). New RNA polymerase sigma factor under *spo0* control in *Bacillus subtilis*. *Proc. Natl. Acad. Sci. USA* **83**, 9438–9442.
(31) Dubnau, E., Weir, J., Nair, G., Carter, L. and Moran, C. (1988). The *Bacillus* sporulation gene *spo0H* codes for σ^{30} (σ^{H}). *J. Bacteriol.* **169**, 1181–1191.
(32) Rong, S., Rosenkrantz, M. and Sonenshein, A. L. (1986). Transcriptional control of the *spoIID* gene of *Bacillus subtilis*. *J. Bacteriol.* **165**, 771–779.
(33) Clarke, S., Lopez-Diaz, I. and Mandelstam, J. (1986). Use of *lacZ* gene fusions to determine the dependence pattern of the sporulation gene *spoIID* in *spo* mutants of *Bacillus subtilis*. *J. Gen. Microbiol.* **132**, 2987–2994.
(34) Labell, T. L., Trempy, J. E. and Haldenwang, W. G. (1987). Sporulation-specific σ factor σ^{29} of *Bacillus subtilis* is synthesized from a precursor protein, P^{31}. *Proc. Natl. Acad. Sci. USA* **84**, 1784–1788.
(35) Kenney, T. J. and Moran, C. P. E. (1987). Organization and regulation of an operon that encodes a sporulation-essential sigma factor in *Bacillus subtilis*. *J. Bacteriol.* **169**, 3329–3339.
(36) Errington, J., Fort, P. and Mandelstam, J. (1985). Duplicated sporulation genes in bacteria. *FEBS Lett.* **188**, 184–188.
(37) Sandman, K., Kroos, L., Cutting, S., Youngman, P. and Losick, R. (1988). Identification of

the promoter for a spore coat gene in *Bacillus subtilis* and studies on the regulation of its induction at a late stage of sporulation. *J. Mol. Biol.* **200**, 461–473.

(38) Kroos, L. and Kaiser, D. (1987). Expression of many developmentally regulated genes in *Myxococcus* depends on a sequence of cell interactions. *Genes and Development* **1**, 840–854.

(39) Errington, J. and Mandelstam, J. (1986). Use of a *lacZ* fusion to study the dependence pattern of sporulation operon *spoIIA* in wild-type *Bacillus subtilis* and in *spo* mutants. *J. Gen. Microbiol.* **132**, 2987–2989.

(40) Zuberi, A. R., Moir, A. and Feavers, I. M. (1987). The nucleotide sequence and gene organization of the *gerA* spore germination operon of *Bacillus subtilis* 168. *Gene* **51**, 1–11.

Chapter **12**

Multicellular Development in Myxobacteria

DALE KAISER

I. INTRODUCTION

Myxobacteria lie on the boundary between uni- and multicellular organisms. They grow and divide as separate cells, yet they constitute a primitive multicellular organism whose cells feed in multicellular units, and which, in times of starvation, assemble compact and regular structures of about 100 000 cells. These structures, called fruiting bodies, have a variety of shapes, four of which are illustrated in Fig. 1. Evidently the shapes are under genetic control because each species produces characteristic fruiting bodies. How are their forms determined? Since myxobacteria are simple cells, perhaps one can answer this question in molecular terms.

Myxobacteria live by secreting hydrolytic enzymes with which they degrade particulate organic matter in the soil (1). When their only nutrient in culture

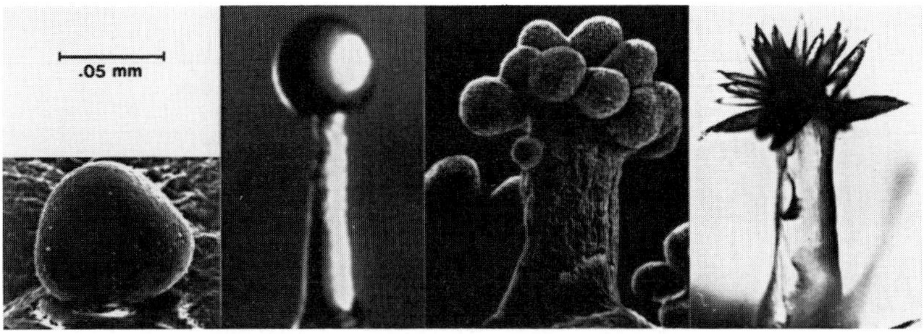

Figure 1. Fruiting bodies of four species of myxobacteria. Left to right: *Myxococcus xanthus*, *Myxococcus stipitatus*, *Stigmatella aurantiaca*, and *Chondromyces apiculatus*. These structures each contain about 100 000 cells. *M. xanthus*, courtesy of J. Kuner. *M. stipitatus* and *C. apiculatus* reproduced from (3); copyright Syndics of the Cambridge University Press. *S. aurantiaca* reproduced from (4); copyright American Society for Microbiology.

is a polymeric substance, casein, their rate of growth is found to increase with cell density (2). Thus, cells feed cooperatively and the association of cells in groups allows them to feed more efficiently. Perhaps the advantage of cooperative feeding has driven their evolution of aggregates in the form of fruiting bodies. When food is again available, the multicellular nature of a fruiting body ensures that a new cycle of growth will be started by a community of cells. Myxobacteria move by gliding on surfaces, which also facilitates their feeding and permits cells to move over each other when they construct a fruiting body (5).

In the history of life on earth, multicellular organisms have evolved from unicells on more than 15 independent occasions (6). Myxobacteria may represent one of the earliest attempts to build a multicellular organism, and their age is estimated at two billion years (7). Myxobacteria lend themselves to experimental investigation of the multicellular state and the cell interactions that support it. Such interactions are essential for their multicellular development. However, their development is gratuitous; it is not necessary for the growth of a cell or a population of cells. Mutants that fail to form multicellular aggregates can be propagated indefinitely by maintaining cells in the growth phase. This chapter describes the use of genetics to investigate the coordinate regulation of large numbers of myxobacterial cells.

II. FRUITING BODY DEVELOPMENT FOLLOWS A PROGRAMME

Despite the relative simplicity of a fruiting body, the progressive and reliable creation of structure and order from initial disorder, which is the hallmark of

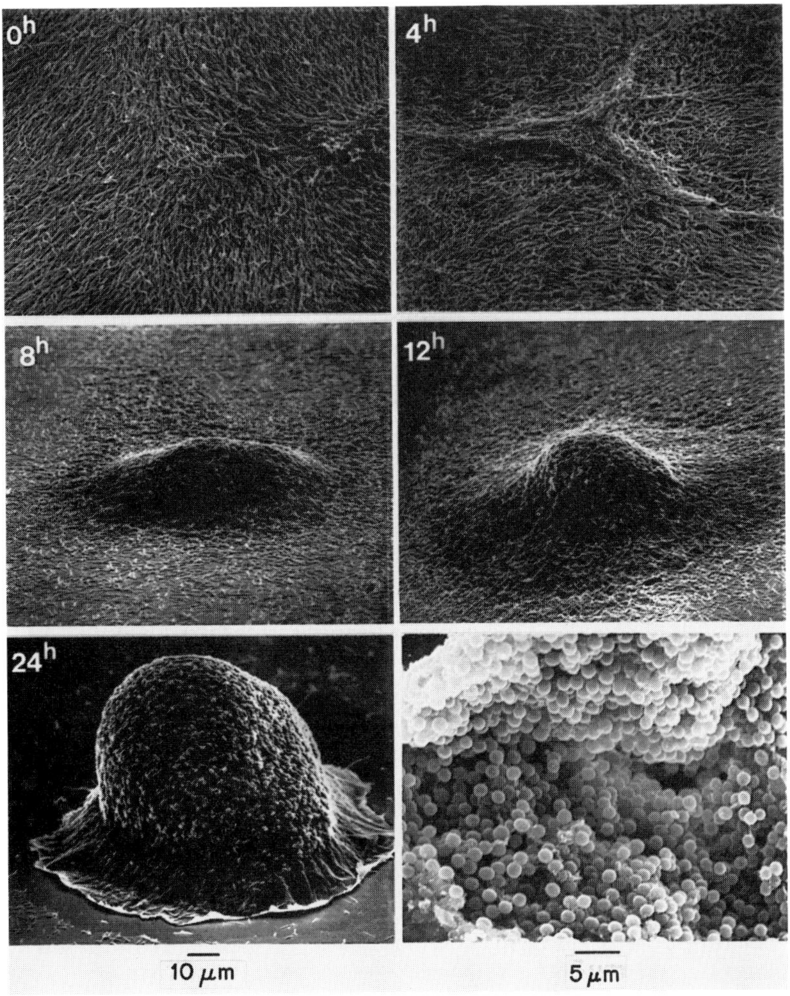

Figure 2. Fruiting body development in *M. xanthus*. Development was initiated at 0 hours by replacing nutrient medium with a buffer devoid of a usable carbon or nitrogen source. The lower right frame shows a fruiting body which has split open, revealing spores inside. This frame is three times the magnification of the others. Scanning electron microscopy by J. Kuner. Reproduced from (8); copyright 1985 Cold Spring Harbor Laboratory.

embryonic development, is clearly evident. Fig. 2 illustrates the process in *Myxococcus xanthus*. Growing cells (0 hours) are rod-shaped. When nutrients are removed, the cells stop growing and initiate development. Centres of aggregation appear almost simultaneously across an entire culture plate at 4 hours. At first these centres are asymmetric, but as more cells accumulate, they become circular mounds (8 and 12 hours). By 24 hours, about 100 000 cells have accumulated and the mound has become steep-sided. Then, within the mound, some cells lyse while others become myxospores. Perhaps the lysing cells provide materials for the sporulating cells. Sporulation itself is a primitive cellular differentiation. The rod-shaped cell becomes a spherical myxospore with a thick wall and resistance to heat and desiccation. Myxospores are less resistant than endospores (Chapter 11), probably because myxospores have one set of cell membranes while endospores have two sets. The regular mound of myxospores, about 0.05 mm high, covered with a skin of lipid, polysaccharide and protein, is the fruiting body of *M. xanthus*. The fruiting bodies in Fig. 1 allow comparison between those of *M. xanthus* and three other myxobacteria. The morphogenetic pathways leading to the more complex fruiting bodies pass through initial stages similar to those of *M. xanthus*. Instead of the immediate transition to sporulation seen with *M. xanthus*, cells of the other species then continue to move, organizing stalks and branches before they too form spores (4, 9).

A number of biochemical events occur along with the morphological development. New proteins are synthesized by the developing culture on a reproducible timetable. The pattern of synthesis of more than 30 proteins, identified as bands separated by gel electrophoresis, is observed to change during development (10). New antigens appear on the cell surface, for which development-blocking monoclonal antibodies have been obtained (11, 12). At 6 hours, a spore coat protein, protein S, begins to be synthesized. This protein has been purified and characterized (13), crystallized (14), and two related genes have been cloned (15). At 10 hours, a protein with lectin ability called protein H, starts to appear (13). This has also been purified and characterized (13), and its gene cloned and sequenced (16). The orderliness and reproducibility of the morphological and biochemical changes suggests that the programme may be encoded in the genome.

III. OPERON FUSIONS EXPOSE A PROGRAMME OF DIFFERENTIAL GENE EXPRESSION

To examine the programme, a transposable reporter of developmentally-regulated gene expression was constructed. The reporter is based on transposon Tn*5* (17), which can be introduced into *M. xanthus* by specialized transduction with phage P1. Coliphage P1 adsorbs to many Gram-negative bacteria, including *Myxococcus,* presumably by attachment to cell-surface lipopolysaccharide as in *Escherichia coli* (18). Adsorbed P1 can inject its DNA into *Myxococcus*, but the phage DNA is unable to replicate, or to establish lysogeny. Thus, when *M. xanthus* is infected with P1::Tn*5* (Fig. 5), transposition of Tn*5* to the host genome yields a kanamycin-resistant strain (19), and P1 is lost. Tn*5* transposes to sites that are random, on the kilobase scale, in *Myxococcus*.

In constructing the reporter, a promoterless *lacZ* gene was added to Tn*5*. (The use of operon fusions was pioneered by Casadaban with the specialized transducing phage Mud*lac* (20).) The structure of Tn*5 lac* is shown in Fig. 3. A promoterless *lacZ* gene has been inserted into a position which

(a) does not interfere with Tn*5* transposition or the expression of the gene for kanamycin resistance, and

(b) is orientated so that transcription initiated from a cell promoter, outside the Tn*5 lac* DNA, proceeds through the gene as shown in Fig. 3. There are also translation stop signals ahead of the *lacZ* gene in all three reading frames. Consequently transcriptional fusions, but not protein fusions, are made.

New developmentally-regulated genes have been located by random insertions of Tn*5 lac* into *Myxococcus*. Among 2374 Tn*5 lac* insertion strains, 548 produced β-galactosidase during growth, indicating that Tn*5 lac* had inserted in the appropriate orientations in transcription units active during growth. Assuming that Tn*5 lac* inserts with equal frequency in both orientations, one would estimate that twice 548, or about half the insertions isolated, were in transcriptionally active regions. This implies that one-half the *Myxococcus* genome is transcribed during growth in laboratory conditions. Of the 2374 strains, 94 appeared to produce more β-galactosidase during starvation-induced fruiting body development than during growth, and in them *lacZ* is

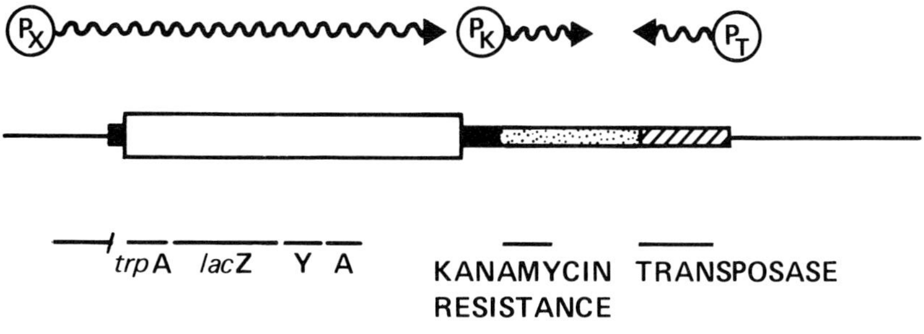

Figure 3. Structure of Tn*5 lac* (21). Transcripts (wavy lines) are shown above the DNA, originating: from P_k, the promoter of the kanamycin-resistance gene; from P_t, the promoter of the transposase gene; and from P_x, the promoter of a transcript into which coding sequence Tn*5 lac* has inserted in the correct orientation to make a transcriptional fusion. Tn*5 lac* consists of sequences from IS*50* left (solid), the *trp-lac* fusion segment (open), the central region of Tn*5* (stippled), and IS*50* right (hatched). Proteins (solid lines) shown below the DNA include a truncated polypeptide (leftmost) encoded by the gene into which Tn*5lac* has inserted. Reproduced from (8); copyright 1985 Cold Spring Harbor Laboratory.

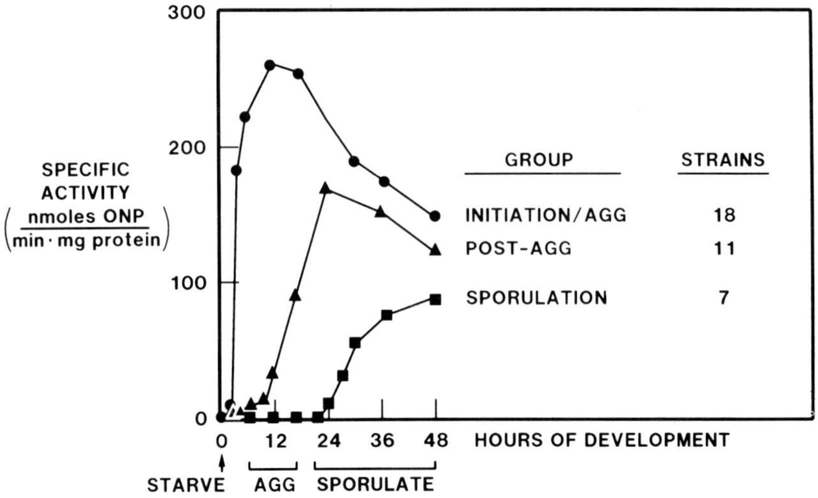

Figure 4. β-Galactosidase synthesis in developing Tn*5lac* strains. The specific activity of β-galactosidase in sonic extracts of cells harvested at different times during development is shown for three representative strains. Reproduced from (8); copyright 1985 Cold Spring Harbor Laboratory. AGG = aggregate; ONP = *o*-nitrophenol (measures β-galactosidase activity).

thus developmentally regulated. (In the rest of this chapter the term "development" will refer to starvation-induced fruiting body development.) In 36 of the 94 strains, the developmental strains were found to be strongly regulated in that β-galactosidase activity increased during development by factors of 3–300.

The time courses of strongly developmentally-regulated β-galactosidase synthesis in all 36 of these strains have been examined. Three are shown for illustration in Fig. 4. In one of these strains, β-galactosidase increases about 30 min after starvation initiates development. In the second, the increase begins around 12 hours. The third strain accumulates β- galactosidase late, at around 24 hours while spores are forming. As shown in Fig. 4, in the sample of 36 strongly regulated Tn*5lac* strains, 18 increase β-galactosidase during the first 8 hours, the period of initiation and preparation for aggregation. Another 11 strains show their increased synthesis between 8 and 16 hours, during or after the period of aggregation. And seven strains show increases during the period (16–24 h) of sporulation and have their highest activity inside spores.

Most of the 36 Tn*5lac* insertions indicated in Fig. 4 are in different regions of the *Myxococcus* genome, because the restriction maps of the DNA adjacent to these insertions do not resemble each other (22). However in five cases, two or more insertions do have related restriction maps and also express

β-galactosidase at about the same time and level during development, suggesting that these pairs are in the same transcript. Taking these restriction map similarities into account, at least 29 different transcription units are represented by these 36 Tn*5lac* insertions. The timing of gene expression for these 29 units, and for proteins S and H as well, clearly demonstrate a programme of differential gene expression.

IV. CELL INTERACTIONS COORDINATE THE PROGRAMME OF FRUITING BODY DEVELOPMENT

Intercellular coordination of fruiting body development is a major problem for the large number of cells that participate in forming a fruiting body. Such coordination would seem to require the passage of regulatory signals between cells. To look for intercellular signals, mutants were sought which were developmentally defective, but could be rescued by added wild-type cells. The idea was that wild-type cells would provide a missing signal, allowing mutants that were defective in signal production to develop normally. Fifty-one such mutants were isolated by Hagen *et al.* (23).

On their own, these mutants have reduced capacity to form spores in fruiting bodies. However, when they are mixed with wild-type cells, they sporulate at practically wild-type levels. Moreover, some pairs of mutant strains can complement each other's defects for sporulation and fruiting body formation. Pairs of mutants which fail to complement in mixtures belong to the same (extracellular) complementation group. In this way, four complementation groups (A, B, C and D) were recognized in the initial set of 51 mutants. If these complementations are to be explained by signals, there may be four different signals.

A connection between cell interactions and the genetic programme that regulates fruiting body development was revealed in the developmental phenotypes of the A, B, C and D groups of complementable mutants. In addition to a deficiency in sporulation, these mutants are each defective at stages in normal development before sporulation. For example, none of the four groups of mutants lyse, yet in wild-type cells lysis is practically complete by 20 hours, while sporulation normally starts at 20 hours. In addition, mutants of complementation groups A and D fail to synthesize proteins S or H, and mutants of group B have reduced levels of both these proteins. These observations suggest that the synthesis of proteins S and H, lysis, and sporulation are regulated by a dependent sequence of gene expressions, and that sporulation is at the end of the sequence. Mutants blocked early in the sequence fail to sporulate, not because they are defective in proper sporulation, but because they fail to execute earlier steps upon which sporulation depends. According to this view, the groups cease development at different stages in the sequence, mutants of groups A, B and D stopping before mutants of group C, for example.

V. MUTANTS OF GROUPS A, B, C AND D DIFFER GENETICALLY

To examine the genes behind the four groups of mutants, it is necessary to return to transposon Tn*5* in its original unmodified form and to describe how that transposon can be used to locate and manipulate genes in *Myxococcus*. Briefly, it is possible to identify an insertion of transposon Tn*5* near any gene or set of interesting mutations. In this way, Tn*5* provides a selectable marker (kanamycin-resistance) for strain construction, localized mutagenesis, chromosome mapping and for gene cloning. Some of these insertions occur within genes, inactivating their function (24). How an insertion of Tn*5* that is near a particular mutation can be isolated from a library of many random Tn*5* insertions is illustrated in Fig. 5. The general principle behind this scheme is discussed by Masters (25).

Linked insertions of Tn*5* have been found near to mutations belonging to each of the four groups. With these Tn*5* insertions as reference points, the different groups are found to have mutations in different genetic loci. Group A loci are called *asg* (for "A signal"); similarly groups B, C and D loci are called *bsg, csg* and *dsg* respectively. The mutations in 20 different group A mutants have been mapped to three different *asg* loci, *asgA, asgB*, and *asgC*. Alleles at each locus are identified by similar frequencies of cotransduction with a neighbouring Tn*5*. None of the three pairs of *asg* loci are cotransducible by the generalized transducing myxophages Mx4 or Mx8. Therefore they are separated from each other by more than 0.9% (about 50 kb) of the *Myxococcus* chromosome, because the DNA molecules of Mx4 and Mx8 represent that fraction of the length of the *Myxococcus* chromosome (26, 27, 28). Five group B mutations define the *bsgA* locus linked to another insertion of Tn*5* (29). Several *bsg* mutations are known not to be linked to *bsgA* by cotransduction, and so at least one other *bsg* locus must exist. All 11 group C mutations analysed map to a single *csg* locus (30); two group D mutations map to one *dsg* locus (31). To summarize, mutations at different loci or different sets of loci are responsible for group A, B, C and D mutant phenotypes.

VI. EXPRESSION OF β-GALACTOSIDASE FROM *lac* FUSION STRAINS DEPENDS ON THE PRODUCTS OF THE *asg, bsg, csg* AND *dsg* GENES

The transcription units to which *lacZ* has been fused were identified because *lacZ* expression was induced by starvation. Are these units dependent on the

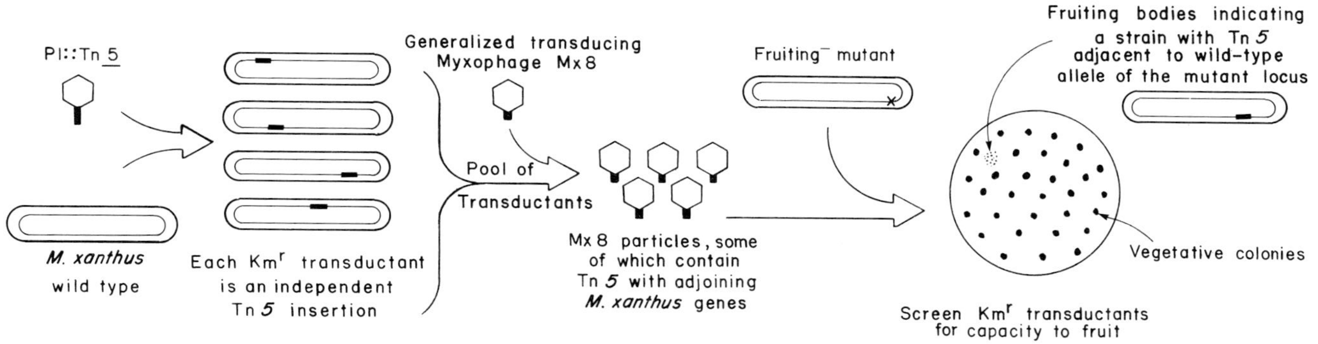

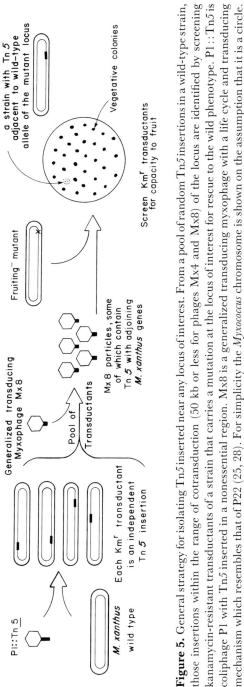

Figure 5. General strategy for isolating Tn*5* inserted near any locus of interest. From a pool of random Tn*5* insertions in a wild-type strain, those insertions within the range of cotransduction (50 kb or less for phages Mx4 and Mx8) of the locus are identified by screening kanamycin-resistant transductants of a strain that carries a mutation at the locus of interest for rescue to the wild phenotype. P1 : : Tn*5* is coliphage P1 with Tn*5* inserted in a nonessential region. Mx8 is a generalized transducing myxophage with a life cycle and transducing mechanism which resembles that of P22 (25, 28). For simplicity the *Myxococcus* chromosome is shown on the assumption that it is a circle. Reproduced from (19).

same regulatory circuit that controls fruiting body development? If they are, then like fruiting body development, β-galactosidase production should depend on the *asg, bsg, csg* and *dsg* genes. Also their dependence should reflect the station of each transcription unit in the regulatory circuit. A test of 21 different *lac* fusions in isogenic asg^+ and *asg* mutant strains showed that three were A-independent and 18 were A-dependent. The same 18 *lac* fusion strains were found to be dependent on both *asgA* and *asgB* (an *asgC* mutation has not yet been tested), so these effects are likely to apply to the whole A group. The A-independent *lac* fusion strains are all normally expressed in both asg^+ and asg^- strains during the first three hours of development; the A-dependent strains are all expressed after 1.5 hours (in asg^+ strains), as if A were required between one and three hours in a dependent sequence of gene expressions.

Are A, B, C and D part of the same dependent sequence? The question can be answered using the principle that interrupting an essential step in a dependent sequence prevents the activation of all genes farther along the sequence. The strongly-regulated *lac* fusion strains are expressed at various times throughout the whole of development and thus sample the dependent sequence along its course. If A is required for an early step in the sequence, then expression of β-galactosidase from most of the *lac* fusion strains should be blocked in *asg* mutants. However, the *csg* mutants appear to alter development at a later stage and would be expected to block expression in fewer *lac* fusion strains. It would also follow that the earliest C-dependent strains should be expressed later in development than the earliest A-dependent *lac* fusion strain. As with the *asg* mutations, a *csg* mutation affects expression from some *lac* fusion strains but not others. Nine strains were found to be C-independent; this represents almost half the 19 strains tested, and a larger fraction than the three out of 19 found in *asg* mutants. Concerning the time of β-galactosidase expression, C seems to be required after about six hours of development, several hours later than A. On the basis of timing and number of signal-dependent *lac* fusion strains, C does act later than A. But, are A and C required by the same regulatory sequence, or by different converging sequences that happen to be synchronized by the same biological clock (32)?

If A and C are part of the same regulatory sequence, with the *lac* fusion strains distributed along it, then because A is earlier than C, any transcription unit which is independent of A should also be independent of C. This conclusion is not necessary if the two are parts of different sequences, even if they are synchronized by the same clock. The experimental results are summarized in the diagram of Fig. 6, showing which particular *lac* fusion strains are affected by A and which by C. It is evident that:

(a) all A-independent strains are C-independent;
(b) all C-dependent strains are A-dependent; and
(c) some A-dependent strains are C-independent.

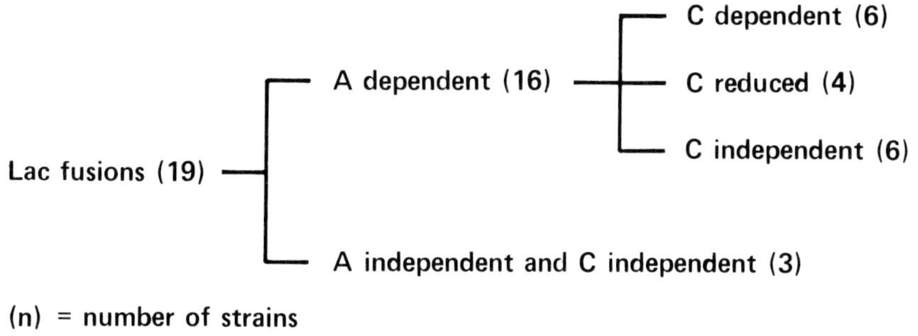

Figure 6. Observed relations between responses to A- and C-factors of a set of 19 *lac* fusions. Numbers of strains in each class are given in parentheses.

These results are those expected if A were early and C later in the same sequence. By similar arguments B can also be placed in the sequence with A and C. Fig. 7 represents the observed degree of dependence of a set of 35 transcriptional fusions on (factors) A, B and C. Plotted against the time at which each fusion would normally express β-galactosidase during development is the degree of its factor-dependence for expression. All the transcription units are at least partially dependent on B. Three transcription units are independent of A and C. Moving upward in the diagram, complete dependence on A is next, then a partial dependence on C, then dependence on B becomes complete, and at the highest level there is complete C-dependence. The diagram shows that dependences on A, B and C are cumulative, as expected in a linear dependent sequence.

Three branches are introduced in Fig. 7 to accommodate pairs of *lac* fusion strains that express β-galactosidase at about the same time, but have different factor-dependences. However, most of the transcription units detected by *lac* fusions are on the uppermost line at each time, the "main" sequence. The regulatory circuit can be represented to a first approximation as a single dependent sequence in which the extracellular factors A, B and C are necessary. Existing data also place factor D in this sequence, but since not all *lac* fusion strains have been examined with a *dsg* mutation, D has not yet been added to the diagram.

VII. A-FACTOR AND C-FACTOR ACTIVITIES CAN BE FOUND IN CELL EXTRACTS

Strains that carry a fusion of a developmentally-regulated promoter to *lacZ* can be used as a bioassay for one of the factors. Recall that the developmental

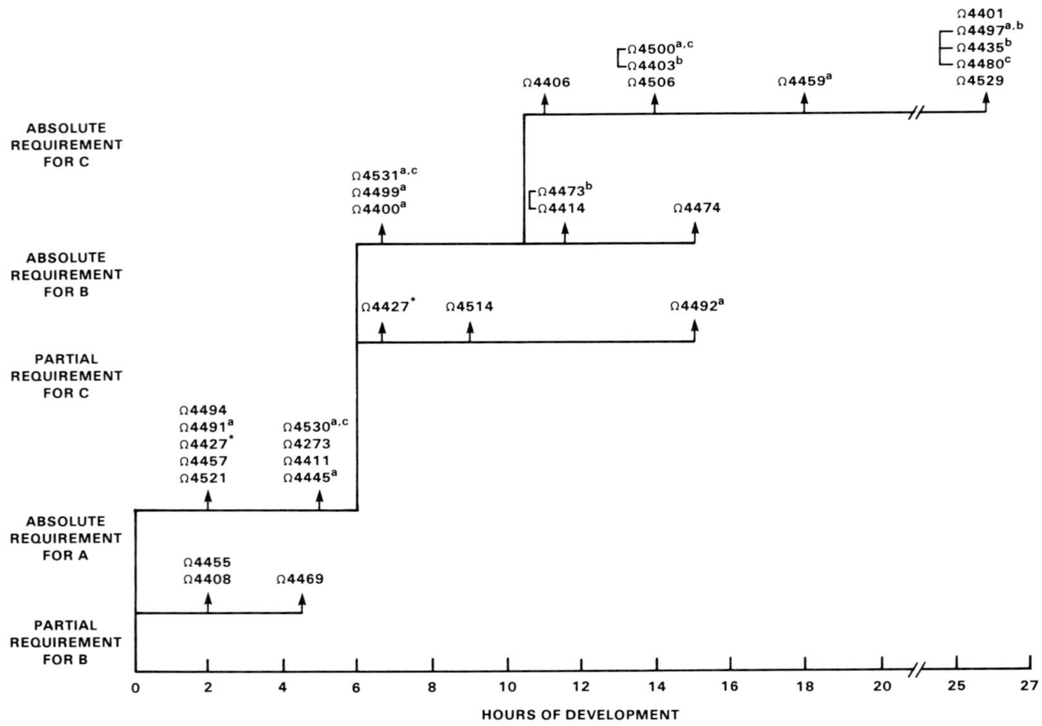

Figure 7. Dependence of gene expression on A-, B- and C-factors. Each *lac* fusion (Ω followed by a four digit number) is shown above an arrow to indicate the time at which it is expressed. Positions on the y-axis indicate the factor-dependence of expression for each *lac* fusion. Requirements accumulate upward. Superscript a, b, or c indicates that the Tn*5 lac* insertion has not been tested in the A^-, B^- or C^- mutant, respectively. Reproduced from (33); copyright 1987 Cold Spring Harbor Laboratory.

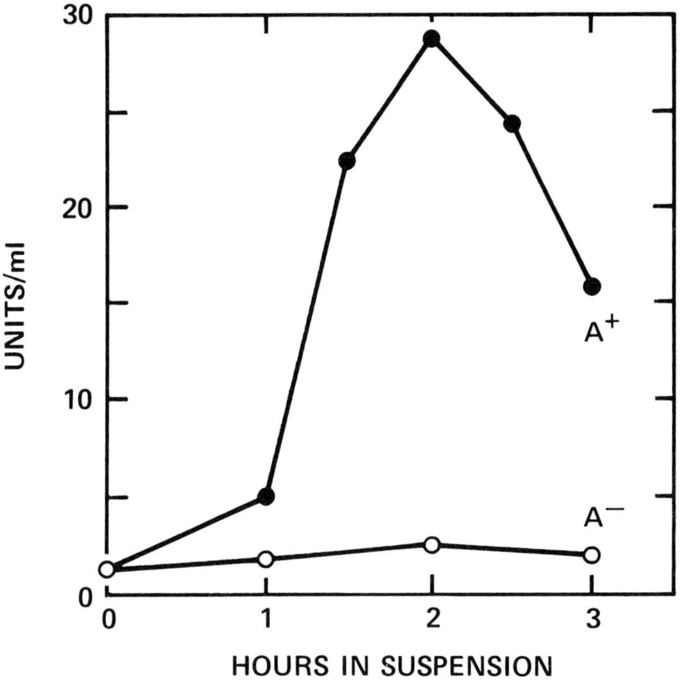

Figure 8. Release of material with A-factor activity from wild-type (A^+) and *asg* mutant (A^-) cells being starved to induce development (34).

defects of A, B and C mutants are corrected by the addition of wild-type cells. To test whether the failure of an A-dependent *lac* fusion strain to express β-galactosidase when it carries an *asg* mutation is due to the absence of extracellular A-factor*, wild-type cells were added to supply the missing substance.

In all seven A-dependent *lac* fusion strains tested, enzyme synthesis was restored by addition of wild-type cells, as illustrated in Fig. 8. Replacing whole wild-type cells with extracts and other materials provides a bioassay for A-factor. Using this A-factor assay, activity can be found in a wash fluid from asg^+ cells. *asg* mutants release much smaller amounts of activity, as shown in Fig. 8. The activity appears in washes made between 1 and 2 hours after the start of development, which agrees with the earliest time (1.5 hours) at which β-galactosidase appears in any of the A-dependent *lac* fusion strains. Crude A-factor is nondialysable and is inactivated by heat.

* Note that this A-factor is distinct from similarly named factors active in other bacteria and other circumstances, such as the A-factor in *Streptomyces* (Chapter 7) and the factor A in *Staphylococcus*.

In a similar fashion, a C-dependent *lac* fusion strain that carries a *csg* mutation can be used to assay C-factor. Material with C-factor activity has been found in sonicated extracts of developing wild type cells, but not from *csg* mutants. Crude C-factor is also heat labile and non-dialysable.

VIII. THE *asg, bsg, csg* AND *dsg* GENE LOCI CAN BE ISOLATED

Interesting, Tn*5* insertions and their adjacent *Myxococcus* genes can be cloned in *E. coli*. It is advantageous for this purpose to use a vector that can be propagated in *E. coli*, where the cloned fragment can be readily manipulated and can be transferred back to *Myxococcus* to examine gene function. The action of such a vector which homes specifically to a Tn*5* inserted in the chromosome is described in Fig. 9. Because cotransduction with Tn*5* does not indicate whether the locus of interest lies to the left or the right of the transposon, it is necessary to clone DNA separately to the left and to the right of Tn*5*. The two clones are obtained as a natural consequence of the fact that the two IS*50* sequences of Tn*5* are orientated in opposite directions and the incoming vector is as likely to recombine with IS*50* right as with IS*50* left. Of course, the Tn*5* must be near enough to the locus so that no site sensitive to the restriction enzyme chosen for cloning lies between Tn*5* and the locus. The figure shows capture of a locus on a 12 kb *Eco*RI fragment. The 15.5 kb fragment from the other side serves as a control when the clones are returned to *Myxococcus* to test their ability to restore wild-type function to the mutant strain. In this way the *asgA, asgB, asgC, csg* and *dsg* genes have been cloned, so that they can now be analysed by molecular techniques.

These vectors have also been constructed to allow phage P1 to transduce them from *E. coli* to *Myxococcus* at high frequency. For this purpose, the vectors

---

Figure 9. Cloning DNA sequences linked to an insertion of Tn*5*. Tn*5* inserted in *Myxococcus* is bounded on the left by IS*50* left and on the right by IS*50* right (shown as open boxes), which are orientated in opposite directions (arrows). (A) Recombination between an IS*50* sequence located on the cloning vector and one of the two chromosomal copies of IS*50* integrates the vector into the chromosome, as shown in (B). Digestion with the restriction enzymes B, C or R produces a DNA molecule containing the vector and adjacent *Myxococcus* DNA up to the first B-, C- or R-sensitive site. This DNA molecule forms a circle, shown in (C), by spontaneous cohesion of its two ends, which are complementary because they were produced by the same staggered cutting restriction enzyme. P1*inc* sequences in the cloning vector encode plasmid incompatibility and permit specialized transduction by P1 as described in Figure 10. Km^r, resistance to kanamycin; Tc^r, resistance to tetracycline; ori, ColE1 origin of DNA replication. Reproduced from (35); copyright Springer-Verlag Berlin, Heidelberg 1987.

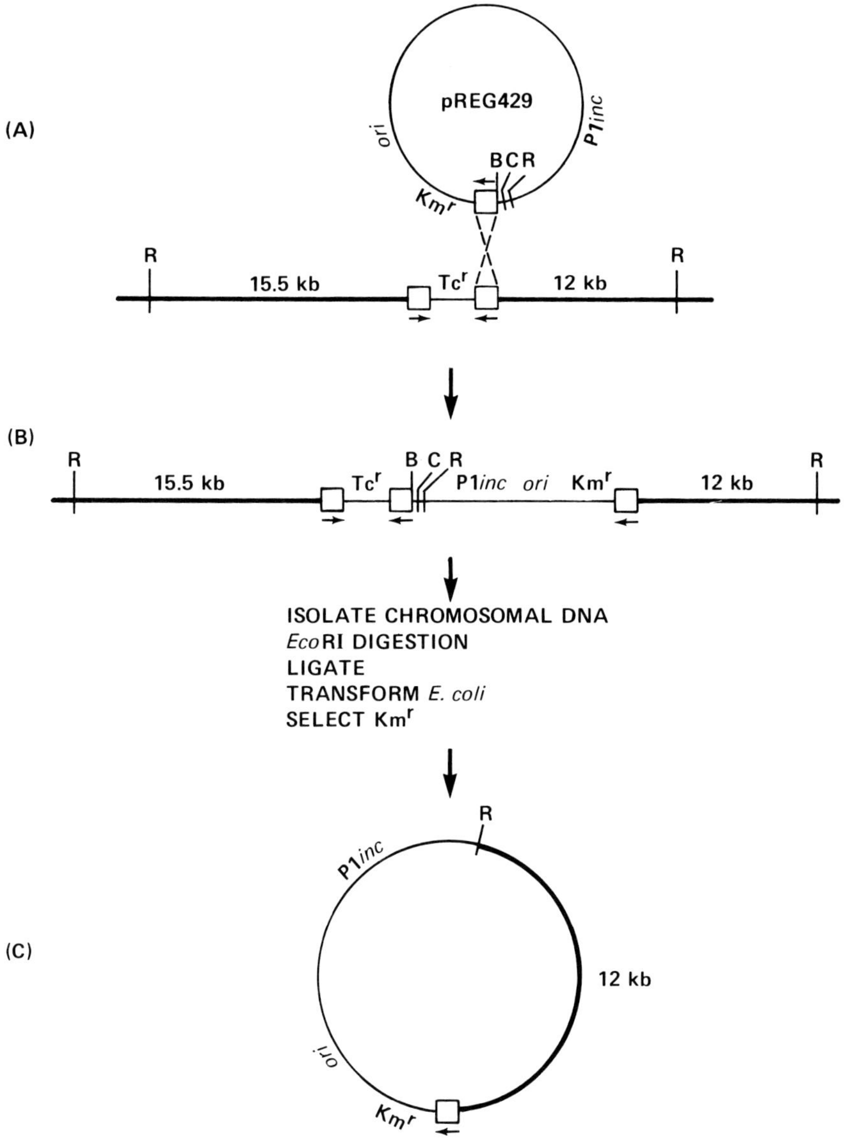
(A)
pREG429
ori
P1inc
BCR
Kmr
R
15.5 kb
Tcr
12 kb
R
(B)
R
15.5 kb
Tcr
B C R
P1inc ori
Kmr
12 kb
R
ISOLATE CHROMOSOMAL DNA
EcoRI DIGESTION
LIGATE
TRANSFORM E. coli
SELECT Kmr
R
P1inc
(C)
12 kb
ori
Kmr

Figure 9.

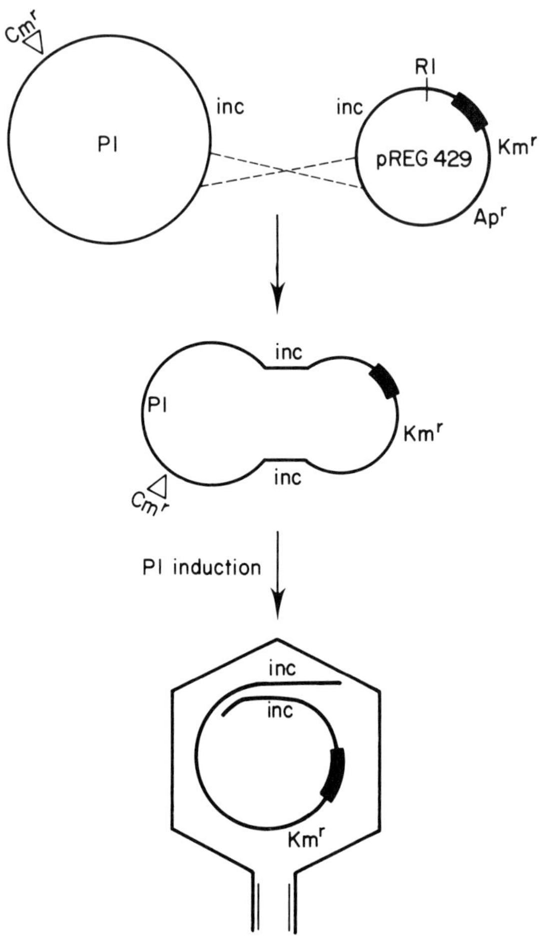

Figure 10. Specialized transduction of plasmid pREG429 by phage P1. When an infecting P1, marked by resistance to chloramphenicol (Cm^r), establishes lysogeny in an *E. coli* strain carrying pREG429, a cointegrate between P1 and the plasmid is formed by recombination in their homologous P1-incompatibility DNA sequences. P1 cannot establish itself as an independent plasmid because the cells were P1-incompatible at the time of infection. Upon induction to lytic growth, P1 begins packaging DNA within the P1 segment of the cointegrate and most of the resulting phage particles contain the entire pREG429 plasmid but an incomplete P1 chromosome. Drawing courtesy of K. Stephens.

contain a piece of phage P1 DNA. As illustrated in Fig. 10, the P1 DNA in the vector allows a superinfecting P1 genome to lysogenize the host by forming a P1::pREG429 cointegrate. The P1 particles which emerge from a cointegrate P1 lysogen have specialized transducing activity for the plasmid. When a P1 phage stock brings a plasmid with its cloned segment back into *Myxococcus*, either by specialized transduction as just described, or by generalized transduction (24), the plasmid DNA reforms a circle in the *Myxococcus* recipient by homologous recombination between duplicate regions (shown aligned in Fig. 10). However, a plasmid such as pREG429 with a ColE1 origin of replication fails to replicate autonomously in *Myxococcus*. K_m^r transductants can nevertheless be produced if a single crossover between the cloned *Myxococcus* DNA and its homologue in the chromosome integrates the cloned segment. These integrants have a duplication of the cloned segment with a copy of the vector located at the vector::insert joint.

A duplication strain is useful as a partial diploid: the phenotype of a heterozygous partial diploid shows which allele is dominant. In this way *asg*, *csg* and *dsg* alleles have been shown to be recessive to their corresponding wild alleles, consistent with their being defective in signal production. In addition, genetic complementation, effectively *in trans*, between two mutant alleles can be examined in such partial diploids. All the point mutant *csg* alleles belong to one cistron by this gene complementation test (36), and both *dsg* alleles belong to one cistron (31). If two crossovers, rather than one, take place between the cloned *Myxococcus* DNA segment and its chromosomal homologue, then recipient sequences are replaced by donor sequences and all vector sequences are lost. Replacements are used to construct deletion strains (37), strains with transposon insertions at particular sites (30), and strains with genetically engineered mutations (29).

IX. OVERVIEW AND PROSPECTS

The genetics of interactions between cells can be explored by extracellular complementation between two types of cells in a mixture. Wild-type cells restore a wild phenotype to cells of an interaction-defective mutant with which they are mixed. Substances released by cells or located on their surface may contribute to this complementation. If rescuing cells can be replaced by factors extracted from those cells, the route to a biochemical analysis of the interaction is open (34, 38, 39).

Extracellular complementation has precedents in classical studies of *Drosophila* mosaics. Cells of the vermilion and cinnabar eye colour mutants became phenotypically wild-type when accompanied by wild-type tissue in a mosaic or transplanted fly (40, 41). Mutants like vermilion and cinnabar were called "nonautonomous" and the behaviour of these nonautonomous mutants in the presence of other cell types provided data on the pathway of synthesis of eye pigments.

Mixtures of bacterial cells are analogues of mosaics or organ and tissue transplants. The Hagen fruiting mutants are nonautonomous since they are rescued by wild-type cells. The behaviour of the Hagen mutants in the presence of other cell types provides data on the regulatory sequence in fruiting body development. Rescue of such mutants by wild-type cells implies that the mutants are defective in factor production, not in factor reception. Extracellular complementation tests between pairs of Hagen mutants divide them into four types. Each type has a different developmental phenotype and corresponds to a different locus or set of loci.

A regulated programme of fruiting body development is evident in the progressive changes of morphology and in the timetable of coordinate gene expression. The properties of the Hagen mutants show that:

(1) the programme requires several different extracellular factors, possibly four;
(2) the factors are required at different stages of development, because mutants that block their production (or release) cease development at different stages;
(3) the developmentally regulated *lacZ* transcriptional fusion strains respond to factors A, B and C as predicted by the branched dependent pathway shown in Fig. 7.

A second example of extracellular complementation in *Myxococcus* can be found in the stimulatable motility mutants, suggesting cell interactions among growth-phase cells. When certain pairs of nonmotile mutants are mixed together, one or both become temporarily motile (42, 43, 44). Six different stimulatable (i.e. extracellularly complementable) types were identified by mixing cells. Each of the six types was found to correspond to a different genetic locus. All mutations at each one of these loci, including Tn*5*-insertion mutations, produce stimulatable strains. Thus, stimulation is locus- and not allele-specific. Any strain, whether motile or not, can complement a particular stimulatable mutant, provided only that the strain has the wild-type allele of the stimulatable locus. The different stimulation types thus reveal different specificities in the corresponding cell interactions.

In *Myxococcus*, cell-to-cell interactions are accessible to experimental study, which has only begun. The structure and function of the products of the stimulatable motility loci remain to be found, and the chemical natures of factors A, B, C and D remain to be defined. The pathway of fruiting body development needs to be worked out, and the regulatory elements that control it need to be identified.

These objectives can be met because the genome of *Myxococcus* consists of only 5700 kb of unique DNA (27); a size that should permit saturation mutagenesis for particular functions. Because myxobacteria are (distantly) related to the enteric bacteria (45), coliphage P1 can transduce them, plasmid RP4, which has sex-factor activity and can mobilize chromosome transfer in other bacteria, can be transferred to them (46) and transposon Tn*5* can move within their DNA. Developmentally-regulated promoters have been identified (47). With these techniques and the tools of genetics and biochemistry, it may be possible to understand fruiting body development in molecular detail.

References

(1) Rosenberg, E. (1984). "Myxobacteria. Development and cell interactions". Springer-Verlag, New York.

(2) Rosenberg, E., Keller, K. H. and Dworkin, M. (1977). Cell density-dependent growth of *Myxococcus xanthus* on casein. *J. Bacteriol.* **129**, 770–777.

(3) Dworkin, M. (1973). Cell-cell interactions in the myxobacteria. *Symp. Soc. Gen. Microbiol.* **23**, 125–142 and plates.

(4) Grilione, P. L. and Pangborn, J. (1975). Scanning electron microscopy of fruiting body formation by myxobacteria. *J. Bacteriol.* **124**, 1558–1565.

(5) Reichenbach, H. and Dworkin, M. (1981). The order Myxobacterales. *In* "The Prokaryotes" (M. P. Starr *et al.*, ed.), pp. 328–355. Springer-Verlag, Berlin, Heidelberg.

(6) Whittaker, R. H. (1969). New concepts of kingdoms of organisms. *Science* **163**, 150–160.

(7) Kaiser, D. (1986). Control of multicellular development: *Dictyostelium* and *Myxococcus*. *Ann. Rev. Genet.* **20**, 539–566.

(8) Kaiser, D., Kroos, L. and Kuspa, A. (1985). Cell interactions govern the temporal pattern of *Myxococcus* development. *Cold Spring Harbor Symposia on Quantitative Biology* **50**, 823–830.

(9) Qualls, G. T., Stephens, K. and White, D. (1978). Morphogenetic movements and multicellular development in the fruiting myxobacterium, *Stigmatella aurantiaca*. *Devel. Biol.* **66**, 270–274.

(10) Inouye, M., Inouye, S. and Zusman, D. R. (1970). Gene expression during development of *Myxococcus xanthus*: pattern of protein synthesis. *Devel. Biol.* **68**, 579–591.

(11) Gill, J. S. and Dworkin, M. (1986). Cell surface antigens during submerged development of *Myxococcus xanthus* examined with monoclonal antibodies. *J. Bacteriol.* **168**, 505–511.

(12) Gill, J. S., Jarvis, B. W. and Dworkin, M. (1987). Inhibition of development in *Myxococcus xanthus* by monoclonal antibody 1604. *Proc. Natl. Acad. Sci. USA* **84**, 4505–4508.

(13) Zusman, D. R. (1984). Developmental program of *Myxococcus xanthus*. See reference (1), pp. 185–213.

(14) Inouye, S., Inouye, M., McKeever, B. and Sarma, R. (1980). Preliminary crystallographic data for protein S, a development-specific protein of *Myxococcus xanthus*. *J. Biol. Chem.* **255**, 3713–3714.

(15) Inouye, S., Ike, Y. and Inouye, M. (1983). Tandem repeat of the genes for protein S, a development-specific protein of *Myxococcus xanthus*. *J. Biol. Chem.* **258**, 38–40.

(16) Romeo, J. M., Esmon, B. and Zusman, D. R. (1986). Nucleotide sequence of the myxobacterial haemagglutinin gene contains four homologous domains. *Proc. Natl. Acad. Sci. USA* **83**, 6332–6336.

(17) Berg, D. E. (1977). Insertion and excision of the transposable kanamycin resistance

determinant Tn*5*. *In* "DNA Insertion Elements, Plasmids and Episomes" (A. J. Bukhari, J. A. Shapiro and S. L. Adhya, eds.), pp. 205–218. Cold Spring Harbor, Cold Spring Harbor Laboratory, New York.
(18) Lindberg, A. A. (1973). Bacteriophage receptors. *Ann. Rev. Microbiol.* **27**, 205–241.
(19) Kuner, J. M. and Kaiser, D. (1981). Introduction of transposon Tn*5* into *Myxococcus* for analysis of developmental and other nonselectable mutants. *Proc. Natl. Acad. Sci. USA* **78**, 425–429.
(20) Casadaban, M. J. and Cohen, S. N. (1979). Lactose genes fused to exogenous promoters in one step using a Mu-lac bacteriophage: *in vivo* probe for transcriptional control sequences. *Proc. Natl. Acad. Sci. USA* **76**, 4530–4533.
(21) Kroos, L. and Kaiser, D. (1984). Construction of Tn*5 lac*, a transposon that fuses *lacZ* expression to exogenous promoters, and its introduction into *Myxococcus xanthus*. *Proc. Natl. Acad. Sci. USA* **81**, 5816–5820.
(22) Kroos, L., Kuspa, A. and Kaiser, D. (1986). A global analysis of developmentally regulated genes in *Myxococcus xanthus*. *Devel. Biol.* **117**, 252–266.
(23) Hagen, D. C., Bretscher, A. P. and Kaiser, D. (1978). Synergism between morphogenetic mutants of *Myxococcus xanthus*. *Devel. Biol.* **64**, 284–296.
(24) O'Connor, K. A. and Zusman, D. R. (1983). Coliphage P1-mediated transduction of cloned DNA from *Escherichia coli* to *Myxococcus xanthus*: use for complementation and recombinational analyses. *J. Bacteriol.* **155**, 317–329.
(25) Masters, M. (1985). Generalized transduction. *In* "Genetics of Bacteria" (J. Scaife, D. Leach and A. Galizzi, eds.), pp. 197–215. Academic Press, London.
(26) Martin, S., Sodergren, E., Masuda, T. and Kaiser, D. (1978). Systematic isolation of transducing phages for *Myxococcus xanthus*. *Virol.* **88**, 44–53.
(27) Yee, T. and Inouye, M. (1982). Two-dimensional DNA electrophoresis applied to the study of DNA methylation and the analysis of genome size in *Myxococcus xanthus*. *J. Mol. Biol.* **154**, 181–196.
(28) Orndorff, P., Stellwag, E., Starich, T., Dworkin, M. and Zissler, J. (1983). Genetic and physical characterization of lysogeny by bacteriophage Mx8 in *Myxococcus xanthus*. *J. Bacteriol.* **154**, 772–779.
(29) Gill, R. E. and Cull, M. G. (1986). Control of developmental gene expression by cell-to-cell interactions in *Myxococcus xanthus*. *J. Bacteriol.* **168**, 341–347.
(30) Shimkets, L. J. and Asher, S. J. (1988). Use of recombination techniques to examine the structure of the *csg* locus. *Mol. Gen. Genet.*, **211**, 63–71.
(31) Cheng, Y. and Kaiser, D. (1987). Manuscript in preparation.
(32) Hartwell, L. H., Culotti, J., Pringle, J. R. and Reid, G. J. (1974). Genetic control of the cell division cycle in yeast. *Science* **183**, 46–51.
(33) Kroos, L. and Kaiser, D. (1987). Expression of many developmentally regulated genes in *Myxococcus* depends on a sequence of cell interactions. *Genes and Devel.* **1**, 840–854.
(34) Kuspa, A., Kroos, L. and Kaiser, D. (1986). Intercellular signalling is required for developmental gene expression in *Myxococcus xanthus*. *Devel. Biol.* **117**, 267–276.
(35) Stephens, K. and Kaiser, D. (1987). Genetics of gliding motility in *Myxococcus xanthus*: molecular cloning of the *mgl* locus. *Mol. Gen. Genet.* **207**, 256–266.
(36) Shimkets, L. J., Gill, R. E. and Kaiser, D. (1983). Developmental cell interactions in *Myxococcus xanthus* and the *spoC* locus. *Proc. Natl. Acad. Sci. USA* **80**, 1406–1410.
(37) Komano, T., Furuichi, T., Teintze, M., Inouye, M. and Inouye, S. (1984). Effects of deletion of the gene for the development-specific protein S on differentiation in *Myxococcus xanthus*. *J. Bacteriol.* **158**, 1195–1197.
(38) Shimkets, L. J. and Kaiser, D. (1982). Murein components rescue developmental sporulation of *Myxococcus xanthus*. *J. Bacteriol.* **152**, 462–470.

(39) Jansen, G. R. and Dworkin, M. (1985). Cell-cell interactions in developmental lysis of *Myxococcus xanthus*. *Devel. Biol.* **112**, 194–202.
(40) Sturtevant, A. H. (1920). The vermilion gene and gynandromorphism. *Proc. Soc. Exper. Biol. Med.* **17**, 70–71.
(41) Ephrussi, B. and Beadle, G. W. (1935). La transplantation des disques imaginaux chez la Drosophile. *C.R. Acad. Sci. (Paris)* **201**, 98–99.
(42) Hodgkin, J. and Kaiser, D. (1977). Cell-to-cell stimulation of movements in nonmotile mutants of *Myxococcus*. *Proc. Natl. Acad. Sci. USA* **74**, 2938–2942.
(43) Hodgkin, J. and Kaiser, D. (1979). Genetics of gliding motility in *Myxococcus xanthus* (Myxobacterales): Genes controlling movements of single cells. *Mol. Gen. Genet.* **171**, 167–176.
(44) Hodgkin, J. and Kaiser, D. (1979). Genetics of gliding motility in *Myxococcus xanthus* (Myxobacterales): two gene systems control movement. *Mol. Gen. Genet.* **171**, 177–191.
(45) Woese, C. R. (1987). Bacterial evolution. *Microbiol. Rev.* **51**, 221–271.
(46) Breton, A. M., Jaoua, S. and Guespin-Michel, J. R. (1985). Transfer of plasmid RP4 to *Myxococcus xanthus* and evidence for its integration into the chromosome. *J. Bacteriol.* **161**, 523–528.
(47) Inouye, S. (1984). Identification of a development-specific promoter of *Myxococcus xanthus*. *J. Mol. Biol.* **174**, 113–120.

Section **IV**

Bacterial Adaptations to Animal Pathogenicity

The development of any bacterial disease—or its limitation or spontaneous cure—involves the expression of large numbers of genes in both pathogen and host. Many of these are not uniquely dedicated to pathogenicity or disease resistance. Thus, obtaining a complete understanding of the disease process is a daunting task. Nevertheless it is possible and useful to identify specific attributes of a bacterium which play a major role in pathogenicity. Molecular genetics is a powerful tool in this undertaking, especially when it allows the cloning of potential pathogenicity determinants and their reintroduction in mutated form into the bacterium to assess the effects of the mutations on pathogenicity. This may give surprising results; thus neither the haemolysin nor the neuraminidase of *Vibrio cholerae*, which were obvious candidates for virulence determinants, turned out to be important for virulence (Chapter 15). When the pathogen is difficult or impossible to grow *in vitro* (*Mycobacterium tuberculosis* and *M. leprae* being good examples: see Hopwood *et al., British Medical Bulletin,* 1988, **44**, 528–546) cloning and expression of its genes in a readily handled surrogate bacterial host such as *E. coli* or *Streptomyces* may lead to the identification of individual antigens which become candidates for further study as components of pathogenicity.

We have selected four examples where molecular genetics has already revealed interesting information about the organization and expression of genes involved in pathogenicity. In Chapters 13 and 14, the basis for a remarkable variation in surface properties of pathogenic *Neisseria* and *E. coli* strains is described. The organisms apparently go to great lengths to control the quantitative expression of genes for pili and other surface proteins, which are essential for colonization of the host but may be antigenic liabilities after colonization, as well as to ensure variation in the type of protein expressed, by using one of a variety of switching mechanisms. These include recombination between "silent" and "expression" gene copies, "flip-flop" of an invertible promoter-containing DNA segment, and variation in the number of multiple repeats of a few bases leading to frameshifts that allow or prevent expression of a messenger RNA. Evolution of such devices presumably reflects the power of the mammalian immune system to recognize and destroy invading bacteria, and the consequent premium on its evasion by the pathogen.

Enterotoxin production by the causal agent of cholera, described in Chapter 15, provides a further example of a highly evolved component of virulence, in this case a toxin beautifully adapted to interact with host cell receptors so as to enter the cells and modify their physiology in the direction of ion and water efflux. Finally, Chapter 16 describes the lengths to which some bacterial pathogens have been forced to go to secure a supply of vital iron and so invade the tissues and fluids of the animal body which are normally kept almost free of available iron and so are difficult niches for bacteria to inhabit.

Chapter **13**

The Molecular Basis of Antigenic Variation in Pathogenic Neisseria

J. R. SAUNDERS

I. INTRODUCTION

The Gram-negative bacteria, *Neisseria gonorrhoeae* (the causative agent of gonorrhoea) and *Neisseria meningitidis* (the principal cause of bacterial meningitis) are important human pathogens. These species are able to colonize their hosts by adhering to the surface of mucosal epithelia. This adhesion is mediated by bacterial adhesins; the most important of which are pili (fimbriae) and outer membrane proteins (1). Pili are filamentous protein structures which project from the bacterial surface and bind to glycoprotein receptors on the surface of epithelial cells. Pili are major pathogenicity determinants in *N. gonorrhoeae* and non-piliated variants show dramatically reduced virulence. Pili are also found on the vast majority of *N. meningitidis* isolates from patients with meningococcal disease and play a role in adhesion of meningococci to human cells *in vitro*. Principal outer membrane protein II (P.II or opacity protein) is a further gonococcal adhesin and virulence determinant. The antigenic and biophysical properties of both pilin and P.II vary within and between different strains. This chapter is concerned with the genetic basis for antigenic diversity in these major surface proteins that pathogenic *Neisseria* present to their exterior environment.

II. DIVERSITY AND VIRULENCE

A. Diversity is Expressed Phenotypically as Phase and Antigenic Variation

Modulation of pilus and P.II expression in *Neisseria* is subject both to phase variation in which piliation or P.II character is reversibly or irreversibly switched on or off, and to antigenic variation in which the qualitative nature of the protein produced is altered. Similar antigenic variations and phase transitions are encountered in a variety of pathogenic bacteria (2). Phase variation in *N. gonorrhoeae* is recognized by observing alterations in colonial morphology when the organism is cultured on clear agar media. Piliated (P^+) gonococci generally produce small, dense colonies with discrete edges (1, 3). In contrast, non-piliated (P^-) variants produce larger more diffuse colonies. The rate of phase variation is sufficiently high in some strains of *N. gonorrhoeae* that P^- sectors can be observed within otherwise P^+ colonies. P.II phase variation is recognized by a change between opaque (Op^+) colonies which express one or more P.II species and transparent (Op^-) colonies which express none (4). Variation in *N. meningitidis* is less apparent since capsular polysaccharide tends to obscure colonial form differences.

B. How is Phase Variation of Adaptive Value?

Pili and P.II proteins make gonococci adhesive but it is advantageous for the bacterium to produce variants that do not express one or other of these structures. This would allow desorption from the surface of one cell and subsequent attachment to another after resynthesis of the relevant protein(s). Furthermore, adhesins may be necessary only for the initial stages of infection, later to become a liability. Thus gonococci producing some P.II species are more susceptible to phagocytosis by human polymorphonuclear leukocytes than their Op^- variants (5) and organisms that do not produce P.IIs may be more virulent and invasive (6, 7). Modulation between an on and off state by a genetic switching system is therefore advantageous in population terms to maintain the organism within the host in a variety of forms and allow adaptation to different environments (2).

C. How is Antigenic Variation of Adaptive Value?

Variation in major surface antigens is of great value to organisms that propagate in a single host species. Pili may comprise 1% or more of total protein in *N. gonorrhoeae* and present a considerable antigenic stimulus to the host. Consequently, the host produces specific antibodies to pili, many of which will block adhesion, and the organism will ultimately be cleared from the body. Nevertheless repeated gonorrhoeal infections with different antigenic types are possible. Furthermore, organisms expressing different pilus types are found during the course of infection of a single patient or of a series of contacts infected from the same source (8, 9). Less evidence is available for meningococci, although it is clear that variation occurs in the properties of pilins and membrane proteins produced by *N. meningitidis* during single infections (10). Antigenic variation allows avoidance of host defences by presenting a sequence of different antigenic stimuli to the immune system. Pilus and P.II variants of *N. gonorrhoeae* also exhibit differential adhesiveness to different human cell lines (1). This implies that variation in surface antigens also allows colonization of different anatomical sites and tissues during infection.

D. The Structure and Function of Pilins

Pili from gonococci and meningococci are each composed of a repeating helical filamentous array of about 10 000 pilin subunits (11). These pilins have apparent molecular weights of approximately 15 000–22 000 (1, 9, 10, 12, 13).

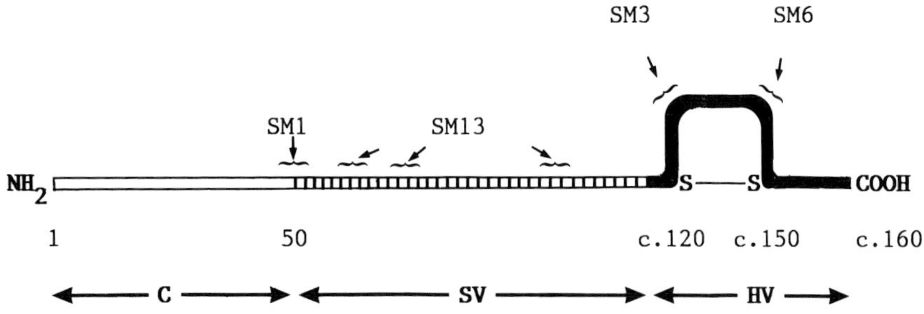

Figure 1. Gross structure of gonococcal pilin showing variable domains and epitopes for type-specific and cross-reacting monoclonal antibodies. C, constant region; COOH, carboxy terminus; S—S, cysteine residues; HV, hypervariable region; NH_2, amino terminus; SV, semivariable region; SM1, epitope for cross-reacting monoclonal antibody SM1; SM3, SM6, SM13, epitopes for type-specific SM3, SM6 and SM13 monoclonal antibodies respectively. Numbers refer to amino acid residues in mature pilin.

The mature pilin molecule is about 160 amino acid residues long (Fig. 1). The first 53 amino acids at the N-terminus are conserved in all gonococcal pilins examined (9, 13–17). There are homologies between this constant (C) region and the equivalent domain of pili from *N. meningitidis* and other bacteria such as *Pseudomonas aeruginosa* (9). The central portion of pilin stretching from residue 54 to a conserved region centred on the first cysteine residue (cys1) at about position 120 constitutes the semivariable (SV) region. Comparison of variant pilin sequences shows that this region is typified by tracts that contain amino acid substitutions embedded in tracts of conserved sequence. The remaining region that stretches from cys1 through a second short conserved region flanking the second cysteine residue at about position 150 (cys2) to the carboxyl terminus is highly variable (HV). Comparisons of different pilins reveal numerous substitutions, deletions and insertions of amino acids in this region. The HV region represents the immunodominant portion of pilin; most antibodies in antisera raised against intact pili are directed against this region (1, 17).

The monoclonal antibody SM1 reacts with pili from all gonococci and from most meningococci (1, 10, 18). *N. meningiditis* pili can be divided into those that are like gonococcal pili (class I) and those that lack the SM1 epitope (class II) (12, 18). It appears that meningococcal strains can produce one or other but not both such pili, although their role in pathogenesis is unclear. The cross-reacting monoclonal antibody SM1 recognizes a weakly immunogenic epitope located in a region of low β-turn potential represented by amino acids 48 to 53 (Fig. 1) of gonococcal pilin (13). Type-specific monoclonal antibodies (such as SM3 and SM6), in contrast, recognize epitopes that lie in the SV and

HV regions, notably but not exclusively between the two cys residues (13, 16, 19). These strongly immunogenic epitopes lie in regions of the pilin polypeptide that are hydrophilic and have high β-turn potential, suggesting that they are located on the outside of mature pilin (11, 13). Type-specific antibodies that recognize these variable domains are known to block adhesion of gonococci to human cells, which implies that these parts of pilin are involved in binding to cell surface receptors (13, 18).

E. The Structure of Protein II Species

Individual gonococci either express no P.II, in which case they produce transparent colonies, or they can produce one or more antigenically distinct P.II species that render the colonies opaque (3, 4, 6). A single organism may produce as many as three distinct P.II species simultaneously in its outer membrane and individual strains are believed to be capable of expressing at least seven different P.II species either separately or in combination. The P.II proteins have apparent molecular weights in the range 24 000 to 30 000. They are heat-modifiable, displaying increases in apparent molecular weight when treated at 100°C. The heat-modifiable class 5 outer membrane proteins of *N. meningitidis* have similar properties and presumably perform an equivalent function (20). Different P.II species, produced by a single strain of *N. gonorrhoeae*, form a family of structurally related proteins that contain a common hydrophobic domain embedded in the outer membrane and a variable, immunodominant hydrophilic domain that is exposed on the outer surface (1, 21).

III. GENETIC MECHANISMS FOR PILUS VARIATION

A. The Gonococcal Genome Contains One (or Rarely Two) Complete Pilin Gene and Many Partial Pilin Sequences

The genome of *N. gonorrhoeae* strain $MS11_{ms}$ contains two distinct pilus-expression (*pilE*) loci (14, 22, 23). These are located about 25 kb apart and each includes the complete pilin coding sequence plus a promoter region and information coding for an N-terminal pre-pilin sequence of seven amino acids not found in mature pilin (23) (Figs. 2 and 3). (The gonococcal pilin promoter, together with those for pilins from several other bacteria such as *P. aeruginosa*, are related structurally to the *Klebsiella pneumoniae ntrA* and *ntrC* promoter sequences (Chapter 6), which may suggest that piliation is regulated by environmental stresses.) Subsequent studies on other variants of strain MS11, for example $MS11_{mk}$, and on unrelated clinical isolates, indicate that a single

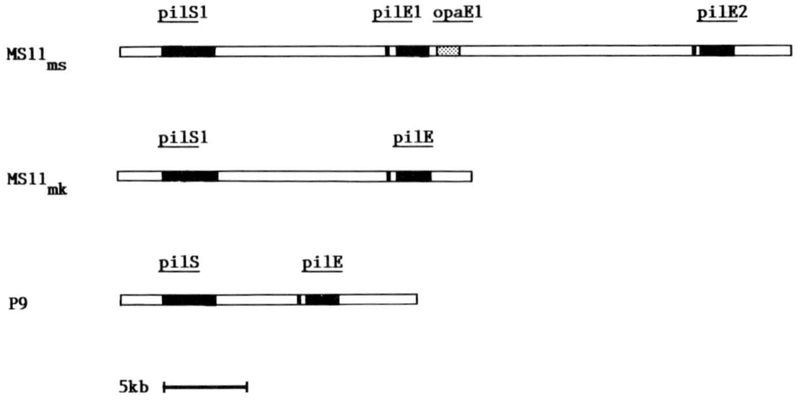

Figure 2. Gross arrangement of *pil* sequences in *Neisseria gonorrhoeae* strains. *pilE*1, *pilE*2, pilin expression loci; *pilS*1, major silent pilin sequence loci; *opaE*1, PII coding locus (shaded). Blocked regions indicate pilin-related DNA sequences.

pilE gene is the normal arrangement in *N. gonorrhoeae* (19, 24, 25). It seems likely that a duplication of *pilE* arose at some stage in the history of strain $MS11_{ms}$, probably as a consequence of the recombinational activity associated with pilus phase and antigenic variation (see below). Gonococci containing two *pilE* loci often show deletion of part of one or the other sequence (14, 22, 24, 26). This might suggest that duplication of *pilE* is an intermediate evolutionary step in generating novel silent pilin sequences (19).

The genome of *N. gonorrhoeae* probably contains 16–18 partial pilin gene sequences in addition to *pilE* (14, 15). These truncated sequences are of varying length, generally spanning all or part of the variable regions and lack both promoter and 5′ terminal sequences (14, 27). The silent *pil* sequences are partly homologous to internal portions of the complete *pilE* gene and each represents a unique resource of variability (16, 27, 28). A cluster of six such

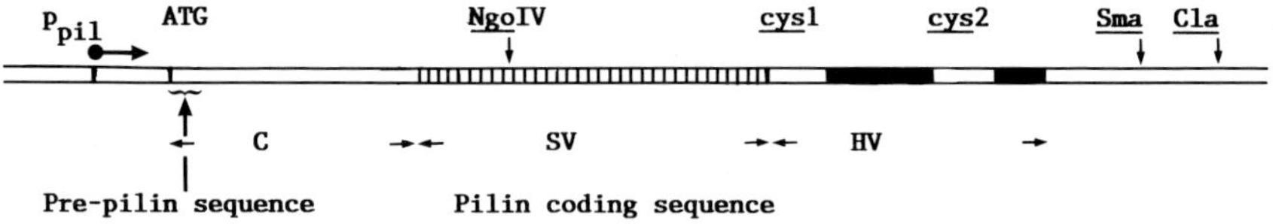

Figure 3. Arrangement of pilin sequences at the *pilE* locus of *N. gonorrhoeae*. ATG, start codon; C, constant coding region; *Cla*, *Cla*I site; *cys*1, *cys*2, conserved regions containing codons for cysteine residues 1 and 2 respectively; HV, hypervariable coding region; *Ngo*IV, location of M.*Ngo*IV modification site; p_{pil}, pilin promoter region; *Sma*, *Sma*I site; SV, semivariable coding region.

partial sequences separated by non-pilin sequences is located in strain $MS11_{ms}$ about 10 kb upstream of *pilE*1 in a silent locus *pilS*1 (23). An equivalent cluster of silent sequences is located about 5 kb upstream of the single *pilE* gene in *N. gonorrhoeae* P9 (28). A single partial pilin sequence is also present about 500 bp upstream of *pilE* in both MS11 and P9 strains (16, 27, 28) (Fig. 2). *N. meningitidis* strains also contain sequences homologous to the gonococcal *pilE* gene regardless of whether they produce class I or class II pili (12, 29, 30). However, there appear to be fewer copies of pilin-related sequences, even in meningococci producing class I pili (12). DNA from commensal neisserial species, with the exception of a few strains of *Neisseria cinerea* and *Neisseria lactamica*, does not hybridize to gonococcal *pil* gene probes (29), emphasizing the importance of pili to the pathogenic *Neisseria* species.

B. The Non-piliated Phenotype is Complex

A complex array of transitions in pilus status is observed when P^+ gonococci are grown *in vitro* (Fig. 4). In some cases the P^+ phenotype is retained but a biochemically different pilin is made which may or may not be antigenically variant. In others, the ability to produce pili is lost. The transition rates from P^+ to P^- and from P^- to P^+ are about 10^{-2} to 10^{-3} and about 10^{-5} per cell per generation respectively (19). Some P^- gonococci give rise to revertants (the P^-r phenotype), whereas others cannot revert (the P^-n phenotype) (24). P^-n variants arise as a consequence of gross chromosomal rearrangements that result in the deletion of 5′-terminal and promoter sequences from *pilE* (14, 19, 24). Similar deletional events are observed in some P^- variants of *N. meningitidis* (16). It is, however, not clear whether P^-n variants would survive *in vivo* since they would be unable to restore the P^+ phenotype except by recombination with exogenous DNA containing the 5′ *pilE* region. (In the case of gonococci carrying two *pilE* loci, it is possible that recombination between one deleted *pilE* locus and an intact *pilE* gene could repair any defect (14, 22).)

P^-r gonococci exhibit no gross chromosomal rearrangements (15, 24). Such variants express pilin-specific mRNA species containing "mutations" that encode truncated or abnormal pilin polypeptides that are not assembled into mature pili (24). Pilin mRNA from some such variants (P^-rp^- phenotype) contains nonsense mutations resulting in synthesis of markedly truncated pilin polypeptides that cannot be detected using anti-pilus antibodies. Such abnormal pilins can arise by a single insertion or deletion in nucleotide sequence that causes premature termination of translation and formation of a polypeptide that lacks the strongly immunogenic epitopes characteristic of the C-terminus. In contrast, other variants (with the P^-rp^+ phenotype) produce pilin mRNA encoding full length pilin that is recognizable immunologically but is nevertheless unable to form pili. Presumably, alterations in the pilin

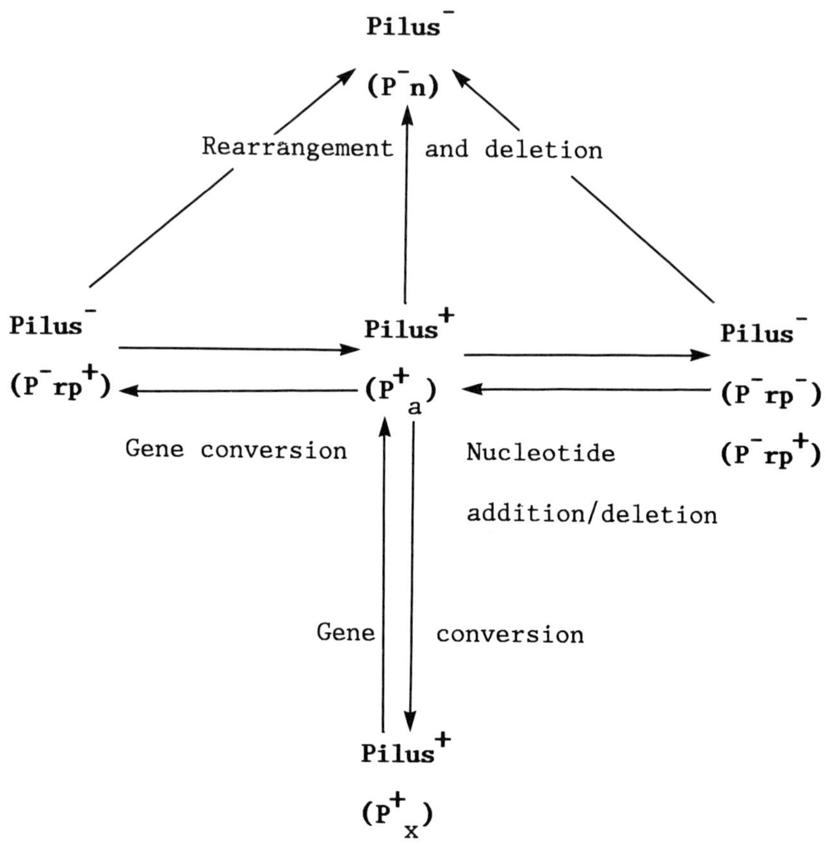

Figure 4. Pilus phenotypes and their interrelationships. a, x pilus antigenic variants; P^+, piliated; P^-, nonpiliated; P^-n, nonreverting pilus-minus; P^-rp^+, reverting, pilin-producing pilus-minus; P^-rp^-, reverting nonpilin-producing pilus-minus.

polypeptide abolish domains necessary for pilus assembly. Reversion of both P^-rp^+ and P^-rp^- to P^+ is accompanied by the reappearance in revertants of pilin mRNA that encodes functionally and immunologically normal pili (24).

C. Gene Conversion Involving a Single *pilE* Locus is Responsible for Generating Structurally-variant Pilins

Changes in amino acid sequence of the SV and the HV regions are responsible for antigenic variation observed in gonococcal pili. The necessary alterations in pilin-coding sequences arise by a gene conversion process (9, 14,

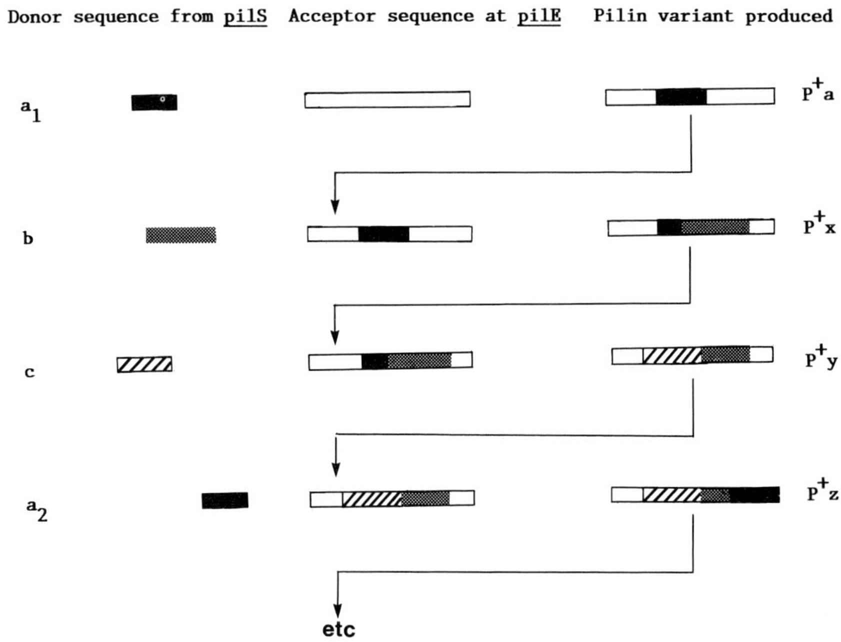

Figure 5. Production of variant pilins by serial gene conversions. Acceptor sequence at *pilE* represents only the SV and HV regions of the pilin gene. Donor sequences a, b, c, etc. represent that part of a particular silent *pil* copy that displaces an equivalent sequence from *pilE* during gene conversion. P^+a, x, y, z, etc. represent genes encoding antigenically and/or physically distinct polypeptides. Subscript numbers on *pilS* donor sequences indicate different lengths of the same donor silent sequence copy.

15). This results from intragenic recombination involving the non-reciprocal exchange of silent sequences, held in storage form at *pilS* and elsewhere, with partially homologous sequences at *pilE*. The incoming copy of silent sequence apparently displaces an analogous tract of the expression locus and the evicted region is presumably discarded. Different silent pilin sequences contain unique tracts of information and their translocation into *pilE* will usually produce chimeric sequences that will encode structually distinct pilins. Simple replacement events can lead to the insertion of a given silent sequence from *pilS*1 into *pilE* (15, 27). However, variation also occurs in the length of silent sequence that may be exchanged. Furthermore, serial gene conversions involving different length tracts of the same or a different silent sequence can lead to the formation of a plethora of distinct pilins (Fig. 5). Pili thus contain a mosaic of polypeptide sequences that reflect the immediate history of such conversions. The silent sequences provide a repertoire of variant sequence information that can be expressed by mixing and matching portions of pilin sequence information. Since gonococcal variants may return to producing one particular pilin variant after producing a series of others, the partial pilin sequences also provide a storage or reference capability.

Phase transition and antigen variation are substantially reduced in recombination-deficient (*recA*$^-$) strains of *N. gonorrhoeae* (19, 24, 31), implying that the displacement process of pilin coding sequences at *pilE* most often involves homologous recombination (or a process having at least one step in common with it). A repeat sequence defined by a 66 bp *Sma*I-*Cla*I restriction fragment is located about 65 bp on the 3′ side of the *pilE* coding region in *N. gonorrhoeae* strains MS11 and P9 (Fig. 3) (14, 23, 25). This sequence is found adjacent to other regions of the gonococcal genome which contain partial pilin sequences and is apparently not deleted in the generation of P$^-$n variants (14, 22). A related sequence (pseudo *Sma*I-*Cla*I) is found downstream of *pil* loci in meningococci (30). These two sequences bear striking homologies to the recognition sequences of several recombinases, most notably the left hand inverted repeat and the left hand end of site I of the *res* site of Tn*3* (28, 30) (Table I). The gonococcal *Sma*I-*Cla*I sequences may therefore be recognized by a recombinase required to initiate, or perhaps resolve, intermediates involved in recombination between *pilE* and silent pilin sequences. Alternatively, these repeats may act as guide sequences to align synapsis between incoming and resident *pil* sequences.

Gonococcal *pilE* and *pilS*1 loci also contain several further classes of recombination sequence (RS) that are unrelated to pilin coding sequences (27) (Table I). All silent pilin sequences in *pilS*1, except copies 2 and 3, are separated by repetitive segments of 39 or 40 bp containing a common 33 bp core and referred to as RS1 (27). RS1 probes hybridize to both *pilE*1 and *pilE*2 loci of strain $MS11_{ms}$ (27) and RS1 sequences have been detected in the region upstream of *pilE* in strain P9 (28). The region of about 425 bp between copies 2 and 3 of *pilS*1 contains RS1 together with two further groups of repeat sequences. These are RS2 which are 49 bp in length and are present as two copies in direct repeat, and RS3, a family of direct and indirect repeats sharing the common core sequence 5′-ATTCCC-3′ or 5′-GGGAAT-3′ where homology between individual members of the family is 9–17 bp (27). Several RS3-related elements are clustered immediately upstream of the *pilE* open reading frame, normally in groups of two or three separated within each group by 3–8 bp (28). Some of the RS sequences are similar to the sequence 5′AGCGAATA-3′ which mediates deletion formation in the gonococcal cryptic plasmid (32), although it is not clear if these sequences are functionally related (Table I). The presence of conserved sequences within the pilin open reading frame (which presumably reflect the critical nature of the corresponding amino acid sequence for producing a functional pilus structure) and the diversity of repeat sequences in and around *pil* loci provide abundant opportunities for homologous recombination. However, the precise role of

Table I

Recombination sequences in *Neisseria* pil regions and their relationship to other recombinase recognition sequences

Salmonella typhimurium Hin right inverted repeat	5′-TTCCTTTTGGAAGGTTTTTGA-3′
	* * * * * * * * * * * * * * * *
Meningococcal Pseudo *Sma*I-*Cla*I repeat	5′-ATTAATCCGGGTGGCTTCCTTTTTAAAGGTTTGCAAGGC-3′
	* * * * * * * * * * *
Gonococcal *Sma*I-*Cla*I repeat	5′-GGGCGGGGTCGTCCGTTCCGGTGGAAATAATATATCGAT-3′
	* * * * * * * * * * * * * * * * * * * * * *
Tn*3* Inverted Repeat	5′-GGGGTCTGACGCT CAGTGGAAGAAAACTCACGTT-3′ (C)
Gonococcal RS1	5′-AAACACCACGCGCCGGATTTCAAACACTTCCAAA-3′
Gonococcal RS2	5′-GGAATCCGGAACGCAAAATCTAAAGAAACCGTTTTACCCGATAAGTTTC-3′
Gonococcal RS3 core sequences	5′-ATTCCC-3′ 5′-GGGAAT-3′
Gonococcal cryptic plasmid deletionogenesis sequence	5′-AGCGAATG-3′

such sequences in gene conversion mechanisms for pilus phase and antigenic variation awaits elucidation.

The pilin-coding region of the gonococcal genome is apparently a recombination hot spot. It is reasonable to presume that gene conversion events are somehow regulated. Gonococci and meningococci produce a surprisingly large variety of methylase activities; many without the corresponding restriction endonucleases (33, 34). Differential methylation by one or more modification methylases might modulate pilin gene conversion processes by interfering directly with recombination or by modulating transcription of some essential product. This would be analogous to *dam*-methylation of promoter sequences in modulating expression of the *cre* gene of *E. coli* phage P1, which encodes a site-specific recombinase (35). The methylase M.*Ngo*IV (recognition sequence 5′-GC^{m}CGGC-3′) has a target site within the pilin structural gene (Fig. 3). It has been suggested that differential methylation of this region could play a role in regulating gene conversion events (14). However, there is no experimental evidence to support this hypothesis.

D. Pilus Phase and Antigenic Variation Arise as a Consequence of the Same Underlying Genetic Mechanism

The generation of P^{-}rp^{+} gonococci, their subsequent reversion to P^{+} and the generation of antigenic pilus variants involve intragenic recombination between *pilE* and silent pilin genes (15). Transfer of the copy 5 silent sequence of *pilS*1 to *pilE* invariably seems to result in creation of a P^{-}rp^{+} phenotype, because the resulting novel pilin genes encode missense polypeptides unable to assemble into pili (15, 19). This raises the possibility that the role of copy 5 is to regulate pilus expression *in vivo*. Pilus expression could be switched off by insertion of copy 5 and subsequently switched on again through eviction of copy 5 sequences by new *pil* sequences encoding assembly-functional pilin (8, 15).

Some reversion of P^{-}r variants to P^{+} can be detected in strains that are deleted for *rec*A. However, this is due to single nucleotide addition or deletion events that bring the pilin mRNA back into a normal translational reading frame (19). The only difference between antigenic variation and reversible phase transition is that in the former, the variant pilin produced is orthodox and can be assembled into pili whereas in the latter, the pilin is altered so as to be non-functional. Some domains of pilin must be involved in structural integrity of the pilus, so it is not surprising that some recombinational events generate combinations of amino acid sequence that are deleterious to pilus structure and function. Deletion events associated with the generation of P^{-}r variants and the duplication of *pilE* that has apparently occurred in some strains may also be aberrations of the recombinational mechanism that produces genetic variation at the pilin locus.

IV. GENETIC MECHANISMS FOR P.II VARIATION

A. P.II Variation Occurs Independently of Pilus Variation

Rates of phase transition between Op^+ and Op^- are about 2×10^{-3} per colony-forming unit per generation in *N. gonorrhoeae* (36). Loss of Op expression can be accompanied by changes in pilin molecular weight in some strains (37). At least one P.II coding locus (*opaE*1) is located close to the *pilE*1 gene of *N. gonorrhoeae* strain $MS11_{ms}$ (Fig. 2) (21, 23). Since this region of the genome is apparently a hot-spot for recombination, it is not surprising that there may be occasional concomitant variation of piliation and opacity characters. However, there is no general correlation between piliation status and production of P.II.

B. Antigenic and Phase Variation Involves the Switching of Translational Reading Frame at a Series of Distinct *opa* loci

The gonococcal genome contains at least nine *opa* expression loci which encode distinct P.II polypeptides (21, 38). This is consistent with the maximum observed number of separate P.II species produced by one strain (6). Cloning and sequencing of a number of *opa* loci from *N. gonorrhoeae* MS11 indicates that each *opa* gene is intact and constitutively transcribed but the functional P.II protein is not necessarily expressed (39). The 5′ region of each *opa* gene contains series of identical pentameric pyrimidine units called coding repeats (CR) with the sequence 5′-CTCTT-3′. A variable number of CR units lie in that part of the gene encoding the N-terminal hydrophobic leader peptide of the P.II polypeptide. This leader is required for transport to the outer membrane and is presumably removed by a signal peptidase post-translationally (Fig. 6). The number of pentameric repeat units varies from seven to 28 in the leader peptide region of each *opa* gene (39). Which of the heterogeneous population of *opa* mRNA species is translated to produce functional P.II polypeptides depends on the precise number of CR units present at any *opa* locus. When the leader region contains certain numbers of CR units, e.g. nine, the DNA sequence coding for the amino-terminus of mature P.II will be in frame with the leader peptide and a functional P.II polypeptide will be produced. However, the addition or deletion of one pentameric CR unit to or from the DNA sequence will oblige translation of the P.II coding sequence in

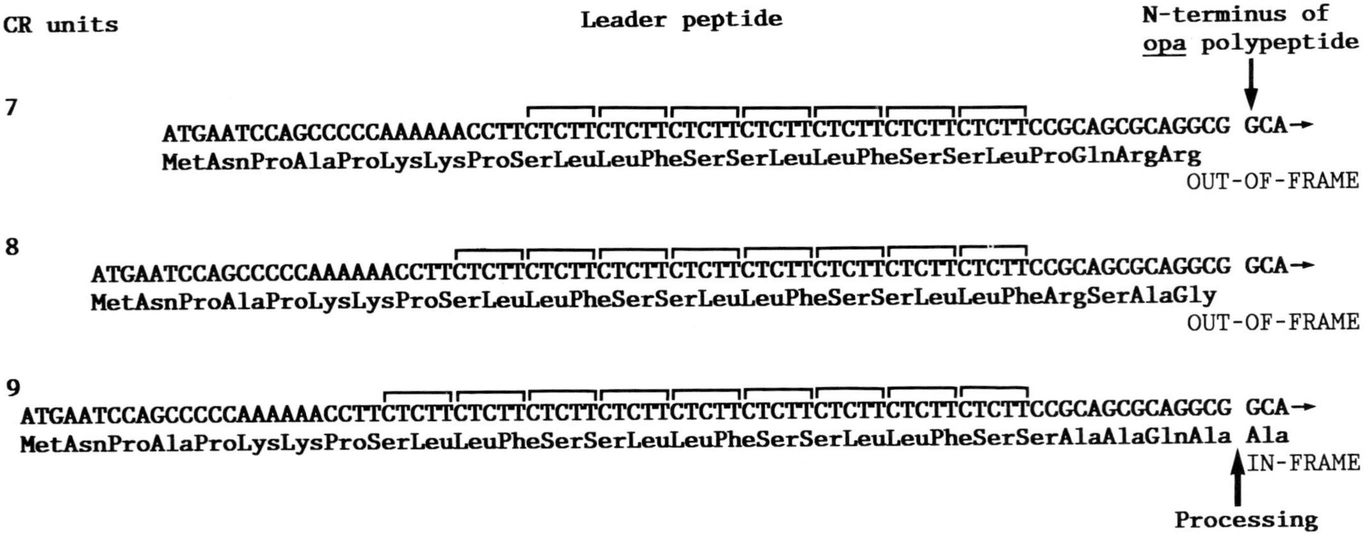

Figure 6. Control of P.II expression by modulation of translational reading frame in the leader peptide region of the *opa* gene sequence. Only 5′ *opa* gene sequences encoding the NH_2-terminus of the unprocessed polypeptide are shown; CR, pentameric 5′-CTCTT-3′ repeats; numbers indicate the degree of repetition of CR in three cases; processing arrow represents position of cleavage by signal peptidase.

either of the two alternative reading frames, both of which produce a non-functional polypeptide (Fig. 6). Hence, a phase change will be observed (Fig. 7). Presumably, the number of CR units in each *opa* gene is altered as a consequence of spontaneous deletional and/or insertional recombination events or by slippages during replication through the pyrimidine-rich regions that these units create (39).

There are no differences in the processing of P.II proteins expressed from *opa* genes with different numbers of CR units. However, it is possible that variation in the length and nature of the leader peptide could influence transport and hence the final properties of P.II when inserted in the outer membrane (39).

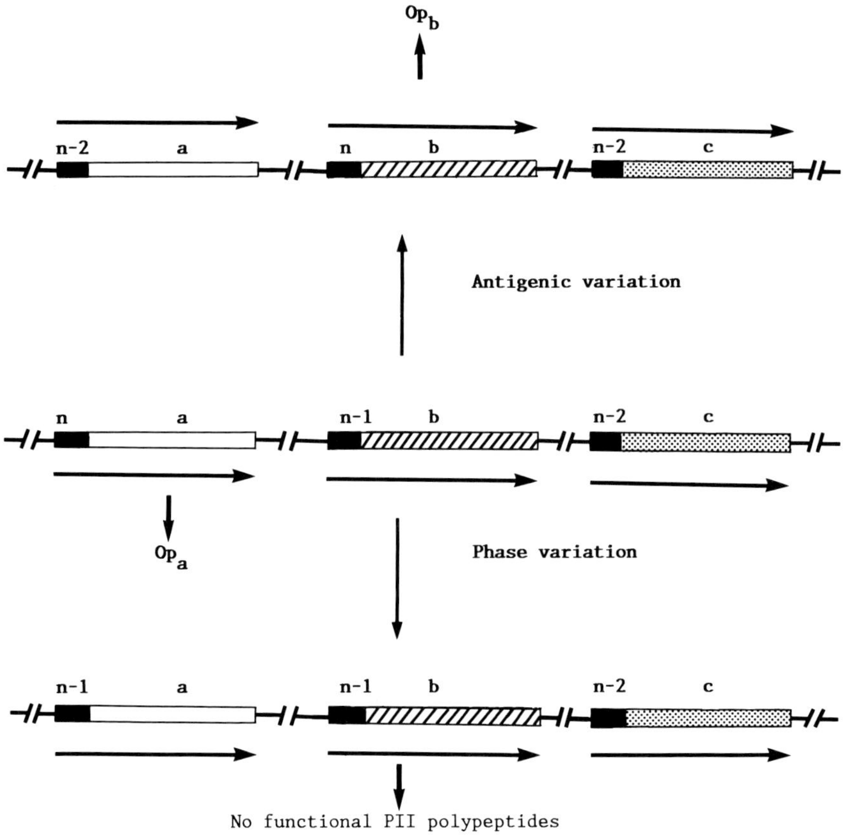

Figure 7. Phase and antigenic variation by differential activation of Neisserial *opa* genes. a, b, c represent the coding sequences for distinct P.II species located in a series of physically separate *opa* loci. $Op_{a,b}$ represent antigenically distinct and functional P.II polypeptides. n = the number of CR units in the leader peptide region that will keep the *opa* polypeptide in-frame.

All *opa* loci capable of expression are linked to CR units. Therefore it seems likely that antigenic variation is caused mainly by sequential expression of different loci, depending on the number of these repeats (Fig. 7). P.II variation thus involves control of the translational reading frame of a series of separate *opa* expression loci (*opaE*) encoding antigenically distinct polypeptides. Gene conversion events (analogous to those of pilus expression) that do not involve CR units have been observed in the structural coding regions of *opaE* loci. Hybridization studies indicate that the sequence of an *opa* donor locus can be duplicated during the replacement of a corresponding variant sequence at an acceptor locus (39). However, many such conversions seem to be silent and do not alter the expression of P.II.

C. Common Mechanism for P.II Variation in Pathogenic *Neisseria*

The genomes of *N. meningitidis* and *Neisseria lactamica* (which is a commensal that can cause infections) also contain gonococcal *opa*-related (*opr*) sequences (29, 39). They contain four and two *opr* sequences respectively (38). Thus, there is less potential for variation in the P.II proteins of these species. However, regulatory features found in gonococcal *opa* genes are conserved in both species; these include constitutive expression of *opr* genes, pentameric 5′-CTCTT-3′ CR units in the 5′ region coding for the hydrophobic leader, variation in the number of CR units at most *opr* loci and association between the number of such units and expression of functional outer membrane protein. Apart from the variable regions, there is also considerable homology between coding sequences for the gonococcal (*opa*) and meningococcal (*opr*) outer membrane proteins. It seems likely, therefore, that the mechanism for switching *opa*/*opr* sequences has been conserved since it regulates an important pathogenicity determinant.

V. CONCLUSIONS

Pathogenic *Neisseria* species have evolved separate mechanisms for modulating the antigenic nature of pili and P.II species. The genome of any one strain contains ample information to account for the observed variability in these important proteins. Exact mechanisms involved in the gene conversion events that manoeuvre previously silent sequences into expression sites remain to be elucidated. It is possible that all the molecular switching events simply involve DNA transactions within a single copy of the neisserial chromosome. However, it is also possible that gene duplications and deletions could involve recombination between two separate chromosomes present in the partner cells of the diplococcus characteristic of these bacteria (19). Furthermore, donor gene sequences could be acquired by interstrain transformation following uptake of chromosomal DNA released by other neisserial cells (33). Both gonococci and meningococci are not only highly competent for transformation but also generally autolyse readily (40). (Interestingly, gonococci only become naturally competent for transformation when they are piliated. However, there is no evidence that pili play a direct role in DNA uptake.) Thus the pathogenic *Neisseria* spp. have both endogenous and exogenous sources of genetic information that might be utilized in generating diversity. Exploitation of the potential for varying two crucial surface components is a significant contributor to the success of these bacteria as agents of human disease.

Acknowledgements

John Heckels, Iain Nicolson, Tony Perry, Maggie So and John Swanson are thanked for generously providing information prior to publication.

References

(1) Heckels, J. E. (1986). Gonococcal antigenic variation and pathogenesis. *In* "Antigenic Variation in Infectious Diseases" (T. H. Birkbeck and C. W. Penn, eds.), pp. 77–94. Society for General Microbiology/IRL Press, Oxford.

(2) Saunders, J. R. (1986). Genetic basis of phase and antigenic variation in bacteria. *In* "Antigenic Variation in Infectious Diseases" (T. H. Birkbeck and C. W. Penn, eds.), pp. 57–76. Society for General Microbiology/IRL Press, Oxford.

(3) Swanson, J. and Barrera, O. (1983). Gonococcal pilus subunit size heterogeneity correlates with transition in colony piliation phenotype, not with changes in colony opacity. *J. Exp. Med.* **158**, 1459–1472.

(4) Swanson, J. (1982). Colony opacity and protein II compositions of gonococci. *Infect. Immun.* **37**, 359–368.

(5) Virji, M. and Heckels, J. E. (1986). The effect of protein II and pili on the interaction of *Neisseria gonorrhoeae* with human polymorphonuclear leukocytes. *J. Gen. Microbiol.* **132**, 503–512.

(6) Barritt, D. S., Schwalbe, Klapper, D. G. and Cannon, J. G. (1987). Antigenic and structural differences among six proteins II expressed by a single strain of *Neisseria gonorrhoeae*. *Infect. Immun.* **55**, 2026–2031.

(7) Sparling, P. F., Cannon, J. G. and So, M. (1986). Phase and antigenic variation of pili and outer membrane protein II of *Neisseria gonorrhoeae*. *J. Infect. Dis.* **153**, 196–201.

(8) Swanson, J., Robbins, K., Barrera, O., Corwin, D., Boslego, J., Ciak, J., Blake, M. and Koomey, J. M. (1987). Gonococcal pilin variants in experimental gonorrhoea. *J. Exp. Med.* **165**, 1344–1357.

(9) Hagblom, P., Segal, E., Billyard, E. and So, M. (1985). Intragenic recombination leads to pilus antigenic variation in *Neisseria gonorrhoeae*. *Nature* **315**, 156–158.

(10) Tinsley, C. R. and Heckels, J. E. (1986). Variation in the expression of pili and outer membrane protein by *Neisseria meningitidis* during the course of meningococcal infection. *J. Gen. Microbiol.* **132**, 2483–2490.

(11) So, M., Billyard, E., Deal, C., Getzoff, E., Hagblom, P., Meyer, T. F., Segal, E. and Tanner, J. (1985). Gonococcal pilus: genetics and structure. *Curr. Top. Microbiol. Immunol.* **118**, 13–28.

(12) Perry, A. C. F., Hart, C. A., Nicolson, I. J., Heckels, J. E. and Saunders, J. R. (1987). Interstrain homology of pilin gene sequences in strains of *Neisseria meningitidis* that express markedly different antigenic pilus types. *J. Gen. Microbiol.* **133**, 1409–1418.

(13) Nicolson, I. J., Perry, A. C. F., Virji, M., Heckels, J. E. and Saunders, J. R. (1987). Localization of antibody binding sites on pili of *Neisseria gonorrhoeae* by DNA sequencing of variant pilin genes. *J. Gen. Microbiol.* **133**, 825–833.

(14) Segal, E., Hagblom, P., Seifert, H. S. and So, M. (1986). Antigenic variation of gonococcal pilus involves assembly of separated silent gene segments. *Proc. Natl. Acad. Sci. USA* **83**, 2177–2181.

(15) Swanson, J., Bergstrom, B., Robbins, K., Barrera, O., Corwin, D. and Koomey, M. (1986). Gene conversion involving the pilin structural gene correlates with pilus^{+} $\langle - \rangle$ changes in *Neisseria gonorrhoeae*. *Cell* **47**, 267–276.

(16) Swanson, J., Robbins, K., Barrera, O. and Koomey, J. M. (1987). Gene conversion variations generate structurally-distinct pilin polypeptides in *Neisseria gonorrhoeae*. *J. Exp. Med.* **165**, 1016–1029.

(17) Schoolnik, G., Fernandez, R., Tai, J. Y., Rothbard, J. and Gotschlich, E. C. (1984). Gonococcal pili: primary structure and receptor binding domain. *J. Exp. Med.* **159**, 1351–1370.

(18) Virji, M. and Heckels, J. E. (1983). Antigenic cross reactivity of *Neisseria* pili: investigations with type- and species-specific monoclonal antibodies. *J. Gen. Microbiol.* **129**, 2761–2768.

(19) Swanson, J. (1987). Genetic mechanisms responsible for changes in pilus expression by gonococci. *J. Exp. Med.* **165**, 1459–1479.

(20) Poolman, J. T., Hopman, C. T. P. and Zanen, H. C. (1985). Colony variants of *Neisseria meningitidis* strain 2996 (B:26:P1.2): influence of class 5 outer membrane proteins and lipopolysaccharides. *J. Med. Microbiol.* **19**, 203–209.

(21) Stern, A., Nickel, P., Meyer, T. F. and So, M. (1984). Opacity determinants of *Neisseria gonorrhoeae*: gene expression and chromosomal linkage to the gonococcal pilus gene. *Cell* **37**, 447–456.

(22) Segal, E., Billyard, E., So, M., Storzbach, S. and Meyer, T. F. (1985). Role of chromosomal rearrangement in *Neisseria gonorrhoeae* pilus phase variation. *Cell* **40**, 293–300.

(23) Meyer, T. F., Billyard, E., Haas, R., Storbach, S. and So, M. (1984). Pilus genes of *Neisseria gonorrhoeae*: chromosomal organization and DNA sequence. *Proc. Natl. Acad. Sci. USA* **81**, 6110–6114.

(24) Bergstrom, S., Robbins, K., Koomey, M. and Swanson, J. (1986). Piliation control mechanisms in *Neisseria gonorrhoeae*. *Proc. Natl. Acad. Sci. USA* **83**, 3890–3894.

(25) Nicolson, I. J., Perry, A. C. F., Heckels, J. E. and Saunders, J. R. (1987). Genetic analysis of variant pilin genes from *Neisseria gonorrhoeae* P9 cloned in *Escherichia coli*: physical and immunological properties of encoded pilins. *J. Gen. Microbiol.* **133**, 553–561.

(26) Meyer, T. F., Mlawer, N. and So, M. (1982). Pilus expression in *Neisseria gonorrhoeae* involves chromosomal rearrangement. *Cell* **30**, 45–52.

(27) Haas, R. and Meyer, T. F. (1986). The repertoire of silent pilus genes in *Neisseria gonorrhoeae*: evidence for gene conversion. *Cell* **44**, 107–115.

(28) Perry, A. C. F., Nicolson, I. J. and Saunders, J. R. (1987). Structural analysis of the *pilE* region of *Neisseria gonorrhoeae* P9. *Gene*. **60**, 85–92.

(29) Aho, E. L., Murphy, G. L. and Cannon, J. G. (1987). Distribution of specific DNA sequences among pathogenic and commensal *Neisseria* species. *Infect. Immun.* **55**, 1009–1013.

(30) Perry, A. C. F., Nicolson, I. J. and Saunders, J. R. (1988). *Neisseria meningitidis* strain C114

contains silent, truncated pilin genes that are homologous to *Neisseria gonorrhoeae pil* sequences. *J. Bacteriol.* **170**, 1691–1697.

(31) Koomey, J. M. and Falkow, S. (1987). Cloning of the *recA* gene of *Neisseria gonorrhoeae* and construction of gonococcal *recA* mutants. *J. Bacteriol.* **169**, 790–795.

(32) Hagblom, P., Korch, C., Jonsson, A. B. and Normark, S. (1986). Intragenic variation by site-specific recombination in the cryptic plasmid of *Neisseria gonorrhoeae. J. Bacteriol.* **167**, 231–237.

(33) Korch, C., Hagblom, P. and Normark, S. (1983). Sequence-specific DNA modification in *Neisseria gonorrhoeae. J. Bacteriol.* **155**, 1324–1332.

(34) Sullivan, K. M., MacDonald, H. J. and Saunders, J. R. (1987). Characterization of DNA restriction and modification activities in *Neisseria* species. *FEMS Microbiol. Lett.* **44**, 389–393.

(35) Sternberg, N., Sauer, B., Hoess, R. and Abremski, K. (1986). Bacteriophage P1 *cre* gene and its regulatory region. Evidence for multiple promoters and for regulation by DNA methylation. *J. Mol. Biol.* **187**, 197–212.

(36) Meyer, L. W. (1982). Rates of *in vitro* changes in gonococcal colony opacity phenotypes. *Infect. Immun.* **37**, 481–485.

(37) Salit, I. E., Blake, M. and Gotschlich, E. C. (1980). Intrastrain heterogeneity of gonococcal pili is related to opacity colony variance. *J. Exp. Med.* **151**, 716–725.

(38) Stern, A. and Meyer, T. F. (1987). Common mechanism controlling phase and antigenic variation in pathogenic neisseriae. *Mol. Microbiol.* **1**, 5–12.

(39) Stern, A., Brown, M., Nickel, P. and Meyer, T. F. (1986). Opacity genes in *Neisseria gonorrhoeae*: control of phase and antigenic variation. *Cell* **47**, 61–71.

(40) Sparling, P. F. (1966). Genetic transformation of *Neisseria gonorrhoeae* to streptomycin-resistance. *J. Bacteriol.* **92**, 1364–1371.

Chapter **14**

Adhesins of Pathogenic **Escherichia coli**

G. HINSON AND P. H. WILLIAMS

I. INTRODUCTION

Escherichia coli is a facultatively anaerobic species adapted to long-term commensal carriage by animals as part of normal intestinal microbial communities. Most strains lead symbiotic existences and are quite harmless, but a minority possess characteristics which result in a variety of infectious diseases in their hosts. Such characteristics, termed virulence determinants, comprise factors which promote growth of the pathogens beyond that achieved by benign strains; sometimes such growth occurs in normally sterile regions of the body, and subsequent damage to the host causes the symptoms of disease. The course of an infection is determined by a combination of bacterial and host factors (such as the immune system), and is best viewed as a dynamic host-parasite interaction. This chapter concerns the genetic analysis of one such factor that is important in the initial stages of infection; the ability of pathogenic bacteria to adhere to specific host tissues prior to colonization. In order to rationalize research effort, individual virulence determinants and host systems are generally studied in isolation. However, we must remember the complex nature of the overall disease process and the involvement of several other factors besides colonization ability.

II. BACTERIAL ADHERENCE TO ANIMAL TISSUES

A. Adherence is Mediated by Adhesins

Pathogenic bacteria have elaborate structures (adhesins) that enable them to adhere tightly to living surfaces of their hosts in order to resist shearing forces that tend to remove them. Adhesins are usually proteins, although polysaccharide capsules, in addition to their antiphagocytic properties (being poorly immunogenic, they physically mask other strong antigens such as outer membrane proteins from the immune system) may also have a role in forming microcolonies on dental and epithelial surfaces. Cells of cariogenic *Streptococcus mutans*, for example, are usually embedded in a matrix of exopolysaccharides stuck firmly to the tooth surface. For the purposes of this chapter, capsular adherence mechanisms will be ignored in favour of the highly specific interactions promoted by proteinaceous adhesins which are clearly associated with pathogenesis.

Electron microscopic examination of most pathogenic isolates of *E. coli* reveals fine fibrils 1–2 μm long and up to 9 nm in diameter, radiating from the bacterial surface (Fig. 1). These structures, called fimbriae or pili (although the latter term is best reserved for conjugative structures such as F-pili with no known role in adherence to animal surfaces), carry the adherence properties of the whole bacteria. Once isolated, fimbriae will adhere to the same substrates as the bacteria from which they were extracted. Isolated fimbriae behave as large structures in the region of 10^6 to 10^7 daltons in non-denaturing conditions. However, each antigenic type of fimbria is principally a non-covalent aggregate of a characteristic kind of protein subunit, whose molecular weights range from 10 000 to 25 000. Genetic data on certain adhesins (see below) point to the presence of smaller numbers of additional proteins which possess the actual adhesive properties of whole fimbriae, but these are not normally visible by sodium dodecyl sulphate (SDS) polyacrylamide gel electrophoresis of purified fimbrial preparations. Furthermore, a few adhesins do not seem to be associated with fimbrial structures at all and so are called "nonfimbrial" or "afimbrial". These may perhaps be irregular aggregates without a clearly defined shape (sometimes appearing in the electron microscope as structures similar to polysaccharide capsules) or simply fibrils which, because of their small diameters (often less than 3 nm), are poorly resolved by normal staining techniques. There appear to be no significant differences between the fimbrial and nonfimbrial adhesins in terms of their function and genetics, so the distinction is of doubtful practical value.

B. Adherence is Highly Specific

Individual adhesin types interact with particular components of a whole range of molecules exposed on animal epithelia. The interaction between bacterial adhesins and host tissues exhibits quite remarkable specificity and has been likened to antibody-antigen recognition; indeed, adhesins are usually classified according to their adherence specificities in *in vitro* assays, the standard method being haemagglutination. Blood samples from a selection of animal species are mixed with bacterial cultures or adhesin preparations and visible agglutination of the erythrocytes denotes a positive reaction. Particular adhesins produce characteristic patterns of reaction and non-reaction with the various blood types. Although haemagglutination tests do not actually assay adherence to the tissues at which colonization occurs *in vivo*, their simplicity and reproducibility makes them diagnostically valuable. More exacting assays examine adherence to tissue samples such as intestinal biopsies (1), frozen tissue sections thawed before use (2), voided urinary epithelial cells (3), and sometimes cultured cell lines (4).

Adherence can be inhibited by providing an excess of the receptor component free in solution rather than bound to a surface, presumably because of competition for a limited number of adhesin active sites. As far as is known at present, all receptors are carbohydrate-containing polymers. Thus, in order to characterize the receptor specificity of a novel adhesin, haemagglutination or other assays may be performed in the presence of putative inhibitors, normally simple sugars or sugar compounds. An early finding was that adherence mediated by type I or common fimbriae, present on the majority of both pathogenic and benign *Enterobacteriaceae*, can be inhibited by D-mannose (5). These fimbriae are therefore also given the alternative designation "mannose sensitive" (MS), while all other adhesins are defined as "mannose resistant" (MR). The abundance of mannose-containing epithelial cell surface components and the lack of correlation with pathogenesis suggest that type I fimbriae, in contrast to MR adhesins, should probably not be regarded as virulence determinants, although some experiments indicate a possible function in mediating adherence of uropathogenic *E. coli* to the lower urinary tract (6). However, of all the adhesin genetic determinants, those for type I fimbriae are probably the best understood and serve as a model system.

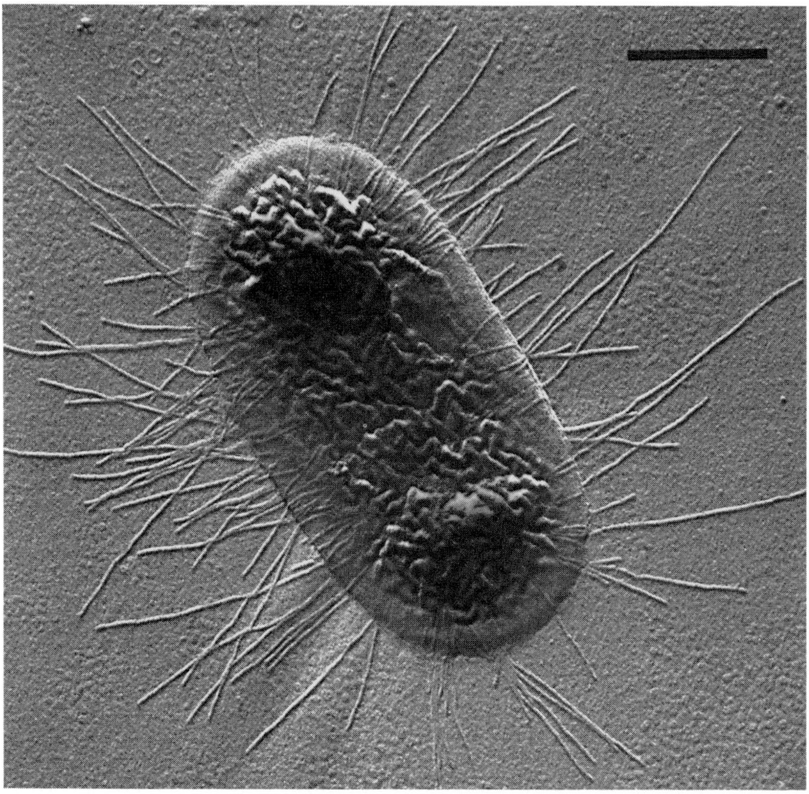

Figure 1. (a)

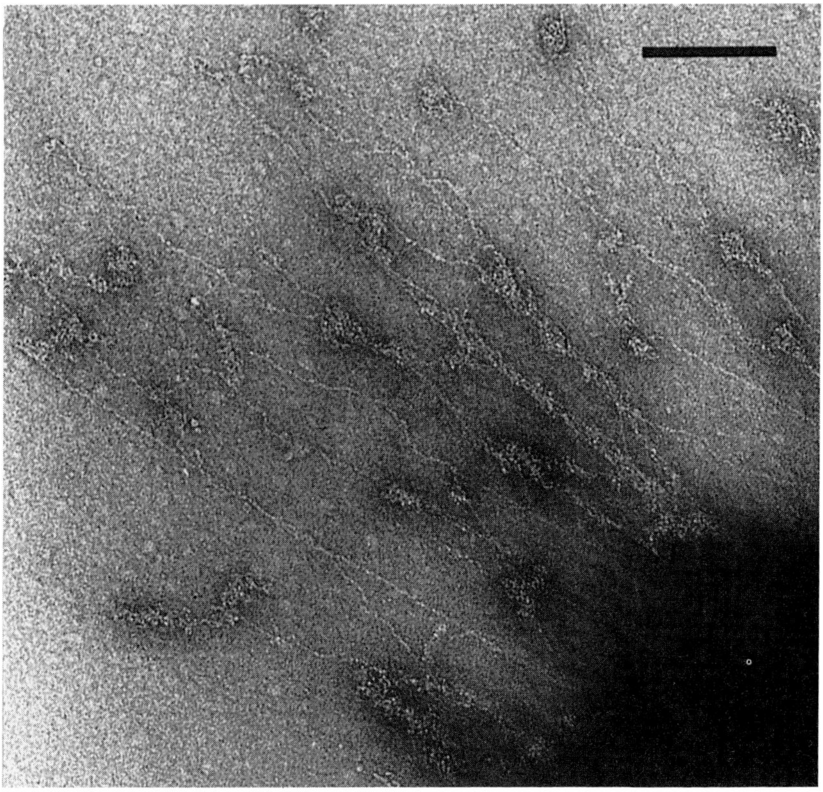

(b)

Figure 1. (a) Electron micrograph of a human diarrhoeal ETEC isolate bearing CS1 fimbriae (one of the CFA/II group of adhesins) dried onto a carbon support and shadowed with carbon and platinum. The rigid, rod-like CS1 fimbriae, approximately 7 nm in diameter and up to about 1 μm long, are easily visible radiating from the bacterial surface. Scale bar = 500 nm. (b) Electron micrograph of part of a laboratory strain of *E. coli* carrying a plasmid with an insert from the chromosome of strain 469-3, a human infantile enteritis-associated *E. coli* isolate. "Fibrillar" 469-3 MR fimbriae are rather more difficult to demonstrate than the rod-like types but can be seen in this enlarged section as apparently flexible, 2–3 nm diameter fibrils after negative staining with uranyl acetate. Scale bar = 200 nm. (Both micrographs courtesy of Dr. S. Knutton.)

Name	Genes	Location	Map	Reference
Type 1	pil	Chromosome	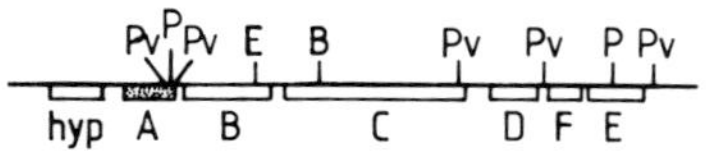	36
	fim	Chromosome	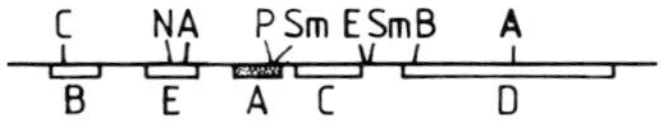	10
K88ab	fae	Plasmid	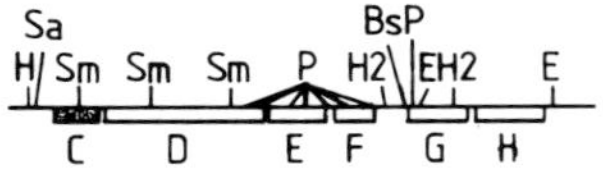	16
K99	fan	Plasmid	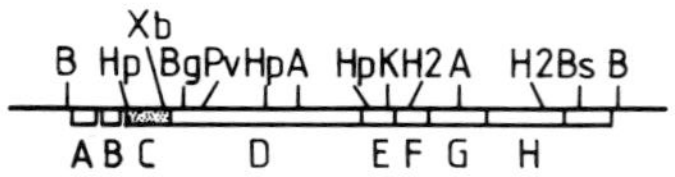	16,22
CS3		Plasmid	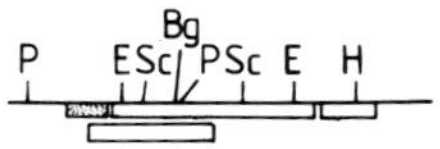	7

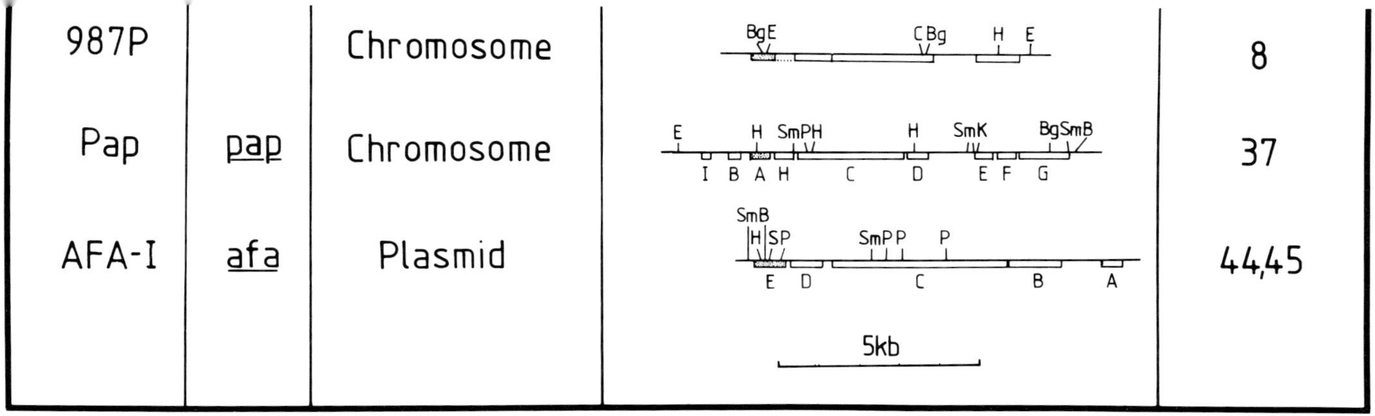

Figure 2. Partial restriction maps of a selection of cloned fimbrial determinants, aligned where possible according to the positions of genes encoding the subunits (stippled) and large proteins. Many of these restriction maps are incomplete because only certain regions of the DNA shown were fully mapped (often, particular restriction fragments were subcloned and then mapped). Restriction enzyme abbreviations given above the lines are:—

A = *Ava*I	B = *Bam*HI	Bg = *Bgl*II	Bs = *Bst*EII	C = *Cla*I	E = *Eco*RI
H = *Hind*III	Hp = *Hpa*I	H2 = *Hind*II	K = *Kpn*I	N = *Nru*I	P = *Pst*I
Pv = *Pvu*II	S = *Sal*I	Sa = *Sau*3AI	Sc = *Sca*I	Sm = *Sma*I	Xb = *Xba*I

Boxes below the lines represent map positions of individual genes. Their names are given by the single letters below the boxes appended to the genetic designations *fim* or *pil* for type 1 "fimbriae" or "pili" (two systems cloned independently), *fae* for "fimbrial adhesin 88" (K88), *fan* for "fimbrial adhesin 99" (K99), *pap* for "pili associated with pyelonephritis" (also called F13) or *afa* for "afimbrial adhesin".

Host tissue specificities attributable to most microbial adhesins dramatically affect the ability of a pathogen to colonize its hosts—and of course without colonization, an infection cannot normally be established. Pathogenic strains infect only a few animal species, and within any host they tend preferentially to colonize particular body regions. In fact, pathogenic *E. coli* strains can each express up to about four different adhesins, often with distinct receptor specificities, raising the possibility of colonizing a limited number of different host sites provided the corresponding receptors are available.

C. Adhesins are Encoded by Complex Gene Clusters

Adhesin genetic systems are carried either on plasmids or on the chromosome. They may be linked to determinants specifying other virulence factors such as haemolysin, enterotoxin or capsule biosynthesis. Apart from subunits which form the fimbrial (or other) structures, additional proteins are clearly required for export and assembly of functional adhesive structures on the bacterial surface (7). Furthermore, the level of adhesin expression is regulated according to environmental and cellular conditions, implying the presence of regulatory genes (8). Thus, the overall size of a typical adhesin determinant is in the region of 8 kilobases (kb), comprising about six genes in addition to that for the subunit protein, and control regions (Fig. 2). Most adhesin gene clusters are uninterrupted by other genetic information, although Colonization Factor Antigen I (CFA/I) fimbriae of human enterotoxigenic *E. coli* (ETEC) are encoded by two distinct regions on plasmids of about 89 kb. Region 1 occupies 6 kb and is closely linked to the genes specifying heat-stable enterotoxin (ST), whereas region 2, 37 kb away and just 2 kb in size, is required for export of CFA/I fimbrial subunits from the cell (9). Many other *E. coli* adhesin determinants have been successfully cloned on single DNA fragments of less than 50 kb in cosmid or plasmid vectors in standard laboratory strains, and the inserts then reduced to about 10 kb without apparent loss of function.

III. ADHESIN GENETICS

So many different adhesins have been cloned from pathogenic *E. coli* isolates that we have chosen topics of particular interest with suitable illustrative examples, rather than attempting to summarize the entire field.

A. Adhesin Expression is Finely Regulated

The major subunit proteins of most adhesins are synthesized in quite large amounts and, of course, are transported across the cell membranes. Thus, it is not surprising to find that expression is quantitatively regulated at a level which avoids overstressing cellular energetic and biosynthetic resources and does not disrupt the cell envelope. Regulatory mechanisms include promoters of different activities allied to particular structural genes, and diffusible repressors and activators whose expression modulates expression of the subunit and other genes. Adhesin synthesis also responds to certain conditions in the bacterial environment such as temperature (see below). Finally, adhesin expression is subject to qualitative control; individual bacteria can switch between expressing and non-expressing "phases" apparently at random ("phase variation", see below).

Expression of type I fimbriae is quantitatively controlled by *fimE* (also known as *hyp*) which specifies a repressor for type I subunit expression (10), and *fimB*, specifying a positive regulator (11, 12; Fig. 3). Thus, *fimE/hyp* mutants are hyperfimbriated (i.e. they express far more fimbriae per cell than do wild-type strains) whereas *fimB* mutants express fewer fimbriae than normal. Presumably as a consequence of the abnormally high biosynthetic

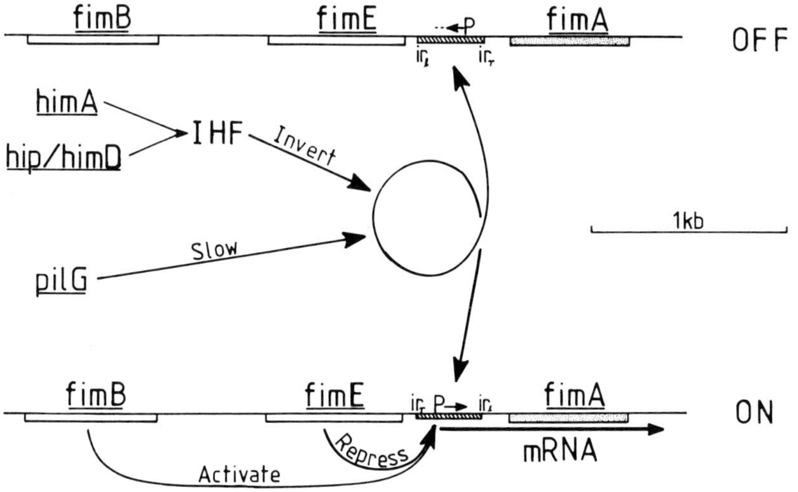

Figure 3. Phase variation of Type 1 fimbriae, encoded at 98 minutes on the *E. coli* chromosomal map, is achieved by inversion of a 314 bp DNA segment (hatched) bounded by 9 bp inverted repeat sequences designated left (ir_l) and right (ir_r). In the "off" orientation, the promoter (P) for subunit gene *fimA* (stippled) reads away from *fimA*, preventing transcription of the gene, whereas after inversion, the promoter initiates transcription of the subunit gene. The two-subunit "Integration Host Factor" (IHF) system, encoded by genes *himA* at map position 38 min and *hip/himD* at 20 min, is required for inversion; conversely, *pilG* at 27 min slows the inversion rate. Subsequent expression of *fimA* is quantitatively regulated by the combined actions of activator gene *fimE* (also known as *hyp*) and repressor gene *fimB*.

load, hyperfimbriated cells also exhibit severely restricted growth relative to nonfimbriated variant cells and wild-type strains, especially when other factors such as composition of the medium reduce the normal growth rate. This may explain the apparent catabolite repression of type I fimbrial expression by glucose, called "pseudocatabolite repression" by Eisenstein and Dodd (13). Glucose added to the culture medium appears to suppress fimbriation directly, whereas in fact it promotes more rapid growth of nonfimbriated cells (which arise by phase variation) than is the case when glucose is absent, and is thus a selective population effect rather than true genetic control. The glucose effect is not abolished by altering the concentration of cyclic adenosine-monophosphate (cAMP), nor in mutant strains deficient in adenyl cyclase, as would be expected of normal catabolite repression in *E. coli* (14).

In general, genes other than those encoding the fimbrial subunits are expressed relatively poorly. The products of such accessory genes are often not visible in protein preparations, even using minicell expression systems. This may be because they have less efficient promoters or, as in the case of K99 fimbriae on ETEC strains isolated from farm animals, because of partial termination of transcription (15). A region of dyad symmetry between the upstream subunit gene (*fanC*) and downstream accessory gene (*fanD*) in K99 plasmids possibly acts as a partially active terminator sequence by allowing stem-loop formation in the mRNA (16); alternatively, it may be involved in temperature-dependent expression (see below).

Gene fusions were used to demonstrate the activity of regulators in the F13 fimbrial system (also known as Pap—pili associated with pyelonephritis) cloned from human uropathogenic *E. coli* (UPEC). Briefly, the promoter region upstream of *papA* (the major subunit gene) was fused with a large part of the *lacZ* gene carried by a specialized transducing phage (17). β-galactosidase activity in the fusion strain, corresponding to *lacZ* mRNA abundance and hence to the level of transcription from the subunit promoter, depends on the activity of two positive regulator genes, *papB* and *papI*.

B. Environmental Signals Control Adhesin Expression

Expression of most, if not all, MR adhesins is maximal at 37°C (mammalian body temperature) but minimal or completely absent during growth below 20°C. Low temperature could be a natural signal that bacteria have been excreted from their host animals; non-expression of adhesins while outside the body would therefore conserve biosynthetic resources. The actual regulatory mechanisms remain largely unstudied apart from two systems, Pap and 469-3 MR fimbriae. Thermoregulation of cloned Pap fimbrial genes was demonstrated to involve a transcriptional component by gene fusions with *lacZ*; activity of the promoter for *papB*, which itself is an activator for *papA* (subunit gene) expression, depends on the growth temperature (18). Temperature-dependent transcription of adhesin genes has also been observed more directly by hybridization of a probe from the cloned 469-3 MR adhesin determinant to total bacterial RNA samples extracted after growth at different temperatures (19). In both strain 469-3 and a laboratory strain harbouring the cloned adhesin genes, the level of transcription was much greater after growth at 37°C than at 18°C.

These results discounted suggestions that, in these systems at least, gross alterations to the microbial physiology (including reduced membrane fluidity at lower temperatures) influence export or assembly of fimbriae as they affect certain other biosynthetic mechanisms such as secretion of K1 capsule (20). However, it is not known precisely how growth temperature causes such transcriptional regulation. An attractive possibility would be the involvement of temperature-sensitive (heat-labile) diffusible repressors of adhesin expression. In the type I fimbrial system, *fimE/hyp* could potentially encode such a repressor; mutants overproduce fimbriae, so the gene evidently encodes a repressor, but if, in addition, the repressor were heat-labile, one might predict a rare class of heat-stable mutation which would repress type I fimbriae at all growth temperatures. Such mutants have not been reported. Similarly, cold-sensitive activators of adhesin expression might be proposed. Here one would predict a major class of mutant repressed for type I fimbriation, and perhaps a minor class of heat-stable activator mutant capable of fimbrial expression at any temperature. Again, *fimB* may be a candidate for such a gene although its putative temperature regulatory function has not been reported. In fact, *fimE/hyp* and *fimB* may be involved in phase variation of type I fimbriae (see below).

Besides the growth temperature, other aspects of the bacterial physiology and environment can also influence adhesin production. For example, production of K99 fimbriae by animal ETEC isolates or laboratory strains harbouring K99 plasmids depends on a range of growth conditions, including alanine concentration and pH, as well as the bacterial growth rate (21, 22), but the mechanisms remain unknown. Another example concerns the three "Coli Surface" (CS) antigens—components of the CFA/II adhesin group on human ETEC strains—that are differentially expressed according to the biotype of bacteria harbouring CFA/II plasmids (23). Initial experiments showed that when CFA/II plasmids were mobilized into *E. coli* strains of certain serotypes (e.g. 06:K15:H16 or H^-), CS1 was expressed only in rhamnose non-fermenting strains and CS2 only in rhamnose-fermenting biotypes; CS3 biosynthesis was independent of the rhamnose fermentative phenotype. Subsequently, CFA/II production was genetically separated from the rhamnose phenotype (24), but it does seem to depend on some general feature of the recipient cell physiology and is under investigation.

C. "Random" Control: Phase Variation

Many pathogenic bacteria are able to repress adhesin synthesis completely for a quasi-random period, the mean length of which depends on the system studied. This process is called phase variation. It follows that non-expressing cells can temporarily evade the host animal immune defences directed against those particular adhesin antigens. Although they are also liable to detach from the substrate, they can often later resume expression and perhaps adhere to new sites, such as previously inaccessible areas exposed by the sloughing of epithelial cells, or distal regions of the same or other tissues. Phase variation causes a supposedly genetically homogeneous culture (a clone) to exhibit phenotypic variability, the extent of which depends on the rate of phase change (typically 1 in 10 000 cells per generation but is highly dependent on the particular strain studied). Slowly-changing strains tend to grow as relatively pure cultures in one or other phase, whereas fast-changing strains tend to produce mixed cultures containing similar numbers of both types of phase variants. Even though individual cells are limited to either of the two phases, the presence of only a few adherent cells in a population is sufficient to give haemagglutination; largely nonfimbriate cultures containing small proportions of fimbriate cells may thus be mistaken for fully fimbriated cultures.

Phase variation of type I fimbriae is achieved by a "flip-flop" change of orientation of a 314 bp invertible DNA segment flanked by 9 bp inverted repeat sequences, identified from the DNA sequence (25). The segment contains a promoter for the operon containing *fimA* (or *pilA*), the subunit gene; in one orientation the promoter initiates transcription of the downstream adhesin genes, while in the alternative orientation it reads the "wrong" way and expression is prevented (Fig. 3). The inversion mechanism was initially compared to that controlling flagellar phase in *Salmonella typhimurium*. There, a site-specific DNA recombinase, specified by the *hin* gene located within the invertible region, controls the inversion event (26). Complementation studies and sequence analysis, however, did not find significant similarities between the type I invertible region and *hin*, nor other such genes.

Interestingly, amino acid sequences predicted from the related DNA sequences of *fimB* and *fimE* genes were found to be similar to those of certain bacteriophage integrase proteins on comparison with over 2600 protein sequences in a database (27). Integrase proteins are phage-encoded, diffusible factors which act in conjunction with additional host cell-specified *trans*-active factors on DNA regions at specific integration sites during lysogeny; in phage lambda, *int* encodes the integrase and "Integration Host Factor" (IHF) is the corresponding host cell-encoded component whose two subunits are specified by *himA* and *hip*/*himD* loci at 38 and 20 minutes on the *E. coli* genetic map. Transcription of subunit gene *fimA* (measured by β-galactosidase expression from *fimA-lacZ* fusion plasmids) was found to be locked in either the on or off phase in IHF-deficient mutant recipients, but complementation of the IHF deficiency with corresponding wild-type *him* alleles relieved the blockage and restored the rate of phase change to the previous level. Thus, *fimB* and *fimE* may not encode simple activators and repressors of type I fimbrial expression as was originally thought (see above), but are implicated in altering the orientation of the invertible region. A further complication to the control of type I fimbrial phase is the fact that the inversion rate is slowed by the product of a gene, *pilG*, located at 27 minutes on the *E. coli* K-12 chromosome (28). The phenotype of *pilG* mutant strains is confusing because, by increasing the inversion rate, cultures contain more equal mixtures of on and off phase variants and appear to have an intermediate level of fimbrial expression. Clearly, control of type I fimbrial expression is a multifactorial process.

Double immunolabelling experiments, using two different fimbrial-specific antisera conjugated to red or green fluorescent labels, seemed to imply that single bacterial cells of multiply-fimbriate human UPEC strains express only one fimbrial antigen at a time (29). A small but significant proportion of cells were double-labelled. These were thought to be cells caught in the process of changing between antigenic types. However, closer examination of the data suggests that the proportion was close to what might have been expected by chance coincidence of two independently controlled antigens. In *E. coli*, therefore, there is little clear evidence of the mutually exclusive expression of different subunit genes picked from a store of silent genes that is found in antigenic variation in *Neisseria* strains (see Chapter 13). Phase variation might perhaps be viewed as a primitive version of true antigenic variation since, although different cell surface antigens are expressed by phase variants, the range of alternatives is limited. Different substrate specificities may be as important to the bacteria as antigenic diversity.

D. Export of Fimbrial Components Involves Many Proteins

Few proteins are truly exported from *E. coli* and the transport mechanisms are poorly understood. There have only been a few cases where adhesin export systems have been studied in detail. While it is clear that adhesin subunits cross the cytoplasmic membrane by "conventional" signal-sequence mediated transport (30), transfer through the outer membrane appears to involve specific interactions with transport proteins in the periplasmic space and integral outer membrane proteins (15, 16, 31).

Expression of K88 fimbriae by comprehensive sets of point and deletion mutants and subclones has been analysed in minicells and by cellular fractionation methods in order to identify the export functions of individual gene products (15, 32). Loss of an 81 kD outer membrane protein blocks export and assembly of fimbriae. In mutants defective for synthesis of the 81 kD

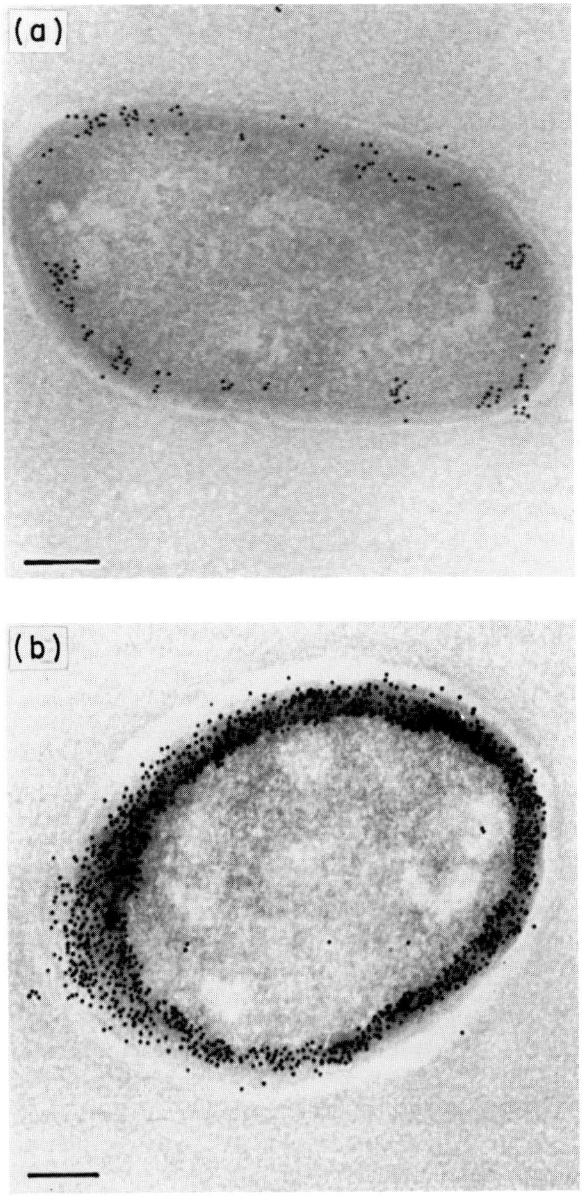

Figure 4. Electron micrograph of thin cryosections of 01 : K^- : K99 *E. coli* cells grown at 18°C (a) and transferred from 18°C to 37°C for 30 min (b), labelled with anti-K99 antiserum and secondary antiserum conjugated to 10 nm gold particles (33). Scale bar = 200 nm. (Micrographs courtesy of Prof. F. K. de Graaf.)

protein, 26 kD protein subunits accumulate in the periplasm and no fimbriae form; this is consistent with the idea that the large membrane protein acts as a transport pore. The 27 kD protein associates with fimbrial subunits in the periplasm, and may prevent them from self-assembling until they are outside the cell; ionic interactions are implied since subunits are acidic proteins while the 27 kD protein is basic. The 17 kD protein is interesting since it appears to modify fimbriae while in the periplasm; mutants lacking the protein do express surface-located fimbriae but they are incapable of mediating adherence to normal substrates. In addition, fimbrial subunits are exported more slowly by these mutants than by wild-type strains; subunits accumulate in the periplasmic spaces, so the 17 kD protein seems to be involved in transport as well as catalysis to the adherent phenotype.

In an elegant series of experiments, Roosendaal *et al.* (33) investigated the synthesis of K99 fimbriae by electron microscopy of bacteria prepared from cultures shifted at various times from 18°C (subunit expression largely repressed) to 37°C (full expression). Sections were reacted with anti-K99 antibodies and subsequently with secondary antibodies conjugated to electron-refractive colloidal gold spheres (Fig. 4). The approach is similar to pulse-chase radiolabelling since the gold particles (and therefore the majority of the K99 subunits) could be readily identified at specific cellular locations in the process of being exported. Following synthesis at the cytoplasmic membrane, the subunits were sequentially transferred through the periplasm to the outer membrane and cell exterior. In conjunction with deletion and transposon mutagenesis, it should now be possible to identify with greater certainty the export functions of the complex K99 system.

E. How Are Fimbriae Assembled?

Fimbrial subunits are very hydrophobic proteins, so isolated fimbriae tend to clump together in aqueous solutions. Furthermore, following dissociation of certain isolated fimbriae and subsequent removal of the denaturant, some fimbria-like structures arise spontaneously. So, does the normal assembly of fimbriae on the bacterial surface occur purely by spontaneous noncovalent subunit-to-subunit interactions or does it involve other assembly proteins? Evidence from those few systems that have been analysed in any detail suggests that subunit-to-subunit interactions are sufficient to cause assembly of native fimbriae. Unlike flagella, axial holes seen in some fimbriae under the electron microscope are not sufficiently wide to allow fimbrial subunits to pass along inside growing fimbriae and attach distally. Immunogold-electron microscopic analysis of the assembly of type I fimbriae has shown that newly exported fimbrial subunits are added basally rather than, for example, "diffusing" along growing fimbriae to their tips (34). Complete assembly of

K99 and type I fimbriae on the bacterial surface takes up to an hour at 37°C (33, 34).

Interestingly, the mean lengths of Pap and type I fimbriae are influenced respectively by PapH and PilF assembly terminator proteins which are normally present at about one hundredth of the concentration of the subunits (35, 36). The relative expression of these proteins affects the fimbrial lengths; greater expression of the subunits (for example, as a result of mutational inactivation of *papH, pilF* or *fimE/hyp*, or by cloning the subunit genes on multicopy vectors) leads to longer fimbriae, whereas greater expression of PapH or PilF shortens them, perhaps by competitive inhibition of subunit polymerization. The PapH protein is additionally involved in anchoring fimbriae to the cell since half or more of the total fimbrial antigen is released from *papH* mutants into the growth medium, as opposed to about a tenth of the total produced by wild-type cells.

F. Fimbrial Subunits Can Carry Distinct Adhesive Proteins

The adhesive components of some fimbriae (including type I, S, $F7_2$ and Pap systems) are known to be minor proteins carried by the fimbrial structures (18, 37, 38). Deletion or mutation of the "true" adhesin gene alone can generate strains which express nonadhesive fimbriae; conversely, inactivation of just the major subunit genes prevents expression of fimbrial structures yet does not necessarily prevent mutant cells adhering in *in vitro* adherence assay systems (and perhaps even *in vivo*). The adherence of Pap fimbriae to digalactoside-containing receptors, for instance, is encoded in an adherence cluster of three genes (*papE, papF* and *papG*) separate from the other Pap genes (Fig. 2) and transcribed from a different promoter. Subclones carrying *papA* (major subunit), *papH* (subunit-like, function unclear), *papC* (outer membrane fimbrial attachment protein or transport pore) and *papD* (probably an export protein), but lacking *papI* and *papB* (regulators) and the adherence cluster (*papE* to *papG*), produce surface-located fimbrial threads but do not mediate haemagglutination or other digalactoside-specific binding. The same "fimbriate but nonadherent" phenotype is caused by transposon insertions in the promoter region for the adherence cluster. Mutants lacking the ability to express either *papA* or *papE* (but not both) while expressing *papG* are capable of digalactoside-specific adherence since *papG* encodes the receptor binding protein which functions so long as the major subunit (*papA* product) or an alternative structural subunit (*papE* product) are expressed (38).

True adhesin proteins have been visualized principally at the tips of native S-(sialyl-specific) and P-fimbriae by immunogold-electron microscopy using monoclonal anti-adhesin antibodies (39, 40). It has been suggested that other minor proteins may be involved in determining the final tip-location of the *papG* adhesin on Pap fimbriae by linking the adhesin and subunit proteins together in a defined structure.

IV. EVOLUTIONARY PERSPECTIVES

A. Homology or Similarity?

We would like to stress the distinction between strictly homologous systems evolved from common origins, and similar systems (even those which cross-hybridize) whose origins are unknown. There is a wealth of information concerning supposed relatedness amongst *E. coli* adhesins and, on comparison, many antigenically distinct adhesins appear genetically similar to various degrees. For example, the restriction maps of most P-fimbrial determinants share common sites and genes from different P systems complement (41), while the subunit DNA sequences of K88 variants K88ab, K88ac and K88ad (42) and those of type I and type I-like fimbriae (43) are nearly identical within each group. *E. coli* strains independently isolated from a child with severe enteritis and a patient with a urinary tract infection express different MR adhesins, yet they are encoded by cross-hybridizing genetic systems with similar restriction maps (1, 44, 45; Fig. 5). Amino- and carboxy-terminal amino acid sequences of a number of adhesin subunits show some similarities while the central portions often appear more disparate (46). Nearly identical systems are commonly assumed to be truly homologous, but how have they diverged, and from where did the less similar adhesins originate?

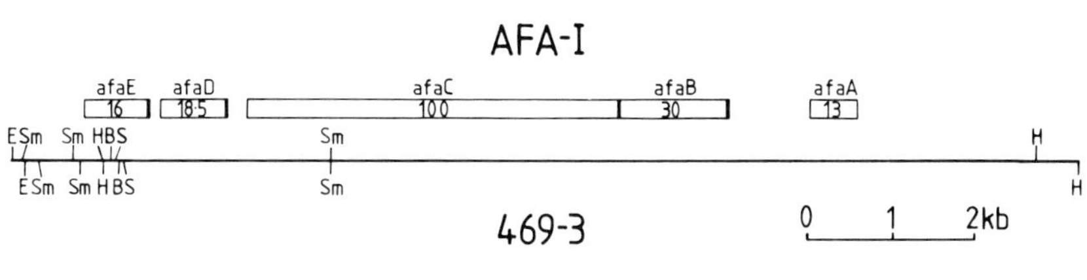

Figure 5. Restriction maps of the determinants encoding biosynthesis of AFA-I (upper) and 469-3 (lower) MR adhesins, aligned for comparison (1, 44, 45). Boxes above the line indicate map positions of AFA-I genes as implied from lengths of DNA required to code for the molecular weights of protein products (marked inside each box) identified by minicell analysis; heavier end markings show signal sequence coding regions. Restriction enzyme sites are annotated as in Figure 2.

B. Possible Evolutionary Mechanisms

We can only speculate on the evolutionary mechanisms tending to conserve certain features while allowing others to diverge. However, opposing selective pressures arising from immunogenicity and functional requirements probably determined the direction and rate of adhesin evolution. Fimbriae are surface located antigens whose exposure, along with capsular and other surface antigens, renders a strain liable to attack by host immunological defences. The presence of multiple fimbrial determinants in pathogenic strains (perhaps generated by conjugal transfer of fimbrial determinants from other strains or by duplication of existing systems) leads to an inherent redundancy. In the presence of at least one functional fimbrial genetic system, others could potentially undergo mutations that change their structure, specificity or expression without eliminating the ability of a strain to colonize its host. It is conceivable that, given sufficient time, a fimbrial system may diverge sufficiently to generate entirely novel antigenic determinants by accumulating such mutations. Strains expressing variant adhesins might be selectively advantaged relative to the wild-type cells since novel substrate specificities might permit colonization of different tissues. Also, it takes some time for the host to develop an immune response to a novel antigen, whereas previously encountered antigens are more rapidly dealt with by the products of immune "memory cells". This situation would not apply directly in very young animals since their immune defences are immature, but there is evidence that immune protection against fimbrial antigens is acquired passively from the mothers' colostrum and milk, and of course, the mothers could present antibodies only to previously encountered fimbriae.

As discussed earlier, synthesis and export of adhesins require considerable cellular resources and can dramatically affect the growth rate of a strain under certain conditions; the complex genetic mechanisms that control adhesin synthesis probably evolved to allow full expression but prevent excessive production. Clearly, adhesins must produce positive selective pressures to counteract the biosynthetic costs incurred, and the ability to mediate adherence to specific host tissues is the most obvious function. Several adhesins are now known to comprise separate fimbrial (structural) and adherence proteins. Furthermore, complementation studies have proved that at least some genes in related fimbrial determinants can complement each others' deficiencies *in vitro* (41). It seems quite possible that the more antigenic parts of adhesin systems, namely the well-expressed fimbrial subunits, may thus diverge quite rapidly without the need to maintain highly specific adherence to host epithelia so long as they simply retain their fundamental structural properties and can still interact with the actual adhesins. Similarly, the adhesins may perhaps evolve novel specificities provided they retain the ability to interact with the structural subunits.

This evolutionary hypothesis may be tested by extending detailed comparisons of different adhesin systems to include accessory as well as subunit genes. The model predicts that functionally important sequences of many adhesins should retain more similarities than do strong antigenic determinants, especially those in the major subunit genes, and therefore has implications in the search for suitable components for vaccines against various infectious bacterial diseases.

References

(1) Hinson, G., Knutton, S., Lam-Po-Tang, M. K.-L., McNeish, A. S. and Williams, P. H. (1987). Adherence to human colonocytes of an *Escherichia coli* strain isolated from severe infantile enteritis: molecular and ultrastructural studies of a fibrillar adhesin. *Infect. Immun.* **55**, 393–402.

(2) Korhonen, T. K., Parkkinen, J., Hacker, J., Finne, J., Pere, A., Rhen, M. and Holthofer, H. (1986). Binding of *Escherichia coli* S-fimbriae to human kidney epithelium. *Infect. Immun.* **54**, 322–327.

(3) Svenson, S. B. and Källenius, G. (1983). Density and localization of P-fimbriae-specific receptors on mammalian cells: fluorescence-activated cell analysis. *Infection* **11**, 6–12.

(4) Scaletsky, I. C. A., Silva, M. L. M. and Trabulsi, L. R. (1984). Distinctive patterns of adherence of enteropathogenic *Escherichia coli* to HeLa cells. *Infect. Immun.* **45**, 534–536.

(5) Duguid, J. P. and Old, D. C. (1980). Adhesive properties of Enterobacteriaceae. *In* "Bacterial Adherence, Receptors and Recognition", Series B, Vol. 6 (E. H. Beachey, ed.), pp. 185–217. Chapman and Hall, London.

(6) Hultgren, S. J., Porter, T. N., Schaeffer, A. J. and Duncan, J. L. (1985). Role of type I pili and effects of phase variation on lower urinary tract infections produced by *Escherichia coli*. *Infect. Immun.* **50**, 370–377.

(7) Boylan, M., Coleman, D. C. and Smyth, C. J. (1987). Molecular cloning and characterization of the genetic determinant encoding CS3 fimbriae of enterotoxigenic *Escherichia coli*. *Microbial Pathogen.* **2**, 195–209.

(8) de Graaf, F. K. and Klaasen, P. (1986). Organization and expression of genes involved in the biosynthesis of 987P fimbriae. *Mol. Gen. Genet.* **204**, 75–81.

(9) Willshaw, G. A., Smith, H. R., McConnell, M. M. and Rowe, B. (1985). Expression of cloned plasmid regions encoding colonization factor antigen I (CFA/I) in *Escherichia coli*. *Plasmid* **13**, 8–16.

(10) Klemm, P. (1986). The *fim* genes of *Escherichia coli* K12: aspects of structure, organization and expression. *In* "Protein-Carbohydrate Interactions in Biological Systems", FEMS symposium no. 31 (D. L. Lark, ed.), pp. 47–50. Academic Press Ltd., London.

(11) Orndorff, P. E. and Falkow, S. (1984). Identification and characterization of a gene product that regulates type 1 piliation in *Escherichia coli*. *J. Bacteriol.* **160**, 61–66.

(12) Klemm, P. (1986). Two regulatory *fim* genes, *fimB* and *fimE*, control the phase variation of type I fimbriae in *Escherichia coli*. *EMBO J.* **5**, 1389–1393.

(13) Eisenstein, B. I. and Dodd, D. C. (1982). Pseudocatabolite repression of type I fimbriae of *Escherichia coli*. *J. Bacteriol.* **151**, 1560–1567.

(14) Epstein, W., Rothman-Denes, L. B. and Hesse, J. (1975). Adenosine 3′:5′-cyclic monophosphate as mediator of catabolite repression in *Escherichia coli*. *Proc. Natl. Acad. Sci. USA* **72**, 2300–2304.

(15) Mooi, F. R. and de Graaf, F. K. (1985). Molecular biology of fimbriae of enterotoxigenic *Escherichia coli*. *Curr. Top. Microbiol. Immunol.* **118**, 119–138.

(16) Mooi, F. R., Roosendaal, B., Oudega, B. and de Graaf, F. K. (1986). Genetics and biogenesis of the K88ab and K99 fimbrial adhesins. *In* "Protein-Carbohydrate Interactions in Biological Systems", FEMS symposium no. 31 (D. L. Lark, ed.), pp. 19–26. Academic Press Ltd., London.

(17) Båga, M., Göransson, M., Normark, S. and Uhlin, B. E. (1985). Transcriptional activation of a Pap pilus virulence operon from uropathogenic *Escherichia coli*. *EMBO J.* **4**, 3887–3893.

(18) Uhlin, B. E., Båga, M., Göransson, M., Lindberg, F. P., Lund, B., Norgren, M. and Normark, S. (1985). Genes determining adhesin formation in uropathogenic *Escherichia coli*. *Curr. Top. Microbiol. Immunol.* **118**, 163–178.

(19) Williams, P. H. and Hinson, G. (1987). Temperature-dependent transcriptional regulation of expression of fimbriae in an *Escherichia coli* strain isolated from a child with severe enteritis. *Infect. Immun.* **55**, 1734–1736.

(20) Nichol, C. P., Davis, J. H., Weeks, G. and Bloom, M. (1980). Quantitative study of the fluidity of *Escherichia coli* membranes using deuterium magnetic resonance. *Biochem.* **19**, 451–457.

(21) Isaacson, R. E. (1980). Factors affecting expression of the *Escherichia coli* pilus K99. *Infect. Immun.* **28**, 190–194.

(22) de Graaf, F. K., Krenn, B. E. and Klaasen, P. (1984). Organization and expression of genes involved in the biosynthesis of K99 fimbriae. *Infect. Immun.* **43**, 508–514.

(23) Smyth, C. J. (1982). Two mannose-resistant haemagglutinins on enterotoxigenic *Escherichia coli* of serotype 06:K15:H16 or H^- isolated from travellers' and infantile diarrhoea. *J. Gen. Microbiol.* **128**, 2081–2096.

(24) Boylan, M. and Smyth, C. J. (1985). Mobilization of CS fimbriae-associated plasmids of enterotoxigenic *Escherichia coli* of serotype 06:K15:H16 or H^- into various wild-type hosts. *FEMS Microbiol. Lett.* **29**, 83–89.

(25) Abraham, J. M., Freitag, C. S., Clements, J. R. and Eisenstein, B. I. (1985). An invertible element of DNA controls phase variation of type I fimbriae of *Escherichia coli*. *Proc. Natl. Acad. Sci. USA* **82**, 5724–5727.

(26) Zieg, J. and Simon, M. (1980). Analysis of the nucleotide sequence of an invertible controlling element. *Proc. Natl. Acad. Sci. USA* **77**, 4196–4200.

(27) Eisenstein, B. I., Sweet, D. S., Vaughn, V. and Friedman, D. I. (1987). Integration host factor is required for the DNA inversion that controls phase variation in *Escherichia coli*. *Proc. Natl. Acad. Sci. USA* **84**, 6506–6510.

(28) Spears, P. A., Schauer, D. and Orndorff, P. E. (1986). Metastable regulation of type I piliation in *Escherichia coli* and isolation and characterization of a phenotypically stable mutant. *J. Bacteriol.* **168**, 179–185.

(29) Nowicki, B., Rhen, M., Väisänen-Rhen, V., Pere, A. and Korhonen, T. K. (1984). Immunofluorescence study of fimbrial phase variation in *Escherichia coli* KS71. *J. Bacteriol.* **160**, 691–695.

(30) Randall, L. L. and Hardy, S. J. S. (1984). Export of protein in bacteria. *Microbiol. Rev.* **48**, 290–298.

(31) Dougan, G., Dowd, G. and Kehoe, M. (1983). Organization of K88ac-encoded polypeptides in the *Escherichia coli* cell envelope: use of minicells and outer membrane protein mutants for studying assembly of pili. *J. Bacteriol.* **153**, 364–370.

(32) Mooi, F. R., Wijfjes, A. and de Graaf, F. K. (1983). Identification and characterization of precursors in the biosynthesis of the K88ab fimbria of *Escherichia coli*. *J. Bacteriol.* **154**, 41–49.

(33) Roosendaal, B., van Bergen en Henegouwen, P. M. P. and de Graaf, F. K. (1986). Subcellular localization of K99 fimbrial subunits and effect of temperature on subunit synthesis and assembly. *J. Bacteriol.* **165**, 1029–1032.

(34) Lowe, M. A., Holt, S. C. and Eisenstein, B. I. (1987). Immunoelectron microscopic analysis of elongation of type I fimbriae in *Escherichia coli*. *J. Bacteriol.* **169**, 157–163.

(35) Båga, M., Norgren, M. and Normark, S. (1987). Biogenesis of *Escherichia coli* Pap pili: PapH, a minor pilin subunit involved in cell anchoring and length modulation. *Cell* **49**, 241–251.
(36) Maurer, L. and Orndorff, P. E. (1987). Identification and characterization of genes determining receptor binding and pilus length of *Escherichia coli* type I pili. *J. Bacteriol.* **169**, 640–645.
(37) Normark, S., Lindberg, F., Lund, B., Båga, M., Ekbäck, G., Göransson, M., Mörner, S., Norgren, M., Marklund, B.-I. and Uhlin, B.-E. (1986). Minor pilus components acting as adhesins. *In* "Protein-Carbohydrate Interactions in Biological Systems", FEMS symposium no. 31 (D. L. Lark, ed.), pp. 3–12. Academic Press Ltd., London.
(38) Lund, B., Lindberg, F., Marklund, B.-I. and Normark, S. (1987). The PapG protein is the α-D-galactopyranosyl-(1 $\rightarrow$ 4)-β-D-galactopyranose-binding adhesin of uropathogenic *Escherichia coli*. *Proc. Natl. Acad. Sci. USA* **84**, 5898–5902.
(39) Moch, T., Hoschützky, H., Hacker, J., Kröncke, K.-D. and Jann, K. (1987). Isolation and characterization of the α-sialyl-β-2,3-galactosyl-specific adhesin from fimbriated *Escherichia coli*. *Proc. Natl. Acad. Sci. USA* **84**, 3462–3466.
(40) Lindberg, F., Lund, B., Johansson, L. and Normark, S. (1987). Localization of the receptor-binding protein adhesin at the tip of the bacterial pilus. *Nature* **328**, 84–87.
(41) van Die, I., van Megen, I., Zuidweg, E., Hoekstra, W., de Ree, H., van den Bosch, H. and Bergmans, H. (1986). Functional relationships among the gene clusters encoding $F7_1$, $F7_2$, F9, and F11 fimbriae of human uropathogenic *Escherichia coli*. *J. Bacteriol.* **167**, 407–410.
(42) Dykes, C. W., Halliday, I. J., Read, M. J., Hobden, A. N. and Harford, S. (1985). Nucleotide sequences of four variants of the K88 gene of porcine enterotoxigenic *Escherichia coli*. *Infect. Immun.* **50**, 279–283.
(43) van Die, I., van Geffen, B., Hoekstra, W. and Bergmans, H. (1984). Type IC fimbriae of a uropathogenic *Escherichia coli* strain: cloning and characterization of the genes involved in the expression of the IC antigen and nucleotide sequence of the subunit gene. *Gene* **34**, 187–196.
(44) Labigne-Roussel, A. F., Lark, D., Schoolnik, G. and Falkow, S. (1984). Cloning and expression of an afimbrial adhesin (AFA-I) responsible for P blood group-independent, mannose resistant hemagglutination from a pyelonephritic *Escherichia coli* strain. *Infect. Immun.* **46**, 251–259.
(45) Labigne-Roussel, A. F., Schmidt, M. A., Walz, W. and Falkow, S. (1985). Genetic organization of the afimbrial adhesin operon and nucleotide sequence from a uropathogenic *Escherichia coli* gene encoding an afimbrial adhesin. *J. Bacteriol.* **162**, 1285–1292.
(46) Moseley, S. L., Dougan, G., Schneider, R. A. and Moon, H. W. (1986). Cloning of chromosomal DNA encoding the F41 adhesin of enterotoxigenic *Escherichia coli* and genetic homology between adhesins F41 and K88. *J. Bacteriol.* **167**, 799–804.

Chapter **15**

Genetic Studies of Enterotoxin and Other Potential Virulence Factors of Vibrio cholerae

RONALD K. TAYLOR

I. INTRODUCTION

Vibrio cholerae, the causitive agent of cholera, has proved to be an ideal organism in which to analyse virulence, because of its amenability to a variety of genetic techniques and the immediate potential application of these findings to improved vaccine development. Cholera remains a serious epidemic disease in various parts of the world, and a suitable anti-cholera vaccine has not yet been developed. *V. cholerae* is a Gram-negative curved rod that is highly motile by means of a single polar flagellum. The vibrios that cause epidemic cholera have been divided into two biotypes; Classical and El Tor, based on haemolysin production and several biochemical tests. Cholera pathogenesis involves oral ingestion through contaminated water or food, survival during passage through the acid barrier of the stomach, colonization of the upper small intestine with concomitant toxin production, and ultimate dissemination in a watery diarrhoea. Throughout this process, the bacterium elaborates a number of proven and potential virulence determinants that help it to reach, adhere to, and colonize the intestinal epithelial layer. Some of the virulence factors, such as cholera toxin, have been extremely well characterized while others, for example those involved in colonization, have just recently been defined.

This chapter describes the methods and results of genetic manipulations used in *V. cholerae* to increase understanding of its pathogenic mechanisms, including the role of enterotoxin and other virulence factors in pathogenesis, as well as the application of genetics to the development of improved cholera vaccines.

II. CHOLERA TOXIN GENES

A. Toxin Structure

Cholera holotoxin is composed of two types of subunits, A and B, in a ratio of 1:5 (1, 2). The A subunit becomes enzymatically active upon proteolytic cleavage (3). The resulting amino-terminal peptide, A_1, catalyses the transfer of the ADP-ribose moiety from NAD to the G_s regulatory component of mammalian adenylate cyclase, thereby increasing cyclase activity (4, 5). This causes an increase of intracellular cAMP concentration, resulting in ion and water efflux and so leading to the watery diarrhoea characteristic of cholera (6). The five B subunits are responsible for binding of the toxin to its eukaryotic cell surface receptor, ganglioside GM_1, and subsequently allowing entry of the toxin into the cell (7). In addition, the B subunits are required for the extracellular secretion of the toxin from the bacterium (8).

B. Mapping the Structural Genes

Classical genetic mapping can be carried out in *V. cholerae* in much the same way as in *Escherichia coli*, utilizing chromosomal transfer by conjugation, and generalized transduction (9, 10). Some strains of *V. cholerae* carry the plasmid sex factor, P, which is analogous to the F fertility factor of *E. coli* (11). The P-factor is present at one copy per cell (12) and can mediate chromosomal transfer at a 10^{-5} to 10^{-6} frequency (13). Although it has been utilized to map the *V. cholerae* chromosome (14), the P-factor cannot form Hfr strains and the frequency of chromosomal transfer is too low to allow routine, efficient mapping. This problem was overcome by providing homology between the P-factor and various positions in the chromosome. This was first accomplished using transposon Tn*1*, with one copy in the P-factor and others distributed around the chromosome, thus forming a transposon-facilitated recombination system (Tfr) with the direction of transfer depending on the orientation of Tn*1* on the P-factor (13). The technique has also been extended using chromosomal insertions made with the Mu-like mutator phage VcA1 and using the P-factor derivative pSJ15 which harbours a defective VcA1 prophage (15). The Tfr technique has been used to map the cholera toxin structural gene operon (*ctxAB*) in both biotypes. While most El Tor strains carry a single *ctxAB* operon between the *his* and *nal* loci, the El Tor strain RV79 contains two copies of the *ctx* genetic element tandemly duplicated in this location (16, 17). Classical strains, exemplified by 569B, contain two copies of the operon, one mapping in the same position as the El Tor copies and the second copy distant on the chromosome (16). The structural genes do not map near other genes that affect toxin production. Most recently, a new set of P-factors has been constructed to facilitate the mapping of Tn*5* or Tn*10* chromosomal insertions (18). Fine structure mapping is facilitated using the generalized transducing phage CP-T1 (19).

C. Cloning and Sequence Analysis of the Toxin Genes

The high degree of structural homology between the A and B subunits of heat-labile toxin (LT) of *E. coli* and the subunits of cholera toxin (20, 21) is also reflected by the hybridization under reduced stringency of LT gene probes to DNA sequences of *V. cholerae* (22), but not of *ctx* deletion mutants (23). These probes were used to screen and isolate *ctx*$^+$ genomic clones from several *V. cholerae* strains. Interestingly, toxin is produced at barely detectable levels from the *E. coli* strains carrying the *ctx*$^+$ plasmids, suggesting positive regulation in *V. cholerae* (Section V). It also remains cell-associated in *E. coli* instead of being extracellular as when produced by *V. cholerae* (24, 25). In fact, both LT and CT are periplasmic in *E. coli* and extracellular in *V. cholerae*, suggesting that *V. cholerae* has a machinery for transport of these toxins to the extracellular environment (8). Additional evidence of such a transport mechanism is provided by a *V. cholerae* mutant defective in transport of both toxins (26).

DNA restriction and sequence analysis of clones from the various strains reveals that the toxin genes form an operon with *ctxA* promoter-proximal (17, 24, 27–30). Since the toxin molecule has five B subunits for every A subunit, the regulation maintaining this ratio must be post-transcriptional. In fact, a gene fusion placing *ctxB* product expression under the translational signals of *ctxA* results in a nine-fold lower amount of B subunit, indicating that the ratio may be regulated by the efficiency of translational initiation (27).

Sequence comparisons further revealed the expected high degree of similarity between *ctx* genes of various strains, except for the number of tandem repeats of a heptamer sequence in the promoter region (Section V). In addition, the sequence revealed a 75% identity of *ctxAB* with the similarly organized *eltAB* genes encoding the heat-labile toxin of enterotoxigenic *E. coli*. It has been suggested that the toxin genes, which are plasmid-encoded in *E. coli* (31) instead of chromosomal as in *V. cholerae*, originated in *V. cholerae* and were acquired by *E. coli* (32). Pertinent to this theory is the location of the *ctx* locus on a genetic element adjacent to a repetitive DNA sequence, RS1, which has been shown to transpose (17, 33).

D. Structure of the *ctx* Genetic Element

Southern analysis using the cloned *ctxAB* region as a hybridization probe revealed that toxinogenic *V. cholerae* strains of both biotypes all carry a region of homologous DNA adjacent to the structural genes that is absent in non-toxinogenic isolates. This DNA completes a genetic element within which lies the *ctx* operon (34–36). The *ctx* element has a substructure similar to that found in composite transposons (Chapter 2). A 2.7 kb repetitive sequence, RS1, is found adjacent to and upstream of the 4.3 kb core region within which lies *ctxAB*. RS1 also forms the junction between tandem duplications of the *ctx* genetic element and sometimes occurs downstream as well, always in the same orientation (34, 35) (Fig. 1). These direct repeats can then flank the 4.3 kb core region which contains *ctx*. RS1 can itself transpose (33) and is also involved in a *recA*-dependent recombination process leading to the duplication or further amplification of tandem repeats of the *ctx* element (33) (Fig. 2). A similar mechanism could result in deletion of the core region of the element and give rise to non-toxinogenic strains that appear identical except for loss of this DNA. Amplification has been shown to occur on repeated passage of *V. cholerae* strains *in vivo* (34). The *in vivo* selective advantage may be due to increased toxin production or to other factors encoded within the core region (37).

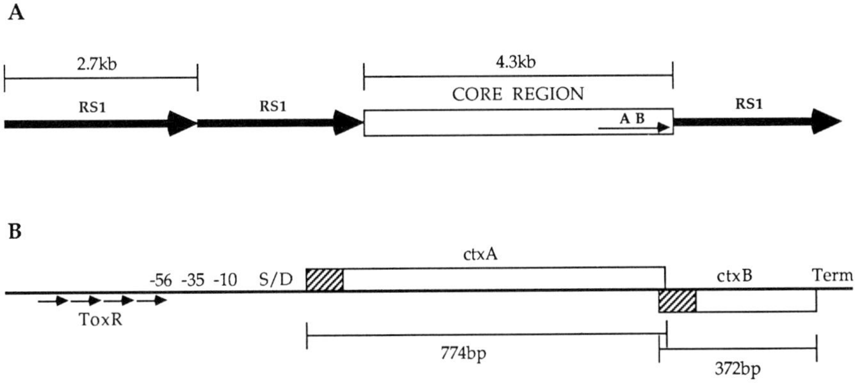

Figure 1. (a) One of the chromosomal arrangements seen for the *ctx* genetic element with its flanking RS1 sequences (34). A copy of RS1 is always found upstream of the core region and can occur in other configurations than the one shown here, depending on the strain and number of copies of the element. The core region contains the toxin structural genes, *ctxA* and *ctxB*, which form an operon. (b) A more detailed view of the *ctxAB* operon including the sites required for its expression and the series of tandem repeats where ToxR interacts to modulate it. S/D, Shine-Dalgarno sequence; Term, transcription terminator.

E. Engineered *ctx* Mutants

The detailed knowledge of the structure of the *ctxAB* locus has been applied to the construction of more effective cholera vaccine strains. Early *ctxA* candidates for live, oral vaccines were isolated after chemical mutagenesis (38, 39) or phage VcA1-induced deletions (40). These strains suffered from the possibility of reversion in the first case, the lack of being precisely defined (which could result in unanticipated properties), or poor colonization ability. New strains are based on defined *ctx* deletions constructed *in vitro*, with their eventual recombination into the homologous position(s) on the *V. cholerae* chromosome. The *in vitro* constructions have involved deletion of the desired region and insertion of an antibiotic resistance gene between the deletion endpoints (deletion/substitution). The resistance gene provides selection for the chromosomal recombinant. A common method to accomplish this involves the use of two plasmids of the same incompatibility group, first developed for *Rhizobium* (41) but since applied to *Vibrio* (27) and a variety of other Gram-negative bacteria, where it has been useful not only for constructing recombinants, but also for selecting transpositions. Mekalanos and coworkers used this technique, as shown in Fig. 3, to construct *ctx* deletion mutants of a Classical strain, 0395. Similar strategies involving recombination of *in vitro* constructed deletions, or tetracycline resistant deletion/substitutions with subsequent

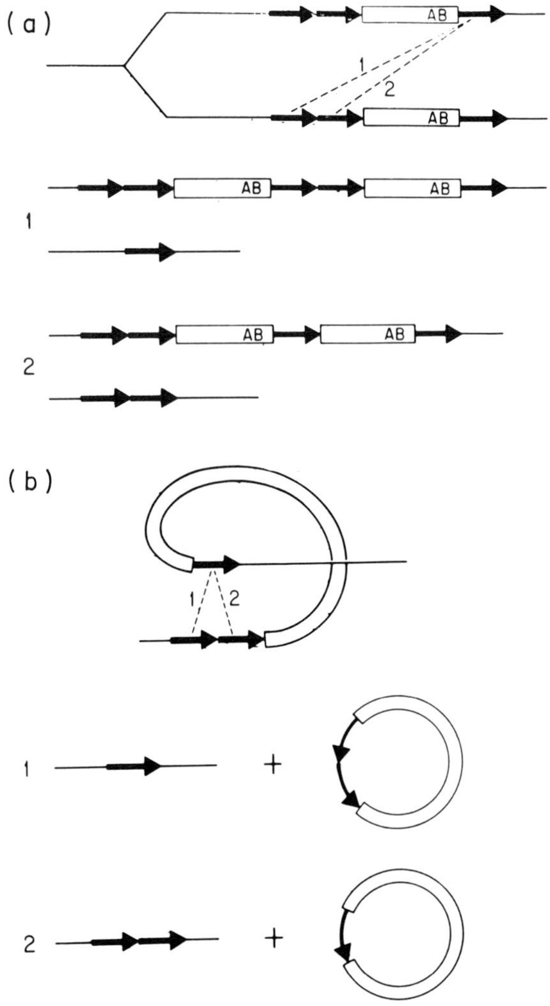

Figure 2. Models for the generation of *ctx* deletion and amplification events. (a) Schematic representation of unequal crossover events between RS1 sequences. The *ctx* region of a typical El Tor strain, which carries a total of three copies of RS1 (arrows) flanking a core region (open box) containing the *ctxAB* operon, is shown. Unequal homologous recombination could occur at two different positions to produce a tandem duplication of the *ctx* region. For position 1, the products are a 9.7 kb tandem duplication and, if reciprocal, a deletion that retains one RS1 copy at the deletion junction. For position 2, the products are a 7 kb tandem duplication and a deletion that retains two tandem RS1 copies at the deletion junction. (b) Schematic representation of possible *ctxAB* operon deletion events *via* intramolecular recombination. The example presented here shows a *ctx* core region flanked by one RS1 at one end and two RS1 copies at the opposite end. Recombination between the RS1 copies would result in one (event 1) or two (event 2) RS1 copies remaining on the chromosome and the formation of a circular segment of DNA consisting of a core region with either two or one copy of RS1, respectively. Reprinted from reference (33) with permission.

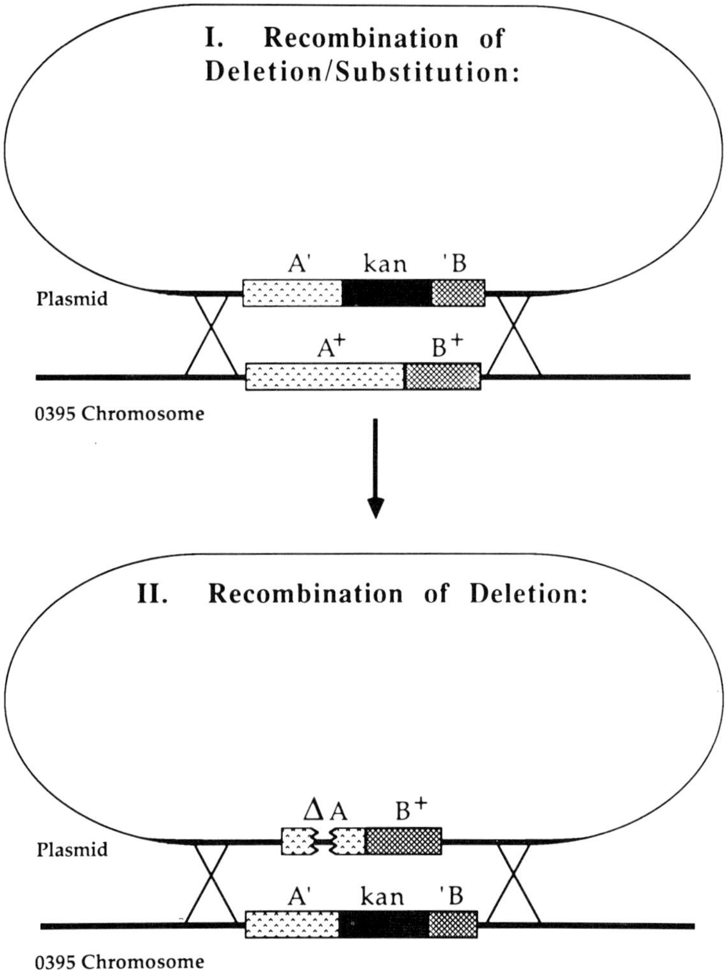

Figure 3. An example of the strategy of using marker exchange to place a precise deletion into the chromosome. I. A kanamycin-resistance gene was inserted into the *ctx* deletion and this construction was cloned into the incompatibility group P plasmid pRK290 which carries tetracycline resistance. After mobilization into *V. cholerae* a second *incP* plasmid, pHI1 encoding gentamicin resistance, was mated in with selection maintained for kanamycin-resistance as well as the incoming plasmid. This selected for loss of pRK290 and resultant recombination, or marker exchange, of the kanamycin-resistant construction into the chromosome. Repeated growth in the presence of kanamycin resulted in homogenotization into both *ctx* copies, as recognized by Southern analysis. II. After curing of the gentamicin plasmid, a pRK290 carrying the ultimately desired unsubstituted *ctxA* deletion mutation was introduced. Upon growth in the absence of kanamycin, mutants that became sensitive were found at a frequency of 0.001. In these, both *ctxA* deletion/substitution copies had been replaced by the unsubstituted deletion (27).

subsequent selection for tetracycline-sensitive colonies, have resulted in the construction of vaccine candidates of both El Tor and Classical strains by Kaper and coworkers (42, 43).

Analysis of these strains in human volunteers showed that, while the defined toxin gene deletion strains cause much less diarrhoea than the wild-type isogenic parent strains, there is still a mild diarrhoea in a significant percentage of the patients (44). Thus the strains will have to be further attenuated for other virulence factors before they are useful as vaccines. Studies in rabbits suggest that the live strains are far superior to dead whole cell preparations in eliciting immunity; ability of the vaccine strains to colonize was the most consistent indication of their ability to stimulate production of antibodies that conferred immunity to subsequent *V. cholerae* challenge (45).

III. ADHERENCE AND COLONIZATION

A. Production of Fimbriae

Fimbriae (pili) are bacterial surface appendages that mediate adherence of several bacterial species to eukaryotic cell surfaces and promote colonization (46). Recently, detailed molecular and genetic studies have identified a role for *V. cholerae* fimbriae in colonization. It now appears that *V. cholerae* serotype 01 can elaborate at least two fimbrial types under different growth conditions. One of these is composed of a repeating 16 kD major subunit protein and is produced when *V. cholerae* is grown in TCG medium (47). There are no genetic studies on this type of fimbria and its role in colonization is yet to be proven. A second type of fimbria is composed of a 20.5 kD subunit and is expressed under different growth conditions from the first. It has been termed TCP (for toxin coregulated pilus) because its expression parallels that of cholera toxin production in a range of growth conditions (48). Interestingly, the N-terminal 25 amino acid residue region of the mature subunit pilin protein, termed TcpA, is highly homologous to a group of pilin proteins, called *N*-methyl-phenylanine (NMePhe) pilins because of a modification of their amino terminus, which are already implicated in the adhesion of several bacterial species to their hosts (48).

B. Identification of *tcpA* using Tn*phoA*

In order to establish the role of TCP in colonization, and to understand further the mechanisms of its regulation and function, gene fusions to the *tcpA* gene were isolated in *V. cholerae* 0395 which produces a large amount of the fimbrial subunit. To facilitate the isolation of such gene fusions, as well as fusions to other genes that encode surface or secreted proteins that might be implicated in virulence mechanisms, the Tn*phoA* gene fusion transposon vector was utilized. Tn*phoA* is a derivative of Tn*5* that can be used to create fusions between target genes and *phoA*, the gene for *E. coli* alkaline phosphatase (49). Such gene fusions encode hybrid proteins composed of an enzymatically potentially active carboxy-terminal portion of alkaline phosphatase fused in-frame to an amino-terminal portion of a target gene product. The critical feature in the use of Tn*phoA* to generate gene fusions is that alkaline phosphatase is active only when exported from the cytoplasm, normally mediated by its signal sequence. The portion of *phoA* encoding its signal sequence was deleted in the construction of Tn*phoA*, so that active hybrid proteins are formed only if Tn*phoA* inserts downstream of export information encoded within the target gene. Since fimbrial subunits and many other proteins implicated in bacterial virulence mechanisms contain signal sequences or other export information, Tn*phoA* can be used to enrich for insertion mutations in the corresponding genes.

Shuttle vectors were developed to introduce Tn*phoA* into *V. cholerae* and other Gram-negative bacteria to facilitate the genetic studies of such determinants (50). Fig. 4 shows how these vectors have been used to isolate Tn*phoA*

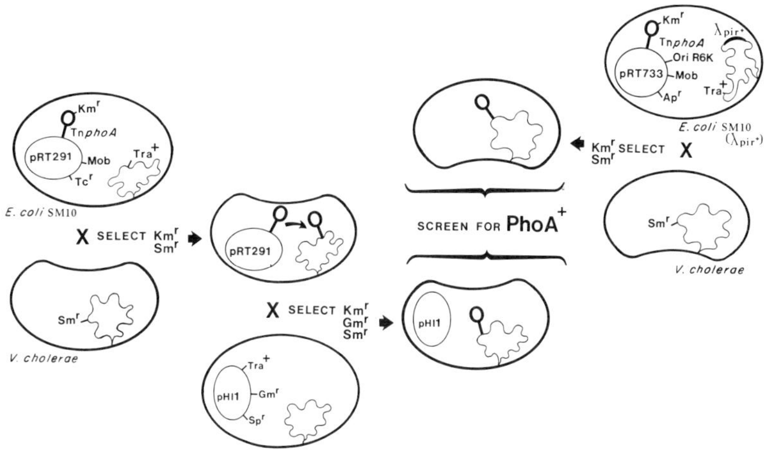

Figure 4. Two methods for the delivery and selection for transposition of Tn*phoA* in *V. cholerae*. On the left, plasmid pRT291, which carries Tn*phoA*, is mobilized from an *E. coli* donor strain (oval) into a *V. cholerae* recipient (crescent). A second plasmid, pHI1, of the same incompatibility group is mated into the strain while maintaining selection to retain the transposon. Resulting colonies that are blue on agar containing the chromogenic substrate 5-bromo-4-chloro-3-indolylphosphate (XP) harbour active gene fusions that have resulted from the transposition of Tn*phoA* from the vector plasmid into a gene that encodes a putative secreted protein. The method shown on the right utilizes a "suicide" plasmid vector that cannot replicate after mobilization into *V. cholerae* or other bacteria that do not carry the λpir prophage that provides transacting replication functions. Selecting for the presence of Tn*phoA* in the recipient yields colonies containing transpositions from the plasmid in a single step (50).

insertions in the *V. cholerae* 0395 chromosome. To find those within the *tcpA* structural gene, fusion-containing strains showing the appropriate regulation of alkaline phosphatase activities were isolated and examined for loss of surface pili and the TcpA protein. Properties of the insertions, such as the kanamycin resistance contributed by Tn*5*, facilitated their cloning and DNA sequencing to prove that the insertions were in *tcpA* and to characterize the gene further.

C. TCP is a Colonization Factor

The wild-type 0395 and the *tcpA*::Tn*phoA* mutant provided an isogenic pair of strains with which to test the role of TCP in host colonization by *V. cholerae*. Using an infant mouse cholera model, the LD_{50} was found to increase from 4×10^3 inoculated bacteria for the wild-type strain to 8×10^8 for the mutant. This was similar to the LD_{50} seen for *ctx* toxin structural gene deletion mutants, demonstrating the central role of TcpA in *V. cholerae* pathogenesis (48). To determine if this role was related to colonization, the ability of the mutant strain to compete with the wild-type strain in the mouse was tested. The competitive index, which is the ratio of the number of mutant bacteria to the number of wild-type bacteria recoverable from an infant mouse one day after intragastral coinoculation at a ratio of 1, was found to be 0.002. This demonstrated a definite defect in the ability of the mutant strain to colonize compared to the isogenic parent (48). Subsequent studies in human volunteers using a toxin-negative strain and a toxin-negative, *tcpA* double mutant supported these results. These studies showed a lack of symptoms and an inability to recover *V. cholerae* colonies from those patients receiving the double mutant, as compared to mild diarrhoea and significant colonization by the *tcpA*$^+$ strain (51). Thus the properties of defined insertion mutations within *tcpA* have established the role of TCP in colonization and virulence.

B. TCP Biogenesis

The Tn*phoA* approach was further exploited to identify several genes involved in TCP biogenesis, either through mutagenesis of the cloned chromosomal region adjacent to *tcpA*, or by characterizing further mutations located throughout the chromosome. Detailed analysis of the location and orientation of the insertions by DNA hybridization with Tn*phoA* probes, and determination of the size and number of respective hybrid proteins by immunoblot analyses with antibody to alkaline phosphatase, have identified nine genes that contribute to TCP biogenesis. Most of these are linked to *tcpA* and lie adjacent to each other immediately downstream of it in the same

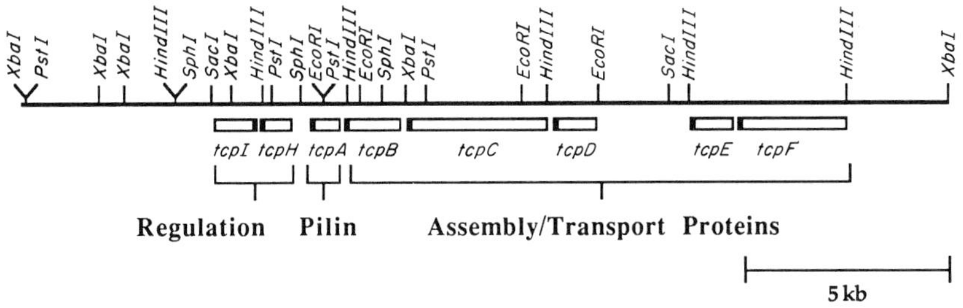

Figure 5. Organization of the gene cluster responsible for TCP pilus biogenesis. The functions of the respective gene products are shown underneath the map. Locations of the genes were determined by mapping the sites of insertion of Tn*phoA* that expressed active hybrid proteins resulting in altered pilus assembly or expression, in conjunction with analysing the sizes of the resulting hybrid proteins (52, 53). The thick vertical bar at the beginning of each gene represents a putative signal sequence for protein export, the presence of which is implied by the isolation of an active *phoA* fusion.

orientation, as shown in Fig. 5 (52, 53). Mutants with lesions in genes *tcpB–F* make TcpA pilin but do not express intact pili on their surface, implicating these gene products in pilus assembly or transport of intermediates to the growing pilus. Two other linked genes, *tcpH* and *tcpI*, appear to encode positive and negative regulators, respectively, of pilus expression. An additional unlinked gene, *tcpG*, has been identified. A *tcpG* insertion mutant strain exhibits pili of normal number and appearance on its surface, yet the strain is avirulent and defective in colonization. The *tcpG* product may therefore be a minor pilus component located at the tip or along the stalk which mediates adhesion in an analogous manner to that of the adhesin proteins of the Type I or PAP pili of uropathogenic *E. coli* strains (54, 55; Chapter 14), or it may moderate TCP function through subtle modifications of TcpA.

E. Additional Adherence Factors

Haemagglutination has long been used to measure the ability of an organism to adhere to eukaryotic cells, and is a property of TCP and other fimbriae. *V. cholerae* elaborates a variety of additional haemagglutinins under different growth conditions or at different phases of the growth curve; their actions are inhibited by a variety of carbohydrates that perhaps mimic the structure of the eukaryotic cell receptor (56, 57). Several genetic approaches have been undertaken to study their potential role in colonization. A gene (*sha*) required for one of them has been mapped and cloned from an El Tor strain. This gene confers the appropriate haemagglutination phenotype, which is not inhibited by D-mannose or L-fucose, on *E. coli* strains that carry it (58, 59). The requirement of this gene product for virulence has not been determined. An additional haemagglutination-defective mutant of the El Tor biotype has been isolated by Tn*5* mutagenesis. The mutation renders the strain unable to agglutinate human type-O erythrocytes and results in decreased ability to colonize the ilea of adult rabbits (60), thus supporting the role of haemagglutinins as virulence factors that help mediate colonization.

Recently, a different genetic screening has identified an additional putative colonization factor found in strains of the Classical biotype. This screen made extensive use of Tn*phoA* mutagenesis in *V. cholerae* coupled with in-depth investigation of the resulting fusions which showed regulated expression consistent with their being part of the ToxR controlled virulence regulon (Section V). A cluster of *toxR*-regulated genes, different from *tcpA*, was identified, mutations which reduce the ability of the parent strain, 0395, to colonize the intestines of infant mice. The property encoded by this gene cluster has been denoted accessory colonization factor (ACF) to describe its role in cholera pathogenesis (61). Thus, it is becoming increasingly clear that adhesion and colonization by *V. cholerae* is a complex process involving multiple factors.

IV. OTHER POTENTIAL VIRULENCE FACTORS

A. Outer Membrane Proteins

Insertional inactivation and deletion are being widely applied to assess the roles of various gene products in virulence. The 25 kD major outer membrane protein, OmpV, a major immunogenic component on the surface of *V. cholerae* (62), has been tested in this manner. Tn*phoA* insertions were isolated in the *ompV* gene and the resulting mutant strains tested in infant mice in a manner analogous to the *tcpA* mutants. The mutants were not significantly affected in their virulence properties with respect to LD_{50} or competitive index, suggesting that, while OmpV is recognized antigenically, it probably has no specific role in *V. cholerae* pathogenesis (48).

A second outer membrane protein, OmpU, has been implicated as a possible virulence determinant. This 38 kD protein is coregulated with cholera toxin and TCP production. The genes encoding all of these proteins appear to be part of a regulon controlled by ToxR (Section V). There are, as yet, no mutants defective in the production of OmpU to assess its role in virulence.

B. Haemolysin

While both *V. cholerae* biotypes contain haemolysin gene sequences, only El Tor strains exhibit haemolysis. The haemolysin structural gene, *hlyA*, has been cloned from El Tor strains and shown to encode an 80 kD protein (63, 64). To determine the role of *hlyA* in the virulence of either El Tor or Classical strains an internal deletion was constructed. Recombinant mutants showed no effects on virulence (65).

C. Neuraminidase

V. cholerae secretes a neuraminidase that acts on several eukaryotic cell surface gangliosides, converting them to ganglioside GM_1 which is the cholera toxin receptor. Cloning and sequencing of *nanH*, the neuraminidase structural gene, have revealed that this gene is probably part of an operon (66). To investigate the obvious potential role of the *nanH* gene product in virulence, kanamycin-resistance insertion mutations were introduced into the *V. cholerae* chromosome utilizing lambda vectors. This was accomplished using broad-host-range plasmid vectors that carry the *lamB* gene, encoding the *E. coli* lambda receptor protein. These allow for the expression of this protein in the outer membrane of several Gram-negative bacteria including *V. cholerae*, thus conferring on them the ability to adsorb lambda phage (67, 68). The phage DNA is injected, but no lysis occurs because of the absence of host factors required for lambda phage growth and DNA replication. If the injected lambda DNA carries a kanamycin-resistance determinant inserted into the cloned *nanH*, selection for kanamycin-resistant colonies will yield mutants with the insertion recombined (by double crossing-over) into the chromosome (if no homology is provided, this strategy can be used for transposon mutagenesis). The recombinant *nanH* mutants thus isolated do not show a dramatic loss of virulence (66). Thus, while the neuraminidase is not essential for disease, it may play a more subtle role.

D. Additional Factors

Genes that encode other potential virulence determinants, or enzymes involved in their biosynthesis, are being identified and isolated at a rapid pace. Such products include a variety of proteases, O-antigens, haemagglutinins, chemotactic sensors, and motility organelles. Some of these factors are of additional interest since they are produced not only by virulent O1 strains, but also by non-O1 *V. cholerae*. While the wider distribution of such factors may preclude them from being classified as discriminating pathogenic factors, they may nevertheless be required (but not sufficient) for virulence. One such factor is the soluble haemagglutinin/protease that has now been well characterized from both O1 and non-O1 isolates (69). Of special interest is the finding that many environmental and human isolates of *V. cholerae* produce small amounts of a Shiga-like toxin (related to that of *Shigella dysenteriae*) (70). Live, oral

V. cholerae vaccine candidate strains that do not produce this toxin are markedly less reactogenic in volunteers than those that do (65). However, an isogenic pair of shiga-like toxin positive and negative strains has not yet been constructed to determine if this additional toxin is the only factor contributing to the reactogenicity. With the application of the aforementioned techniques the roles of these macromolecules in pathogenesis and immunity will quickly be assessed.

V. REGULATION OF VIRULENCE GENE EXPRESSION

A. ToxR is a Positive Regulator of *ctx* Expression

A toxin is not expressed constitutively by *V. cholerae,* but varies with environmental conditions such as pH, aeration, osmolarity, and amino acids present in the growth medium, as well as the growth phase (17, 71). The first toxin-negative mutants of *V. cholerae* carried lesions, not in the structural genes, but in a regulatory gene called *tox* (72, 73). We now know that these mutations were isolated preferentially because of the duplication of the *ctx* genetic element in most strains. The *tox* locus has been studied in detail and is at least partly responsible for the regulation of toxin expression in response to environmental conditions.

When the *ctx* genes were cloned in *E. coli,* it was observed that toxin production was very low, as expected if a positive regulatory element of *V. cholerae* was required for maximal expression (24). Utilizing a *ctx-lacZ* transcriptional fusion, additional clones were isolated which overcame the expression defect, indicating that the clones encoded an activator of *ctx* transcription (36). When mobilized back into *V. cholerae tox* mutants, the clones restored toxin production. The gene responsible for activating toxin expression was designated *toxR.* Further mapping studies, as well as Northern analysis of *ctx* RNA levels, support the idea that *tox* and *toxR* are the same and that their regulation of toxin expression is at the level of transcription (74). The cloned region of DNA has been studied in detail. The *toxR* gene encodes a protein of 32.5 kD that shares little primary amino acid sequence homology with most known DNA-binding proteins (75). It does, however, share extensive homology with several transcriptional activators involved in bacterial transmembrane regulatory systems that respond to environmental stimuli (76 and Chapter 2).

B. ToxR is Required for the Expression of *tcpA* and Other Genes

The DNA sequence of *toxR* allowed for the construction of specific insertional mutations within the gene to prevent its expression. In this case, a new technique was used so that the mutations could be recombined into the *V. cholerae* chromosome via homologous recombination using only a single manipulation. A mobilizable, suicide delivery plasmid was developed which, when it contains the internal fragment of a gene, results in integration of the plasmid sequences within the gene, causing loss of expression of the corresponding protein (71: see also Fig. 5 of Chapter 7). This plasmid, designated pJM703.1, is also used as the basis of one of the Tn*phoA* delivery systems shown in Fig. 4.

The protein profiles from the strains carrying the *toxR* insertions showed dramatic effects on the levels of several proteins in addition to the lack of toxin production. Most strikingly, TcpA synthesis is shut off and there is a flip in the expression of two major outer membrane proteins, OmpT and OmpU; OmpU disappears and OmpT is expressed instead (71, 75). This same strategy has been used to demonstrate that *toxR* also regulates the *acf* gene cluster involved in colonization (61) (Section III, E). It therefore seems that *ToxR* in *V. cholerae*, like the *vir* gene product of *Bordetella* (77), controls a regulon of virulence-determining genes.

C. Mechanism of ToxR Regulation

Perhaps the most striking feature of the ToxR sequence is the presence of a hydrophobic stretch of amino acids indicative of a membrane-spanning domain in the carboxy-terminal third of the molecule. With the development of Tn*phoA* has come its use in studying the topology of membrane proteins (49). This tool was used to help demonstrate that ToxR spans the inner membrane of the bacterial cell, with the amino-terminus in the cytoplasm and the carboxy-terminus in the periplasm (75). This location provides a mechanism for the regulation of gene expression in response to environmental stimuli, as depicted in Fig. 6. The ToxR-PhoA hybrid proteins were still capable of activating expression of ToxR-regulated genes, even though this regulation was altered under some conditions, indicating a role of the periplasmic domain in this regulation. Since the hybrid proteins were still capable of activating transcription, cytoplasmic membrane fractions were isolated from an *E. coli* strain carrying the *toxR-phoA* containing plasmid and used in a gel retardation assay (78) to detect binding of the protein to the *ctx* promoter. This assay, in combination with a series of *ctx* promoter deletions, demonstrated that ToxR, in its transmembrane bound state, binds to the DNA sequence TTTTGAT found in multiple, tandem, direct repeats in the *ctx* promoter, thereby activating its expression. If the protein was solubilized from the membrane, it no longer bound DNA. The topology of ToxR suggests that the DNA-binding domain is toward its amino-terminus, which is where it shares homology with the other transcriptional activator proteins involved in transmembrane regulatory systems (76). It remains to be seen if the promoters

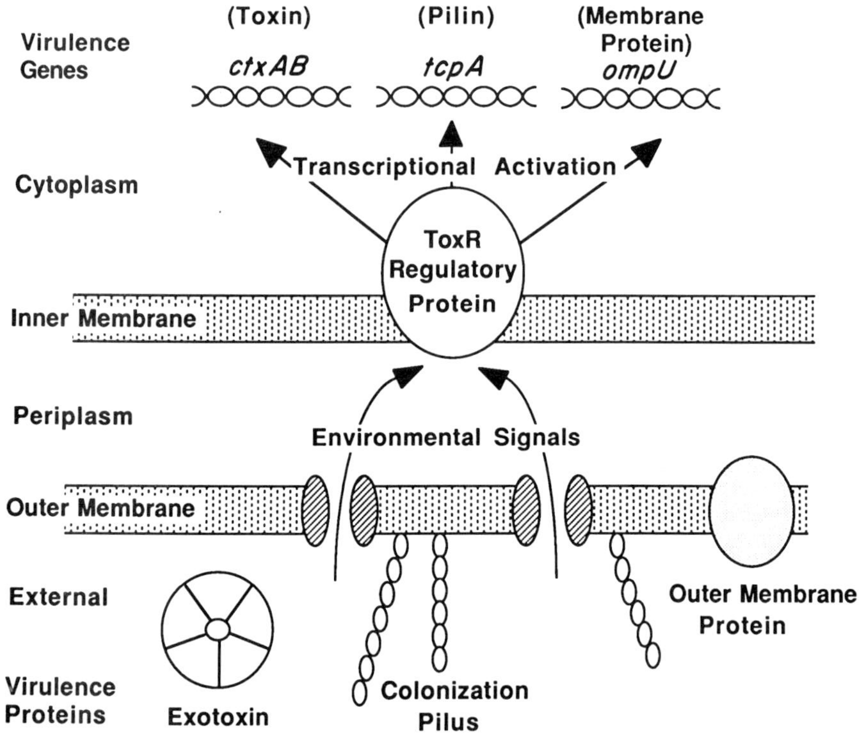

Figure 6. A model for the basis on which ToxR mediates the expression of a regulon of virulence-associated genes in response to environmental stimuli. ToxR is an integral inner membrane protein situated with its carboxy-terminus accessible to conditions in the periplasm and its amino-terminus in the cytoplasm where it interacts directly with the promoter of at least one of the transcriptional units it regulates, the *ctxAB* operon (75).

of other ToxR-regulated genes share homology with the *ctx* promoter and whether ToxR directly binds to these promoters.

While the membrane location of ToxR is consistent with its role in modulating gene expression in response to environmental stimuli, ToxR may not act independently. The product of a second gene, *toxS*, which is located in the same transcriptional unit and only four base pairs downstream of *toxR*, modulates its action. The product of this gene is an inner membrane protein as well, but with the bulk of its structure on the periplasmic side. The phenotype exhibited by *toxS* insertion mutations is a decrease in the expression of ToxR-regulated genes under some growth conditions. The role of the ToxS protein may be to stabilize ToxR in a dimerized conformation since *toxS* mutants that express ToxR-PhoA hybrid proteins exhibit a ToxS-independent phenotype (PhoA assumes an active conformation as a dimer) (79).

VI. PERSPECTIVES

It is clear that the application of the tools of modern molecular genetics to studies of *V. cholerae* is quickly unravelling the fundamental mechanisms that underlie cholera pathogenesis. The ability to isolate defined transposon induced mutations and move them between strains to construct isogenic sets specifically altered in various putative virulence determinants has made it possible to assess their role and focus further detailed studies on those of the most immediate importance. In addition, the ability to go on to isolate the genes of interest, determine their sequences and identify their products and subsequently to manipulate them in a variety of ways and introduce the altered genes back into *V. cholerae* has created a means to address the mechanisms by which confirmed virulence determinants function. These findings will have broad implications not only in further development of vaccines against cholera and other enteric diseases, but also on the mechanisms by which gene expression is influenced in response to environmental conditions.

References

(1) Gill, D. M. (1976). The arrangement of subunits in cholera toxin. *Biochemistry* **15**, 1242–1248.

(2) Lospalluto, J. J. and Finkelstein, R. A. (1972). Chemical and physical properties of cholera exo-enterotoxin (choleragen) and its spontaneously formed toxoid (choleragenoid). *Biochim. Biophys. Acta* **257**, 158–166.

(3) Mekalanos, J. J., Collier, R. J. and Romig, W. R. (1979). Enzymatic activity of cholera toxin. II. Relationships to proteolytic processing, disulphide bond reduction, and subunit composition. *J. Biol. Chem.* **254**, 5855–5861.

(4) Cassel, D. and Pfeuffer, T. (1978). Mechanism of cholera toxin action: Covalent modification of the guanyl nucleotide binding protein of the adenylate cyclase system. *Proc. Natl. Acad. Sci. USA* **75**, 2669–2673.

(5) Gill, D. M. and Meren, R. (1978). ADP-ribosylation of membrane proteins catalysed by cholera toxin: basis of the activation of adenylate cyclase. *Proc. Natl. Acad. Sci. USA* **75**, 3050–3054.

(6) Field, M. (1980). Intestinal secretion and its stimulation by enterotoxins. *In* "Cholera and Related Diarrheas" (O. Ouchterlony and J. Holmgren, eds.), pp. 46–52. Karger, Basel.

(7) Cutrecasas, P. (1973). Gangliosides and membrane receptors for cholera toxin. *Biochemistry* **12**, 3558–3566.

(8) Hirst, T. R., Sanchez, J., Kaper, J. B., Hardy, S. J. S. and Holmgren, J. (1984). Mechanism of toxin secretion by *Vibrio cholerae* investigated in strains harbouring plasmids that encode heat-labile enterotoxins of *Escherichia coli*. *Proc. Natl. Acad. Sci. USA* **81**, 7752–7756.

(9) Kaper, J. B. and Baldini, M. M. Genetics. *In* "Cholera" (D. Barua and W. B. Greenbough III, eds.), Plenum Medical Publ., New York (In press).

(10) Guidolin, A. and Manning, P. A. (1987). Genetics of *Vibrio cholerae* and its bacteriophages. *Microbiol. Rev.* **51**, 285–298.

(11) Bhaskaran, K. (1958). Genetic recombination in *Vibrio cholerae*. *J. Gen. Microbiol.* **19**, 71–75.

(12) Datta, A., Parker, C. D., Wohlhieter, J. A. and Baron, L. S. (1973). Isolation and characterization of the fertility factor P of *Vibrio cholerae*. *J. Bacteriol.* **113**, 763–771.

(13) Sublett, R. D. and Romig, W. R. (1981). Transposon-facilitated recombination in classical biotypes of *Vibrio cholerae*. *Infect. Immun.* **32**, 1132–1138.

(14) Parker, C, Gauthier, D., Tate, A., Richardson, K. and Romig, W. R. (1979). Expanded linkage map of *Vibrio cholerae*. *Genetics* **91**, 191–214.

(15) Johnson, S. R. and Romig, W. R. (1981). *Vibrio cholerae* conjugative plasmid pSJ15 contains transposable prophage dVcA1. *J. Bacteriol.* **146**, 632–638.

(16) Sporecke, I., Castro, D. and Mekalanos, J. (1984). Genetic mapping of *Vibrio cholerae* enterotoxin structural genes. *J. Bacteriol.* **157**, 253–261.

(17) Betley, M. J., Miller, V. L. and Mekalanos, J. J. (1986). Genetics of bacterial enterotoxins. *Ann. Rev. Microbiol.* **40**, 577–605.

(18) Newland, J. W., Green, B. A. and Holmes, R. K. (1984). Transposon-mediated mutagenesis and recombination in *Vibrio cholerae*. *Infect. Immun.* **45**, 428–432.

(19) Ogg, J. E., Timme, T. L. and Alemohammad, M. M. (1981). General transduction in *Vibrio cholerae*. *Infect. Immun.* **31**, 737–741.

(20) Dallas, W. S. and Falkow, S. (1980). Amino acid sequence homology between cholera toxin and *Escherichia coli* heat-labile toxin. *Nature* **288**, 499–501.

(21) Spicer, E. K., Kavanaugh, W. M., Dallas, W. S., Falkow, S., Konigsberg, W. H. and Schafer, D. E. (1981). Sequence homologies between A subunits of *Escherichia coli* and *Vibrio cholerae* enterotoxins. *Proc. Natl. Acad. Sci. USA* **78**, 50–54.

(22) Moseley, S. L. and Falkow, S. (1980). Nucleotide sequence homology between the heat-labile enterotoxin gene of *Escherichia coli* and *Vibrio cholerae* deoxyribonucleic acid. *J. Bacteriol.* **144**, 444–446.

(23) Kaper, J. B., Moseley, S. L. and Falkow, S. (1981). Molecular characterization of environmental and nontoxigenic strains of *Vibrio cholerae*. *Infect. Immun.* **32**, 661–667.

(24) Pearson, G. D. N. and Mekalanos, J. J. (1982). Molecular cloning of *Vibrio cholerae* enterotoxin genes in *Escherichia coli* K12. *Proc. Natl. Acad. Sci. USA* **79**, 2976–2980.

(25) Gennaro, M. L., Greenaway, P. J. and Broadbent, D. A. (1982). The expression of biologically active cholera toxin in *Escherichia coli*. *Nucleic Acids Res.* **10**, 4883–4890.

(26) Neill, R. J., Ivins, B. E. and Holmes, R. K. (1983). Synthesis and secretion of the plasmid-coded heat-labile enterotoxin of *Escherichia coli* in *Vibrio cholerae*. *Science* **221**, 289–291.

(27) Mekalanos, J. J., Swartz, D. J., Pearson, G. D. N., Hartford, N., Groyne, F. and de Wilde, M. (1983). Cholera toxin genes: nucleotide sequence, deletion analysis and vaccine development. *Nature* **306**, 551–557.

(28) Gennaro, M. L. and Greenaway, P. J. (1983). Nucleotide sequences within the cholera toxin operon. *Nucleic Acids Res.* **11**, 3855–3861.

(29) Lockman, H. A., Galen, J. E. and Kaper, J. B. (1984). *Vibrio cholerae* enterotoxin genes: nucleotide sequence analysis of DNA encoding ADP-ribosyltransferase. *J. Bacteriol.* **159**, 1086–1089.

(30) Lockman, H. and Kaper, J. B. (1983). Nucleotide sequence analysis of the A2 and B subunits of *Vibrio cholerae* enterotoxin. *J. Biol. Chem.* **258**, 13722–13726.

(31) Smith H. W. and Halls, S. (1968). The transmissible nature of the genetic factor in *Escherichia coli* that controls enterotoxin production *J. Gen. Microbiol.* **52**, 319–334.

(32) Yamamoto, T., Gojobori, T. and Yokota, T. (1987). Evolutionary origin of pathogenic determinants in enterotoxigenic *Escherichia coli* and *Vibrio cholerae* O1. *J. Bacteriol.* **169**, 1352–1357.

(33) Goldberg, I. and Mekalanos, J. J. (1985). Effect of a *recA* mutation on cholera toxin gene amplification and deletion events. *J. Bacteriol.* **165**, 723–731.

(34) Mekalanos, J. J. (1983). Duplication and amplification of toxin genes in *Vibrio cholerae*. *Cell* **35**, 253–263.

(35) Mekalanos, J. J. (1985). Cholera toxin: genetic analysis, regulation and role in pathogenesis. *Curr. Top. Microbiol. Immunol.* **118**, 97–118.

(36) Miller, V. L. and Mekalanos, J. J. (1984). Synthesis of cholera toxin is positively regulated at the transcriptional level by *toxR*. *Proc. Natl. Acad. Sci. USA* **81**, 3471–3475.
(37) Pierce, N. F., Kaper, J. B., Mekalanos, J. J. and Cray, W. C. (1985). Role of cholera toxin in enteric colonization by *Vibrio cholerae* O1 in rabbits. *Infect. Immun.* **50**, 813–816.
(38) Finkelstein, R. A., Vasil, M. L. and Holmes, R. K. (1974). Studies on toxinogenesis in *Vibrio cholerae*. I. Isolation of mutants with altered toxinogenicity. *J. Infect. Dis.* **129**, 117–123.
(39) Honda, T. and Finkelstein, R. A. (1979). Selection and characteristics of a *Vibrio cholerae* mutant lacking the A (ADP-ribosylating) portion of the cholera enterotoxin. *Proc. Natl. Acad. Sci. USA* **76**, 2052–2056.
(40) Mekalanos, J. J., Moseley, S. L., Murphy, J. R. and Falkow, S. (1982). Isolation of enterotoxin structural gene deletion mutations in *Vibrio cholerae* induced by two mutagenic vibriophages. *Proc. Natl. Acad. Sci. USA* **79**, 151–155.
(41) Ruvkun, G. B. and Ausubel, F. M. (1981). A general method for site-directed mutagenesis in prokaryotes. *Nature* **289**, 85–88.
(42) Kaper, J. B., Lockman, H., Baldini, M. M. and Levine, M. M. (1984). Recombinant nontoxinogenic *Vibrio cholerae* strains as attenuated cholera vaccine candidates. *Nature* **308**, 655–658.
(43) Kaper, J. B., Lockman, H., Baldini, M. and Levine, M. M. (1984). Recombinant live oral cholera vaccine. *Bio/Technology* **2**, 345–349.
(44) Levine, M. M., Kaper, J. B., Herrington, D., Losonsky, G., Morris, J. G., Clements, M. L., Black, R. E., Tall, B. and Hall, R. (1988). Volunteer studies of deletion mutants of *Vibrio cholerae* O1 prepared by recombinant techniques. *Infect. Immun.* **56**, 161–167.
(45) Pierce, N., Cray, W. C., Jr., Kaper, J. B. and Mekalanos, J. J. (1988). Determinants of immunogenicity and mechanisms of protection by virulent and mutant *Vibrio cholerae* O1 in rabbits. *Infect. Immun.* **56**, 142–148.
(46) Beachey, E. H. (1981). Bacterial adherence: adhesin-receptor interactions mediating the attachment of bacteria to mucosal surfaces. *J. Infect. Dis.* **143**, 325–345.
(47) Ehara, M., Ishibashi, M., Ichinose, Y., Iwanaga, M., Shimodori, S. and Naito, T. (1987). Purification and partial characterization of *Vibrio cholerae* O1 fimbriae. *Vaccine* **5**, 283–288.
(48) Taylor, R. K., Miller, V. L., Furlong, D. B. and Mekalanos, J. J. (1987). Use of *phoA* gene fusions to identify a pilus colonization factor coordinately regulated with cholera toxin. *Proc. Natl. Acad. Sci, USA* **84**, 2833–2837.
(49) Manoil, C. and Beckwith, J. (1985). Tn*phoA*: a transposon probe for protein export signals. *Proc. Natl. Acad. Sci. USA* **82**, 8129–8133.
(50) Taylor, R. K., Manoil, C. and Mekalanos, J. J. (1988). Broad-host-range vectors for delivery of Tn*phoA*: use in genetic analysis of secreted virulence determinants of *Vibrio cholerae. J. Bacteriol.* (In press.)
(51) Herrington, D. A., Losonsky, G., Hall, R., Mekalanos, J. J., Taylor, R. K. and Levine, M. M. (1988). Evaluation of *toxR* and *tcpA* mutant strains of *Vibrio cholerae* O1 in volunteers. *In* "Proceedings of the Twenty-Third Joint Conference on Cholera" National Institutes of Health, Bethesda, MD.
(52) Taylor, R., Shaw, C., Peterson, K., Spears, P. and Mekalanos, J. (1988). Safe, live *Vibrio cholerae* vaccines? *Vaccine* **6**, 151–154.
(53) Shaw, C. E., Peterson, K. M., Mekalanos, J. J. and Taylor, R. K. Genetic studies of *Vibrio cholerae* TcpA pilus biogenesis. *In* "Advances in Research on Cholera and Related Diarrhoeas, Vol. 6" (Y. Takeda and R. B. Sack, eds.), Martinus Nijhoff, Hague (In press).
(54) Lindberg, F., Lund, B., Johansson, L. and Normark, S. (1987). Localization of the receptor-binding protein adhesin at the tip of the bacterial pilus. *Nature* **328**, 84–87.
(55) Abraham, S. N., Goguen, J. D., Sun, D., Klemm, P. and Beachey, E. H. (1987). Identification of two ancillary subunits of *Escherichia coli* type I fimbriae using antibodies against synthetic oligopeptides of *fim* gene products. *J. Bacteriol.* **169**, 5530–5536.

(56) Jones, G. (1980). The adhesive properties of *Vibrio cholerae* and other *Vibrio* species. *In* "Bacterial Adherence" (E. H. Beachey, ed.), pp. 219–249, Chapman and Hall, London.
(57) Hanne, L. F. and Finkelstein, R. A. (1982). Characterization and distribution of the haemagglutinins produced by *Vibrio cholerae*. *Infect. Immun.* **36**, 209–214.
(58) Franzon, V. L. and Manning, P. A. (1986). Molecular cloning and expression in *Escherichia coli* K12 of the gene for a haemagglutinin from *Vibrio cholerae*. *Infect. Immun.* **52**, 279–294.
(59) Van Dongen, W. M. A. M. and de Graaf, F. K. (1986). Molecular cloning of a gene coding for a *Vibrio cholerae* haemagglutinin. *J. Gen. Microbiol.* **132**, 2225–2234.
(60) Finn, T. N., Reiser, J., Germanier, R. and Cryz, S. J., Jr. (1987). Cell-associated haemagglutinin-deficient mutant of *Vibrio cholerae*. *Infect. Immun.* **55**, 942–946.
(61) Peterson, K. M. and Mekalanos, J. J. (1988). Characterization of the *Vibrio cholerae* ToxR regulon: identification of novel genes involved in intestinal colonization. *Infect. Immun.* (In press.)
(62) Stevenson, G., Leavesley, D. I., Lagnado, C. A., Heuzenroeder, M. W. and Manning, P. A. (1985). Purification of the 25 000 dalton *Vibrio cholerae* outer membrane protein and the molecular cloning of its gene: *ompV*. *Eur. J. Biochem.* **148**, 385–390.
(63) Goldberg, S. L. and Murphy, J. R. (1984). Molecular cloning of the haemolysin determinant from *Vibrio cholerae* El Tor. *J. Bacteriol.* **160**, 239–244.
(64) Manning, P. A., Brown, M. H. and Heunzenroeder, M. W. (1984). Cloning of the structural gene (*hly*) for the haemolysin of *Vibrio cholerae* El Tor strain 017. *Gene* **31**, 225–231.
(65) Kaper, J. B., Mobley, H. L. T., Michalski, J. M., Herrington, D. A. and Levine, M. M. Recent advances in developing a safe and effective live oral attenuated *Vibrio cholerae* vaccine. *In* "Advances in Research on Cholera and Related Diarrhoeas, Vol. 6" (Y. Takeda and R. B. Sack, eds.), Martinus Nijhoff, Hague (In press).
(66) Galen, J. E., Vimr, E. R., Lawrisuk, L. and Kaper, J. B. Cloning, sequencing, mutation, and expression of the gene, *nanH*, for *Vibrio cholerae* neuraminidase. *In* "Advances in Research on Cholera and Related Diarrhoeas, Vol. 6" (Y. Takeda and R. B. Sach, eds.), Martinus Nijhoff, Hague (In press).
(67) DeVries, G. E., Raymond, C. K. and Ludwig, R. A. (1984). Extension of bacteriophage host range: selection, cloning, and characterization of a constitutive receptor gene. *Proc. Natl. Acad. Sci. USA* **81**, 6080–6084.
(68) Harkki, A., Hirst, T. R., Holmgren, J. and Palva, E. T. (1986). Expression of the *Escherichia coli lamB* gene in *Vibrio cholerae*. *Microb. Pathog.* **1**, 283–288.
(69) Honda, T., Booth, B. A., Boesman-Finkelstein, M. and Finkelstein, R. A. (1987). Comparative study of *Vibrio cholerae* non-O1 protease and soluble haemagglutinin with those of *Vibrio cholerae* O1. *Infect. Immun.* **55**, 451–454.
(70) O'Brien, A. D., Chen, M. E., Holmes, R. K., Kaper, J. B. and Levine, M. M. (1984). Environmental and human isolates of *Vibrio cholerae* and *Vibrio parahaemolyticus* produce a *Shigella dysenteriae* 1 (shiga)-like cytotoxin. *Lancet* **2**, 77–78.
(71) Miller, V. L. and Mekalanos, J. J. (1988). A novel suicide vector and its use in construction of insertion mutations: osmoregulation in *Vibrio cholerae* of outer membrane proteins and virulence determinants requires *toxR*. *J. Bacteriol.* **170**, 2575–2583.
(72) Holmes, R. K., Vasil, M. L. and Finkelstein, R. A. (1975). Studies on toxinogenesis in *Vibrio cholerae*. III. Mutants *in vitro* and in experimental animals. *J. Clin. Invest.* **55**, 551–560.
(73) Vasil, M. L., Holmes, R. K. and Finkelstein, R. A. (1975). Conjugal transfer of a chromosomal gene determining production of enterotoxin in *Vibrio cholerae*. *Science* **187**, 849–850.
(74) Miller, V. L. and Mekalanos, J. J. (1985). Genetic analysis of the cholera toxin positive regulatory gene *toxR*. *J. Bacteriol.* **163**, 580–585.
(75) Miller, V. L., Taylor, R. K. and Mekalanos, J. J. (1987). Cholera toxin transcriptional activator ToxR is a transmembrane DNA binding protein. *Cell* **48**, 271–279.

(76) Ronson, C. W., Nixon, B. T. and Ausubel, F. M. (1987). Conserved domains in bacterial regulatory proteins that respond to environmental stimuli. *Cell* **49**, 579–581.
(77) Weiss, A. A. and Falkow, S. (1984). Genetic analysis of phase change in *Bordetella pertussis*. *Infect. Immun.* **43**, 263–269.
(78) Fried, N. G. and Crothers, B. M. (1983). CAP and RNA polymerase interactions with the *lac* promoter: binding stoichiometries and long range effects. *Nucl. Acids Res.* **11**, 141–158.
(79) Peterson, K. M., Taylor, R. K., Miller, V. L., DiRita, V. J. and Mekalanos, J. J. Coordinate regulation of virulence determinants in *Vibrio cholerae*. *In* "Proceedings of the Centenary Symposium of the Pasteur Institute" (M. Schwartz, ed.), Elsevier, Amsterdam (In press).

Chapter **16**

Iron Scavenging in the Pathogenesis of Escherichia coli

PETER H. WILLIAMS *and* MARK ROBERTS

I. INTRODUCTION

A. Iron and Infection

The tissues and fluids of an animal's body are rich media for the growth of bacteria. The fact that they are nevertheless normally sterile is due to the successful application of a range of defensive mechanisms, from the simple physical barriers of epithelial surfaces, through mechanical clearing or flushing (such as coughing or micturition), to the sophisticated targeted responses of serum bacteriolysis and phagocytosis in which the immune system of the host plays a key role. A more general response to bacterial infection, however, is immediate reduction of the amount of iron in the blood plasma, which deprives invading microorganisms of an essential growth requirement. This effect, known as "hypoferraemia of infection", results from reduced intestinal absorption of iron and increased iron storage in the liver induced by interleukin 1, a small protein released into the plasma from macrophages within an hour or two of inoculation with microorganisms or with endotoxin.

The ability of the host to withhold iron from infecting bacteria is called, somewhat inaccurately, "nutritional immunity". It can be overcome experimentally by adding excess iron salts to bacterial inocula in animal infections. Furthermore, certain human clinical conditions which lead to raised levels of serum iron render patients very sensitive to even a small number of invading bacteria; hyperferraemia as a result of liver disease or haemolytic anaemia, or indeed simply from the accidental introduction of iron into the body, is frequently complicated by secondary bacterial infection. Conversely, processes which contribute to nutritional immunity, thereby decreasing the availability of iron to invading bacteria, significantly enhance the ability of the host organism to resist bacterial infections (1).

B. The Trouble with Iron

Virtually all living cells, both pro- and eukaryotic, require iron. It is a component of a number of enzyme systems, such as the cytochromes, catalases, and ribonucleotide reductase, in which the generation of a redox potential by interconversion of ferric (FeIII) and ferrous (FeII) forms is an essential feature. Iron is very abundant in the biosphere; however, in the strongly oxidizing atmosphere of the earth, at least since the emergence of photosynthetic organisms, iron exists predominantly as FeIII, which at physiological pH values is virtually insoluble. As such it would be effectively unavailable to biological systems were it not for the evolution of a variety of iron-specific ligands capable of supplying essential ferric iron as a soluble complex.

In an animal body, most extracellular iron is complexed with transferrin and lactoferrin, glycoproteins with molecular weights about 75–80 000 and two similar but not identical high affinity binding sites for FeIII. Transferrin is synthesized in the liver and secreted into the serum and lymph, from which it supplies iron to cells throughout the body. It also supplies iron to the bone marrow for haemoglobin synthesis, and when highly saturated it may donate iron to the liver and spleen for intracellular storage in ferritin. Lactoferrin has a higher affinity for iron than transferrin, and greater stability in acidic conditions; it is synthesized and secreted by mucosal epithelial cells of the respiratory, gastrointestinal and urogenital tracts, and is also present in polymorphs (associated with lysozyme in secondary granules) where it may be important in bactericidal activity (2). In a healthy individual, transferrin and lactoferrin are only partially saturated with iron; in human serum, for example, only about 30% of transferrin iron-binding sites are normally occupied. The concentration of "free" ferric ions in the serum is therefore orders of magnitude lower than that required to support bacterial growth. Superimpose on this the hypoferraemic response, and we see that the body has a powerful and very versatile defence against the establishment of bacterial infections.

C. Bacterial Iron Metabolism

Nevertheless pathogenic bacteria of a variety of species are able to proliferate successfully *in vivo* despite conditions of extreme iron stress imposed by nutritional immunity. The mechanisms by which they sequester iron for growth are many and various, at least in chemical and biochemical detail, but mechanistically they fall into just two classes—those that depend on siderophores, and those that do not.

Siderophores are low molecular weight iron-specific ligands secreted by microorganisms in response to iron stress. They solubilize free ferric iron, or scavenge it from soluble complexes such as transferrin or ferritin, and deliver it to the secreting organism by interacting with siderophore-specific receptors on the cell surface. At least 100 siderophores of bacterial (both Gram-positive and Gram-negative) and fungal origin have already been isolated and well characterized. They vary enormously in chemical structure, but virtually all have three bidentate groups (usually catechol or hydroxamate: Fig. 1) positioned to form an intramolecular hexadentate complex that fits the six octahedrally-directed valence bonds of FeIII. The two enterobacterial siderophores enterochelin and aerobactin, which are considered in detail in this chapter, are examples of the two main chemical classes of microbial iron-specific ligands.

Some pathogenic bacteria use mechanisms for iron uptake that do not involve siderophores. For example, *Neisseria gonorrhoeae* and *Neisseria meningitidis* (but not nonpathogenic *Neisseria* spp.) express surface receptors that specifically bind lactoferrin and transferrin, from which iron can be removed directly in an energy-dependent manner. However, although mutants have been isolated which are unable to utilize these host proteins as iron sources, it has not yet been possible to identify which of several iron-regulated outer membrane components is involved.

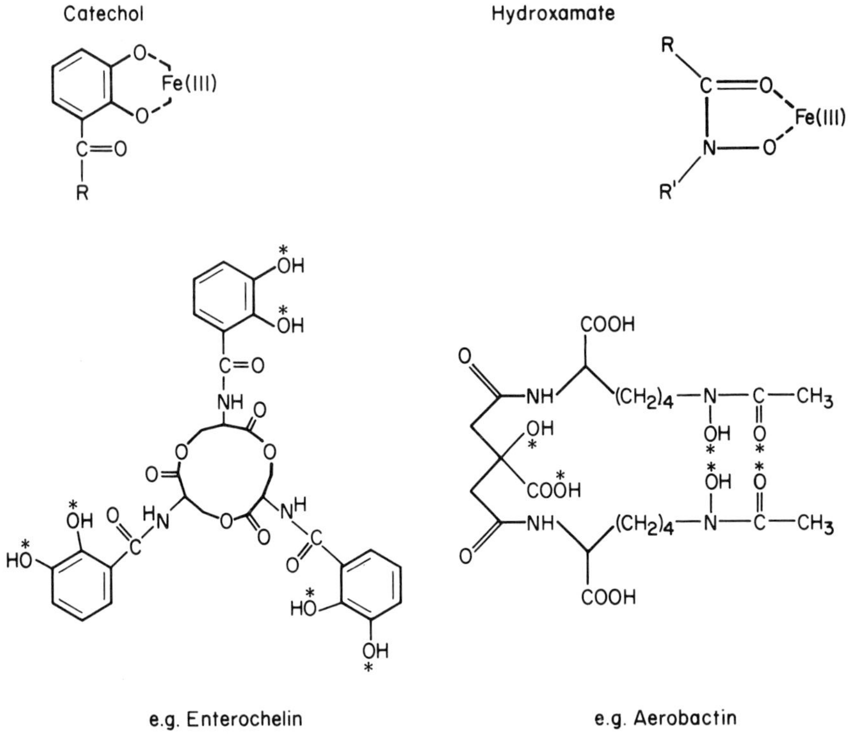

Figure 1. The two main chemical types of bidentate ligand systems of microbial siderophores exemplified by the catechol enterochelin, a cyclic triester of 2,3-dihydroxybenzoyl serine, and the hydroxamate aerobactin, a conjugate of two N^6-acetyl-N^6-hydroxylysine moieties and citric acid. Asterisks indicate groups involved in hexadentate complex formation with Fe III.

II. ENTEROBACTERIAL IRON UPTAKE SYSTEMS

A. The Involvement of Plasmids

With the rapid development in the early 1970s of techniques for the detection and analysis of plasmids, it became clear that several important virulence determinants of pathogenic strains of *Escherichia coli* were encoded in genes carried on plasmids; enterotoxin production and the expression of protein adherence antigens by isolates from human and animal diarrhoeal disease are classic examples. Although *E. coli* is generally regarded as a gut-living organism, some strains exert pathogenic effects at extraintestinal sites. In a search for plasmid-associated virulence determinants among such invasive strains, Williams Smith discovered a plasmid in a strain isolated from an outbreak of bacteraemia in chickens which, when transferred by conjugation, enhanced the lethality of the recipient bacterial strain for chickens and mice in experimental infections (3, 4). This characteristic, which was due to a greater ability to survive in the blood or peritoneal fluid of infected animals rather than to any toxic activity, was linked to a plasmid that specified production of colicin V. Moreover, the majority of isolates from bacteraemia of man and several domestic animals expressed colicin V, and among selected strains conjugal transfer of the ColV plasmid always rendered transconjugants more virulent, while curing (detected as loss of colicin V production) was correlated with lowered lethality for experimental animals in all cases.

Several ColV plasmid-associated features have been implicated in pathogenicity, including enhanced ability to adhere to isolated mouse intestinal epithelium, perhaps promoted by the slightly increased hydrophobicity of a bacterial surface covered with conjugative pili (5, 6), and resistance to the bactericidal effects of serum (7). It is also possible that colicin V itself plays a role in pathogenesis; for instance, colicin V in combination with endotoxin increases vascular permeability sufficiently to permit bacterial entry into the bloodstream (8), and may also have a slight depressive effect on peritoneal macrophage activity (9). In each case, however, the importance of these phenomena in invasive infections has not been conclusively proved.

It is the ability of $ColV^+$ bacteria to overcome nutritional immunity imposed by the host hypoferraemic response which is the most powerful contribution of the plasmid to pathogenicity. Extensive epidemiological evidence indicates that a ColV plasmid-specified high affinity iron uptake system based on the siderophore aerobactin is an important and widespread virulence determinant among the Enterobacteriaceae. This system alone, separated from other plasmid functions by molecular cloning, is sufficient to restore full virulence in animal infections to a clinical strain whose potency has been reduced by the loss of its resident ColV plasmid.

B. Discovery of the Aerobactin System

The role of ColV plasmids in iron metabolism of pathogenic strains was first demonstrated by Williams (10), who showed that plasmid-containing strains have a clear selective advantage over plasmid-free derivatives for growth *in vitro*, in media supplemented with transferrin or other iron-chelating agents, and *in vivo* in the bodies of experimentally infected animals. Significantly, however, the advantage of carrying a ColV plasmid is completely overcome by the presence of excess iron in growth media or the addition of iron salts to bacterial inocula for experimental infections.

In conditions of iron stress strains of *E. coli*, whether or not they are pathogenic, secrete the catechol siderophore enterochelin (Fig. 1) and synthesize an outer membrane receptor protein for ferric-enterochelin, the *fepA* gene product. Mutants of *E. coli* K-12 defective in the biosynthesis of enterochelin can grow only if a high concentration of free iron is available (allowing passive nonspecific uptake by an ill-defined low affinity system) or if utilizable iron chelators such as sodium citrate or ferrichrome are supplied. The presence in enterochelin-deficient mutants of ColV plasmids from various sources, however, overcomes dependence on excess iron or exogenous chelators, and represses the synthesis of the *fepA* protein and other iron-regulated components of the outer membrane, indicating the operation in ColV$^+$ strains of an additional high affinity iron transport system (10). In the case of one particular plasmid, ColV-K30, point mutations affecting iron transport resulted in loss of the virulence-enhancing property of the wild-type plasmid (11). Two classes of mutants were identified, those in which siderophore synthesis was affected (designated *iuc*, for iron uptake chelator) and others which were unable to use exogenously supplied siderophore (designated *iut*, for iron uptake transport). The product of *iuc* gene activity was identified as aerobactin (Fig. 1), a hydroxamate siderophore that had previously been described in culture supernatants of a strain of *Aerobacter aerogenes* (12, 13). Apo-aerobactin has a molecular mass of 565; it binds a single atom of FeIII through its two hydroxamic acid groups and a carboxyl and hydroxyl group in the central citric acid moiety. (At very high iron concentrations, a second ferric ion may bind to carboxyl groups in the lysine residues of the molecule (14), but it is unlikely that such a salt would be significant in nature.) The *iut* gene product is an outer membrane protein with an apparent molecular mass of 74 000, which is the receptor for ferric-aerobactin and for a bacteriocin called cloacin DF13 synthesized by strains of *Enterobacter cloacae* (15, 16). *E. coli* is normally insensitive to cloacin, but the presence of a ColV plasmid carrying the aerobactin system renders a strain cloacin-sensitive because of the presence of the 74 kD *iut* protein. Cloacin-resistant mutants cannot utilize aerobactin as a source of iron.

III. MOLECULAR GENETICS

A. Cloning

Cloacin-susceptibility has proved an invaluable tool at several stages in the molecular analysis of the aerobactin system. In the initial cloning of restriction fragments of plasmid ColV-K30, Bindereif and Neilands (17) screened transformants harbouring recombinant plasmids for sensitivity to cloacin as an indicator of the expression of the receptor protein. One plasmid that conferred cloacin-sensitivity, pABN1 (Fig. 2), also carried the complete

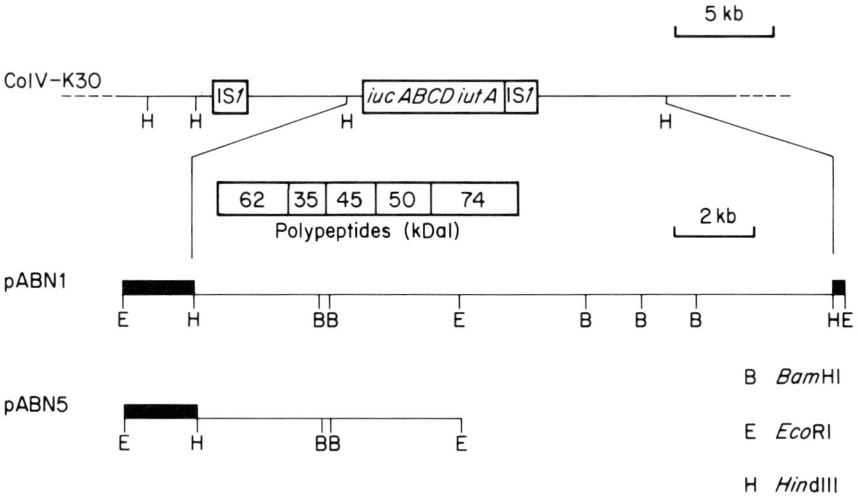

Figure 2. Important plasmids in the genetic analysis of the aerobactin system of *E. coli*. The five-gene cluster (*iucA–D* and *iutA*) of the prototype plasmid ColV-K30 is flanked by IS*1* elements in reverse orientation (24). pABN1 comprises a 16.3 *Hin*dIII fragment containing the entire gene cluster cloned into a multicopy vector pP*lac* (17); the positions of the five genes and the sizes of their products are indicated in boxes above the partial restriction map of pABN1 (18). pABN5 was derived by *Eco*RI digestion and relegation of pABN1 (17). Note that analogous plasmids have also been described by other groups.

coding capacity for biosynthesis of aerobactin, including an iron-regulated promoter that allowed expression only in conditions of iron stress. A strain harbouring the derivative plasmid pABN5 also synthesized aerobactin in an iron-regulated fashion, but was resistant to cloacin. Subsequent work with conventional minicell and maxicell expression systems established that pABN1, and similar constructs from other laboratories, express four polypeptides in addition to the 74 kD receptor protein. For three of these, there is general agreement on their molecular sizes (approximately 35 000, 50 000 and 62 000) but the fourth has proved more difficult to characterize (estimated molecular weights of 45 000 and 62 000 have been suggested for it) because of unexplained variability in its level of expression (18–20).

B. Mapping

Evidence that four gene products are required for the biosynthesis of aerobactin came from detailed transposon mapping (18). In pABN1 deriva-

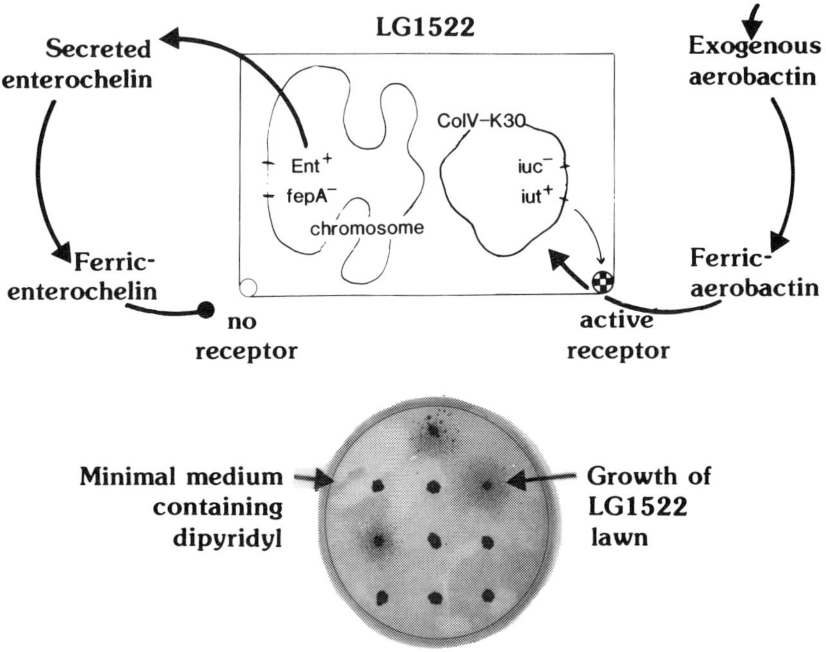

Figure 3. Aerobactin bioassay. *E. coli* K-12 strain LG1522 makes enterochelin (Ent^+) but cannot utilize it because it lacks the enterochelin receptor (*fepA*); moreover it carries a ColV-K30 plasmid in which aerobactin biosynthesis is inactivated by a point mutation (*iuc*) but in which the activity of the aerobactin receptor gene (*iutA*) is unaffected. LG1522 therefore cannot grow in iron-restricted conditions (e.g. defined minimal medium containing the iron chelator α,α'-dipyridyl at 200 μM) unless supplied with aerobactin. Isolates inoculated onto plates seeded with LG1522 are identified as aerobactin producers if they support a halo of growth of the test strain around the point of inoculation.

tives defective in the synthesis of aerobactin, as determined in a simple plate bioassay (Fig. 3), the positions of transposon insertion mapped to a 5.5 kb region corresponding to most of the insert DNA in pABN5 (Fig. 2). All such mutants were still sensitive to cloacin DF13, indicating that transposon insertion into the biosynthesis genes did not abolish expression of the receptor gene. Conversely, transposon insertions that resulted in cloacin-resistance had no effect on aerobactin biosynthesis; sites of insertion mapped to a 2 kb region immediately adjacent to the defined biosynthesis region.

By comparing translation products in minicells and maxicells of several pABN1 mutants and derivatives, the order of genes in the aerobactin system was determined (Fig. 2). Moreover the observation of putative truncated forms of some of the polypeptides in these analyses suggests that the overall direction of transcription is from left to right on the map as drawn. We intend in this chapter to follow Neilands' suggested nomenclature for the individual genes of the aerobactin system, in which *iucA-D* represent the biosynthesis genes in map order and *iutA* is the receptor gene (20). Inevitably, however, genetic nomenclature is confusing. Braun and his coworkers (19) favour the descriptive genotype assignments *aerA-D* for the aerobactin biosynthesis genes (such that *aerA* = *iucD*, *aerB* = *iucB*, *aerC* = *iucC* and *aerD* = *iucA*), but unfortunately the individual designations of the genes bear no relationship either to their order in the gene cluster or, as we shall see, to the sequence of enzyme activities in the biosynthetic pathway.

C. An Aerobactin Transposon?

The aerobactin system is a widespread characteristic of species of the Enterobacteriaceae, particularly among isolates from human and animal disease (Table I). It may be located on plasmids (ColV or otherwise), or on the chromosome where in some strains it may be present in more than one copy. A reasonable explanation for such ubiquity is that the aerobactin genes form part of a transposable element, although transposition has never been demonstrated. Nevertheless the finding that the aerobactin genes in many species are flanked by IS*1* elements in reverse orientation supports the

Table I
Ubiquity of the aerobactin system among the *Enterobacteriaceae*

Species	Genetic location	Flanking IS*1* elements	References
E. coli from human extraintestinal infections	Usually chromosomal	+	21, 22, 23
E. coli from animal extraintestinal infections	Plasmids (usually ColV)	+	24,25
Enteroinvasive *E. coli*	Chromosome	+	26
Enteropathogenic *E. coli*	Plasmids (not ColV) or chromosome	+ / −	26
Shigella spp. (not *dysenteriae* type I)	Chromosome	+	27
Salmonella spp. from gastroenteritis	Chromosome	−	28
Salmonella spp. from extraintestinal infections	Plasmids	+	29
Klebsiella spp.	Plasmids	−	30, 31
Enterobacter spp.	Plasmids	+	30
Yersinia spp. (not *pseudotuberculosis* or *enterocolitical*)	?	?	32

hypothesis, although the absence of either or both of these in some chromosomally-located systems suggests that a putative transposon may have been inactivated by mutations accumulated in the absence of selective pressures for transposition.

IV. BIOCHEMICAL GENETICS

A. Lysine and Citric Acid are Precursors of Aerobactin

Aerobactin consists of two N^6-acetylated, N^6-hydroxylated lysine residues linked by peptide bonds to a citric acid residue (Fig. 1). With the discovery that four gene products are involved in aerobactin biosynthesis, it was logical to suppose that two specific enzymes might modify lysine and two more might effect condensation of the product with a citric acid moiety; however, the actual assignment of functions was possible only when a variety of subclones and deletion derivatives of pABN1 and similar plasmids could be accurately mapped with respect to the genes that they carried. In Table II, we show a selection of the most pertinent data from a great deal of careful biochemical analysis carried out in several laboratories. These data, and the facts summarized below, point to the involvement of the products of the four genes *iucA-D* in the following biosynthetic pathway for aerobactin:

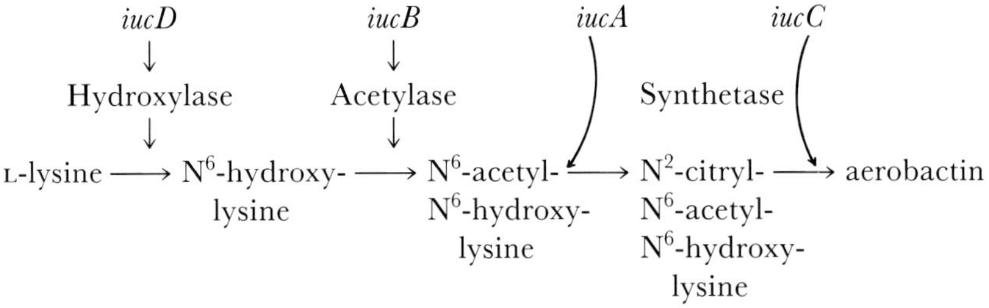

B. The *iucD* Gene Product

Strains harbouring plasmid clones of the *iucD* gene alone excrete large amounts of N^6-hydroxylysine in conditions of iron stress. Conversely, no putative intermediates in aerobactin biosynthesis were detected with strains carrying plasmids that lacked only *iucD* of the four *iuc* genes (19, 20). These observations indicate that the 50 kD *iucD* protein is in fact a hydroxylase, which acts on lysine in the first step of the biosynthetic pathway, and confirm previous enzymological studies of aerobactin production in *Aerobacter aerogenes*. They conflict, however, with the data of Ford *et al.* (33) who fed ^{14}C-lysine to

Table II
Biochemical genetics of aerobactin biosynthesis

Genotype	Activity
iuc $A^+B^+C^+D^+$	Aerobactin synthesized *in vivo*
iuc D^+	N^6-hydroxylysine synthesized *in vivo*
iuc $A^+B^+C^+$	No intermediates observed
iuc $A^+B^+C^+D^-$::Tn	Lysine converted to acetyl-lysine *in vivo*
iuc A^+B^+	N^6-hydroxylysine converted to N^6-acetyl-N^6-hydroxylysine in the presence of acetyl-CoA in cell free extracts
iuc A^+	No acetylation of N^6-hydroxylysine *in vitro*
iuc $B^+C^+D^+$	N^6-acetyl-N^6-hydroxylysine synthesized *in vivo*
iuc $A^+B^+C^-D^+$ (small deletions or duplications)	N^2-citryl-N^6-acetyl-N^6-hydroxylysine synthesized *in vivo*

All clones and mutants were derived from pABN1 (itself a derivative of ColV-K30) or from an equivalent plasmid containing the cloned aerobactin system from ColV-K311.

strains carrying an *iucD* transposon insertion mutant of pABN1 and observed the production of a radiolabelled compound that was identified unequivocally as acetyl-lysine. This suggests that hydroxylation catalysed by the *iucD* gene product follows acetylation in the synthesis of the aerobactin intermediate N^6-acetyl-N^6-hydroxylysine. Neither Gross *et al.* (19) nor de Lorenzo *et al.* (20) looked for N^6-acetyl-lysine in culture media of strains defective in aerobactin biosynthesis functions; nevertheless, since substrate specificities of individual enzymes tend to favour a model in which hydroxylation is the first step in aerobactin biosynthesis, this is the pathway presented here.

C. The *iucB* Gene Product

A plasmid containing *iucA* and *iucB* specifies an activity which, in cell-free extracts, acetylates hydroxylysine in the presence of acetyl-CoA (20). Since a plasmid expressing *iucA* alone does not, clearly the 35 kD *iucB* gene product is an acetylase (N^6-hydroxylysine: acetyl-CoA N^6-transacetylase).

D. Aerobactin Synthetase

Strains carrying pABN5-like plasmids in which part of the *iucA* gene has been deleted secrete N^6-acetyl-N^6-hydroxylysine, suggesting that *iucA* mutants are blocked in the third reaction of the biosynthetic pathway. Furthermore, plasmids with functional *iucA* (as well as *iucB* and *iucD*) but in which *iucC* was inactivated by either deletion or duplication of a small restriction fragment promoted the accumulation of N^2-citryl-N^6-acetyl-N^6-hydroxylysine, a compound in which only one modified lysine residue is attached to citric acid (19, 34). Thus, assembly of two lysine derivatives with citric acid to form aerobactin is a two-step reaction involving the enzyme products of first the *iucA* and then the *iucC* genes. Aerobactin synthetase of the *A. aerogenes* strain from which the siderophore was first isolated has an estimated molecular size of over 100 000, and exhibits two distinct catalytic activities, one specific for citric acid and the other capable of activating derivatives of lysine for subsequent linkage by peptide bonds. These reactions consume four molecules of ATP for each molecule of aerobactin formed (35). Assuming that the enzyme is similar in the two bacterial species, it is therefore possible that the aerobactin synthetase of *E. coli* strains carrying ColV plasmids is a complex of at least two subunits, the *iucA* and *iucC* gene products.

V. REGULATION

A. Transcriptional Regulation in Response to Iron Availability

It is characteristic of bacterial iron transport mechanisms that they are expressed only in conditions of iron deprivation. In the first systematic study of regulation of the aerobactin system, Braun and Burkhardt (36) isolated derivatives of plasmid ColV-K311 into which phage *Mud*(Ap,*lac*) was inserted such that the synthesis of hydroxamate material was undetectable, and expression of β-galactosidase (from the phage *lacZ* gene) was modulated by the availability of iron. Depending on the particular strain tested, the level of β-galactosidase synthesized in iron-limited conditions was 6 to 30 times higher than when iron was freely available, the variability probably being due to trivial differences in growth rates leading to different levels of iron deficiency within the cells.

Although these studies were limited by the consequences of mutational inactivation of the very means of transporting iron into the cells, one important observation was that β-galactosidase was constitutively expressed in these fusion plasmids in so-called *fur* (ferric uptake regulation) mutants of *E. coli* K-12. Thus the aerobactin system, like several genes and gene clusters scattered throughout the genome, appears to be under the negative transcriptional control of a common regulatory element, the *fur* gene product (37). The *fur* gene, which maps at about 15.5 minutes on the *E. coli* K-12 chromosome (38), has been cloned and sequenced; the deduced product is a 17 kD polypeptide with an unusually high histidine content (39). Coupling the cloned *fur* gene to the *E. coli recA* promoter-operator sequence permits overexpression on induction with nalidixic acid and the product is then readily purified to homogeneity by binding to zinc iminodiacetate agarose and subsequent elution with histidine (40). The Fur protein binds divalent metal ions; thus the normal corepressor for generalized repressor activity on various iron-regulated systems would be FeII iron rather than the virtually insoluble unbound FeIII form, although divalent ions of manganese, cobalt and some other metals are also effective.

Measurement of mRNA levels by dot-blot hybridization confirmed that expression of all the genes of the aerobactin system is normally regulated by the availability of iron in the external medium, and demonstrated directly that control is exerted at the level of transcription (41). In a strain carrying ColV-K30, transcription of the *iuc* genes is induced about 30-fold from a low basal level when iron stress is imposed. With plasmid pABN1, the basal level when iron is freely available is about 20 times higher than with the parent plasmid. It is enhanced a further ten-fold by iron stress to an overall induced level some six times higher than the maximum level of transcription of the biosynthesis genes of plasmid ColV-K30. These results show that pABN1 contains *cis*-acting regulatory sequences, and suggests that at high copy-number there may be partial titration of the chromosomally determined regulatory gene product. Transcription of the *iutA* gene of both ColV-K30 and pABN1, on the other hand, is induced only three-fold from comparatively high basal levels; since in conditions of iron stress all five genes of the aerobactin cluster are transcribed at a higher level, presumably as a single polycistronic message from a major promoter upstream of the *iucA* gene (Fig. 4a), it appears that *iutA* has its own minor unregulated promoter whose activity is swamped by induced transcription from the major *fur*-regulated promoter. No obvious promoter region was found by sequence analysis of the region immediately upstream of the *iutA* gene; nevertheless, transposon insertion into the *iuc* genes completely inactivates aerobactin biosynthesis but merely reduces receptor activity (18).

B. The Major Regulatory Region of the Aerobactin Operon

In vitro transcription experiments using purified *E. coli* RNA polymerase and plasmid pABN5 as template supported the idea that all five genes of the aerobactin system could be transcribed as a single unit, and placed the site of initiation of transcription to the left of the gene cluster as shown in Fig. 4a (42). Use of a 700 bp *Hin*dIII-*Sal*I fragment from the extreme left hand end of pABN5 in the *in vitro* transcription system generated a major transcript 260 nucleotides in length and a minor product of approximately 320 nucleotides, indicating that the aerobactin system has two promoters about 60 bp apart. However, S1 mapping using total RNA purified from bacteria harbouring ColV-K30 showed that, while the major promoter (P1) was also active *in vivo*, there was no evidence for *in vivo* activity of the minor upstream promoter (P2)

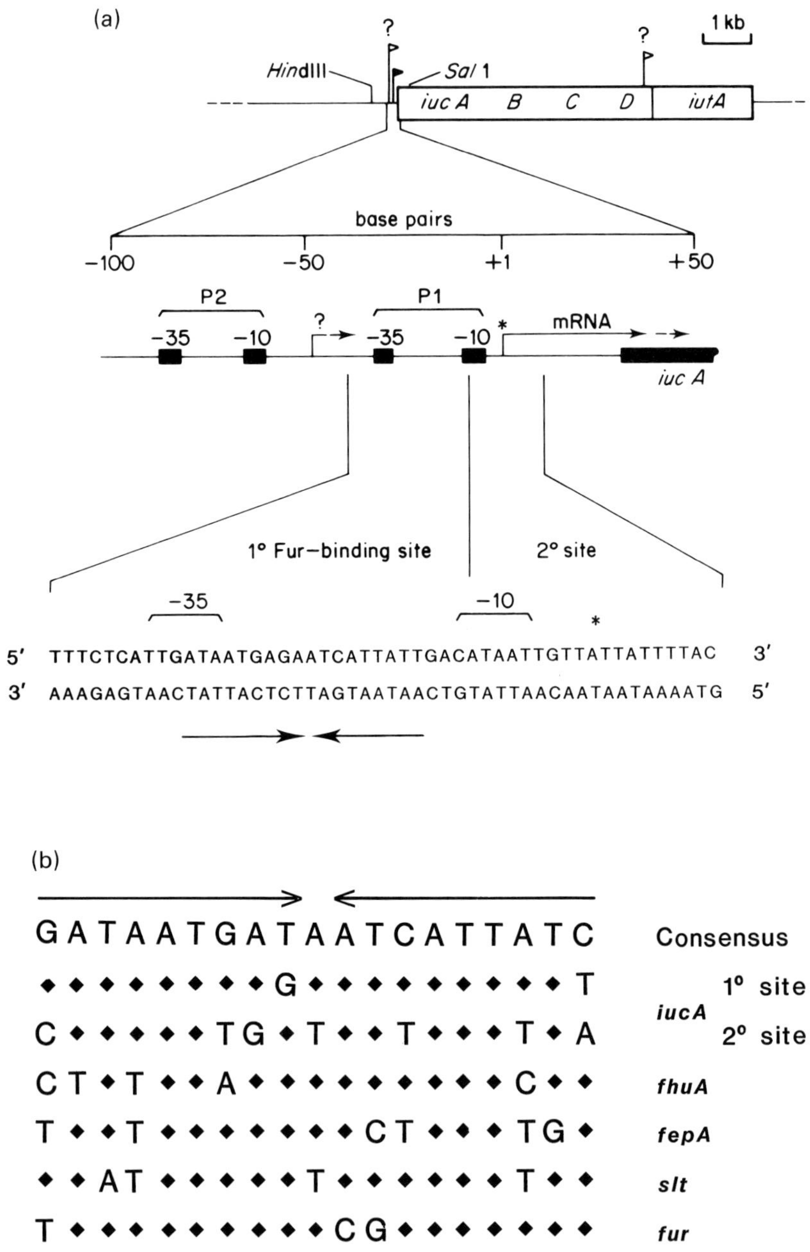
(a)
?
?
1 kb
HindIII
Sal 1
iuc A
B
C
D
iutA
base pairs
−100
−50
+1
+50
P2
P1
−35
−10
?
−35
−10
*
mRNA
iuc A
1° Fur−binding site
2° site
−35
−10
*
5′ TTTCTCATTGATAATGAGAATCATTATTGACATAATTGTTATTATTTTAC 3′
3′ AAAGAGTAACTATTACTCTTAGTAATAACTGTATTAACAATAATAAAATG 5′
(b)
GATAATGATAATCATTATC Consensus
◆◆◆◆◆◆◆◆G◆◆◆◆◆◆◆◆◆T iucA 1° site
C◆◆◆◆◆TG◆T◆◆T◆◆◆T◆A iucA 2° site
CT◆T◆◆A◆◆◆◆◆◆◆◆◆C◆◆ fhuA
T◆◆T◆◆◆◆◆◆◆CT◆◆◆TG◆ fepA
◆◆AT◆◆◆◆◆T◆◆◆◆◆◆T◆◆ slt
T◆◆◆◆◆◆◆◆◆CG◆◆◆◆◆◆◆ fur

Figure 4.

at different levels of iron stress. The role of the minor promoter in regulation of the aerobactin system therefore remains unclear. Sequencing of the 700 bp *Hin*dIII-*Sal*I fragment (Fig. 4a) revealed that the −10 and −35 regions of P1 each differ only in one nucleotide from the consensus sequences of *E. coli* promoters, with the distance between the two regions (17 nucleotides) being optimal for a strong promoter.

DNaseI protection experiments using purified Fur protein and a divalent metal as corepressor indicated that there are two contiguous sites for repressor binding with different lengths and affinities (43). The primary binding site extends 31 bp from just upstream of the −35 region of P1 to the upstream edge of the −10 region (Fig. 4a); at higher protein concentrations the next 19 bp downstream are protected, and at very high levels the upstream region as far as and beyond P2 may also be protected, but without defined boundaries of protection. The primary binding site for the Fur protein in the aerobactin system shows a certain degree of homology with the secondary binding site immediately downstream (Fig. 4b). Moreover, comparison with regions upstream of other iron-regulated genes, and indeed with the promoter of the *fur* gene itself, suggests an AT-rich consensus sequence comprising a 9 bp unit repeated in reverse orientation with a single base pair between (44–47). Deviations from the consensus sequence tend to retain, or even increase, the proportion of AT pairs, and an AT-TA dinucleotide at equivalent positions in the inverted repeats is, so far at least, invariant.

C. Complexities of Regulation

There is elegant simplicity in a regulatory model that involves a common repressor, the Fur protein, and its corepressor FeII, acting at many operator sites on the chromosome and on plasmids to control a variety of iron-regulated

Figure 4. (a) The operator-promoter region of the aerobactin operon. Transcriptional analyses indicate a major promoter (P1) for the operon and a minor promoter (P2) which functions only *in vitro* (42); there may also be an unregulated promoter for the receptor gene only (41). Asterisks indicate the initiation point for transcription; from P1 there is a leader sequence of about 30 bases before the translation initiation codon of the *iucA* gene. Purified *fur* repressor binds to two adjacent sites spanning P1 (43). (b) Consensus sequence in the putative regulatory regions of the iron-regulated genes *iucA* (the aerobactin operon, primary and secondary *fur* binding sites, 43), *fhuA* (which encodes a receptor for the fungal siderophore ferrichrome, 44), *fepA* (which encodes the enterochelin receptor, 45), *slt* (the Shiga-like toxin type I operon of *E. coli*, 46) and *fur* (which encodes the general repressor for iron-regulated systems, 47). Only bases that differ from the consensus sequence are shown.

systems. Inevitably, however, consideration of overall iron metabolism in a bacterial cell introduces layers of complexity which are as yet only very poorly understood. It is likely, for instance, that bacteria accumulate intracellular pools of FeII iron which both supply iron-activated enzymes and regulate the mechanisms by which the metal is transported into the cell (48). Williams and Carbonetti (49) used streptonigrin, a drug whose bactericidal activity depends on the presence of iron, as a probe of iron pool sizes. Cells grown in iron-rich medium accumulate iron and are very sensitive to killing by streptonigrin. However, when bacteria that could make neither aerobactin nor enterochelin were shifted to iron-limited medium, rapid depletion of the iron pool to support active growth was accompanied by increasing resistance to the drug; uptake of iron complexed to externally supplied citrate immediately restored sensitivity. However, starved bacteria that depended on enterochelin for growth remained streptonigrin-sensitive when shifted from iron-rich to iron-limited media, presumably as the iron pool was continually replenished by release of iron. A strain that made aerobactin, however, rapidly became streptonigrin-resistant during active growth in iron-limited conditions, suggesting that aerobactin in some way supplies iron directly to various enzymes that require it, rather than releasing it into the pool from which its withdrawal for essential functions would be concentration-dependent.

A model that fits these observations is shown in Fig. 5. What it does not explain is that in strains that make both aerobactin and enterochelin, there is differential regulation of the two systems, such that, as mentioned previously,

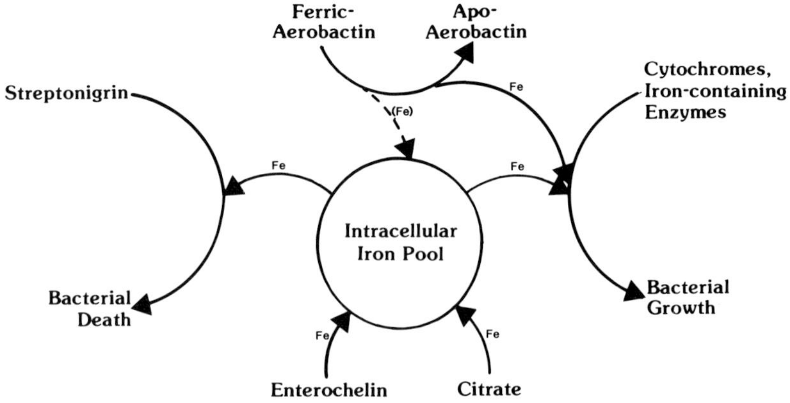

Figure 5. Compartmentalization of aerobactin iron. Enterochelin- and citrate-mediated iron uptake systems contribute to an intracellular iron pool which supplies iron-activated enzymes and catalyses the bactericidal activity of streptonigrin. Aerobactin withholds its iron from streptonigrin, delivering most of it directly to proteins essential for bacterial growth.

the presence of a ColV plasmid encoding the aerobactin system results in repression of the enterochelin receptor (*fepA*) gene in conditions where it would be induced in a plasmid-free strain. Moreover, mutants (with chromosomal mutations) exist which are constitutive for expression of the aerobactin system but display normal inducibility of enterochelin synthesis. Thus, if we postulate that aerobactin supplies iron directly to the Fur protein so that the enterochelin system should be repressed, we also need to suggest that the repressor has a lower avidity for the aerobactin operator (which may be further reduced by mutation) so that adequate synthesis of this siderophore can still occur. Whether this is due simply to sequence differences at the respective operators, which change their affinity for the Fur protein, or whether the aerobactin system requires additional specific cofactors remains to be seen.

VI. EPILOGUE

In the contest between pathogenic microorganisms and the defence mechanisms of their victims, competition for available iron is pivotal to the outcome of the infection. It is likely, for instance, that in the very early stages of an invasive *E. coli* infection, when the number of invading pathogens is low, the ability to synthesize aerobactin immediately to provide adequate iron for rapid initiation of bacterial proliferation is a critical factor in the development of disease symptoms. However, it has recently become clear that for some pathogens the elaboration of toxins, haemolytic activity and possibly adherence structures may also depend on the availability of iron. It is ironic (!) indeed that the defensive hypoferraemic response of the host may be a trigger for the expression of a variety of genes whose products are directly involved in pathogenesis.

The aerobactin system is the best understood of all iron-regulated genes in bacteria, and it provides a useful model whose principles may be applied to other systems in which genetical and biochemical details are harder to come by. But aerobactin is also important in practical terms because of its crucial role in the virulence of many species of enteric bacteria, and any means of interfering with its activity *in vivo* may therefore form the basis of a highly selective control of a variety of diseases in man and domestic animals.

References

(1) Weinberg, E. D. (1984). Iron withholding: a defence against infection and neoplasia. *Physiological Rev.* **64**, 65–102.

(2) Aisen, P. and Listowsky, I. (1980). Iron transport and storage proteins. *Ann. Rev. Biochem.* **49**, 357–393.

(3) Williams Smith, H. (1974). A search for transmissible pathogenic characters in invasive strains of *Escherichia coli*: the discovery of a plasmid-controlled toxin and a plasmid-controlled lethal character closely associated, or identical, with colicine V. *J. Gen. Microbiol.* **83**, 95–111.
(4) Williams Smith, H. and Huggins, M. B. (1976). Further observations on the association of the colicine V plasmid of *Escherichia coli* with pathogenicity and with survival in the alimentary tract. *J. Gen. Microbiol.* **92**, 335–350.
(5) Clancy, J. and Savage, D. C. (1981). Another colicin V phenotype: in vitro adhesion of *Escherichia coli* to mouse intestinal epithelium. *Infect. Immun.* **32**, 343–352.
(6) Tewari, R., Smith, D. G. and Rowbury, R. J. (1985). Effect of ColV plasmids on the hydrophobicity of *Escherichia coli*. *FEMS Microbiol. Lett.* **29**, 245–249.
(7) Binns, M. M., Davies, D. L. and Hardy, K. G. (1979). Cloned fragments of the plasmid ColV, I-K94 specifying virulence and serum resistance. *Nature* **279**, 778–781.
(8) Ozanne, G., Mathieu, L. G. and Baril, J. P. (1977). Production of colicin V *in vitro* and *in vivo* and observations on its effects in experimental animals. *Infect. Immun.* **17**, 497–503.
(9) Ozanne, G., Mathieu, L. G. and Baril, J. P. (1977). Production de colicines V et V_2 *in vitro* et *in vivo*: étude de leur action inhibitrice sur la phagocytose par des macrophages peritoneaux. *Rev. Can. Biol.* **36**, 307–316.
(10) Williams, P. H. (1979). Novel iron uptake system specified by ColV plasmids: an important component in the virulence of invasive strains of *Escherichia coli*. *Infect. Immun.* **26**, 925–932.
(11) Williams, P. H. and Warner, P. J. (1980). ColV plasmid-mediated, colicin V-independent iron uptake system of invasive strains of *Escherichia coli*. *Infect. Immun.* **29**, 411–416.
(12) Gibson, F. and Magrath, D. I. (1969). The isolation and characterization of a hydroxamic acid (aerobactin) formed by *Aerobacter aerogenes* 62-I. *Biochim. Biophys. Acta.* **192**, 175–184.
(13) Warner, P. J., Williams, P. H., Bindereif, A. and Neilands, J. B. (1981). ColV plasmid-specified aerobactin synthesis by invasive strains of *Escherichia coli*. *Infect. Immun.* **33**, 540–545.
(14) Braun, V. (1981). *Escherichia coli* cells containing the plasmid ColV produce the iron ionophore aerobactin. *FEMS Microbiol. Lett.* **11**, 225–228.
(15) Grewal, K. K., Warner, P. J. and Williams, P. H. (1982). An inducible outer membrane protein involved in aerobactin-mediated iron transport by ColV strains of *Escherichia coli*. *FEBS Lett.* **140**, 27–30.
(16) van Tiel-Menkveld, G. J., Oudega, B., Kempers, O. and de Graaf, F. K. (1981). The possible involvement of the cloacin DF13 receptor protein in the hydroxamate-mediated uptake of iron by *Enterobacter cloacae* and *Escherichia coli* (ColV). *FEMS Microbiol. Lett.* **12**, 373–380.
(17) Bindereif, A. and Neilands, J. E. (1983). Cloning of the aerobactin-mediated iron assimilation system of plasmid ColV. *J. Bacteriol.* **153**, 1111–1113.
(18) Carbonetti, N. H. and Williams, P. H. (1984). A cluster of five genes specifying the aerobactin iron uptake system of plasmid ColV-K30. *Infect. Immun.* **46**, 7–12.
(19) Gross, R., Engelbrecht, F. and Braun, V. (1985). Identification of the genes and their polypeptide products responsible for aerobactin synthesis by pColV plasmids. *Mol. Gen. Genet.* **201**, 204–212.
(20) de Lorenzo, V., Bindereif, A., Paw, B. H. and Neilands, J. B. (1986). Aerobactin biosynthesis and transport genes of plasmid ColV-K30 in *Escherichia coli* K12. *J. Bacteriol.* **165**, 570–578.
(21) Bindereif, A. and Neilands, J. B. (1985). Aerobactin genes in clinical isolates of *Escherichia coli*. *J. Bacteriol.* **161**, 727–735.
(22) Carbonetti, N. H., Boonchai, S., Parry, S. H., Väisänen-Rhen, V., Korhonen, T. K. and Williams, P. H. (1986). Aerobactin-mediated iron uptake by *Escherichia coli* isolates from human extraintestinal infections. *Infect. Immun.* **51**, 966–968.
(23) Valvano, M. A., Silver, R. P. and Crosa, J. H. (1986). Occurrence of chromosome- or plasmid-mediated aerobactin iron transport systems and hemolysin production among clonal groups of human invasive strains of *Escherichia coli* K1. *Infect. Immun.* **52**, 192–199.

(24) Waters, V. L. and Crosa, J. H. (1986). DNA environment of the aerobactin iron uptake system genes in prototypic ColV plasmids. *J. Bacteriol.* **167**, 647–654.
(25) Linggood, M. A., Roberts, M., Ford, S., Parry, S. H. and Williams, P. H. (1987). Incidence of the aerobactin iron uptake system among *Escherichia coli* isolates from infections of farm animals. *J. Gen. Microbiol.* **133**, 835–842.
(26) Roberts, M., ParthaSarathy, S., Lam-Po-Tang, M. K. L. and Williams, P. H. (1986). The aerobactin iron uptake system in enteropathogenic *Escherichia coli*: evidence for an extinct transposon. *FEMS Microbiol. Lett.* **37**, 215–219.
(27) Lawlor, K. M. and Payne, S. M. (1984). Aerobactin genes in *Shigella* spp. *J. Bacteriol.* **160**, 266–272.
(28) McDougall, S. and Neilands, J. B. (1984). Plasmid- and chromosome-coded aerobactin synthesis in enteric bacteria: insertion sequences flank operon in plasmid-mediated systems. *J. Bacteriol.* **159**, 300–305.
(29) Colonna, B., Nicoletti, M., Visca, P., Casalino, M., Valenti, P. and Maimone, F. (1985). Composite IS*1* elements encoding hydroxamate-mediated iron uptake in FI*me* plasmids from epidemic *Salmonella* spp. *J. Bacteriol.* **162**, 307–316.
(30) Krone, W. J. A., Koningstein, G., de Graaf, F. K. and Oudega, B. (1985). Plasmid-determined cloacin DF13-susceptibility in *Enterobacter cloacae* and *Klebsiella edwardsii*, identification of the cloacin DF13/aerobactin outer membrane receptor proteins. *Antonie van Leeuwenhoek* **51**, 203–218.
(31) Nassif, X. and Sansonetti, P. J. (1986). Correlation of the virulence of *Klebsiella pneumoniae* K1 and K2 with the presence of a plasmid encoding aerobactin. *Infect. Immun.* **54**, 603–608.
(32) Stuart, S. J., Prpic, J. K. and Robins-Browne, R. M. (1986). Production of aerobactin by some species of the genus *Yersinia*. *J. Bacteriol.* **166**, 1131–1133.
(33) Ford, S., Cooper, R. A. and Williams, P. H. (1986). Biochemical genetics of aerobactin biosynthesis in *Escherichia coli*. *FEMS Microbiol. Lett.* **36**, 281–285.
(34) de Lorenzo, V. and Neilands, J. B. (1986). Characterization of *iucA* and *iucC* genes of the aerobactin system of plasmid ColV-K30 in *Escherichia coli*. *J. Bacteriol.* **167**, 350–355.
(35) Appanna, V. D. and Viswanatha, T. (1986). Effects of some substrate analogs on aerobactin synthetase from *Aerobacter aerogenes* 62-1. *FEBS Letters* **202**, 107–110.
(36) Braun, V. and Burkhardt, R. (1982). Regulation of the ColV plasmid-determined iron(III)-aerobactin transport system in *Escherichia coli*. *J. Bacteriol.* **152**, 223–231.
(37) Hantke, K. (1981). Regulation of ferric iron transport in *Escherichia coli* K12: isolation of a constitutive mutant. *Mol. Gen. Genet.* **182**, 288–292.
(38) Bagg, A. and Neilands, J. B. (1985). Mapping of a mutation affecting regulation of iron uptake systems in *Escherichia coli* K12. *J. Bacteriol.* **161**, 450–453.
(39) Schaffer, S., Hantke, K. and Braun, V. (1985). Nucleotide sequence of the iron regulatory gene *fur*. *Mol. Gen. Genet.* **200**, 110–113.
(40) Bagg, A. and Neilands, J. B. (1987). Ferric uptake regulation protein acts as a repressor, employing iron(II) as a cofactor to bind the operator of an iron transport operon in *Escherichia coli*. *Biochem.* **26**, 5471–5477.
(41) Roberts, M., Leavitt, R. W., Carbonetti, N. H., Ford, S., Cooper, R.A. and Williams, P. H. (1986). RNA-DNA hybridization analysis of transcription of the plasmid ColV-K30 aerobactin gene cluster. *J. Bacteriol.* **167**, 467–472.
(42) Bindereif, A. and Neilands, J. B. (1985). Promoter mapping and transcriptional regulation of the iron assimilation system of plasmid ColV-K30 in *Escherichia coli* K12. *J. Bacteriol.* **162**, 1039–1046.
(43) de Lorenzo, V., Wee, S., Herrero, M. and Neilands, J. B. (1987). Operator sequences of the aerobactin operon of plasmid ColV-K30 binding the ferric uptake regulation (*fur*) repressor. *J. Bacteriol.* **169**, 2624–2630.
(44) Coulton, J. W., Mason, P., Cameron, D. R., Carmel, G., Jean, R. and Rode, H. N. (1986).

Protein fusions of β-galactosidase to the ferrichrome-iron receptor of *Escherichia coli* K12. *J. Bacteriol.* **165**, 181–192.

(45) Lundrigan, M. D. and Kadner, R. J. (1986). Nucleotide sequence of the gene for the ferrienterochelin receptor FepA in *Escherichia coli*. *J. Biol. Chem.* **261**, 10797–10801.

(46) Calderwood, S. B., Auclair, F., Donohue-Rolfe, A., Keusch, G. T. and Mekalanos, J. J. (1987). Nucleotide sequence of the Shiga-like toxin genes of *Escherichia coli*. *Proc. Natl. Acad. Sci. USA* **84**, 4364–4368.

(47) Schaffer, S., Hantke, K. and Braun, V. (1985). Nucleotide sequence of the iron regulatory gene *fur*. *Mol. Gen. Genet.* **200**, 110–113.

(48) Williams, R. J. P. (1982). Free manganese(II) and iron(II) cations can act as intracellular cell controls. *FEBS Letters* **140**, 3–10.

(49) Williams, P. H. and Carbonetti, N. H. (1986). Iron, siderophores, and the pursuit of virulence: independence of the aerobactin and enterochelin iron uptake systems in *Escherichia coli*. *Infect. Immun.* **51**, 942–947.

Section V

Bacteria that Interact with Plants as Parasites or Symbionts

Members of a group of related genera in the α subdivision of the purple bacteria have evolved special relationships with higher plants in which plant tissue is induced to proliferate to give rise to nodules or benign tumours which become the habitat for specialized derivatives of the bacteria or some of their genes. One of the recent successes of bacterial genetics has been the understanding of the roles of genes carried by *Agrobacterium* species or rhizobia (including members of the new genera *Rhizobium*, *Bradyrhizobium* and *Azorhizobium*) in the genetic transformation of plant tissue in crown gall or hairy root disease on the one hand, or in root nodule formation and nitrogen fixation on the other (Chapters 18 and 19). While bacterial chromosomal genes undoubtedly play a role in determining the outcome of the interaction between the bacterium and its host plant (those for exopolysaccharide production by rhizobia being examples), remarkably, the genes that determine the formation of galls or root nodules, as well as the specialized biochemical functions manifested in them—opine and plant hormone production, or nitrogen fixation, respectively—all form parts of large plasmids whose possession by a relatively unspecialized Gram-negative rod converts it into a unique plant parasite or symbiont.

The molecular understanding of crown gall was greatly simplified by the circumscribed nature of the disease. Other bacterial diseases of plants approach in complexity those of animals described in Section IV and are correspondingly difficult to understand. In vertebrates, specific responses of the immune system to infection may often help to pinpoint particular attributes of the pathogen that are likely to be relevant to the development of disease. The absence of an immune system from plants removes this possibility (although evidence is growing that plants do possess other kinds of specific defence system that may be triggered in response to infection). The best approach may then be to isolate bacterial mutations that affect pathogenicity, without prior assumptions about mechanisms, as has proved fruitful in the study of *Xanthomonas campestris* (Chapter 17).

Chapter 17

Pathogenicity of **Xanthomonas** *and Related Bacteria Towards Plants*

M. J. DANIELS

I. INTRODUCTION

The principal groups of plant pathogens are viruses, fungi and bacteria. Bacterial pathogens generally flourish in warm, moist climates and consequently they are much less important for agriculture in the temperate regions than in the tropics. Perhaps for this reason, a relatively small proportion of the literature of plant pathology is concerned with bacteria. However those investigators whose interests lie in the area of fundamental rather than

Table I
The principal groups of bacterial plant pathogens

Genus	Types of disease
Gram-negative:	
Agrobacterium	Hypertrophy (galls)
Erwinia	Soft rot, wilting, necrosis
Pseudomonas	Leaf spot, hypertrophy, soft rot, wilting
Xanthomonas	Leaf spot, necrosis
Gram-positive:	
Clavibacter	Canker, wilting
Curtobacterium	Wilting
Rhodobacterium	Hypertrophy

applied aspects of pathology have regarded bacteria as ideal "model" pathogens. This has been particularly true in recent years as molecular genetics has been applied to analysing pathogenicity.

The principal groups of pathogenic bacteria fall into a small number of genera (Table I). *Erwinia* spp. are closely related to *Escherichia coli* and have the advantage that much of the genetic methodology of *E. coli* can be applied to them with little modification. *Curtobacter*, *Clavibacter* and *Rhodobacterium* (formerly classified with the genus *Corynebacterium*) are the only Gram-positive genera; members of this group are more difficult to investigate and there have been few attempts to study them genetically. The remaining three genera, *Agrobacterium* (the subject of Chapter 18), *Pseudomonas* and *Xanthomonas* are members of the family *Pseudomonadaceae* and have been studied extensively. The examples given in this chapter will be drawn mostly from research on *Xanthomonas campestris*, the subject of investigations in the author's laboratory, but much of what follows applies to other xanthomonads and pseudomonads. A discussion of the taxonomy of these organisms is inappropriate here, but certain points of nomenclature are important because they describe the specificity of the pathogens for their hosts. Members of the genus *Xanthomonas*, which consists exclusively of plant pathogens, are yellow-pigmented, motile, aerobic, Gram-negative rods. Five species are currently recognized, of which one, *X. campestris*, is divided into about 120 pathovars, defined as groups distinguishable from one another with certainty only by plant host range. Typically a given pathovar can infect only a small number of closely-related plant species. Pathovars are often further divided into races which show specificity for certain genotypes or cultivars of a single plant species. *Pseudomonas syringae* also provides examples of subspecific division into pathovars and races.

X. campestris pathovar *campestris* (hereafter *X. c. campestris*) causes black rot of crucifers (1), one of the most serious diseases of brassica crops worldwide (Fig. 1). Infection of a crop usually begins with contaminated seed. As the seedling

Figure 1. Black rot symptoms produced by *X.c. campestris* about two weeks after inoculation via small wounds at the leaf margins. The bacteria migrate inwards through the vascular system and as the populations increase at each point the characteristic progression of symptoms of chlorosis (chlorophyll destruction), vein blackening (deposition of melanin-like pigments) and rotting is seen. The photograph also shows the effect of humidity on disease symptoms; the leaf on the left was maintained at low ambient humidity while that on the right was covered with a plastic bag to give high humidity.

emerges and grows the bacteria colonize the surface of the plant. Plant pathogenic bacteria (in contrast to fungi) do not possess mechanisms for breaching the outer protective layers of plants, and penetration depends on natural openings or wounds, produced for example by storm or frost damage. In the case of *X. c. campestris*, the epiphytic bacteria gain access to the internal tissues via hydathodes at the leaf margins (structures which function in water exudation or guttation). They then migrate through the vascular system and cause the characteristic black rot lesions. Not all pathogens are capable of systemic invasion or rotting of tissue. Many cause only localized tissue necrosis, which may nonetheless cause significant loss of yield through destruction of photosynthetic tissue, or render fruits unattractive and unsaleable. Others may enter the vascular system and interfere with water movement to cause vascular wilt symptoms. Multiplication invariably takes place in the intercellular spaces of plants (Fig. 2).

Pathogenicity in the wide sense is a complex phenomenon. In addition to the obvious manifestations of damage to the plant, pathogenic bacteria have to be able to gain access and grow on and in plant tissues (non-pathogens are generally unable to do this), and they have to be able to resist the host's chemical defence mechanisms. Little is known of the biochemical basis of these properties. In view of the power of genetic and molecular biological approaches for unravelling complex biological processes it is not surprising that many laboratories have recently begun to study the molecular biology of host-pathogen interactions. It is worth noting that the plant pathologist enjoys certain advantages over his colleagues who study human and animal pathogens, in that he is able to infect and sacrifice unlimited numbers of clonal host organisms without being restricted by ethical or economic factors. Since many aspects of genetic research involve handling large numbers of individuals, this can be an important factor.

II. STRATEGIES AND TECHNIQUES FOR STUDYING THE GENETICS OF PATHOGENICITY

A. Genetic Methodology

Genetic manipulations with Gram-negative bacterial plant pathogens are straightforward (2). The organisms are easily cultured in simple, defined

Figure 2. *Xanthomonas* growing in intercellular spaces in leaves. (a) Scanning electron micrograph showing bacteria (B) between plant cells (P). Scale bar = 5 μm. (b) Thin section though similar material; (W) indicates plant cell walls. Note that the bacteria seem to be surrounded by poorly-staining matrix material, perhaps polysaccharide. Scale bar = 1 μm. Micrographs by courtesy of Dr I. N. Roberts.

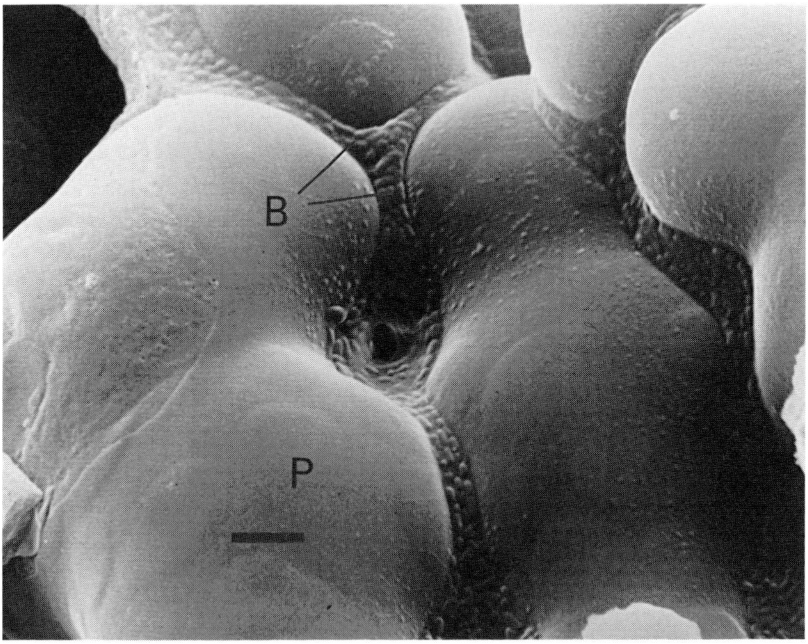

a

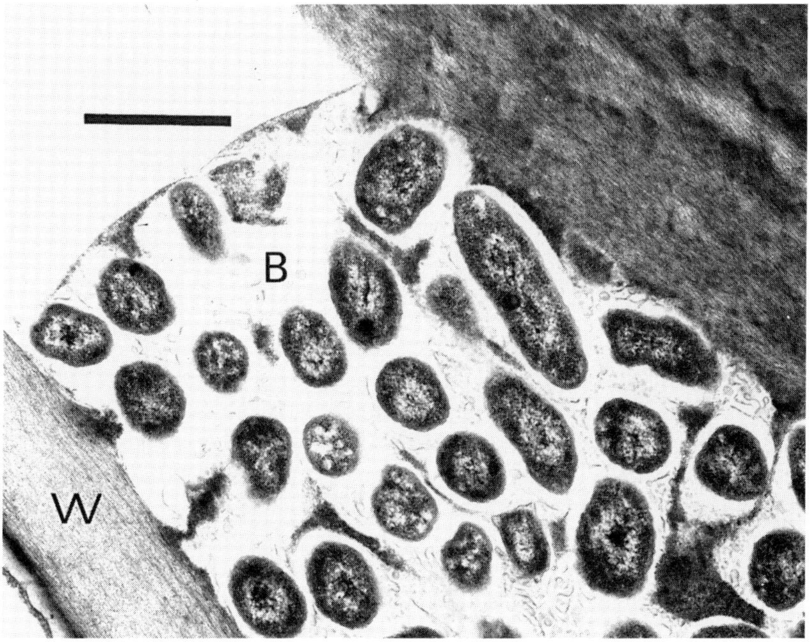

b

laboratory media. Mutations may be induced by chemicals, radiation, or by transposon insertions. Because of the advantages of transposon-induced mutations for subsequent analysis, much effort has been devoted to adapting techniques for various species (3). The basic procedure is to introduce the transposon into the target cell via a vector (usually a plasmid) which can transfer but not replicate, so that any colonies of the target bacterium which express the antibiotic resistance marker of the transposon should have arisen by transposition to the chromosome or another indigenous replicon. The most commonly used transposon is Tn*5*. Experience indicates that different pathogen species, or even different strains of a single species, vary widely in the ease with which Tn*5* insertion mutants can be isolated. *X. campestris* seems to be particularly refractory, probably because the frequency of Tn*5* transposition is about one hundredfold lower than in other bacteria (4). However, a derivative of Tn*1721* has recently been used successfully to mutagenize the organism (5).

Gene mapping may be accomplished by conjugation mediated by the broad-host-range R-factor R68:45 (6). Linkage between markers has been demonstrated in some *P. syringae* pathovars (2) and *X. c. campestris* (P. C. Turner and M. J. Daniels, unpublished). However the exploitation of the technique has been limited because of the current preference for gene cloning methods for characterizing the genome.

Molecular genetic procedures for plant pathogens are similar to those discussed elsewhere in this volume. The use of broad-host-range cloning vectors has been an important feature of recent work. These vectors are derived from plasmids of incompatibility groups P, Q and W and may be transferred by conjugation to virtually all Gram-negative bacteria in which they can replicate. High and low copy-number vectors, including cosmids, promoter-probes and expression vectors have been developed, so that a wide range of experimental manipulations are feasible.

B. Testing Pathogenicity

The design of suitable assays for pathogenicity is one of the greatest challenges in the study of host-pathogen interaction. Although it is desirable that the procedure adopted should mimic closely the natural disease observed in the field, it must be simple, rapid to perform and reliable, because one of the features of genetic experimentation is the screening of large numbers of individuals to detect altered phenotypes.

As noted above, the characteristic progression of symptoms caused by *X. c. campestris* results from entry into intercellular spaces and the vascular system of leaves through hydathodes. This may be simulated by spraying suspensions of bacteria on to plants which have been maintained under warm humid conditions to promote water exudation, i.e. guttation. Thereafter the plants are moved to a drier environment so that the guttation droplets with the bacteria inside are sucked back into the leaf. An alternative, less "natural", technique involves dipping the tip of a leaf into a bacterial suspension and then clipping the leaf margin with scissors to allow the bacteria to gain access to the interior (Fig. 1). Neither of these techniques is suitable for mass screening of thousands of colonies which might be necessary to isolate mutants. Simpler methods can be used to infiltrate leaves with bacterial suspensions from syringes, either with a fine needle to inject into thicker tissue adjacent to the midrib or major veins, or by forcing the suspension through stomata by placing the end of the syringe (without a needle) firmly against the lower leaf surface, supported by a finger on the other side, and applying gentle pressure to the plunger. Perhaps the quickest method is to touch a sharp sterile needle on to a bacterial colony and then to stab it into a seedling (7), which permits up to 1000 colonies to be screened per day. It must be appreciated that the simplified pathogenicity tests by-pass some of the earlier stages of natural infection, so that in using them one is in effect studying only part of the overall process. A new method for *X. c. campestris* combines simplicity with similarity to natural infection. Seeds are soaked in bacterial suspensions and planted out, and as they germinate and grow bacteria colonize the surfaces and eventually infect through natural openings and kill the seedlings (M. J. Daniels, unpublished).

Selection of the most appropriate inoculation technique usually depends on trial and error, because different plants (and even different varieties of a single species) may vary widely in their suitability for each method.

C. Cloning of Genes Encoding Known Suspected Pathogenicity Factors

Physiological studies have led to the identification of numerous factors produced by bacteria which are *a priori* likely to be pathogenicity determinants, including enzymes, phytotoxins, plant growth regulators and extracellular polysaccharides (EPS). In many cases the role, if any, of these substances in the disease process is unclear, and since genetics offers the possibility of demonstrating unequivocally the requirement for a particular factor, there has been much interest in cloning genes which encode or are involved in the production of the various factors.

The starting point for gene cloning exercises is usually a library of wild-type DNA cloned in a broad-host-range cosmid vector, such as pLAFR1 (8). Synthesis of a certain factor directed by cloned DNA can often be detected by simple plate tests. For example, protease production can be detected on agar containing skimmed milk, on which positive colonies are surrounded by a clear zone (9). Phytotoxins produced by bacteria are frequently active against a wide range of target organisms, including bacteria, and toxin-producing colonies may then be detected by inhibition of growth of test bacteria in an agar overlay above the primary colonies (10). EPS production causes colonies on appropriate media to assume a slimy appearance. It is usually a simple matter to detect colonies which produce either more or less than the wild-type. *P. syringae* pv. *savastanoi* produces indoleacetic acid from tryptophan, causing hypertrophy of adjacent plant cells. Mutants defective in production are resistant to α-methyltryptophan and the use of this and other analogues gives a useful tool for investigating the genetics of the pathway (11).

Cosmid libraries are usually maintained in *E. coli* strains. In favourable cases, the genes of interest may be expressed in *E. coli* and production of the factor detected simply by plating the library on indicator media. This simple procedure can usually be employed for *Erwinia* libraries (12), and a protease gene of *X. c. campestris* has been cloned in a similar manner (9). However, most *Xanthomonas* and *Pseudomonas* genes are not expressed in *E. coli*. To clone such genes, the library must first be transferred into a suitable host, either a mutant of the pathogen or a related organism which does not produce the factor. For example *X. c. campestris* genes involved in polysaccharide synthesis were cloned by virtue of the ability of a cosmid to restore EPS production to a non-slimy mutant (13). The related cereal pathogen *X. campestris* pv. *translucens* does not produce cellulase (an endoglucanase which breaks down carboxymethyl-cellulose) or polygalacturonate lyase, and transfer of *X. c. campestris* libraries into *X. c. translucens* permits the identification of clones carrying genes encoding these enzymes (unpublished). Transfer of libraries into the desired recipients by conjugation can be done either *en masse* with pooled library cultures (14), or through large numbers (perhaps hundreds) of matings, each with an individual clone (15, 16). Although the latter procedure seems cumbersome, it may be necessary because individual clones may show widely differing frequencies of transfer, so that the transconjugant pool arising from transfer *en masse* may be a highly biased sample of the library.

D. Cloning of Genes Encoding Unknown Pathogenicity Factors

Unfortunately the number of potential pathogenicity factors which are known and for which simple techniques such as those described in the previous section can be applied is small. These factors are restricted to those that cause visible damage to plants, but pathogenicity in the widest sense probably involves a large number of factors. In general, mutation of genes encoding or involved in the synthesis of the factors will have some effect on the host-pathogen interaction. One method of identifying and cloning such genes is therefore to produce a set of mutants which are in some way altered in pathogenicity, but without obvious defects in growth *in vitro* (such as auxotrophy). This is achieved by the laborious testing of large numbers of colonies of mutagenized bacteria on plants, using pathogenicity tests such as those described in Section B. In the case of *X. c. campestris* mutagenized with nitrosoguanidine or ethylmethanesulphonate, pathogenicity mutants arose at a frequency of about one-tenth the frequency of auxotrophs (7). Mutants selected using a simplified seedling pathogenicity test gave a range of phenotypes when tested subsequently in mature turnip leaves.

Genes defined by such mutations can be cloned by transfer of a wild-type DNA library into each mutant in turn, and then screening individual transconjugants for restoration of function (14). Alternatively, if the mutation is caused by insertion of a transposon, flanking genomic sequences can be excised and cloned from the mutants and used as hybridization probes to locate corresponding clones in a DNA library (17).

E. Genes That Show Differential Expression Following Entry into the Host

A novel strategy has recently been devised to detect pathogenicity genes which does not depend either on prior knowledge of the nature or function of the gene products or on pathogenicity testing (18). It was supposed that a subset of pathogenicity genes will show differential expression when the bacteria enter plant tissues. Such genes must exist; for example polygalacturonate lyase is induced by plant cell wall degradation products, but is not synthesized appreciably *ex planta* in the absence of the inducers. However, intercellular fluids in plants have a composition different from laboratory bacteriological media, so that the nutritional shift between the two media could be expected to cause many changes of expression of "housekeeping" genes, thus giving perhaps a high "noise level" in the search for genes required for pathogenicity. The technique makes use of turnip seedlings treated with chloramphenicol, which is taken up through the roots and accumulates in plant tissues to bacteriostatic concentrations, so that wild-type bacteria are unable to grow and produce disease in the treated plants. A library of *X. c. campestris* DNA fragments was constructed in a broad-host-range promoter probe plasmid with chloramphenicol acetyltransferase as the reporter gene. This library was transferred into the wild-type *X. c. campestris* and individual transconjugants were tested for pathogenicity on chloramphenicol-treated seedlings. Only those clones that contained a promoter functional in the plant environment were chloramphenicol-resistant and gave disease, and when the strains were tested for antibiotic resistance on laboratory media, a proportion were found to show differential expression. The promoter fragments were then used as physical probes to isolate surrounding regions of the genome from a cosmid library.

F. Genes Controlling Host Specificity

Most plant pathogens are able to cause disease only in a tiny fraction of the plant species with which they may come into contact. The basis of host specificity has been a major preoccupation of plant pathologists, and genetic methods have been used for many years to study the phenomenon in some fungal pathogens. Strong supporting evidence for some concepts derived from this work has been obtained by molecular genetic studies of certain *Pseudomonas* and *Xanthomonas* strains. In these cases, strains of a particular pathovar can be classified into races distinguishable by their differential reaction in tester cultivars of the plant host. DNA libraries of one bacterial race are prepared and transferred into another, and the transconjugants are then screened on the tester cultivars for alteration of host specificity (15, 16). Genes cloned in this fashion probably affect specificity only at the level of pathogen race-host cultivar level; little is yet known about determinants of specificity at other levels, e.g. pathovar-plant genus and species.

III. FUNCTION OF SOME PATHOGENICITY GENES

A. Extracellular Enzyme Production

Not all bacterial pathogens produce extracellular enzymes capable of degrading plant cell constituents. Most work has been carried out with soft-rotting *Erwinia* species and more recently with *X. c. campestris*, which can cause spreading, rotting symptoms under favourable conditions. However, many pathogens which cause localized lesions do not appear to produce significant amounts of enzymes such as pectinases (including pectin methylesterase, pectate lyase (polygalacturonate lyase) and polygalacturonase, of which the first two are produced by *X. c. campestris*), proteases and cellulases. Many plant pathogenic fungi produce hydrolases able to break down a wide range of plant cell wall polymers, but bacteria seem to be more limited in their capabilities. Until recently the evaluation of the role of these enzymes in pathogenicity was difficult because mutants defective in production were not available. In the case of pectinases, which have received most attention, attempts to make mutants were frustrated because multiple isozymes encoded by different genes are produced. Genes encoding the enzymes are cloned by plating DNA libraries maintained in hosts capable of expressing the genes on indicator media and screening for colonies that produce the enzyme (Section II.C). While protease is detected on agar containing skimmed milk, positive colonies being surrounded by a clear zone (9), pectinase- and cellulase-positive colonies are surrounded by clear zones on polygalacturonate- or carboxymethylcellulose-containing agar subsequently developed with cetyltrimethylammonium bromide or congo red, respectively. In *E. chrysanthemi* cloning has revealed the presence of five polygalacturonate lyase genes (*pelA–E*). By manipulation of the genes *in vitro* and subsequent introduction of the mutant alleles into the wild-type organism the role of each isozyme in pathogenicity can be studied (19). In *X. c. campestris*, three or four forms of polygalacturonate lyase can be separated chromatographically (20), and the structural genes for two isozymes have been cloned but not yet studied in detail.

The gene(s) encoding *X. c. campestris* extracellular protease are unusual because the enzyme is produced in *E. coli* harbouring the cloned DNA. The DNA was mutagenized with Tn*5* and those clones with transposon insertions that abolished protease production were transferred into wild-type *X. c. campestris*. Cells in which marker exchange of the transposon by recombination from the cloned DNA into the genome has taken place can be readily isolated by introducing a plasmid which is incompatible with the cloning vector while maintaining selection for the transposon (21). This simple technique makes it possible to produce a set of transposon insertions in a common genetic background, and the positions of the mutations can be accurately mapped by restriction analysis of the cloned DNA. Marker exchange of mutations that abolish protease production directed by a single cloned DNA fragment yield *X. c. campestris* strains with no detectable extracellular protease, suggesting that a single small region of the genome carries the structural gene or genes. Surprisingly, the protease-deficient strains appear to retain full pathogenicity when introduced into turnip seedlings and mature leaves, and the bacteria grow in seedlings at the same rate and to the same population levels as wild type strains (9). The role of protease in pathogenicity (if any) is therefore unclear.

Similar approaches have been used to study the cellulase (endoglucanase) of *X. c. campestris* (22). This enzyme is the major extracellular protein produced by the organism. As with the protease gene, marker exchange of the Tn*5* insertions from the cloned DNA into the wild-type genome generates strains with no detectable enzyme activity which nevertheless retain pathogenicity (C. L. Gough, personal communication). Similar results have been obtained with *P. solanacearum* (C. Boucher, personal communication). The apparent lack of a requirement for two plant cell-degrading enzymes in pathogenicity is surprising. However it must be remembered that the common methods used for testing pathogenicity differ from natural infection processes (Section II.B) and the observations of gross symptoms or bacterial growth *in planta* (9) are, at best, incomplete indicators of the normal host-pathogen interaction. More detailed studies may reveal that protease and cellulase are involved in the process.

Certain classes of *X. c. campestris* mutants selected on the basis of reduced pathogenicity to seedlings (7) show pleiotropic defects in extracellular enzyme levels. The mutants have been complemented with cosmid clones from a wild-type DNA library (14), and analysis of the DNA has revealed some novel physiological mechanisms. Mutations in one region of the genome, cloned in the cosmid pIJ3020 (14), lead to loss of the ability to synthesize all known extracellular enzymes that have been tested (amylase, cellulase, polygalacturonate lyase and protease). This region of the genome appears to encode interacting positive regulators of enzyme synthesis, and there is also some evidence for an unlinked negative regulator. These genes may be involved in a complex "global" regulatory network superimposed on local controls. Synthesis of protease is known to be induced by polypeptides, and polygalacturonate lyase by pectin degradation products, but these inducers each control only one enzyme (or family of isozymes). There is currently much interest in multicomponent genetic regulatory systems in bacteria (23) and it will be interesting to determine whether the *X. c. campestris* genes have properties similar to those found in other plant-associated bacteria such as *Agrobacterium* and *Rhizobium*.

A further class of mutant with defects in extracellular enzyme production can be complemented with the cosmid pIJ3000. Mutational analysis of the cloned DNA revealed the presence of two linked clusters of genes together spanning approximately 11 kb of DNA (21) (Fig. 3). All Tn*5* insertions in this region give a phenotype in which amylase, cellulase, polygalacturonate lyase and protease are synthesized at near wild-type levels, but fail to be exported into the extracellular medium. The enzymes accumulate in the periplasmic space (20). Part of the putative export gene cluster has been sequenced and has revealed the presence of three open reading frames, two of which have the

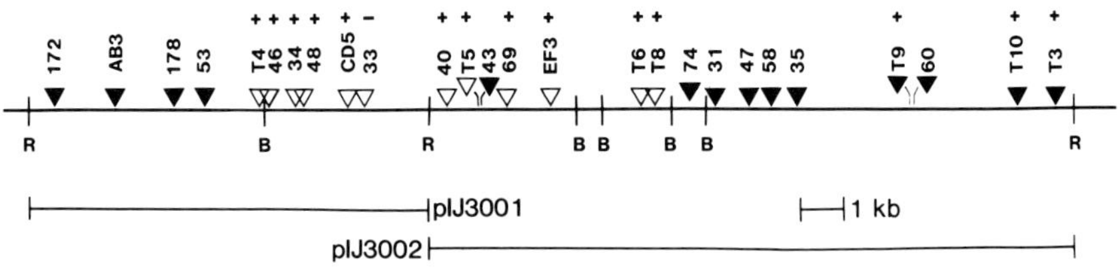

Figure 3. An example of mutational analysis of cloned *X.c. campestris* DNA which reveals the presence of a cluster of linked genes required for pathogenicity. A cosmid clone, pIJ3000, was identified in a genomic library of wild-type DNA by its ability to complement a nonpathogenic mutant. A large number of independent Tn*5* insertions in the cloned DNA were isolated and mapped. Only one insertion (number 33, indicated with a minus sign) abolished the complementation, suggesting that the gene defined by the original mutation lies in this position. However, when all the Tn*5* insertions were transferred by marker exchange to the *X.c. campestris* genome, all those indicated by open triangles, spanning a distance of approx. 11 kb, gave nonpathogenic mutants (21). This gene cluster is required for enzyme export (20). B and R represent sites for *Bam*H1 and *Eco*R1 respectively, and pIJ3001 and pIJ3002 are subclones which have been used in the analysis of the gene cluster.

characteristics of membrane proteins (F. Dums, personal communication). The periplasmic enzymes that accumulate in the export-deficient mutants have molecular weights indistinguishable from those of the extracellular forms exported by wild-type cells, suggesting that cleavage of a portion of the molecule is not involved in passage through the outer membrane.

To summarize, many genes are involved in the production of extracellular enzymes. Structural genes are regulated by both specific and global systems. Presumably secretion genes (24) are required to transfer the proteins across the inner bacterial membrane, and a further set of genes is needed for onward transport into the medium. Although protease and cellulase seem to be relatively unimportant for pathogenicity, polygalacturonate lyase is likely to be critical for symptom development, and hence mutations in the regulatory and export genes which affect this enzyme will abolish pathogenicity.

B. Some General Properties of Pathogenicity Genes

(1) Mutant phenotypes. The problems of testing for pathogenicity have been discussed above (Section II.B). The isolation of mutants altered in pathogenicity by direct screening on plants necessarily involves testing large numbers of individual clones, so that simple inoculation procedures are essential. Unfortunately this may create a model which differs significantly from the natural host-pathogen system and therefore not all pathogenicity genes are likely to be detected in "mutant hunts". Subject to these limitations, mutational analysis has been useful in revealing a range of genes required for pathogenicity. Apart from *X. c. campestris* (7) mutants have been isolated in *P. solanacearum* (25) and *P. syringae* pvs. *phaseolicola*, *pisi* and *syringae* (26–29). Our experience with *X. c. campestris* is typical. While some non-pathogenic mutants show alterations in production of suspected pathogenicity determinants such as extracellular enzymes, the majority show wild-type levels. A range of macroscopic phenotypes can be discerned: some mutants produce no visible response, others produce rapid localized necrosis similar to a hypersensitive response to alien pathogens (30), and others give partial symptoms which spread less rapidly than the wild-type. In many cases, mutants are unable to grow significantly in inoculated plants, but in other cases populations *in planta* may increase as rapidly and to similar levels as the wild-type. Thus growth of pathogens *in planta* is necessary but not sufficient for pathogenicity or, equivalently, pathogenicity can be dissociated from parasitism.

A characteristic of many plant pathogenic pseudomonads and xanthomonads is that when inoculated into a dicotyledonous plant that is not a compatible host, they induce a rapid hypersensitive resistance response (HR, 30). The phenomenon has been studied most extensively in tobacco challenged with *P. syringae* strains, but the general features are common to many other systems. Induction of HR requires a period of about four hours, during which time the pathogen must be metabolically active. Thereafter it can be inactivated without affecting the subsequent course of events. Collapse of plant cells in the inoculated area is detectable after about 8 h, and by 24 h the area is desiccated, resembling parchment. However the effect is strictly confined to the tissue originally inoculated. Inhibitors and polymeric materials produced during HR prevent bacteria spreading and affecting other tissues. A proportion of *P. syringae* and *P. solanacearum* mutants selected as being non-pathogenic to their homologous host plant have been found to be unable to induce HR on heterologous hosts (25–27), and conversely mutants selected primarily as HR-negative are non-pathogenic. These results suggest that both the compatible (disease) and incompatible (HR) interactions depend on some common mechanism(s), the biochemical nature of which remains obscure.

(2) Organization of pathogenicity genes. The clustering of genes encoding pathways is a common feature of prokaryotic genome organization, and there is evidence from several pathogens of clustering of pathogenicity-related genes. Clusters in *X. c. campestris* required for control and export of extracellular enzymes (Fig. 3) have been mentioned above (Section III.A), and in addition there is some clustering of genes required for extracellular polysaccharide biosynthesis (13, 31, 32). Gene clusters of unknown function required for pathogenicity have been found in *P. syringae* pvs. *syringae* and *phaseolicola* and *P. solanacearum*; in the latter organism, the genes in question are carried on a megaplasmid, estimated to have a size of about 1500 kb (33). No general rules have yet emerged about the location of pathogenicity genes on plasmids. In some strains of *P. syringae* pv. *savastanoi* genes for auxin biosynthesis are plasmid-borne, whereas in other strains they are apparently chromosomal (34). Mills and colleagues (35) have shown that *P. syringae* strains harbour plasmids which can integrate into and dissociate from the chromosome, taking pieces of flanking chromosomal DNA of variable size with them, so that there is considerable scope for genomic rearrangements in these organisms. The strain of *X. c. campestris* from which our genetic stocks are derived appears to lack indigenous plasmids and we assume that the genes that we have been studying are chromosomally located. A clear demonstration of this would require classical genetic mapping studies.

It is difficult to estimate what fraction of the genome is required solely for pathogenicity. In *X. c. campestris*, *P. syringae* pv. *phaseolicola* and *P. solanacearum* the proportion of mutants altered only in pathogenicity is about 10% of the proportion that are auxotrophic. If it is assumed that mutagenesis is random, this might imply a number of genes between 10 and 100. Drawing together data from cloning of pathogenicity genes indicates that at least 30 genes have already been identified.

C. Avirulence Genes and Host Specificity

Interest in the genetics of host-pathogen specificity owes much to the need to breed disease-resistant crop plants. Current concepts are derived principally from experiments with fungal pathogens, because until recently it was not possible to analyse plant-pathogenic bacteria genetically. An important contribution to the development of the subject was made by Flor (36) who recognized the need to study in parallel the genetics of both the pathogen and the host. Flor's research was concerned with the interaction of flax (*Linum usitatissimum*) and the rust fungus *Melampsora lini*. At that time *M. lini* could not be cultured *in vitro* (it behaved as an obligate parasite), and consequently the only disease-related genes that could be observed to segregate were those determining ability to colonize and cause disease in particular tester plant lines, i.e. specificity genes. It was found that for every gene in the pathogen determining virulence or avirulence there was a corresponding gene in the host determining resistance or susceptibility. In the host, resistance was conferred by the dominant allele, whereas in the pathogen virulence was recessive, leading to the concept of avirulence genes. It is supposed that the products of the resistance and avirulence genes interact to trigger plant defence mechanisms that restrict pathogen growth. There is genetic evidence for analogous mechanisms in many other plant-fungal systems.

Although there is no direct genetic evidence for avirulence genes in bacteria, several exist as a number of races that interact specifically with plant cultivars in a manner reminiscent of fungi. To test the hypothesis that specificity is determined by avirulence genes, genomic libraries have been transferred into other races, and transconjugants assayed for their reaction on differential tester cultivars. Staskawicz *et al.* (15) found a cosmid clone in a library of race 6 of *P. syringae* pv. *glycinea* which determined the inability to cause disease on certain soybean cultivars, and when the clone, carrying a gene designated *avrA*, was transferred to other races they became unable to cause disease on these cultivars. A strain of *X. c. glycines* normally pathogenic to all soybean lines acquired the specificity of race 6 of *P. s. glycinea* when *avrA* was introduced. Similar avirulence genes have been cloned from other *P. s. glycinea* races (37), and also from *X. c. malvacearum* (16) and *X. c. vesicatoria* (38) (respective hosts: cotton and pepper). The xanthomonads may have some advantages for future work because the genetic basis of resistance of cotton and pepper is better understood than that of soybean. Although some *avr* genes have been extensively studied and sequenced (39), their functions in interacting with plant genes are not understood. Nevertheless the cloning of these genes is an important step forward in plant pathology. Table II summarizes the interaction of an *X. c. vesicatoria avr* gene with a pepper resistance gene.

Table II
Interactions between *Xanthomonas campestris* pv. *vesicatoria* and pepper determined by a bacterial avirulence gene and a plant resistance gene

X. c. vesicatoria strain	Pepper cultivar ECW	Pepper cultivar ECW10R
Pepper race 1	C	C
Pepper race 2	C	HR
Pepper race 1 carrying cloned avirulence gene *avrBs*$_1$ from race 2	C	HR

The near isogenic cultivars ECW and ECW10R (which carries the Bs$_1$ gene for resistance to bacterial spot) can be distinguished by differential susceptibility to pepper race 2 strains of *X. c. vesicatoria*. Both are susceptible to race 1 strains. Introduction of a cloned *avrBs*$_1$ gene from a race 2 strain into race 1 changes the reaction from C (compatible, i.e. spreading disease symptoms) to HR (localized hypersensitive resistance).

avr genes determine specificity at the level of races of a particular pathogen and cultivars of a single plant. It is not known what mechanisms determine wider classes of specificity, e.g. between pathovars and plant genera and species.

IV. CONCLUDING REMARKS

Genetic studies of bacterial pathogens have been pursued for a comparatively short time, but already a wealth of information about pathogenicity mechanisms is beginning to emerge, justifying the faith of those who believe that genetic approaches are unrivalled for analysing complex biological phenomena. Although this chapter has concentrated almost exclusively on aspects of the genetics of pathogens, it must be borne in mind that disease involves specific interactions between at least two organisms (the host and the pathogen), or perhaps more if vectors and competing microbes are considered. Plant hosts probably respond to or resist infection in a number of different ways, and many genes are presumably involved. It may be helpful to consider disease in terms of a programmed developmental pathway of the host-pathogen complex. The complex passes through a number of states, each characterized by specific patterns of metabolic activity and gene expression in host and pathogen. Events occurring in one state will result in the system being driven into the next, and disruption of the normal pathway by mutation of appropriate genes may generate unpredictable interaction phenotypes. This line of argument emphasizes the need to conduct parallel studies of the several partners of the interaction if full understanding is to be achieved. We are approaching the time when precise genetic manipulation of crop plants will become feasible, and disease resistance will be an important property to manipulate. If the potentialities are to be realized, we have to fill in the many gaps in our understanding of the fundamental biological features of plant disease.

References

(1) Williams, P. H. (1980). Black rot: a continuing threat to world crucifers. *Plant Disease* **64**, 736–742.

(2) Chatterjee, A. K. and Vidaver, A. K. (1986). Genetics of pathogenicity factors: application of phytopathogenic bacteria. *Adv. Plant Pathol.* **4**, 1–218.

(3) Mills, D. (1985). Transposon mutagenesis and its potential for studying virulence genes in plant pathogens. *Annu. Rev. Phytopathol.* **23**, 297–320.

(4) Turner, P., Barber, C. and Daniels, M. (1984). Behaviour of the transposons Tn*5* and Tn*7* in *Xanthomonas campestris* pathovar *campestris*. *Mol. Gen. Genet.* **195**, 101–107.

(5) Shaw, J. J., Settles, L. G. and Kado, C. I. (1988). Transposon Tn*4431* mutagenesis of *Xanthomonas campestris* pv. *campestris*: characterization of a nonpathogenic mutant and cloning of a locus for pathogenicity. *Molec. Plant-Microbe Interactions* **1**, 39–45.

(6) Haas, D. and Holloway, B. W. (1976). R-factor variants with enhanced sex factor activity in *Pseudomonas aeruginosa*. *Mol. Gen. Genet.* **144**, 243–251.

(7) Daniels, M. J., Barber, C. E., Turner, P. C., Cleary, W. G. and Sawczyc, M. K. (1984). Isolation of mutants of *Xanthomonas campestris* pv. *campestris* showing altered pathogenicity. *J. Gen. Microbiol.* **130**, 2447–2455.

(8) Friedman, A. M., Long, S. R., Brown, S. E., Buikema, W. J. and Ausubel, F. M. (1982). Construction of a broad-host-range cosmid cloning vector and its use in the genetic analysis of *Rhizobium* mutants. *Gene* **18**, 289–296.

(9) Tang, J-L., Gough, C., Barber, C. E., Dow, J. M. and Daniels, M. J. (1987). Molecular cloning of protease genes from *Xanthomonas campestris* pv. *campestris*: expression in *Escherichia coli* and role in pathogenicity. *Mol. Gen. Genet.* **210**, 443–448.

(10) Staskawicz, B. J. and Panapoulos, N. J. (1979). A rapid and sensitive microbiological assay for phaseolotoxin. *Phytopathology* **69**, 663–666.

(11) Smidt, M. and Kosuge, T. (1979). The role of indole-3-acetic acid accumulation by α-methyl tryptophan resistant mutants of *Pseudomonas savastanoi* in gall formation on oleanders. *Physiol. Plant Pathol.* **13**, 203–214.

(12) Keen, N. T., Dahlbeck, D., Staskawicz, B. and Belser, W. (1984). Molecular cloning of pectate lyase genes from *Erwinia chrysanthemi* and their expression in *Escherichia coli*. *J. Bacteriol.* **159**, 825–831.

(13) Barrère, G. C., Barber, C. E. and Daniels, M. J. (1986). Molecular cloning of genes involved in the biosynthesis of the extracellular polysaccharide xanthan by *Xanthomonas campestris* pv. *campestris*. *Int. J. Biol. Macromol.* **8**, 372–374.

(14) Daniels, M. J., Barber, C. E., Turner, P. C., Sawczyc, M. K., Byrde, R. J. W. and Fielding, A. H. (1984). Cloning of genes involved in pathogenicity of *Xanthomonas campestris* pv. *campestris* using the broad-host-range cosmid pLAFR1. *EMBO J.* **3**, 3323–3328.

(15) Staskawicz, B. J., Dahlbeck, D. and Keen, N. T. (1984). Cloned avirulence gene of *Pseudomonas syringae* pv. *glycinea* determines race-specific incompatibility on *Glycine max* (L.) Merr. *Proc. Natl. Acad. Sci. USA* **81**, 6024–6028.

(16) Gabriel, D. W., Burges, A. and Lazo, G. R. (1986). Gene-for-gene interactions of five cloned

avirulence genes from *Xanthomonas campestris* pv. *malvacearum* with specific resistance genes in cotton. *Proc. Natl. Acad. Sci. USA* **83**, 6415–6419.

(17) Niepold, F., Anderson, D. and Mills, D. (1985). Cloning determinants of pathogenesis from *Pseudomonas syringae* pathovar *syringae*. *Proc. Natl. Acad. Sci. USA* **82**, 406–410.

(18) Osbourn, A. E., Barber, C. E. and Daniels, M. J. (1987). Identification of plant-induced genes of the bacterial pathogen *Xanthomonas campestris* pathovar. *campestris* using a promoter-probe plasmid. *EMBO J.* **6**, 23–28.

(19) Reid, J. L. and Collmer, A. (1988). Construction and characterization of an *Erwinia chrysanthemi* mutant with directed deletions in all of the pectate lyase structural genes. *Molec. Plant-Microbe Interactions* **1**, 32–38.

(20) Dow, J. M., Scofield, G., Trafford, K., Turner, P. C. and Daniels, M. J. (1987). A gene cluster in *Xanthomonas campestris* pv. *campestris* required for pathogenicity controls the excretion of polygalacturonate lyase and other enzymes. *Physiol. Mol. Plant Pathol.* **31**, 261–271.

(21) Turner, P. C., Barber, C. E. and Daniels, M. J. (1985). Evidence for clustered pathogenicity genes in *Xanthomonas campestris* pv. *campestris*. *Mol. Gen. Genet.* **199**, 338–343.

(22) Gough, C. L., Dow, J. M., Barber, C. E. and Daniels, M. J. (1988). Molecular cloning of a cellulase (endoglucanase) gene from *Xanthomonas campestris* pv. *campestris* and study of its role in pathogenicity. (Submitted.)

(23) Ronson, C. W., Nixon, B. T. and Ausubel, F. M. (1987). Conserved domains in bacterial regulatory proteins that respond to environmental stimuli. *Cell* **49**, 579–581.

(24) Pugsley, A. P. and Schwartz, M. (1985). Export and secretion of proteins by bacteria. *FEMS Microbiol. Revs.* **42**, 3–38.

(25) Boucher, C., Barberis, P. A., Trigalet, A. P. and Demery, D. A. (1985). Transposon mutagenesis of *Pseudomonas solanacearum*: isolation of Tn*5*-induced avirulent mutants. *J. Gen. Microbiol.* **131**, 2449–2457.

(26) Anderson, D. M. and Mills, D. (1985). The use of transposon mutagenesis in the isolation of nutritional and virulence mutants in two pathovars of *Pseudomonas syringae*. *Phytopathology* **75**, 104–108.

(27) Lindgren, P. B., Peet, R. C. and Panopoulos, N. J. (1986). Gene cluster of *Pseudomonas syringae* pv. "*phaseolicola*" controls pathogenicity of bean plants and hypersensitivity on non-host plants. *J. Bacteriol.* **168**, 512–522.

(28) Harper, S., Zewdie, N., Brown, I. R. and Mansfield, J. W. (1987). Histological, physiological and genetical studies of the responses of leaves and pods of *Phaseolus vulgaris* to three races of *Pseudomonas syringae* pv. *phaseolicola* and to *Pseudomonas syringae* pv. *coronafaciens*. *Physiol. Mol. Plant Pathol.* **31**, 153–172.

(29) Malik, A. N., Vivian, A. and Taylor, J. D. (1987). Isolation and partial characterization of three classes of mutant in *Pseudomonas syringae* pathovar *pisi* with altered behaviour towards their host, *Pisum sativum*. *J. Gen. Microbiol.* **133**, 2393–2399.

(30) Klement, Z. (1982). Hypersensitivity. *In* "Phytopathogenic Prokaryotes", Vol. 2 (M. S. Mount and G. H. Lacy, eds.), pp. 149–177. Academic Press, New York.

(31) Harding, N. E., Cleary, J. M., Cabanas, D. K., Rosen, I. G. and Kang, K. S. (1987). Genetic and physical analyses of a cluster of genes essential for xanthan gum biosynthesis in *Xanthomonas campestris*. *J. Bacteriol.* **169**, 2854–2861.

(32) Thorne, L., Tansey, L. and Pollock, T. J. (1987). Clustering of mutations blocking synthesis of xanthan gum by *Xanthomonas campestris*. *J. Bacteriol.* **169**, 3593–3600.

(33) Boucher, C., Martinel, A., Barberis, P., Alloing, G. and Zischek, C. (1986). Virulence genes are carried by a megaplasmid of the plant pathogen *Pseudomonas solanacearum*. *Mol. Gen. Genet.* **205**, 270–275.

(34) Comai, L., Surico, G. and Kosuge, T. (1982). Relation of plasmid DNA to indoleacetic acid

production in different strains of *Pseudomonas syringae* pv. *savastanoi*. *J. Gen. Microbiol.* **128**, 2157–2163.

(35) Szabo, L. J. and Mills, D. (1984). Characterization of eight excision plasmids of *Pseudomonas syringae* pv. *phaseolicola*. *Mol. Gen. Genet.* **195**, 90–95.

(36) Flor, H. H. (1955). Host-parasite interaction in flux rust—its genetics and other implications. *Phytopathology* **45**, 680–685.

(37) Staskawicz, B., Dahlbeck, D., Keen, N. and Napoli, C. (1987). Molecular characterization of cloned avirulence genes from race 0 and race 1 of *Pseudomonas syringae* pv. *glycinea*. *J. Bacteriol.* **169**, 5789–5794.

(38) Swanson, J., Kearney, B., Dahlbeck, D. and Staskawicz, B. (1988). Cloned avirulence gene of *Xanthomonas campestris* pv. *vesicatoria* complements spontaneous race-change mutants. *Molec. Plant-Microbe Interactions* **1**, 5–9.

(39) Napoli, C. and Staskawicz, B. (1987). Molecular characterization and nucleic acid sequence of an avirulence gene from race 6 of *Pseudomonas syringae* pv. *glycinea*. *J. Bacteriol.* **169**, 572–578.

Chapter **18**

Tumorigenicity of Agrobacterium *on Plants*

P. J. J. HOOYKAAS

I. INTRODUCTION

A. Tumour Induction on Plants by Agrobacteria

Agrobacteria are Gram-negative soil bacteria, which belong to the *Rhizobiaceae* (1). Bacteria belonging to this family interact with plants in various ways. Rhizobia induce nitrogen-fixing root nodules in leguminous plants (see Chapter 19), while agrobacteria are the causative agents of the neoplastic plant diseases crown gall and hairy root. The crown gall disease is

Figure 1. Crown gall tumours induced by *Agrobacterium tumefaciens* on *Kalanchoe*.

characterized by tumour formation at infected wound sites on the root crown (hence the name crown gall), stem or leaves of plants (Fig. 1). An abundant proliferation of roots from infection sites is typical of the hairy root disease (Fig. 2). The taxonomy of agrobacteria is based on the phytopathogenic character of the bacteria: crown gall inducers belong to the species *Agrobacterium tumefaciens*, those forming hairy roots to the species *Agrobacterium rhizogenes*, while avirulent wild-type strains are called *Agrobacterium radiobacter* (1).

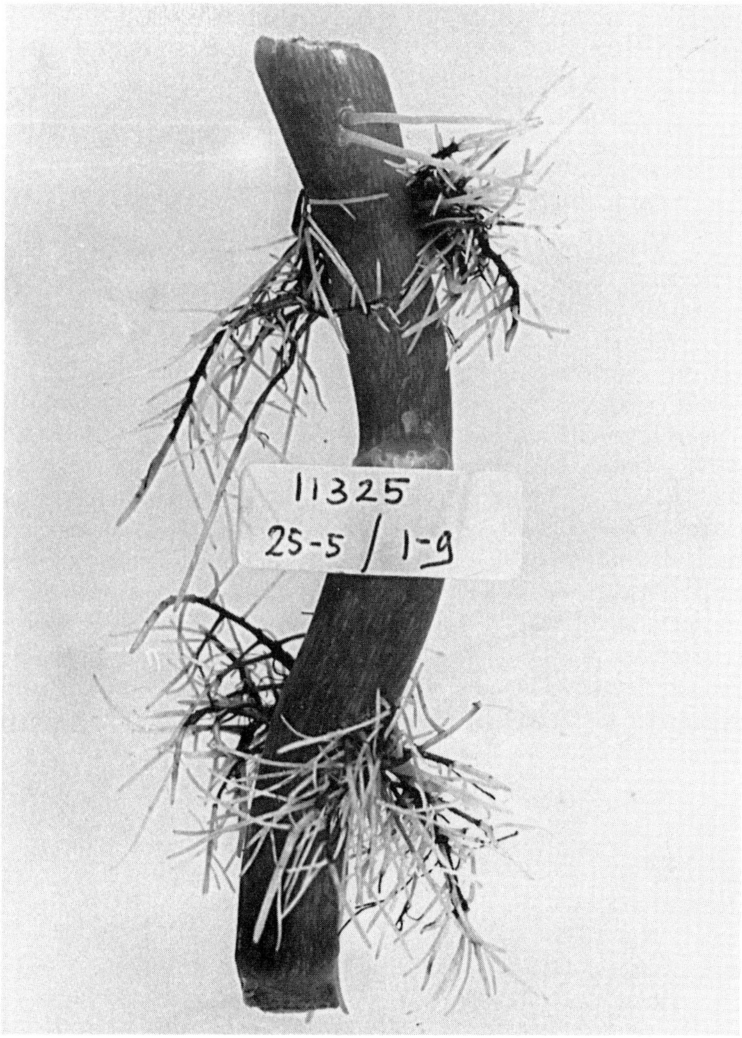

Figure 2. Hairy roots induced by *Agrobacterium rhizogenes* on *Kalanchoe.*

B. Novel Properties of Crown Gall and Hairy Root Cells

Crown gall tissues and hairy roots can be cultured *in vitro.* They share two unusual properties in which they differ from normal plant tissues. The first is that they can proliferate in the absence of phytohormones (auxins and cytokinins), whereas normal plant tissues need the addition of one or both

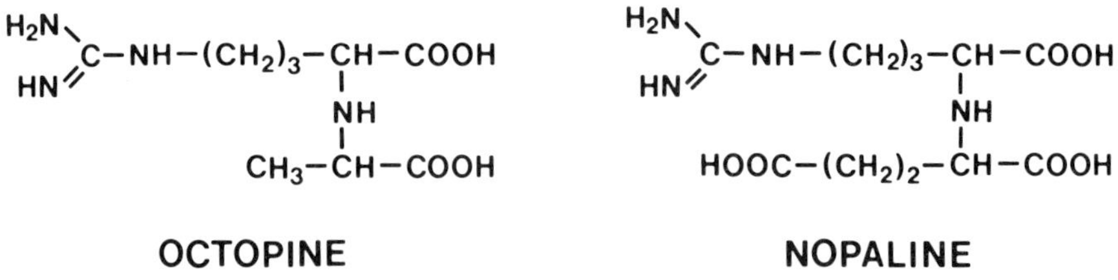

Figure 3. Structural formulae of the opines octopine and nopaline, which are derivatives of arginine linked to pyruvate and α-ketoglutarate, respectively.

classes of phytohormones for growth (2). *In vitro* in the absence of phytohormones crown gall cells grow into calli—amorphous masses of cells, which are somewhat comparable to tumours—whereas hairy root cells differentiate into roots just as they do *in vivo* (3). Besides their ability to proliferate without phytohormones crown gall tissues and hairy roots share a second unique property: they produce and excrete opines (4). Opines are compounds (mostly amino acids linked to small sugars such as pyruvate or α-ketoglutarate) which are formed via enzymes present in crown gall and hairy root cells but not in normal plant cells (Fig. 3). There are several different types of opines. The types of opines produced in the crown galls or hairy roots depend on the particular *Agrobacterium* strain responsible for the infection (5). Thus octopine, nopaline, leucinopine and succinamopine strains of *A. tumefaciens* and agropine and mannopine strains of *A. rhizogenes* are distinguishable.

II. Ti AND Ri PLASMIDS

A. Role of Plasmids in Tumour Induction

More than a decade ago, it was found that certain virulent *A. tumefaciens* strains could be converted into avirulent forms by heat treatment (incubation at 37°C instead of 29°C which is the temperature at which these bacteria are usually grown). In these strains phytopathogenicity is apparently an unstable trait. Interestingly, the avirulent derivatives could regain virulence during mixed infections with virulent strains *in planta* (6). These experimental data suggested the possibility that phytopathogenicity was determined by a transmissible extrachromosomal element. Procedures were soon developed for the detection of such extrachromosomal elements in *Agrobacterium* strains. It turned out that virulent strains did indeed carry a large (about 200 kb) extrachromosomal element (a plasmid), absent from the avirulent derivatives. Avirulent strains which had become virulent in mixed infections with virulent bacteria *in planta* turned out to have received a large plasmid from the virulent bacteria (7). This shows unequivocally that a large plasmid in *A. tumefaciens* contains virulence determinants and it was therefore called a Ti (tumour inducing) plasmid. Similarly, genes present on a large Ri (root inducing) plasmid in *A. rhizogenes* determine the ability of its host to induce the hairy root disease (8).

Since the taxonomy of agrobacteria was originally based on their phytopathogenicity (1), the discovery that this trait is determined by plasmid-borne genes threw doubt on the classification and it was found that quite different types of bacteria were grouped in each of the "species" *A. tumefaciens*, *A. rhizogenes* and *A. radiobacter* as originally defined. More recent classifications are therefore based on characteristics other than phytopathogenicity. Accordingly, agrobacteria can be divided into at least three clusters or biotypes, which might eventually be accepted as different *Agrobacterium* species (9).

B. Can any Bacterium become Tumorigenic after Contact with *Agrobacterium*?

Is it possible to convert any bacterium into a tumour-inducing one by introducing a Ti plasmid into it? Transfer of the Ti plasmid to related bacteria belonging to the *Rhizobiaceae* has been accomplished (10), but transfer to more distant species was impossible because of the inability of the Ti plasmid to replicate in these more distantly related hosts. Eventually, this problem was circumvented by using cointegrates between the Ti plasmid and *inc*P type wide host range plasmids such as RP4 and R772 (11). Root nodule-inducing bacteria (*Rhizobium trifolii*, *Rhizobium leguminosarum*) and leaf nodule inducers (*Phyllobacterium myrsinacearum*) gained tumorigenicity after receipt of such a Ti plasmid cointegrate (10; R. J. M. Van Veen and P. J. J. Hooykaas, unpublished). Other related bacteria such as *Rhizobium meliloti* became only very inefficient crown gall inducers, while more distantly related bacteria such as *Escherichia coli* and *Pseudomonas aeruginosa* remained completely avirulent. It can thus be concluded that the Ti plasmid can confer tumorigenicity only on a selected group of bacterial hosts. This might be explained by a lack of expression of the virulence genes in other bacteria or perhaps by a requirement for certain chromosomal virulence genes, which might be absent in most bacteria, or to the presence of gene products which interfere with the process of tumour induction in certain bacteria.

C. Chromosomal Virulence Genes

A number of chromosomal virulence (*chv*) genes have been identified in *Agrobacterium*. Some of these are essential for virulence on many host plants, while others are necessary for tumour induction only on a few plant species. The genes *chvA* and *chvB* are important for the ability of *Agrobacterium* to attach to plant cell walls, which is an initial and essential step in the tumour induction process (12). Mutants in *chvA* or *chvB* do not produce the exopolysaccharide β-1,2 glucan (G. Cangelosi, personal communication), but a direct role of this polysaccharide in attachment has not been established. The production of cellulose (β-1,4 glucan fibrils) is another trait determined by chromosomal genes (13). Although their production is not essential for virulence, these fibrils act to anchor the bacteria more firmly to the plant cells, which may enhance the efficiency of tumour induction in certain cases. Recently it was found that similar chromosomal virulence genes are also present in *Rhizobium*, and at least some (the counterparts of *chvA* and *chvB*) are essential for the formation of root nodules by *Rhizobium*. Apparently, attachment mediated by the *chvA* and *chvB* gene products is an essential and common step in the processes of tumour induction by *Agrobacterium* and root nodule initiation by *Rhizobium*.

III. T-DNA

A. Identification of *Agrobacterium* Genes Responsible for Oncogenicity and Opine Production

The molecular mechanism underlying plant tumour induction is now known in some detail. A key finding was that plant tumour cells (as well as hairy root cells) carry a (defined) piece of DNA originating from the Ti plasmid (14). This segment of DNA, the T-DNA, was apparently transferred to plant cells at the infection sites during tumour induction. Further research showed that three genes present on the T-DNA are responsible for the conversion of normal plant cells into tumour cells (15). These oncogenicity (*onc*) genes have been characterized in detail: they code for enzymes involved in the production of phytohormones. One of the *onc* genes, called *ipt*, *cyt* or *tmr*, codes for an isopentenyl transferase, an enzyme which catalyses the conversion of AMP into isopentenyl-AMP, a compound with cytokinin activity. The two other *onc* genes, which are called *aux*-1, *tms*-1 or *iaaM* and *aux*-2, *tms*-2 or *iaaH* respectively, code for a monooxygenase and a hydrolase, which together catalyse the conversion of tryptophan via indoleacetamide into indole acetic acid, a compound with auxin activity. From work with plant tissue cultures, it is known that initiation of growth (callus formation) often occurs only if both an auxin and a cytokinin are added to the medium. It is thus easy to understand why the introduction of T-DNA, determining the production of an auxin as well as a cytokinin, into plant cells leads to an unlimited proliferation of these transformed cells, resulting eventually in tumour formation on the

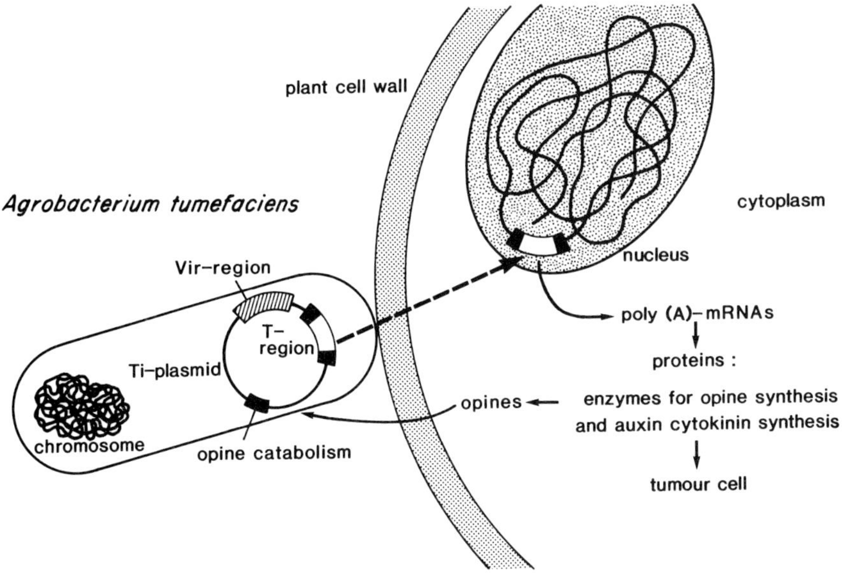

Figure 4. Schematic view of the process of plant tumour induction by *Agrobacterium tumefaciens*.

infected plant (Fig. 4). It also explains why crown gall tissues do not require the addition of phytohormones to the medium for *in vitro* growth.

The molecular basis of hairy root initiation is similar to that of crown gall formation. A (defined) segment of DNA from the Ri plasmid, the Ri T-DNA, is transferred to plant cells during the process of hairy root induction (14). The T-DNA of the Ri plasmid is responsible for the conversion of normal plant cells into hairy root cells. Several genes involved in this process have been identified. These genes are different from the three *onc* genes present in the T-DNA of the Ti plasmid, and their functions are still unknown.

The basis for the production of different types of opines depending on the particular Ti or Ri plasmid that was present in the *Agrobacterium* strain responsible for the infection has also been elucidated. The T-DNA carries, besides three *onc* genes, several other genes, some of which code for enzymes that catalyse the production of specific opines. Different *Agrobacterium* strains sometimes have a different set of these opine synthase genes in their T-DNA.

B. Why are T-DNA Genes Expressed in Plant Cells?

The expression signals of genes in prokaryotes and eukaryotes (plants) are quite different. One may wonder then why a set of bacterial genes (present in the T-DNA) is expressed in plant cells. The reason for this became clear when the 5′ and 3′ regulatory regions surrounding these T-DNA genes were studied in more detail. It was found that these genes contained expression signals that are typical of eukaryotic genes. In the 5′ region sequences such as TATA boxes, CAAT boxes and enhancer-like sequences and in the 3′ region AATAAA boxes were identified (16). Later, the importance of some of these consensus sequences for the expression of T-DNA genes in plant cells became clear from experiments in which specific mutants were used. For instance, the TATA boxes in the 5′ end of the *cyt* gene of the T-DNA were indeed found to be essential for expression. They determined the transcription start sites *in planta*. In the 3′ end of this same gene the AATAAA and GT-rich boxes, which are known to be involved in transcription termination and mRNA-polyadenylation in other eukaryotic genes, were found to be essential for proper gene expression in plants (17).

IV. GENES AND SEQUENCES NECESSARY FOR T-DNA TRANSFER

A. Virulence Genes on the Ti Plasmid

One of the continuing mysteries of the tumour induction process is how the T-DNA is transmitted from *Agrobacterium* to plant cells. Mutagenesis of the T-region of the Ti plasmid revealed that none of the genes in the T-region are involved in T-region transfer (18). When the *onc* genes were mutated, this resulted in non-oncogenic strains, but T-DNA transfer still occurred from such strains, as shown by the presence of markers (e.g. opines) in the infected tissues. The conclusion was therefore that the genes responsible for T-DNA transfer were located elsewhere in the *Agrobacterium* genome. Now we know from transposon mutagenesis that these genes are in fact clustered in a region distinct from but adjacent to the T-region on different Ti and Ri plasmids (Fig. 5). This region is about 40 kb in size and is called the Virulence (Vir) region. Complementation experiments showed that it embraces seven operons in the case of the octopine Ti plasmid which are called *virA*–*virG* (19, 20). From DNA sequence analysis it is now known that they contain one (*virA*, *virG*), two (*virC*, *virE*), four (*virD*) or eleven genes (*virB*). The number of genes present in the *virF* operon is not yet known. Some of these operons (*virA*, *virB*, *virD*, *virG*) are essential for tumour induction, as indicated by the avirulence of mutants on all plant species tested. Mutations in any of the other *vir* operons lead to strains that are either attenuated in virulence or have a restricted host range for tumour induction.

Other types of Ti and Ri plasmids have a Vir region that is similar to that of the octopine Ti plasmid, but subtle differences in their Vir regions may

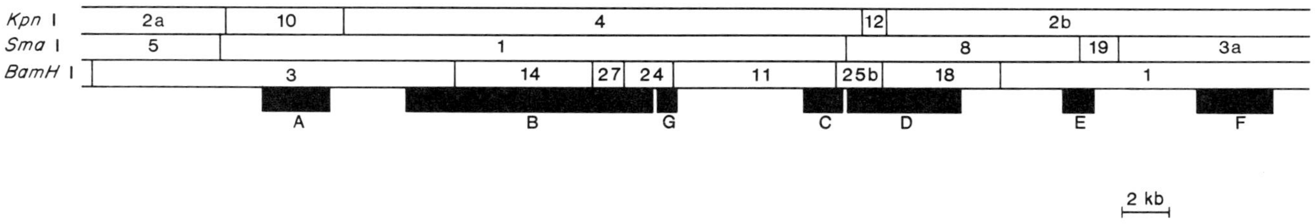

Figure 5. Restriction map of the virulence region of the octopine Ti plasmid. The location of the *vir* operons is shown.

underlie small differences in host range or T-DNA transfer efficiency between different strains. For instance the absence of the *virF* region in nopaline Ti strains may be the reason for their attenuated phenotype on plants such as *Nicotiniana glauca*. Conversely, the presence of an extra *vir* gene (*tzs*), which is responsible for *trans*-zeatin production and secretion in nopaline strains (as compared to octopine strains), may give nopaline strains enhanced virulence on certain plant species (21).

B. Regulation of Expression of the Virulence Genes

With the exception of the *virA* and *virG* genes, the *vir* operons are not transcribed during normal vegetative growth (22). Their expression becomes coordinately induced, however, when their host bacteria sense (wounded) plant tissues or plant cell exudates. Thus the *vir* genes together form a regulon. Apparently, plant cells contain and excrete compounds which can trigger *vir* expression. These compounds were recently purified from tobacco and shown to be plant phenolics such as acetosyringone and α-hydroxyacetosyringone (Fig. 6), which are accumulated in tobacco tissues especially after wounding (23). Interestingly, it has been known for many years that wounding is a prerequisite for tumour induction. Besides the two compounds mentioned above a number of other phenolics have been found to be inducers of the *vir* regulon. Examples are lignin precursors such as sinapinic acid, ferulic acid and coniferyl alcohol. Acetosyringone, however, is still the "strongest" *vir* inducer identified; *i.e.* it induces maximum levels of *vir* expression at the lowest concentration (5 μM).

Fusions between *vir* promoters and indicator genes such as those for β-galactosidase (19) and luciferase (24) have aided enormously studies on the regulation of *vir* gene expression in *Agrobacterium*. With such fusions, it was shown that induction of *vir* gene expression depends on the presence of intact

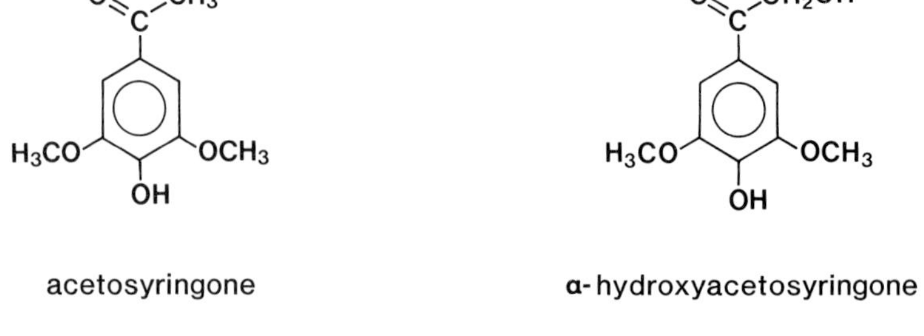

Figure 6. Structure of the *vir*-inducing compounds acetosyringone and α-hydroxy-acetosyringone.

virA and *virG* genes, which themselves have constitutive expression, although *virG* is induced to still higher levels of expression in the presence of plant phenolics (22). Apparently, the products of the *virA* and *virG* genes are involved in the regulation of expression of the other *vir* genes. DNA sequence analysis showed that *virA* codes for a transmembrane protein, while *virG* codes for a soluble protein. The VirA protein resembles proteins such as EnvZ and PhoR in *E. coli*, which are transmembrane sensor proteins; the VirG protein is strongly homologous to proteins such as OmpR and PhoB, which are positive regulators of transcription (25). A model was recently proposed for such two-factor regulatory systems (Fig. 7). In this model, the positive regulator of transcription is inactive until it becomes activated by a "sensor" protein. This "sensor" protein has two domains; the first domain can sense the presence of particular compounds in the medium (in the case of PhoR, phosphate) or conditions (in the case of EnvZ, the osmolarity of the medium), and the second domain has an enzymatic activity with which it can activate the accompanying activator of transcription (for PhoR, the PhoB protein; for EnvZ, the OmpR protein). The "sensor" protein is built in such a way that its catalytic domain becomes active only after its "sensor" domain meets the proper compounds or conditions. In the case of the *ntrB-ntrC* two-factor system the activator NtrC is activated via phosphorylation by the NtrB protein (25). Whether phosphorylation plays a role in any of the other two-factor systems mentioned is not known. From the homologies of the *Agrobacterium virA-virG* genes with the other two-factor regulatory systems, it may be inferred that the VirA protein is a transmembrane sensor protein for plant phenolics such as acetosyringone,

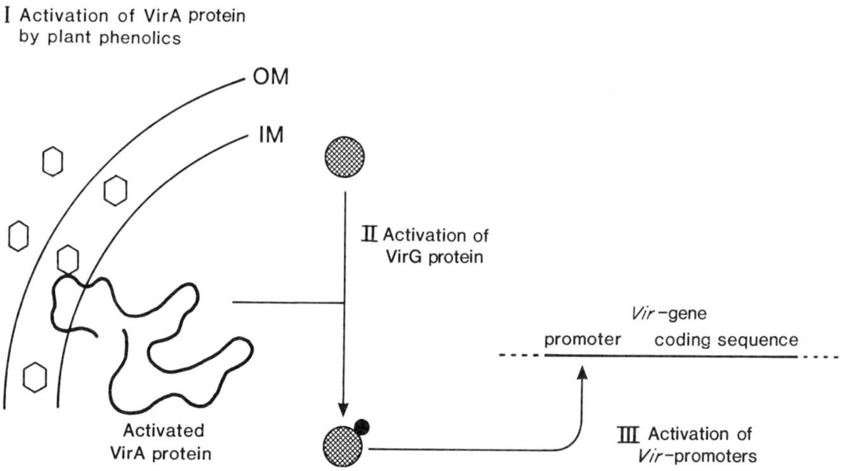

Figure 7. Model for the regulation of the *vir* genes via the regulatory proteins VirA and VirG.

which can activate the VirG activator protein (by phosphorylation?). An activated VirG protein may then act as a positive regulator of the *vir* promoters, leading to a coordinated expression of the *vir* genes in the presence of plant phenolics. Thus the *Agrobacterium* T-DNA transfer system is operative only if the bacterium is in the presence of (wounded) plants.

C. Role of Border Repeats in T-DNA Transfer

It is now clear that *vir* gene products are involved in the transfer of T-DNA to plant cells (see next Section). The question then arises why the T-DNA is transferred alone and not accompanied by any other segments of the Ti plasmid. Several years ago it was found that deletions removing the right end of the T-region led to avirulence of the host bacteria, even if none of the *onc* genes or *vir* genes had been affected (26). Sequence analysis showed that the T-region—as present in the Ti plasmid—is surrounded by an imperfect direct repeat of 24 bp (16). In the avirulent deletion mutants mentioned above the right repeat was missing. Later experiments showed that this right border repeat indeed plays an essential role in the T-DNA transfer process. It might be that the border repeats form the recognition signals for the T-DNA transfer apparatus. As a first experiment to test this idea, the T-region was cloned (B in Fig. 8) and then introduced into an *Agrobacterium* strain carrying a Ti plasmid that lacked the T-region, but still had an intact Vir region ("helper plasmid"; A, Fig. 8). It was found that the separation of T-region and Vir region on two different replicons did not affect virulence (27): strains with such a "binary" system were as virulent as wild-type *Agrobacterium* strains (Fig. 8). This was true even for strains with the T-region integrated in the chromosome of *Agrobacterium* (28). Next it was shown that transfer of any gene to plant cells can occur via the *Agrobacterium* virulence system, provided that this gene is surrounded by the 24 bp border repeats. This principle forms the rationale for the "binary vector system" for plant transformation. Genes of choice are cloned between the 24 bp border repeats in a plasmid that can replicate in *Agrobacterium* and preferably also in *E. coli* ("binary vector"). After transfer to an *Agrobacterium* strain with a helper (Vir-region) plasmid, the genes of choice can be transferred to plant cells as part of the artificial T-region present on the binary vector.

Further experiments have shown that a right border repeat alone, but not a left border repeat alone, is sufficient to allow transfer of the genes of choice to plant cells via the virulence system of *Agrobacterium*. There is, however, in fact no intrinsic difference between left and right border repeats: they are interchangeable without apparent effect on virulence. The explanation for all these findings is that the orientation of the border repeat towards the genes that are

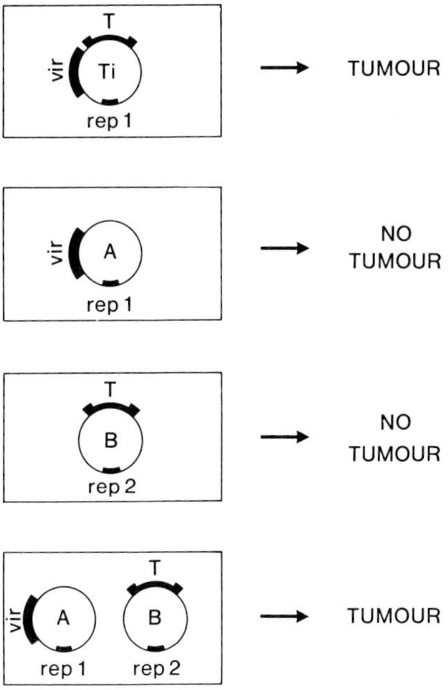

Figure 8. Principle underlying the binary vector system. T, T-DNA flanked by right and left border repeats; *vir*, virulence region; rep1, rep2, different origins of replication.

to be transferred is the significant factor (29). Apparently, T-DNA transfer is an orientated process, which starts at the right border repeat and terminates at the left border repeat. When the left border repeat is missing, termination probably occurs at sequences which resemble a border repeat ("pseudo border repeats"). However, when the right border repeat is missing, the process does not start at all (the left border repeat points in the wrong direction); therefore, strains lacking the right border repeat are avirulent (see also the next Section). Although left and right border repeats have intrinsically identical activity, there is a big difference in the (T-DNA transfer) activity of left and right border fragments. This is due to the presence of a 25 bp sequence called "overdrive" at 13–14 bp from the right border repeat (30). The left border region lacks such a sequence. In the absence of the overdrive sequence near the right border repeat, strains are strongly attenuated in virulence. Recent work has shown that the overdrive sequence is a T-region transfer enhancer, working equally well in all possible orientations and positions with respect to the right border repeat. Increasing the distance between the border repeat and the overdrive sequence from 13 bp up to more than 4300 bp did not affect the activity of this enhancer (31).

V. DIFFERENT STEPS IN THE PROCESS OF TUMOUR INDUCTION

Although the different steps in the process of plant tumour induction by *Agrobacterium* have not all been studied in detail, sufficient data are available to give the following model. The accumulation and multiplication of agrobacteria at wound sites on plants possibly represent the initial events in the process. A role for chemotaxis may be envisaged here. Concomitantly (or subsequently) the *vir* genes on the Ti plasmid may be induced by compounds such as acetosyringone, which are released from wounded plant tissues (23). The bacteria may attach to plant cell walls via their chromosomally encoded attachment system (*chv* genes), and form large aggregates of plant cells and bacteria via cellulose fibrils (13).

Recently it was found that acetosyringone-induced bacteria contain single-stranded DNA molecules representing one particular strand, the bottom strand, of the T-region (32). These molecules, which are called T-strands, are possibly the putative T-DNA intermediates which are transferred to the plant cells during tumour induction. The precise mechanism underlying T-strand production is unknown, but it has been found that nicks (single-stranded DNA breaks) are introduced at a specific position in the bottom strand of each of the 24 bp border repeats via two proteins encoded by the *virD* operon (33). These same VirD proteins are the only Vir proteins involved in T-strand production. Whether host proteins also play a role is not known. Although the mechanism of T-strand formation has not yet been studied, it can easily be envisaged that this occurs via displacement synthesis (Fig. 9). If DNA synthesis, which always proceeds in the $5' \rightarrow 3'$ direction, started at the nick site in the bottom strand of the right border repeat and moved in the direction of the left border repeat, this would eventually lead to the displacement of the bottom strand. Termination of DNA synthesis would then occur at the nick site in the bottom strand of the left border repeat. This model also explains the fact that the right border repeat is essential for tumorigenicity, whereas the left border repeat is not: in the absence of a right border repeat, T-strand formation would not occur at all, while in the absence of a left border repeat, T-strands would be formed terminating at an alternative site. The number of T-strands produced per bacterium is lower if the right border repeat is not accompanied by the overdrive sequence (31). The presence of this sequence apparently enhances virulence by stimulating T-strand production in the bacterium. If the T-region were indeed transferred to plant cells as a single-stranded DNA molecule, this would be identical to what happens during conjugation between Gram-negative bacteria (34). In this latter process, DNA is transmitted from the donor to the recipient in the form of a single-stranded molecule, which is formed in the donor via displacement synthesis, starting at a specific nick site called *oriT* (origin of transfer). Interestingly, indeed, it turns out that the 24 bp

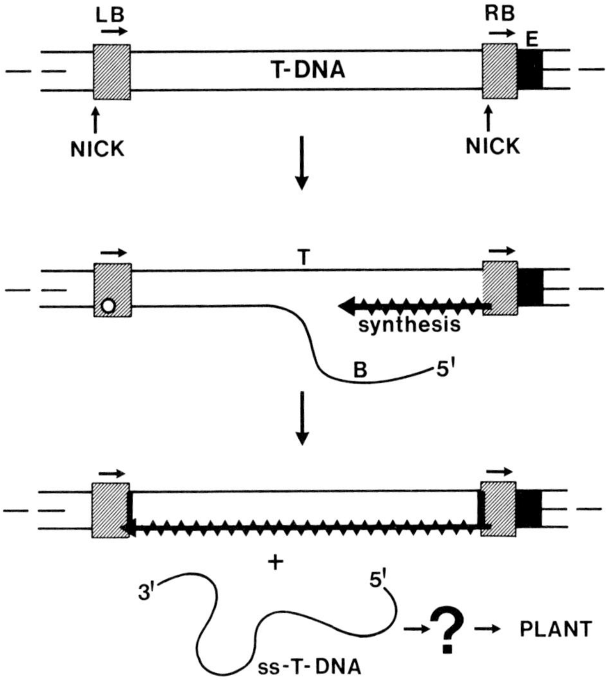

Figure 9. Model for the formation of T-strands in *Agrobacterium* by displacement synthesis. LB, RB left and right border repeats; E, overdrive sequence; T, top strand; B, bottom strand.

right border repeat can in fact be replaced functionally by the *oriT* sequence of an *incQ* plasmid, provided that the genes for the enzymes capable of recognizing and nicking at this *oriT* sequence are also present (35). This finding indicates that the processes of T-DNA transfer and bacterial conjugation may be closely related.

In bacterial conjugation, pili play an important and specific role by constituting a bridge between donor and recipient via which DNA transfer can occur (34). Attachment of *Agrobacterium* to plant cell walls is mediated via products determined by the *chv* genes. However, specific (pilus-like?) structures may well be necessary on the surface of *Agrobacterium* to allow T-DNA transfer to the plant cell. Such structures might be encoded by the *virB* operon, which is absolutely essential for virulence and, according to its DNA sequence, determines ten different proteins, many of which are strongly hydrophobic and, therefore, might be membrane proteins (42). So far, however, a Vir-region-determined pilus or other surface structure has not been described.

At the moment it is not known whether T-strands as such are transferred to the plant cell. It might be that T-strands are converted into another type of T-DNA intermediate before transfer. Anyway, after the T-DNA intermediates enter the plant cell, they must find their way to the nucleus and integrate there in any of the chromosomes. Although the T-DNA may sometimes become severely scrambled or deleted during the transfer-integration process, most often it is inserted into the plant genome in a rather preserved form. Small deletions are usually present at the ends. These are more prominent at the left than at the right ends, in these well preserved T-DNAs (36). Taken together, this suggests that the T-DNA is well protected—especially at its right end—against degradation during its transfer to the plant cell nucleus. Protection might be mediated by specific single-stranded DNA binding proteins, which might of course be determined by some of the *vir* genes. The integration of the T-DNA into the plant genome seems to occur essentially at random; preferred sites or sequences for integration have not been found. Whether integration is mediated by proteins determined by any of the *vir* genes is not known. However, from what is known about the integration of foreign DNA into animal and plant cells it seems likely that plant host proteins catalyse the integration of the T-DNA into the plant genome.

After integration, the T-DNA is expressed as a number of mRNAs, which are translated into a set of proteins reponsible for the novel properties (phytohormone-independent growth and opine synthesis) of the crown gall cells. Although the expression of each of the T-DNA genes is controlled by its own specific 5′ and 3′ regulatory sequences, the insertion site of the T-DNA may strongly influence the expression of some or all of the T-DNA genes. For instance, silencing of the expression of T-DNA genes via DNA methylation has been observed frequently (37).

VI. PROSPECTS FOR APPLICATIONS

The *Agrobacterium* system can now be exploited for the transfer of genes of choice to plant cells. The binary vector system is especially useful for this purpose, because genes can be cloned directly into such a vector in *E. coli* and thereafter transferred to an *Agrobacterium* helper strain carrying the *vir* and *chv* genes. Traits so far transmitted to plants via *Agrobacterium* include herbicide tolerance and resistance to viruses or insects. The range of plants into which *Agrobacterium* can introduce DNA is broader than originally thought. Initially it was believed that the *Agrobacterium* system could be exploited for the genetic engineering of dicotyledons, but not for that of monocotyledons because tumours are not induced on the latter. Recently, however, it was demonstrated that T-DNA transfer and integration can occur in both types of plants (38–40). The lack of tumours on monocotyledons is probably due to the fact that the auxin and cytokinin, which are produced via the T-DNA genes, cannot stimulate tumorous overgrowths in monocotyledons. Therefore, it is likely that the *Agrobacterium* system will become applicable to a vast number of crops in the future. On certain plant species, T-DNA transfer occurs only with a low frequency. For *Arabidopsis*, this is due to the fact that this plant species releases rather limiting amounts of the *vir*-inducing compounds. The addition of acetosyringone during the transformation process led to much enhanced transformation frequencies in this case (41).

Concluding, the situation is now such that the application of *Agrobacterium* for crop improvement is less limited by shortcomings in the vector system than by imperfections in the *in vitro* cell culture technology needed to regenerate the transformed cells into complete fertile plants. Progress in this latter area, however, is rapid and it is expected that in the near future, transgenic plants will be obtained from many crops.

References

(1) Kersters, K. and De Ley, J. (1984). Genus *Agrobacterium. In* "Bergey's Manual of Systematic Bacteriology" (N. R. Krieg, ed.), pp. 244–254. Williams and Wilkins, Baltimore, USA.

(2) Braun, A. C. (1958). A physiological basis for autonomous growth of crown gall tumor cell. *Proc. Natl. Acad. Sci. USA* **44**, 344–349.

(3) Tepfer, M. and Casse-Delbart, F. (1987). *Agrobacterium rhizogenes* as a vector for transforming higher plants. *Microbiol. Sci.* **4**, 24–28.

(4) Tempé, J. and Goldmann, A. (1982). Occurrence and biosynthesis of opines. *In* "Molecular Biology of Plant Tumors" (G. Kahl and J. Schell, eds.), pp. 427–449. Academic Press, New York.

(5) Bomhoff, G., Klapwijk, P. M., Kester, H. C. M., Schilperoort, R. A., Hernalsteens, J. P. and Schell, J. (1976). Octopine and nopaline synthesis and breakdown genetically controlled by a plasmid of *Agrobacterium tumefaciens. Mol. Gen. Genet.* **145**, 177–181.

(6) Kerr, A. (1969). Transfer of virulence between isolates of *Agrobacterium. Nature* **223**, 1175–1176.

(7) Van Larebeke, N., Genetello, C., Schell, J., Schilperoort, R. A., Hermans, A. K., Hernalsteens, J. P. and van Montagu, M. (1975). Acquisition of tumour-inducing ability by non-oncogenic agrobacteria as a result of plasmid transfer. *Nature* **255**, 742–743.

(8) White, F. F. and Nester, E. W. (1980). Hairy root: plasmid encodes virulence traits in *Agrobacterium rhizogenes. J. Bacteriol.* **141**, 1134–1141.

(9) Kerr, A. and Panagopoulos, C. G. (1977). Biotypes of *Agrobacterium radiobacter* var. *tumefaciens* and their biological control. *Phytopath. Z.* **90**, 172–179.

(10) Hooykaas, P. J. J. (1983). Plasmid genes essential for the interactions of agrobacteria and rhizobia with plant cells. *In* "Molecular Genetics of the Bacteria-Plant Interaction" (A. Pühler, ed.), pp. 229–239. Springer Verlag, Berlin.

(11) Hille, J., Hoekema, A., Hooykaas, P. and Schilperoort, R. (1984). Gene organization of the Ti plasmid. *In* "Genes Involved in Microbe-Plant Interactions" (D. P. S. Verma and T. Hohn, eds), pp. 287–309. Springer Verlag, Wien.

(12) Douglas, C. J., Staneloni, R. J., Rubin, R. A. and Nester, E. W. (1985). Identification and genetic analysis of an *Agrobacterium tumefaciens* chromosomal virulence region. *J. Bacteriol.* **161**, 850–860.

(13) Matthysse, A. G. (1987). Characterization of nonattaching mutants of *Agrobacterium tumefaciens*. *J. Bacteriol.* **169**, 313–323.
(14) Bevan, M. W. and Chilton, M.-D. (1982). T-DNA of the Agrobacterium Ti and Ri plasmids. *Ann. Rev. Genet.* **16**, 357–384.
(15) Yang, N.-Y. (1985). Agrobacterium T-DNA oncogenes encode plant growth regulators: are there cellular homologues in plants? *Trends Biotechnol.* **3**, 297–299.
(16) Barker, R. F., Idler, K. B., Thompson, D. V. and Kemp, J. D. (1983). Nucleotide sequence of the T-DNA region from *Agrobacterium tumefaciens* octopine Ti plasmid pTi15955. *Plant Mol. Biol.* **2**, 335–350.
(17) Memelink, J., de Pater, B. S., Hoge, J. H. C. and Schilperoort, R. A. (1987). T-DNA hormone biosynthetic genes; phytohormone and gene expression in plants. *Devel. Genet.* **8**, 321–337.
(18) Leemans, J., Deblaere, R., Willmitzer, L., de Greve, H., Hernalsteens, J. P., van Montagu, M. and Schell, J. (1982). Genetic identification of functions of TL-DNA transcripts in octopine crown galls. *EMBO J.* **1**, 147–152.
(19) Stachel, S. E. and Nester, E. W. (1986). The genetic and transcriptional organization of the *vir* region of the A6 Ti plasmid of *Agrobacterium tumefaciens*. *EMBO J.* **5**, 1445–1454.
(20) Hooykaas, P. J. J. and Schilperoort, R. A. (1984). The molecular genetics of crown gall tumorigenesis. *Advances in Genetics* **22**, 209–283.
(21) Melchers, L. S. and Hooykaas, P. J. J. (1987). Virulence of *Agrobacterium*. *In* "Oxford Surveys of Plant Molecular and Cell Biology", Vol. 4, (B. J. Miflin, ed.), pp. 167–220. Oxford University Press, Oxford.
(22) Stachel, S. E. and Zambryski, P. C. (1986). *vir*A and *vir*G control the plant-induced activation of the T-DNA transfer process of *Agrobacterium tumefaciens*. *Cell* **46**, 325–333.
(23) Stachel, S. E., Messens, E., Van Montagu, M. and Zambryski, P. (1985). Identification of the signal molecules produced by wounded plant cells that activate T-DNA transfer in *Agrobacterium tumefaciens*. *Nature* **318**, 624–629.
(24) Rogowsky, P., Close, T. J. and Kado, C. I. (1987). Dual regulation of virulence genes of *Agrobacterium* plasmid pTiC58. *In* "Molecular Genetics of Plant-Microbe Interactions" (D. P. S. Verma and N. Brisson, eds.), pp. 14–19. M. Nijhoff, Dordrecht.
(25) Ronson, C. W., Nixon, B. T. and Ausubel, F. M. (1987). Conserved domains in bacterial regulatory proteins that respond to environmental stimuli. *Cell* **49**, 579–581.
(26) Ooms, G., Hooykaas, P. J. J., Van Veen, R. J. M., Van Beelen, P., Regensburg-Tuink, A. J. G. and Schilperoort, R. A. (1982). Octopine Ti-plasmid deletion mutants of *Agrobacterium tumefaciens* with emphasis on the right side of the T-region. *Plasmid* **7**, 15–29.
(27) Hoekema, A., Hirsch, P. R., Hooykaas, P. J. J. and Schilperoort, R. A. (1983). A binary plant vector strategy based on separation of *vir* and T-region of the *Agrobacterium tumefaciens* Ti-plasmid. *Nature* **303**, 179–180.
(28) Hoekema, A., Roelvink, P. W., Hooykaas, P. J. J. and Schilperoort, R. A. (1984). Delivery of T-DNA from the *Agrobacterium tumefaciens* chromosome into plant cells. *EMBO J.* **3**, 2485–2490.
(29) Wang, K., Herrera-Estrella, L., Van Montagu, M. and Zambryski, P. (1984). Right 25 bp terminus sequence of the nopaline T-DNA is essential for and determines direction of DNA transfer from *Agrobacterium* to the plant genome. *Cell* **38**, 455–462.
(30) Peralta, E. G., Hellmiss, R. and Ream, W. (1986). *Overdrive*, a T-DNA transmission enhancer on the *Agrobacterium tumefaciens* tumour-inducing plasmid. *EMBO J.* **5**, 1137–1142.
(31) Van Haaren, M. J. J., Sedee, N. J. A., Schilperoort, R. A. and Hooykaas, P. J. J. (1987). Overdrive is a T-region enhancer which stimulates T-strand production in *Agrobacterium tumefaciens*. *Nucleic Acid Res.* **15**, 8983–8997.
(32) Stachel, S. E., Timmerman, B. and Zambryski, P. (1986). Generation of single-stranded

T-DNA molecules during the initial stages of T-DNA transfer from *Agrobacterium tumefaciens* to plant cells. *Nature* **322**, 706–712.

(33) Wang, K., Stachel, S. E., Timmerman, B., Van Montagu, M. and Zambryski, P. C. (1987). Site-specific nick in the T-DNA border sequence as a result of *Agrobacterium vir* gene expression. *Science* **235**, 587–591.

(34) Willetts, N. (1986). Plasmids. *In* "Genetics of Bacteria" (J. Scaife, D. Leach and A. Galizzi, eds.), pp. 165–195. Academic Press, London.

(35) Buchanan-Wollaston, V., Passiatore, J. E. and Cannon, F. (1987). The *mob* and *ori*T mobilization functions of a bacterial plasmid promote its transfer to plants. *Nature* **328**, 172–175.

(36) Jorgensen, R., Snyder, C. and Jones, J. D. G. (1987). T-DNA is organized predominantly in inverted repeat structures in plants transformed with *Agrobacterium tumefaciens* C58 derivatives. *Mol. Gen. Genet.* **207**, 471–477.

(37) Hepburn, A. G., Clarke, L. E., Pearson, L. and White, J. (1983). The role of cytosine methylation in the control of nopaline synthase gene expression in a plant tumour. *J. Mol. Appl. Genet.* **2**, 315–329.

(38) Hooykaas-Van Slogteren, G. M. S., Hooykaas, P. J. J. and Schilperoort, R. A. (1984). Expression of Ti plasmid genes in monocotyledonous plants infected with *Agrobacterium tumefaciens*. *Nature* **311**, 763–764.

(39) Grimsley, N., Hohn, B., Hohn, T. and Walden, R. (1986). "Agroinfection", an alternative route for viral infection of plants by using the Ti plasmid. *Proc. Natl. Acad. Sci. USA* **83**, 3283–3286.

(40) Schäfer, W., Gorz, A. and Kahl, G. (1987). T-DNA integration and expression in a monocot crop plant after induction of *Agrobacterium*. *Nature* **327**, 529–532.

(41) Scheikholeslam, S. N. and Weeks, D. P. (1987). Acetosyringone promotes high efficiency transformation of *Arabidopsis thaliana* explants by *Agrobacterium tumefaciens*. *Plant Mol. Biol.* **8**, 291–298.

(42) Thompson, D. V., Melchers, L. S. Idler, K. B., Schirperoort, R. A. and Hooykaas, P. J. J. (1988). Analysis of the complete nucleotide sequence of the *Agrobacterium tumefaciens virB* operon. *Nucl. Acids Res.* **16**, 4621–4636.

Chapter 19

The Symbiosis Between **Rhizobium** *and Legumes*

A. W. B. JOHNSTON

I. INTRODUCTION

When peas, clovers, soya beans or any other legume seedlings are grown in sterile conditions, there is little or nothing to distinguish their root systems from those of other dicotyledonous plants. Likewise, bacteria comprising the rhizobia have little to single them out morphologically or biochemically from various other Gram-negative soil bacteria. Yet, when these two very different organisms come together, there ensues an interaction of great complexity which culminates in the production of organs, the root nodules, in which large numbers of the bacteria are housed: in this exclusive home, they reduce atmospheric dinitrogen gas to ammonia, a source of nitrogen that the plant uses for subsequent synthesis of other nitrogenous compounds.

The reason why these two ostensibly undistinguished partners engage in this extraordinary and agronomically important interaction is that both contain genes that are expressed only in the presence of the other. For example, legumes make proteins, called nodulins, which are synthesized only in the nodule; the functions of some of these, such as the oxygen-carrying leghaemoglobin and some of the enzymes involved in assimilating ammonia, are known, but the roles of most of the nodulins are unknown. Turning to the bacterial partner, most strains of rhizobia fix nitrogen only in nodules because the nitrogen fixation (*nif*) genes are active at high levels only when the bacteria occupy the nodule. Genes involved in the earlier steps in the interaction are also expressed specifically when the bacteria are exposed to roots of the appropriate host legume.

This chapter will consider the contribution of genetic studies of these bacteria to the understanding of the interaction and will concentrate on the early stages of infection. Chapter 6 deals with genes involved in nitrogen fixation itself. First, however, a brief description of the infection process and also an excursion into the taxonomy of these bacteria are necessary.

A. The Infection Process

Normally, the first visible effect of inoculating rhizobia on legume roots is that young root hairs are induced to curl, twist or become otherwise deformed (Fig. 1a). At the root hair tips, the bacteria enter the plant via an "infection thread", a tube made by the plant which conducts the bacteria down the root hair cell (Fig. 1a, b). Ahead of the growing infection thread, plant cell division is initiated to form an incipient nodule and, as the nodule grows, the infection

Figure 1. Light microscope views of nodule development. Bars = 20 μm. (a) An infected root hair of bean (*Phaseolus vulgaris*). Rhizobia are conveyed from the point of infection (arrow) into the curled root hair by a branched infection thread (IT). Infection threads (seen predominantly in tranverse sections) have penetrated into the underlying cortical cells (arrowheads). The nucleus (N) of the root hair cell is indicated. (b) Release of rhizobia from the infection thread in a pea (*Pisum sativum*) nodule. The infected cells contain numerous undifferentiated *Rhizobium* (R) which are no longer encased by a plant cell wall. An infection thread can be seen traversing several cells at the upper right. (c) Nitrogen-fixing bacteroids in a mature pea nodule. The bacteroids have been stained for nitrogenase activity using immunogold labelling and silver enhancement and therefore appear black. (Photographs courtesy of K. A. Vanden Bosch.)

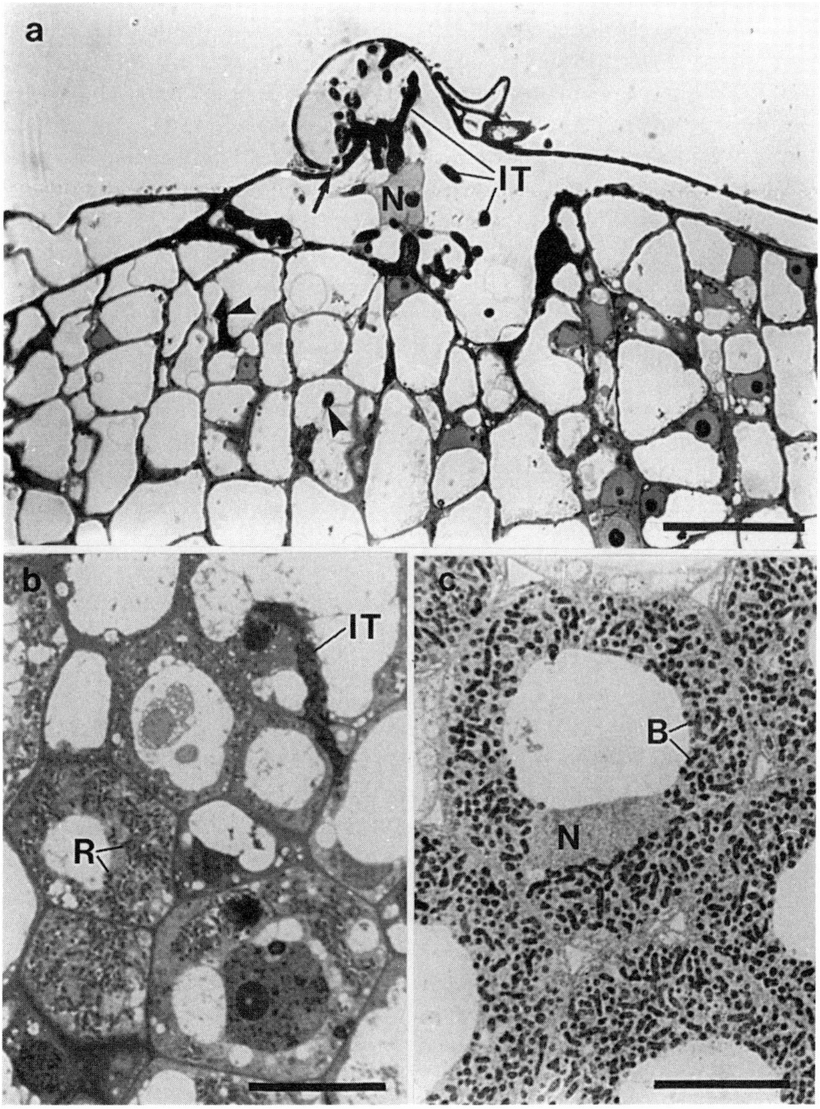

Figure 1.

thread branches and travels between and through the plant cells. The bacteria are then released from the tips of infection threads by a "pinching off" process in which they are surrounded by a plant membrane; these "bacteroid" forms (Fig. 1c) express the ability to fix nitrogen. What precise biochemical mechanisms govern the early steps in the infection process, and allow the bacteria to survive in and to be released from the infection threads and then to be transformed into the morphologically and biochemically distinct bacteroids? At present these questions are still unanswered, but bacterial genes that determine several of these steps have been identified; these studies should form the basis for future work that may elucidate the precise mechanisms determining the nature of the interaction.

B. Host Specificity and Taxonomy of Rhizobia

An important point about the interaction is its specificity, particular host legumes being nodulated only by certain strains or species of bacteria. In addition to the distinction between different strains based on host-range specificity itself are more basic differences between the bacteria, which are used to classify them into three genera: *Rhizobium*, which includes relatively fast-growing strains that tend to have narrow host-ranges in temperate legumes and do not fix nitrogen in free-living culture; *Bradyrhizobium*, whose members have less stringent host-ranges than *Rhizobium* species (some even nodulate the non-legume *Parasponium*); and *Azorhizobium* which includes strains that induce nodules on stems of the tropical tree *Sesbania*. All are in the α group of Gram-negative bacteria (Chapter 1), but bear as little relationship to each other as to other genera in this group; they are united only by their ability to nodulate. Nevertheless, "rhizobia" is a convenient term to encompass the three genera. Although the structure and function of the symbiotic genes are generally similar in the three genera, there are, nonetheless, significant differences in their organization and regulation (1). In *Rhizobium* strains, genes concerned with early steps in nodulation (*nod* genes), in the determination of host-range, and in the ability to fix nitrogen (*fix* and *nif*) are clustered on large "symbiotic" (*Sym*) plasmids (1). Indeed, plasmid-determined host-range specificity defines different biovars (bv.) of the same *Rhizobium* species; thus *R. leguminosarum* strains are divided into bv. *phaseoli*, bv. *trifolii* and bv. *viciae* (which nodulate *Phaseolus*, clover and peas respectively) as a consequence of the host-range and *nod* genes on their Sym plasmids. In contrast, in strains of *Bradyrhizobium*, *nif* and *nod* genes are more dispersed than in *Rhizobium* strains and are chromosomally located. Some strains of *Bradyrhizobium* fix nitrogen at low levels in free-living culture but do not assimilate the ammonia and cannot grow on the fixed nitrogen. A different situation exists in strains of *Azorhizobium*, which fix nitrogen at high levels *in vitro* and can grow on the ammonia that is formed.

II. METHODS FOR IDENTIFYING BACTERIAL GENES INVOLVED IN NODULATION

Genes involved in nodulation or nitrogen fixation have been analysed both physically and functionally in several species of rhizobia. These studies have been facilitated by the availability of wide host-range cloning vectors that can be transferred between *Escherichia coli* and rhizobia (2, 3), so that symbiotic phenotypes conferred by cloned genes can be determined. The introduction of transposon mutagenesis in rhizobia (4, 5) has allowed the isolation and mapping of mutations that affect the symbiosis. These agents have been particularly valuable since, in most cases, mutant strains defective in nodulation or nitrogen fixation have no readily scorable phenotype in free-living culture: having the mutant gene marked by the transposon greatly facilitates the determination of its location. A powerful technique which harnesses the powers of transposons and *in vivo* and *in vitro* genetic techniques involves the mutagenesis with a transposon of previously cloned DNA followed by the introduction, by recombination, of the transposon into the corresponding position in the genome (6). Thus, a region of defined DNA can be intensively mutagenized and the phenotypic effects of the mutations can readily be determined.

With the techniques described above, two complementary approaches have been used to isolate and characterize genes involved in the early stages of the infection process and in the determination of the host-range of a given bacterial strain. In one, mutant strains affected in a phenotype that is readily scorable in bacteria growing in free-living culture are isolated, and the effects of such mutations on nodulation ability are subsequently examined. Thus, it is clear that polysaccharides on the bacterial cell surface are important in the nodulation process (Section III). The second approach involves the direct screening of survivors of a mutagenic treatment for any that are defective in the induction of normal nodules; though more laborious initially, this has the advantage that, potentially, all genes essential for the interaction (but inessential for normal growth) may be identified.

III. POLYSACCHARIDE SYNTHESIS IS IMPORTANT FOR NODULATION

Gram-negative bacteria make various polysaccharides (7) which are in the outer membrane (lipopolysaccharides: LPS) or are exuded from the cell (exopolysaccharides: EPS). *A priori*, it might be predicted that the surface architecture of rhizobia would be important in interactions between the plant and the bacterium. Certainly, the fact that some mutant bacteria, defective in the synthesis of one or other of the various polysaccharides, are also deficient in their symbiotic phenotype supports this view but the picture is not entirely clear, since there are other examples in which mutations that block synthesis of a polysaccharide have no apparent effect on nodulation or nitrogen fixation.

A. β1–2 Linked Glucan

Strains of *Rhizobium* and *Agrobacterium*, which are related to each other and both of which induce cell proliferation in plants, make an unusual cyclic low molecular weight β1–2 glucan EPS. This polymer is needed for the bacteria to infect normally since mutants of *R. meliloti* defective in its production induce abnormal, non-fixing nodules on the host plant, alfalfa (8). Significantly, these nodules contain no bacteria; thus rhizobia can induce these so-called "pseudonodules" even when they have lost the ability to invade the plant. Just as mutants of *Rhizobium* defective in the synthesis of this glucan are affected in their interaction with plants, mutant strains of *Agrobacterium* which fail to make the polymer no longer induce tumours (9, 10). Interestingly, this defect was corrected by the corresponding cloned DNA from *R. meliloti*, confirming the relatedness of the glucan in the two genera (9).

B. Acidic Exopolysaccharide

Colonies of rhizobia have a mucoid appearance (Fig. 2) because they produce large amounts of a high molecular weight acidic EPS. Genetic evidence has shown that this polymer, too, is important for the normal development of nodules. Mutant strains defective in its synthesis (Eps^-) have been isolated in different species of rhizobia. Although some of these mutants are apparently unaffected in symbiotic nitrogen fixation (11), other mutants fail to nodulate (11, 12), and yet others form nodules but fail to fix nitrogen (13, 14, 15). The situation is further complicated by the finding that the effect of a particular Eps^- mutation may depend on the host legume. Thus, certain non-mucoid derivatives of a *Rhizobium* strain with a wide host-range amongst various tropical legumes induced nitrogen-fixing nodules on some hosts, but tumourous, non-fixing nodules on others (16). Similarly, Borthakur *et al.* (12) introduced a particular Eps^- mutation into two biovars of *R. leguminosarum* which differed only in the symbiotic plasmid that they contained and thus nodulated different hosts (see below). In *R. leguminosarum* bv. *phaseoli*, whose host is *Phaseolus* beans, the non-mucoid mutants induced normal nodules on this legume; in contrast, when the same mutant allele was introduced into *R. leguminosarum* bv. *viciae*, which nodulates peas and vetches, the mutant derivatives failed to nodulate. Currently, it is not clear why certain Eps^- mutant strains are unaffected in the symbiotic interaction whereas others fail to initiate or to develop nodules. There is, therefore, considerable scope for

Figure 2. Appearance of colonies of wild-type *Rhizobium leguminosarum* bv. *viciae* (left) and of a mutant that fails to make the acidic EPS and hence is non-mucoid; this mutant fails to nodulate peas.

determining the precise stage at which individual mutations block the synthesis of EPS and for correlating the biochemical basis of the defects with their effects on the symbiosis.

Although *Rhizobium* strains lacking Sym plasmids make normal amounts of the acidic EPS, there are genes on at least one such plasmid which influence the production of the polymer. In a strain of *R. leguminosarum* bv. *phaseoli*, a gene was identified which, when present in multiple copies in cloning vectors, caused strains of *Rhizobium* to be non-mucoid and to lose the ability to nodulate. This gene, termed *psi* (polysaccharide inhibition), specifies a polypeptide which, judged by its deduced sequence, is likely to be inserted into the membrane of the bacterial cell (17, 18). Since bacteroids make little or no EPS (19), the function of *psi* may be to inhibit the production of the polymer and it was suggested that *psi* is normally expressed only in the bacteroids; however, expression of the cloned gene is deregulated and its effect on EPS synthesis can be observed in the free-living state. Transcription of *psi* itself is controlled in *R. leguminosarum* bv. *phaseoli* by a regulatory gene (*psr*) on the Sym plasmid which inhibits transcription of *psi* (18). Thus, although the Sym plasmid does not appear to contain structural genes for the synthesis of the acidic EPS, it does possess at least two genes influencing the regulation of its production.

In *R. meliloti*, several mutants have been isolated which are altered in their staining properties with Calcoflor, a dye that binds to the acidic EPS. Like mutants deficient in production of the cyclic β1–2 glucans, these also induce tumour-like, non-fixing nodules devoid of bacteria (14, 15). These mutations have been mapped to a large plasmid, different from the Sym plasmid (20).

C. Lipopolysaccharide

A third class of polysaccharide required for normal nodulation is the membrane-bound LPS. In a mutant of *R. leguminosarum* bv. *phaseoli* which induced tumourous, non-fixing nodules, this was associated with a defect in LPS synthesis (21).

Although considerable work is required to establish the exact role of the different polymers in the infection process, at least there is a biochemical indication of the type of molecule that is involved. However, the precise basis of the interaction between any of these polymers and plant-specified components has not been fully elucidated. One attractive hypothesis is that plant lectins (proteins that recognize specific sugars) bind to one or other of the *Rhizobium* polysaccharides and thus cause the bacteria to adhere to the root surface (22). Although there is some evidence in favour of this model, in which it has been claimed that the lectin from a particular legume binds specifically to the surface of the rhizobia which infect it, other data have not supported the "lectin hypothesis".

IV. ANALYSIS OF *NOD* GENE FUNCTION

A. Structures of the *nod* Genes

Work in several laboratories, using different strains and species of rhizobia, has been directed towards a detailed description of the structure and regulation of genes that confer host-range specificity and/or govern the earliest steps in nodule development (23). The *nod* genes of different strains of *Rhizobium* and *Bradyrhizobium* turn out to be highly conserved in their structure and function, though, as would be anticipated from the different host-ranges of the bacteria, there are also some significant, species-specific features. The most detailed studies have been made on *R. meliloti* which nodulates alfalfa, *R. leguminosarum* bv. *trifolii* which nodulates clovers, and *R. leguminosarum* bv. *viciae* which nodulates vetches and peas. In all three, *nod* genes are clustered on less than 20 kb of DNA of the Sym plasmid that also carries *nif* genes required for nitrogen fixation (24–26). Transfer of cloned fragments containing this *nod* DNA from a donor strain to strains lacking a Sym plasmid or to different species of *Rhizobium* allows the recipient to form nitrogen-fixing nodules on the legume which was the host of the donor of the *nod* genes (1). The ability to nodulate can even be transferred to *Agrobacterium* by the transfer of the entire Sym plasmid or cloned fragments derived from it, though the nodules so formed develop abnormally and fail to fix nitrogen. Thus, although the Sym plasmid is important, there must be additional genetic information contained in *Rhizobium* which allows it to induce normal nodules (27, 28).

Since *nod* genes are clustered in these strains, they have been subjected to detailed structural and functional analyses. Work in the author's laboratory has been aimed at the description of a cluster of *nod* genes on the Sym plasmid pRL1JI (from a strain of *R. leguminosarum* bv. *viciae*) which confers the ability to nodulate peas and vetches. On this plasmid, the *nod* genes are in a 10 kb region located between two clusters of *nif* genes, one containing *nifH, D* and *K* and the other *nifA* and *B* (29). Transfer of this *nod* DNA to other *Rhizobium* species or to a strain cured of its Sym plasmid conferred on the recipient the ability to carry out normally the early stages of infection of peas. This region was sequenced and ten genes, *nodL, M, E, F, D, A, B, C, I* and *J*, were identified (24–34, Fig. 3). A cloned fragment containing *nodABC*, when transferred to a strain lacking a Sym plasmid, restored the ability to curl root hairs (30); this was consistent with the fact that mutations in these genes abolished root hair curling and nodulation (31). Mutations in *nodD* also abolished root hair curling and nodulation, because of its role in the regulation of other *nod* genes (see below). Mutations in *nodFE* severely inhibited the number of nodules and delayed the time of their appearance but did not interfere with root hair curling; mutations in *nodI* and *nodJ* had only a minor effect, causing a slight reduction in nodule number.

By comparing the deduced sequences of *nod* gene products with those of proteins with known functions, it was found (34) that *nodF* specified a polypeptide similar to the acyl carrier protein, involved in the addition of acetate residues to growing chains of fatty acids and which, in *E. coli*, is required for synthesis of a periplasmically located polysaccharide. It is thus possible that *nodF* is required for synthesizing some lipid or polysaccharide needed for normal infection. The *nodI* product is homologous to a series of related inner membrane proteins that transport various low molecular weight compounds into enteric bacteria, and so *nodI* may be involved in the transport of some compound into or out of the bacteria (33). The product of *nodJ*, the gene immediately downstream of *nodI*, may also be associated with the membrane since it would be extremely hydrophobic (33). The findings on the structure, location and possible function of the various *nod* genes are shown in Table I.

B. Comparison of *nod* Genes in Different Rhizobia

nod genes are conserved in different rhizobia. In *R. meliloti, R. leguminosarum* bv. *trifolii* and *B. japonicum*, genes corresponding to *nodABC* have been identified and, in all three, they are transcribed divergently from a *nodD* gene, though in *B. japonicum, nodA* is preceded by an open reading frame which is absent from the fast-growing strains (35, 36, 37). In addition, *R. leguminosarum* bv. *trifolii* contains genes corresponding to *nodF* and *E* in the same location as

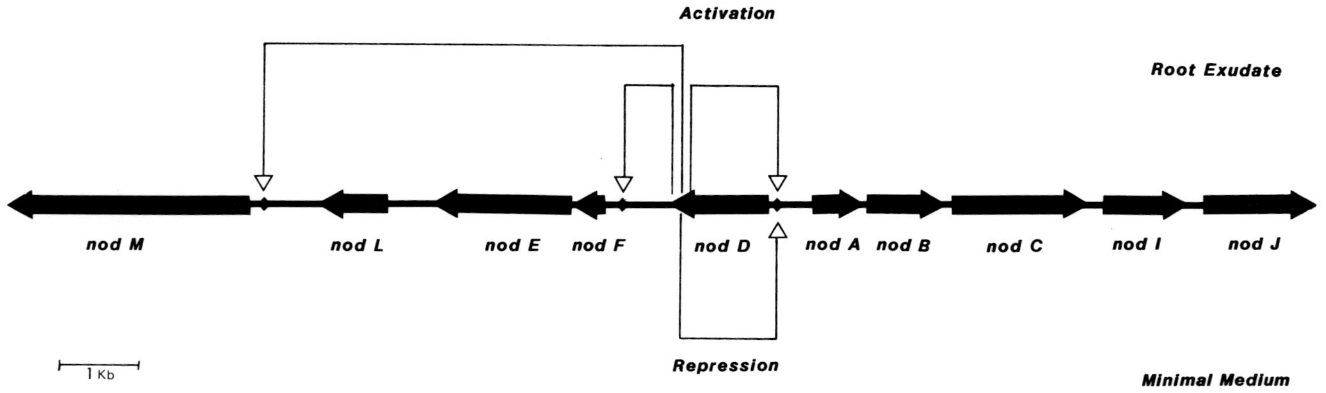

Figure 3. Organization and regulation of transcription of the *nod* gene cluster on the *R. leguminosarum* bv. *viciae* Sym plasmid pRL1JI. The dimensions and the directions of transcription of the ten *nod* genes are shown as thick arrows. The *nodD*-dependent "regulatory loop" is indicated. In the presence of root exudate, or defined *nod* gene inducers, *nodD* activates transcription of other *nod* transcriptional units, its gene product interacting with the *nod* boxes, whose locations are indicated by ◆.

Table I
Features of *nod* genes in different rhizobia

Gene	Phenotype of mutants°	Similarity to other genes	Found in species*	Comments
nodA	Nod^-; Rhc^-	?	1, 2, 3, 4	Cytosolic
nodB	Nod^-; Rhc^-	?	1, 2, 3, 4	—
nodC	Nod^-; Rhc^-	?	1, 2, 3, 4	Membrane-bound
*nodD***	Nod^-; Rhc^- or Nod^+; Rhc^+	*lysR*	1, 2, 3, 4, 5	Regulatory
nodE	Nod^d; Rhc^+		1, 2, 3, 4, 5	Determinant of host-range
nodF	Nod^d; Rhc^+	Acyl carrier protein	1, 2, 3	Determinant of host-range
nodG	Nod^d; Rhc^+	Ribitol dehydrogenase	3	Specific to *R. meliloti*
nodH	Nod^-; Rhc^-	?		Specific to *R. meliloti*
nodI	Nod^d; Rhc^+	Transport proteins	1	—
nodJ	Nod^d; Rhc^+	Membrane-bound?	1	—
nodK	Nod^+; Rhc^+	?	5	Specific to *Bradyrhizobium*
nodL†	Nod^d; Rhc^+	?	1	Determinant of host-range
nodM	Nod^d; Rhc^+	Amidophosphoribosyl-transferase (*purF*)	1	—
nodN	Nod^d; Rhc^+	?	1	Involved in host-range
nodX′	Nod^+ or Nod^-	Membrane-bound	1	Needed to nodulate Afghan peas

°Nod^-, unable to nodulate; Rhc^-, unable to curl root hairs; Nod^d, delayed in the onset of nodulation and reduced in nodule number.

*Genes have been sequence in 1 = *R. leguminosarum* bv. *viciae*; 2 = *R. leguminosarum* bv. *trifolii*; 3 = *R. meliloti*; 4 = *B. parasponiae*; 5 = *R. fredii*.

**In strains (e.g., *R. meliloti*) with multiple *nodD* genes, mutations in individual copies of *nodD* have only a minor effect on nodulation ability but in strains (e.g. *R. leguminosarum* biovars *trifolii* and *viciae*) with a single copy of *nodD*, a mutation in this gene abolishes nodulation ability completely.

′*nodX* is specifically needed for the nodulation of primitive Afghanistan peas by certain strains of *R. leguminosarum* bv. *viciae*. Mutations in *nodX* abolish nodulation of these peas but not of commercial varieties of peas.

†Mutations in *nodL* reduce nodulation of peas, lentils and *Lathyrus* but have little effect on the nodulation of vetches.

on pRL1JI (36). In contrast, *nodF* and *E* of *R. meliloti*, though in the same positions relative to each other, are approximately 13 kb downstream from *nodC* (38). Also, in *R. meliloti*, *nodF* and *E* are flanked by two genes, *nodG* and *nodH*, which are important for nodulation of alfalfa but which have no counterparts in *R. leguminosarum* (38, 39). Of these, *nodG* specifies a protein similar in sequence to ribitol dehydrogenase, but the significance of this remains to be determined (38).

Whereas mutations in the single *nodD* of *R. leguminosarum* bv. *viciae* and *R. leguminosarum* bv. *trifolii* abolish nodulation of their respective hosts (peas and clover), mutations in *nodD* of *R. meliloti* have little effect on nodulation of alfalfa, probably because *R. meliloti* contains multiple copies of *nodD* (40).

An important observation concerning the effects of mutations in *nodF* and *E* in *R. leguminosarum* bv. *trifolii* is that, although they reduce nodulation on the "correct" host, clover, they confer nodulation of "non-host" species such as peas, albeit poorly (41). Thus, these genes are involved in determining host-range, although their sequences are conserved in different biovars. Possibly, the particular version of *nodF* and *nodE* in an individual biovar causes subtle changes to the bacteria which, while facilitating the nodulation of the normal host, prevents nodulation of non-host legumes. A role for *nodF* and *E* in host-range determination is also indicated since strains of one species or biovar containing mutations in these genes are not corrected by corresponding genes from other *Rhizobium* species. In contrast, mutations in *nodD, A, B* and *C* can be functionally equivalent in different biovars. Thus, the nodulation defects of strains carrying mutations in these genes are corrected by cloned DNA with the corresponding gene from different strains (42). The products of *nodA* and *nodC* of *R. meliloti* are in the cytoplasm and the outer membrane respectively of the bacteria (43, 44); since these genes are equivalent in different rhizobia, it is likely that this localization applies to the *nodA* and *nodC* gene products in all rhizobia.

Thus, the fine-scale structures of several genes specifying early stages of the infection process have been identified and some clues concerning their possible functions have been obtained. This information has been crucial in allowing studies on the regulation of these genes to be undertaken.

V. REGULATION OF *NOD* GENE TRANSCRIPTION

A. Host Plant Exudate and *nodD* are Required for the Activation of *nod* Operons

Progress in understanding the regulation of *nod* genes has been facilitated by using gene fusions in which individual *nod* genes plus their upstream regulatory sequences were fused to *lacZ* of *E. coli*. In *R. leguminosarum* bv. *viciae*, three transcription units, *nodABCIJ*, *nodFE* and *nodD*, were identified; of these, only *nodD* is transcribed in cells grown in normal growth media. When root exudate from peas was added to the medium, *nodFE* and *nodABCIJ* were actively expressed, but only when there was a functional copy of *nodD* in the strain (45). In *R. leguminosarum* bv. *viciae*, *nodD* also has an autoregulatory function, repressing its own transcription (45). Similarly, in *R. meliloti* and *R. leguminosarum* bv. *trifolii nodD* is transcribed constitutively and, in the presence of inducer molecules found in legume root exudates, activates other *nod* genes (46, 47). It is not clear, though, why the *nodD* of *R. meliloti* is not autoregulatory (46).

B. Flavonoids are the Inducers in Legume Root Exudates

In several cases, the molecules in legume root exudates which activate *nodD*-dependent induction of the other *nod* genes have been identified. In alfalfa exudate, the flavone luteolin is the most potent inducer of *nod* gene expression for *R. meliloti* (48) and a related molecule, 7,4′-dihydroxyflavone, is the major inducer in clover seedlings (49). Similarly, several flavones and flavanones (Fig. 4) activate *nod* genes of *R. leguminosarum* bv. *viciae* (50). Comparison of the structures of the flavonoids that are inducers with those of closely related molecules that are not shows that hydroxylation at the 7 and the 3′ positions is important for the flavonoids to act as inducers (Fig. 4). The potency of the inducer molecules is remarkable; significant induction occurs at concentrations as low as 100 nM (49, 50). Though flavones and flavanones are widespread in the plant kingdom, legumes may be unusual in excreting them from their roots; we were unable to detect inducer activity in extracts from roots of a range of non-legume plants.

C. Legumes Make Compounds that Antagonize *nod* Gene Induction

It is apparent that plants make other molecules that antagonize the effects of the inducer flavonoids. Certain flavonols, isoflavones and acetophenone analogues, similar in their structure to the inducers (Fig. 5), depress the effects of inducer flavonoids. One of these molecules, acetosyringone, is an inducer of the virulence (*vir*) genes of *Agrobacterium tumefaciens* (51). The effects of acetosyringone on the transcription of the *vir* genes of *A. tumefaciens* and the *nod* genes of *Rhizobium* are likely to have different molecular bases since the mechanism of *vir* gene induction is different from that of the *nod* genes in that two regulatory genes, *virA* and *virG*, are required for the activation of the other *vir* genes and there is no similarity in the sequence of *nodD* with *virA* or *virG* (Chapter 18). Root exudates from peas and from clover contain molecules that inhibit *nod* gene transcription (50; 52). Thus, transcription of *nod* genes is influenced in a complex way and must be determined by the relative concentrations of the inducers and the "anti-inducers" in the rhizosphere of the host. It is possible that these "anti-inducers" act by competing with the inducer molecules, either at the level of uptake or in their proposed interaction with the product of *nodD* (see below).

D. Host-Range and the *nodD* Gene

Initially it appeared that the *nodD* genes of different species of rhizobia were functionally equivalent, since the Nod$^-$ defect on clover of a *nodD* mutant

Flavone Ring Structure

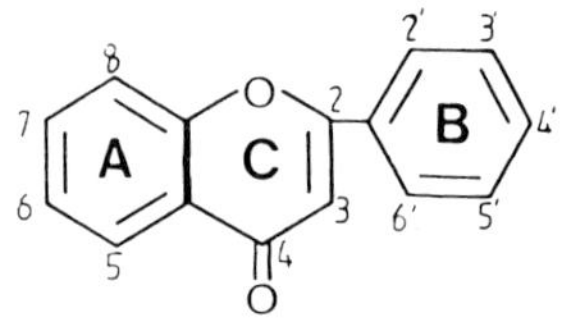

Common Name	Chemical Name	Effect on nod gene expression
Apigenin	5,7,4′-trihydroxyflavone	inducer
Apigenin-7-O-glucoside	5,4′-dihydroxy, 7-O-glucosyl flavone	inducer
Luteolin	5,7,3′,4′-tetrahydroxyflavone	inducer*
–	7,4′-dihydroxyflavone	inducer*

* : major inducers of R.meliloti nod gene expression

Flavanone Ring Structure

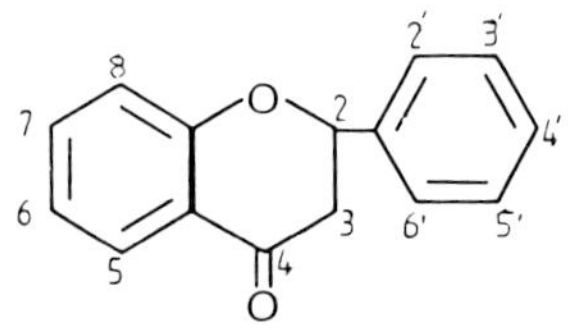

Common Name	Chemical Name	Effect on nod gene expression
Eriodictyol	5,7,3′,4′-tetrahydroxyflavanone	inducer
Hesperitin	5,7,3′-trihydroxy, 4′-methoxy flavanone	inducer

Figure 4. Structures of some flavonoid molecules that are potent inducers of the *nod* genes of *R. leguminosarum* bv. *viciae*. Note that in all cases, inducers have —OH substitutions in the 7-position in the A ring and substitutions at the 3′ and/or 4′ positions in the B ring, and that in one case (apigenin 7-*O*-glucoside) a substitution with a sugar at the 7-position does not interfere with inducing activity.

Isoflavone Ring Structure

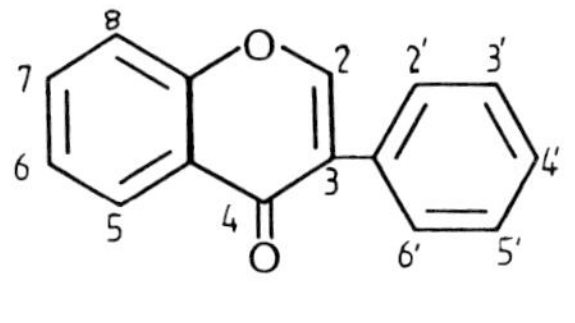

Common Name	Chemical Name	Effect on nod gene expression
Daidzein	7,4'-dihydroxyisoflavone	anti-inducer
Genistein	5,7,4'-trihydroxyisoflavone	anti-inducer

Flavonol Ring Structure

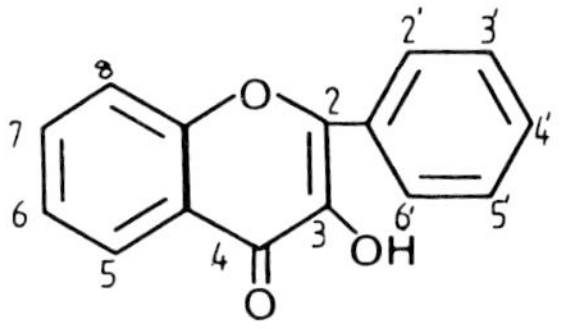

Common Name	Chemical Name	Effect on nod gene expression
Kaempferol	5,7,4'-trihydroxyflavonol	anti-inducer

Figure 5. Structures of flavonoid molecules that antagonize *nod* gene induction. Addition of these molecules to strains of *R. leguminosarum* bv. *viciae* which have been grown in the presence of the inducers shown in Figure 4 causes a significant reduction in the level of *nod* gene induction. Note the structural similarities between these "anti-inducers" and the inducer molecules.

strain of *R. leguminosarum* bv. *trifolii* could be corrected by the introduction of the cloned *nodD* of *R. meliloti* (42). Also, luteolin, the flavone that activates *nod* transcription in *R. meliloti*, is a potent inducer of *nod* genes of *R. leguminosarum* bv. *viciae* and extracts from alfalfa activate *nod* transcription in the latter strain (50). However, the flavanone hesperitin was relatively ineffective for *R. meliloti* (48) whereas it was the most potent inducer identified for the induction of *nod* genes of *R. leguminosarum* bv. *viciae* (50). It is now apparent that such differences in the sensitivities of different *nodD* products to individual molecules can, in most cases, influence host-range specificity; the ability of a wide host-range strain of *Rhizobium* to nodulate the tropical forage legume siratro (*Macroptillium*) is due to its possession of a version of *nodD* which, when transferred to *R. meliloti*, conferred on the recipient the ability to nodulate this host (53). Similarly, the root exudate of red clover activated transcription of *nod* genes of *R. leguminosarum* bv. *trifolii* but not those of *R. meliloti* (54). These observations indicate that the inducer molecules interact directly with the polypeptide product of *nodD*. Despite the similarity in the sequences of the *nodD* genes in different rhizobia, it is clear that there must be significant functional differences between them; presumably the *nodD* gene product of particular strains may "sense" individual inducer molecules that are present in the exudates of the host legumes.

E. The Product of *nodD* Binds to a Regulatory DNA Sequence

A conserved sequence, spanning approximately 35 bp, precedes the *nodFE* and *nodABCIJ* operons in *R. meliloti*, *R. leguminosarum* bv. *trifolii* and *R. leguminosarum* bv. *viciae* which are activated by the *nodD* polypeptide (Fig. 6). This "*nod*-box" (34, 36, 55) also occurs upstream of *nodG* of *R. meliloti* (36) and *nodM* of *R. leguminosarum* bv. *viciae* (J. A. Downie, personal communication), suggesting that these genes are also under the same regulatory control. Transcription of *nodABCIJ* of *R. meliloti* is initiated near the *nod*-box (56) and the simplest explanation for *nod* gene regulation is that the *nodD* gene product interacts with inducer flavonoids and this alters its conformation so that it "opens up" the promoters of genes whose transcription it activates.

There is evidence to support this model. Part of the deduced amino acid sequence of the *nodD* product is similar to that of the *araC* product of *E. coli* (34); *araC* is a regulatory gene whose product, in the presence of arabinose, binds to the promoter of the *araBAD* operon and activates its transcription. *araC* is transcribed divergently from *araBAD* and is autoregulatory; in the presence or absence of arabinose it represses its own transcription. Similarly, *nodD* of *R. leguminosarum* bv. *viciae* is autoregulatory in the presence or absence of flavonoid inducers. More extensive homology was detected between the *nodD* product and the product of *lysR*, a regulatory gene of *E. coli* which is also autoregulatory

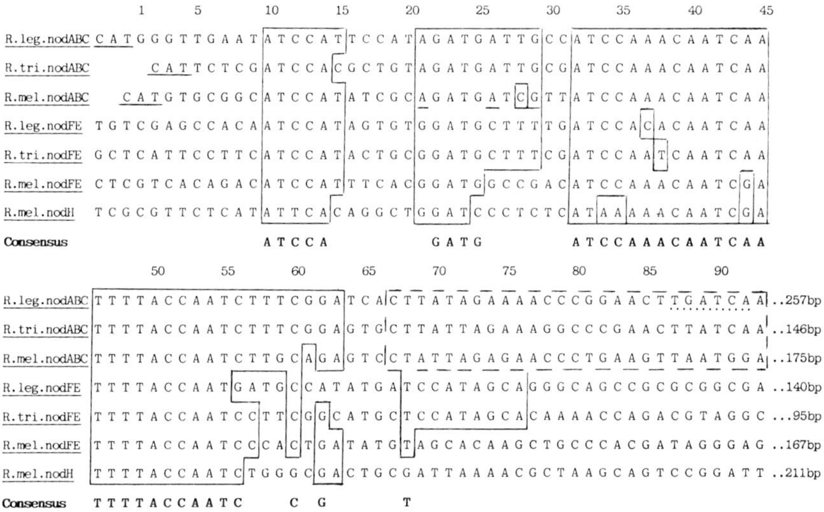

Figure 6. Comparison of *nod*-box sequences preceding transcriptional units whose expression requires *nodD* plus inducer flavonoids. Sequences preceding *nodABC* genes of *R. leguminosarum* bv. *viciae* (*R. leg.*), *R. leguminosarum* bv. *trifolii* (*R. tri.*) and *R. meliloti* (*R. mel.*) and *nodH* of *R. meliloti* were aligned to give maximum similarity. Regions of identity are boxed and a consensus sequence is presented where six or more base pairs are identical. Underlined CAT refer to translational starts (ATG on the other strand) of *nodD* which are transcribed divergently from *nodABC*. The boxes with broken lines indicate regions with a reduced level of similarity in the *nod*-boxes. The numbers at the right of the sequences refer to the distances (in base pairs) to the translational starts of the *nodA*, *nodF* or *nodH* genes. The dotted line refers to a region of diad symmetry that may be involved in the binding of the product of *nodD*.

and, in the presence of diaminopimelate, activates transcription of other genes involved in lysine biosynthesis.

Burn *et al.* (57) isolated several mutant forms of *nodD* of *R. leguminosarum* bv. *viciae* which were altered in their regulatory properties. Some were defective both in autoregulation of *nodD* and in the induction of other *nod* genes in the presence of flavonoid inducers. Others were specifically defective in only one of these two regulatory properties, and others activated transcription even in the absence of inducers. These last mutants were of particular interest since they still responded to the inducer flavonoids; there was "hyperinduction" of *nodABCIJ* and *nodFE* in strains carrying these mutant *nodD* alleles. Further, compounds such as the flavonol kaempferol and the isoflavone genistein, which, in strains carrying wild-type *nodD* are anti-inducers, actually acted as inducers in strains containing this last type of mutant *nodD*. The fact that a single mutation can affect the response to both the inducers and the anti-inducers suggests strongly that both types of molecule interact directly with the product of *nodD*. It is striking that strains of *R. leguminosarum* bv. *viciae* harbouring these "constitutive" mutant forms of *nodD* induced fewer nodules on peas than normal, and that the nodules failed to fix nitrogen. Thus, continued, high-level expression of *nod* genes, far from enhancing the efficacy of the interaction, is detrimental both for nodulation and nitrogen fixation (57).

These "constitutive" forms of *nodD* were due to an amino acid substitution (from asparagine to aspartate) in the carboxy-terminal region of the *nodD* product, suggesting that this region is involved in the interaction with the flavonoids (57). Horvath *et al.* (53) came to the same conclusion; they constructed hybrid *nodD* genes by joining part of *nodD* of *R. meliloti* to part of *nodD* of the wide host-range strain that nodulated siratro (see above) and showed that the 3′ region of *nodD* determines the response (in terms of its ability to induce other *nod* genes) to the particular inducer molecules.

Using the DNA fragment retardation technique (Chapter 3), strong evidence that the *nodD* product binds to specific DNA sequences was obtained (58). A fragment spanning the *nodA–nodD* intergenic region of *R. leguminosarum* bv. *viciae* was labelled and, before electrophoresis, was exposed to cell extracts of this species. If the extract was from a strain that lacked a copy of *nodD*, a single band was obtained following electrophoresis but with extracts from strains with wild-type *nodD*, an additional, more slowly migrating band was found, whether inducer molecules were present or not. Since addition of protease to cell extract prevented the production of this extra band, it apparently comprises a complex between protein(s) in the cell extracts and the labelled DNA. When extract obtained from an *E. coli* strain containing the cloned *nodD* gene was added to the *nod*-box DNA, the retarded band was still obtained, providing strong evidence that this DNA-protein complex includes the *nodD* product itself.

VI. CONCLUSIONS

Considering that the genetic study of rhizobia started relatively recently, there has been real progress in the analysis of these bacteria. For obvious reasons, this has focused on the study of the genes involved in the symbiotic interaction (though it is a pity, perhaps, that we do not know what "normal" promoters of *Rhizobium* look like). As shown in this Chapter and Chapter 6, the most detailed knowledge is available for the two extreme "ends" of the symbiotic interaction: the identification of the early steps in the recognition of the host and the expression of the *nif* genes. Perhaps to a greater extent than with any other group of bacteria, there is available a large body of comparative data on the structure and properties of the "same" genes in different strains. Though this has led to some replication of effort, it has not been entirely wasteful since differences and similarities of corresponding genes in different species can help in determining their function, as has already been done for *nodD*.

In the future, it seems likely that there will be rapid developments in the understanding of the precise mechanisms of how the inducers and anti-inducers exert their effects and of how the *nif* genes are turned on in the nodule. A more challenging task, though, will be to establish physiological and biochemical roles for these genes. For example, the detailed information that is available concerning the structure of *nod* genes must be translated into meaningful biochemical data that will give a clearer molecular picture of the early stages of this complex interaction.

References

(1) Long, S. R. (1984). Genetics of *Rhizobium* nodulation. *In* "Plant Microbe Interactions" (T. Kosuge and E. W. Nester, eds.), Vol. 1, Molecular and Genetic Perspectives, pp. 265–306, Academic Press, New York.

(2) Friedman, A. M., Long, S. R., Brown, S. E., Buikema, W. J. and Ausubel, F. M. (1982). Construction of a broad-host-range cosmid cloning vector and its use in the genetic analysis of *Rhizobium* mutants. *Gene* **18**, 289–296.

(3) Bagdasarian, M., Lurz, R., Ruckert, B., Franklin, F. C. H., Bagdasarian, M. M., Frey, J. and Timmis, K. T. (1981). Specific-purpose plasmid cloning vectors. II. Broad-host-range, high-copy, RSF1010-derived vectors and a host-vector system for gene cloning in *Pseudomonas*. *Gene* **16**, 237–247.

(4) Beringer, J. E., Beynon, J. L., Buchanan-Wollaston, A. V. and Johnston, A. W. B. (1978). Transfer of the drug resistance transposon Tn*5* to *Rhizobium*. *Nature* **276**, 633–634.

(5) Simon, R., Preifer, U. and Pühler, A. (1983). Vector plasmids for *in vivo* and *in vitro* manipulations of Gram-negative bacteria. *In* "Molecular Genetics of the Plant-Microbe Interactions" (A. Pühler, ed.), pp. 98–106, Springer Verlag, Berlin.

(6) Ruvkun, G. and Ausubel, F. M. (1981). A general method for site-directed mutagenesis in prokaryotes. *Nature* **289**, 85–88.

(7) Sutherland, I. W. (1985). Biosynthesis and composition of Gram-negative bacterial extracellular and wall polysaccharides. *Ann. Rev. Microbiol.* **39**, 243–270.

(8) Dylan, T., Ielpi, L., Stanfield, S., Kayshap, L., Douglas, C., Yanofsky, M., Nester, E. W., Helinski, D. R. and Ditta, G. (1986). *Rhizobium meliloti* genes required for nodule development are related to chromosomal virulence genes in *Agrobacterium tumefaciens*. *Proc. Natl. Acad. Sci. USA* **83**, 4403–4407.

(9) Douglas, C. J., Staneloni, R. J., Rubin, R. A. and Nester, E. W. (1985). Identification and genetic analysis of an *Agrobacterium tumefaciens* chromosomal virulence region. *J. Bacteriol.* **161**, 850–860.

(10) Puvanesarajah, V., Schell, F. M., Stacey, G., Douglas, C. J. and Nester, E. W. (1985). A role for β-2 glucan in the virulence of *Agrobacterium tumefaciens*. *J. Bacteriol.* **164**, 102–106.

(11) Sanders, R., Raleigh, E. and Signer, E. (1981). Lack of correlation between extracellular polysaccharide production and nodulation ability in *Rhizobium*. *Nature* **292**, 148–149.

(12) Borthakur, D., Barber, C. E., Lamb, J. W., Daniels, M. J., Downie, J. A. and Johnston, A. W. B. (1986). A mutation that blocks exopolysaccharide synthesis prevents nodulation of peas by *Rhizobium leguminosarum* but not of beans by *Rhizobium phaseoli* is corrected by cloned DNA from the phytopathogen *Xanthomonas*. *Mol. Gen. Genet.* **203**, 320–323.

(13) Chakrovorty, A. K., Zurkowski, W., Shine, J. and Rolfe, B. G. (1982). Symbiotic nitrogen fixation: molecular cloning of *Rhizobium* genes involved in exopolysaccharide synthesis and effective nodulation. *J. Mol. Appl. Genet.* **1**, 585–596.

(14) Finan, T. M., Hirsch, A. M., Leigh, J. A., Johansen, E., Kuldau, G. A., Deegan, S., Walker, G. C. and Signer, E. R. (1985). Symbiotic mutants of *Rhizobium meliloti* that uncouple plant from bacterial differentiation. *Cell* **40**, 869–877.

(15) Leigh, J. A., Signer, E. R. and Walker, G. C. (1985). Exopolysaccharide-deficient mutants of *Rhizobium meliloti* that form ineffective nodules. *Proc. Natl. Acad. Sci. USA* **82**, 6231–6235.

(16) Chen, H., Batley, M., Redmond, J. and Rolfe, B. G. (1985). Alteration of the effective nodulation properties of a fast growing broad-host-range *Rhizobium* due to changes in exopolysaccharide synthesis. *J. Plant Physiol.* **120**, 331–349.

(17) Borthakur, D., Downie, J. A., Johnston, A. W. B. and Lamb, J. W. (1985). *psi*, a plasmid-linked *Rhizobium phaseoli* gene that inhibits exopolysaccharide production and which is required for symbiotic nitrogen fixation. *Mol. Gen. Genet.* **200**, 278–282.

(18) Borthakur, D. and Johnston, A. W. B. (1987). Sequence of *psi*, a gene on the symbiotic plasmid of *Rhizobium phaseoli* which inhibits exopolysaccharide synthesis and nodulation, and demonstration that its transcription is inhibited by *psr*, another gene on the symbiotic plasmid. *Mol. Gen. Genet.* **207**, 149–154.

(19) Tully, R. E. and Terry, M. E. (1985). Decreased exopolysaccharide synthesis by anaerobic and symbiotic cells of *Bradyrhizobium japonicum*. *Plant Physiol.* **79**, 445–450.

(20) Hynes, M. F., Simon, R., Niehaus, K., Labes, M. and Pühler, A. (1986). The two megaplasmids of *Rhizobium meliloti* are involved in the effective nodulation of alfalfa. *Mol. Gen. Genet.* **202**, 356–362.

(21) Vandenbosch, K. A., Noel, K. D., Kaneko, Y. and Newcomb, E. H. (1985). Nodule initiation elicited by noninfective mutants of *Rhizobium phaseoli*. *J. Bacteriol.* **162**, 950–959.

(22) Dazzo, F. B. and Hubbell, D. (1975). Cross-reactive antigens and lectin as determinants of symbiotic specificity in the *Rhizobium*-clover association. *Appl. Env. Microbiol.* **30**, 1017–1033.

(23) Rossen, L., Davis, E. O. and Johnston, A. W. B. (1987). Plant-induced expression of *Rhizobium* genes involved in host specificity and early stages of nodulation. *TIBS* **12**, 430–433.

(24) Downie, J. A., Hombrecher, G., Ma, Q-S., Knight, C. D., Wells, B. and Johnston, A. W. B. (1983). Cloned nodulation genes of *Rhizobium leguminosarum* determine host-range specificity. *Mol. Gen. Genet.* **190**, 359–365.

(25) Schofield, P. R., Ridge, R. W., Rolfe, B. G., Shine, J. and Watson, J. M. (1984). Host-specific nodulation is encoded on a 14 kb DNA fragment in *Rhizobium trifolii*. *Plant Mol. Biol.* **3**, 3–11.

(26) Kondorosi, E., Banfalvi, Z. and Kondorosi, A. (1984). Physical and genetic analysis of a symbiotic region of *Rhizobium meliloti*: identification of nodulation genes. *Mol. Gen. Genet.* **193**, 445–452.

(27) Hooykaas, P. J. J., van Brussel, A. A. N., den Dulk-Ras, H., van Slogteren, G. M. S. and Schilperoort, R. A. (1981). Sym plasmid of *Rhizobium trifolii* expressed in different rhizobial species and *Agrobacterium tumefaciens*. *Nature* **291**, 351–353.

(28) Govers, F., Moerman, M., Downie, J. A., Hooykaas, P. J. J., Franssen, H. J., Louwerse, J., van Kammen, A. and Bisseling, T. (1986). *Rhizobium nod* genes are involved in inducing an early nodulin gene. *Nature* **323**, 564–566.

(29) Downie, J. A., Ma, Q-S., Knight, C. D., Hombrecher, G. and Johnston, A. W. B. (1983). Cloning of the symbiotic region of *Rhizobium leguminosarum:* the nodulation genes are between the nitrogenase genes and a *nifA*-like gene. *EMBO J.* **2**, 947–952.

(30) Rossen, L., Johnston, A. W. B. and Downie, J. A. (1984). DNA sequence of the *Rhizobium leguminosarum* nodulation genes *nodA*, *B* and *C* required for root hair curling. *Nucleic Acids Res.* **12**, 9497–9508.

(31) Downie, J. A., Knight, C. D., Johnston, A. W. B. and Rossen, L. (1985). Identification of genes and gene products involved in nodulation of peas by *Rhizobium leguminosarum*. *Mol. Gen. Genet.* **198**, 255–262.

(32) Downie, J. A., Surin, B. P., Evans, I. J., Rossen, L., Firmin, J. L., Shearman, C. A. and Johnston, A. W. B. (1986). Nodulation genes of *Rhizobium leguminosarum*. *In* "Molecular Genetics of Plant-Microbe Interactions" (D. P. S. Verma and N. Brisson, ed.), pp. 225–228, Martinus Nijhoff, Amsterdam.

(33) Evans, I. J. and Downie, J. A. (1986). The *nodI* product of *Rhizobium leguminosarum* is closely related to ATP-binding bacterial transport proteins; nucleotide sequence of the *nodI* and *nodJ* genes. *Gene* **43**, 95–101.

(34) Shearman, C. A., Rossen, L., Johnston, A. W. B. and Downie, J. A. (1986). The *Rhizobium leguminosarum* gene *nodF* encodes a protein similar to acyl carrier protein and is regulated by *nodD* plus a factor in pea root exudate. *EMBO J.* **5**, 647–652.

(35) Torok, I., Kondorosi, E., Stepkowski, T., Posfai, J. and Kondorosi, A. (1984). Nucleotide sequence of *Rhizobium meliloti* nodulation genes. *Nucleic Acids Res.* **12**, 9509–9523.

(36) Schofield, P. R. and Watson, J. M. (1986). DNA sequence of *Rhizobium trifolii* nodulation genes reveals a reiterated and potentially regulatory sequence preceding *nodABC* and *nodFE*. *Nucleic Acids Res.* **14**, 2891–2903.

(37) Scott, K. F. (1986). Conserved nodulation genes from the non-legume symbiont *Bradyrhizobium* sp. *parasponia. Nucleic Acids Res.* **14**, 2905–2919.

(38) Debelle, F. and Sharma, S. B. (1986). Nucleotide sequence of *Rhizobium meliloti* RCR2011 genes involved in host specificity of nodulation. *Nucleic Acids Res.* **14**, 7453–7472.

(39) Horvath, B., Kondorosi, E., John, M., Schmidt, J., Torok, I., Gyorgypal, Z., Barabas, I., Weineke, U., Schell, J. and Kondorosi, A. (1986). Organization, structure and symbiotic function of *Rhizobium meliloti* nodulation genes determining host specificity for alfalfa. *Cell* **46**, 335–343.

(40) Gottfert, M., Horvath, B., Kondorosi, E., Putnoky, P., Rodriguez-Quinones, F. and Kondorosi, A. (1986). At least two functional *nodD* genes are necessary for efficient nodulation of alfalfa by *Rhizobium meliloti. J. Mol. Biol.* **191**, 411–420.

(41) Djordjevic, M. A., Schofield, P. R. and Rolfe, B. G. (1985). Tn*5* mutagenesis of *Rhizobium trifolii* host-specific nodulation genes results in mutants with altered host-range ability. *Mol. Gen. Genet.* **200**, 463–471.

(42) Fisher, R. F., Tu, J. K. and Long, S. R. (1985). Conserved nodulation genes in *Rhizobium meliloti* and *Rhizobium trifolii. Appl. Env. Microbiol.* **49**, 1432–1435.

(43) Egelhoff, T. T. and Long, S. R. (1985). *Rhizobium meliloti* nodulation genes: identification of *nodDABC* gene products, purification of *nodA* protein and expression of *nodA* in *Rhizobium. J. Bacteriol.* **164**, 591–599.

(44) John, M., Schmidt, J., Weineke, U., Kondorosi, E., Kondorosi, A. and Schell, J. (1985). Expression of the nodulation gene *nodC* of *Rhizobium meliloti* in *Escherichia coli*: role of the *nodC* gene product in nodulation. *EMBO J.* **4**, 2425–2430.

(45) Rossen, L., Shearman, C. A., Johnston, A. W. B. and Downie, J. A. (1985). The *nodD* gene of *Rhizobium leguminosarum* is autoregulatory and in the presence of plant root exudate induces the *nodABC* genes. *EMBO J.* **4**, 3369–3374.

(46) Innes, R. W., Kuempl, P. L., Plazinski, J., Canter-Cramers, H., Rolfe, B. G. and Djordjevic, M. A. (1985). Plant factors induce expression of nodulation and host range genes in *Rhizobium trifolii. Mol. Gen. Genet.* **201**, 426–432.

(47) Mulligan, J. T. and Long, S. R. (1985). Induction of *Rhizobium meliloti nodC* expression by plant exudate requires *nodD. Proc. Natl. Acad. Sci. USA* **82**, 6609–6613.

(48) Peters, N. K., Frost, J. W. and Long, S. R. (1986). A plant flavone, luteolin, induces expression of *Rhizobium meliloti* nodulation genes. *Science* **233**, 977–980.

(49) Redmond, J. W., Batley, M., Djordjevic, M. A., Innes, R. W., Kuempl, P. L. and Rolfe, B. G. (1986). Flavones induce expression of nodulation genes in *Rhizobium. Nature* **323**, 632–634.

(50) Firmin, J. L., Wilson, K. E., Rossen, L. and Johnston, A. W. B. (1986). Flavonoid induction of nodulation genes in *Rhizobium* is reversed by other compounds present in plants. *Nature* **324**, 90–92.

(51) Stachel, S. E., Messens, E., Van Montagu, M. and Zambryski, P. (1985). Identification of the signal molecules produced by wounded plant cells that activate T-DNA transfer in *Agrobacterium tumefaciens*. *Nature* **318**, 624–629.

(52) Djordjevic, M. A., Redmond, J. W., Batley, M. and Rolfe, B. G. (1987). Clovers secrete specific phenolic compounds which either stimulate or repress *nod* gene expression. *EMBO J.* **6**, 1173–1179.

(53) Horvath, B., Bachem, C. W. B., Schell, J. and Kondorosi, A. (1987). Host-specific regulation of nodulation genes in *Rhizobium* is mediated by a plant-signal, interacting with the *nodD* gene product. *EMBO J.* **6**, 841–845.

(54) Spaink, H. P., Wijfellman, C. A., Pees, E., Okker, R. J. H. and Lugtenberg, B. J. J. (1987). *Rhizobium* nodulation gene *nodD* as a determinant of host specificity. *Nature* **328**, 337–339.

(55) Rostas, K., Kondorosi, E., Horvath, B., Simoncsits, A. and Kondorosi, A. (1986). Conservation of extended promoter regions of nodulation genes in *Rhizobium*. *Proc. Natl. Acad. Sci. USA* **83**, 1757–1761.

(56) Fisher, R. F., Brierley, H. L., Mulligan, J. T. and Long, S. R. (1987). Transcription of *Rhizobium meliloti* nodulation genes: identification of a *nodD* transcription initiation site *in vitro* and *in vivo*. *J. Biol. Chem.* **262**, 6849–6855.

(57) Burn, J. E., Rossen, L. and Johnston, A. W. B. (1987). Four classes of mutation in the *nodD* gene of *Rhizobium leguminosarum* biovar *viciae* which affect its ability to autoregulate and/or to activate other *nod* genes in the presence of flavonoid inducers. *Genes and Dev.* **1**, 456–464.

(58) Hong, G.-F., Burn, J. E. and Johnston, A. W. B. (1987). Evidence that DNA involved in the expression of nodulation (*nod*) genes in *Rhizobium* binds to the product of the regulatory gene *nodD*. *Nucleic Acids Res.* **15**, 9677–9691.

Section **VI**

Bacterial Population Genetics

Since this section contains a single chapter, it is hard to think of things that might usefully be said in an introduction to the section that have not already been better said in the chapter itself. Suffice to say that a chapter on bacterial population genetics has not been included in this book merely to justify its title. We really feel that the time is now ripe for a much greater dialogue between molecular bacterial geneticists, who are applying a powerful battery of newly available analytical tools to the study of all kinds of interesting properties of particular bacterial groups, and the small number of geneticists who are thinking of bacteria in populational and evolutionary terms. We hope that the inclusion of both approaches in this book may provide a small extra impetus to the development of this dialogue.

Chapter **20**

The Population Genetics of Bacteria

J. P. W. YOUNG

I. INTRODUCTION

A. What is Population Genetics?

The techniques of molecular biology are having a profound impact in all areas of genetics, and the field of population genetics and evolution is no exception. Molecular biology exerts a unifying influence, providing a common language and currency: the same cloned gene may tell us about the control of gene expression, the consequences of mutation, the structure of proteins, the genetic diversity of populations, the evolutionary relationship between organisms, and so on. To avoid misunderstanding, however, it is necessary to appreciate the very different modes of thought which each discipline brings to bear on this common ground.

Population genetics lies in that corner of the field of genetics that adjoins ecology on the one hand and systematics on the other. It is concerned with genetic variation, its amount and nature, its distribution over space and time, the forces that affect it and its biological significance. This is the raw stuff of evolution. To understand genetic variation we must take into account a wide range of phenomena. We must know about the processes that generate new genotypes: mutation, rearrangements within genomes, and the passage of genes from one individual to another. We must know about the processes that determine the fate of genotypes: selection on the basis of fitness, migration between populations, and the laws of chance. If, for particular populations, we have a sufficient description of all these processes, then we can predict the pattern of genetic variation that will result: the level of genetic polymorphism, linkage disequilibrium, spatial heterogeneity, temporal changes, and so on. Conversely, the observed pattern of variation can tell us about the processes that generated it.

There are three aspects to population genetics: the description of variation in field populations, an assessment of the processes that cause this variation (which may involve both field and laboratory studies), and a body of theory that connects the cause with the effects. The textbook by Futuyma (1) provides a lucid introduction to population genetics, speciation, coevolution, and other concepts raised in this chapter. Some of the terms used in this chapter are defined in Table I.

B. Do Bacteria Have Population Genetics?

Bacteria were largely omitted from the great burgeoning of descriptive population genetics that followed in the 1960s and 1970s from the development of enzyme electrophoresis as a tool for detecting genetic variation in natural populations (2, 3). This was not because bacteria are not amenable to this approach; indeed, a recent spate of publications has shown its value. Rather, the answer lies in the background and expertise of the investigators. The prevailing attitude has been aptly characterized as “find ’em and grind ’em”: samples of organisms were collected from wild populations, and a tissue sample from each individual was homogenized to prepare an extract for electrophoresis. No laboratory culture was required, in contrast to the specialized and unfamiliar (to the population geneticist) techniques needed for the isolation, culture and identification of bacteria.

Bacteria differ from typical higher eukaryotes in many relevant ways: their haploidy, clonal reproduction, vast population sizes and short generation times. The most controversial, however, from the point of view of population

Table I

Glossary of population genetics and related terms

Accessory element: DNA that is inessential for the organism and can over-replicate relative to chromosomal DNA; phages, plasmids, transposons and insertion sequences of bacteria (31; see Section IV.A).

Biotype: A group of bacterial isolates sharing certain biochemical and physiological attributes.

Coevolution: A series of reciprocal evolutionary responses in each of two or more species (1).

Chemostat: A vessel for the continuous culture of microbes in a fixed volume of liquid, with constant addition of new medium and removal of culture. One nutrient is supplied at a low concentration, limiting for growth, and this controls the cell density at a steady state in which growth balances outflow.

Cryptic gene: A phenotypically silent DNA sequence not normally expressed during the life cycle of an individual, but capable of activation as a rare event in a few members of a population by mutation, recombination, insertion of a DNA element, or other genetic mechanisms (40; see Section V).

Drift, genetic: Random fluctuation of allele frequencies from generation to generation in a finite population.

Effective number of alleles: A measure, for one locus, of genetic diversity in a population; the reciprocal of the sum of the squares of allelic frequencies. It is less than the actual number of alleles except in the special case that all alleles are equally frequent.

Effective population size: The number of individuals in a notional "ideal" population of constant size, and with equal fitnesses, etc., that would have the same rate of random drift in allele frequencies as the real population. It is normally less than the real, or "census" population size.

Electrophoretic type (ET): A set of bacterial isolates with the same electrophoretic alleles at all enzyme loci examined (9; see Section II.A).

Epistasis: A synergistic effect, on the phenotype or fitness, of two or more gene loci, whereby alleles at one locus alter the effects of alleles at another.

Fitness: The average contribution of an allele or genotype to succeeding generations.

Heterosis: The superiority of a heterozygote over either homozygote for a particular trait.

Homologous: Of genes, chromosomes or structures, similar in different organisms because they are inherited from a common ancestor. Now often, but inaccurately, used to describe similarity in sequence of nucleic acids or proteins.

Linkage disequilibrium: Nonrandom association of alleles at different loci in a population (the loci need not necessarily be in the same linkage group).

Neutral theory: The theory

(i) that most of the intraspecific variation at the molecular level is essentially neutral, so that polymorphism is maintained by the balance of mutation and drift; and

(ii) that the great majority of evolutionary changes at the molecular level, as revealed by comparative studies of protein and DNA sequences, are caused not by selection but by random drift of selectively neutral or nearly neutral mutants (12; see Section II.B). Also the mathematics that supports these ideas.

Periodic selection: The periodic invasion of an asexual population by clones of higher fitness that arise as one or very few individuals by mutation from an existing clone. Causes "founder effects" that reduce effective population size (14, 18; see Section II.C).

Phylogeny: The evolutionary history, or genealogy, of a group of organisms.

Polymorphism, genetic: The presence of more than one allele at a given locus in a population.

Selection, natural: The differential reproduction of alternative genotypes caused by differences in fitness.

Speciation: The evolutionary process of species formation.

Species, biological: Groups of interbreeding natural populations that are reproductively isolated from other such groups (19; see Section III.A).

genetics, is their range of mechanisms for genetic exchange and rearrangement, which include conjugation, transduction and transformation, plasmids, phages and transposable elements. Much of the current interest in the population genetics of bacteria centres on the role of these phenomena in determining population structure and evolution, but direct evidence of their quantitative significance in nature is still scanty. There is no good reason to expect the balance of all these forces to be the same in all bacteria, for bacteria span an enormous range of genetic distance and of lifestyles. We should not forget that *Escherichia coli* is no more closely related to *Bacillus subtilis* than toads are to toadstools (4).

II. GENETIC VARIATION AND ITS INTERPRETATION

A. What Does Enzyme Polymorphism Tell Us About Bacteria?

Without genetic variation, there can be no population genetics. Molecular techniques now provide many ways of detecting naturally-occurring genetic variation, but the first in widespread use was enzyme electrophoresis, a simple technique for detecting those amino acid substitutions that cause charge changes in enzyme molecules (5). It rapidly became clear that virtually all organisms had abundant intraspecific variation, and levels of genetic diversity were measured in a vast number of animal species, and not a few plants (2, 3). More recently, restriction fragment patterns and DNA sequences have taken the analysis to a further degree of refinement (6).

Bacteriologists have long distinguished between closely-related organisms by techniques based on molecular differences, such as serology (which recognizes differences in antigens), phage typing (based on resistance to certain phages), biotyping (metabolic capabilities) and antibiotic-resistance typing (7). While invaluable in the tracing of clinical outbreaks, these methods have the disadvantage, from the point of view of population genetics, that it is seldom simple to deduce the genotype of an isolate from its phenotype. When population geneticists took an interest in bacteria, they brought their familiar tools with them (Fig. 1). Milkman published a study of enzyme polymorphism in *E. coli* in 1973 (8), but our present knowledge stems largely from work published in the 1980s by Selander and colleagues (7, 9, 10).

Thousands of isolates of *E. coli* have now been examined for variation in a dozen or more enzymes specified by chromosomal loci. Isolates with the same alleles at all loci examined are said to belong to the same electrophoretic type (ET). The main generalizations and conclusions are as follows (7, 10).

(1) *E. coli* has a much higher level of genetic diversity than is found in most eukaryotes; more than 90% of loci are variable, compared with about one third in eukaryotes, and the average effective number of alleles

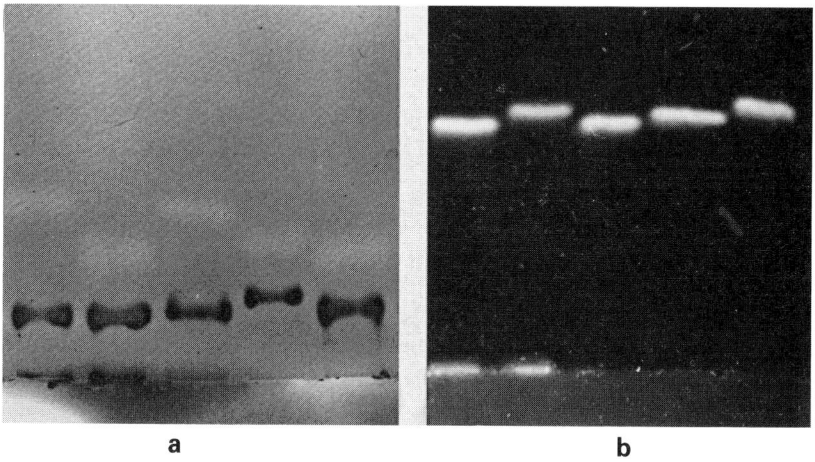

Figure 1. Enzyme polymorphism in *Rhizobium leguminosarum*. Cell extracts from natural isolates are separated by electrophoresis in polyacrylamide gels, which are then stained for enzyme activity to reveal allelic mobility variants of (a) glucose-6-phosphate dehydrogenase (dark bands) and superoxide dismutase (pale bands) and (b) β-galactosidase. (Reproduced with permission from reference 51.)

(diversity at each locus) is about threefold higher. These comparisons beg the question of whether *E. coli* as a taxonomic unit is biologically equivalent to a eukaryotic species; this will be considered in Section III.

(2) There are very strong associations (linkage disequilibria) between loci. This almost certainly means that the rate of chromosomal recombination is very low, so that the population structure is basically clonal.

(3) The majority of *E. coli* isolates fall into one of three clusters based on multilocus similarity, which are postulated to represent three major evolutionary lineages (Fig. 2).

(4) Isolates of *Shigella* species are not distinct from *E. coli*, but fall into two of the three clusters. The diversity within *Shigella* is much less than in *E. coli* as a whole.

(5) Isolates from pathogenic infections are not a random sample from the normal faecal flora; most belong to one of a relatively small number of "specialist" ETs. Although very closely related genotypes are sometimes isolated from different animal species, the overall proportions of different ETs vary greatly from one host group to another. These observations indicate that the ETs show ecological specializations.

(6) Most of the genetic diversity occurs locally; differences between localities contribute rather little to the total variation. Almost identical ETs have been isolated from geographically distant samples; and the

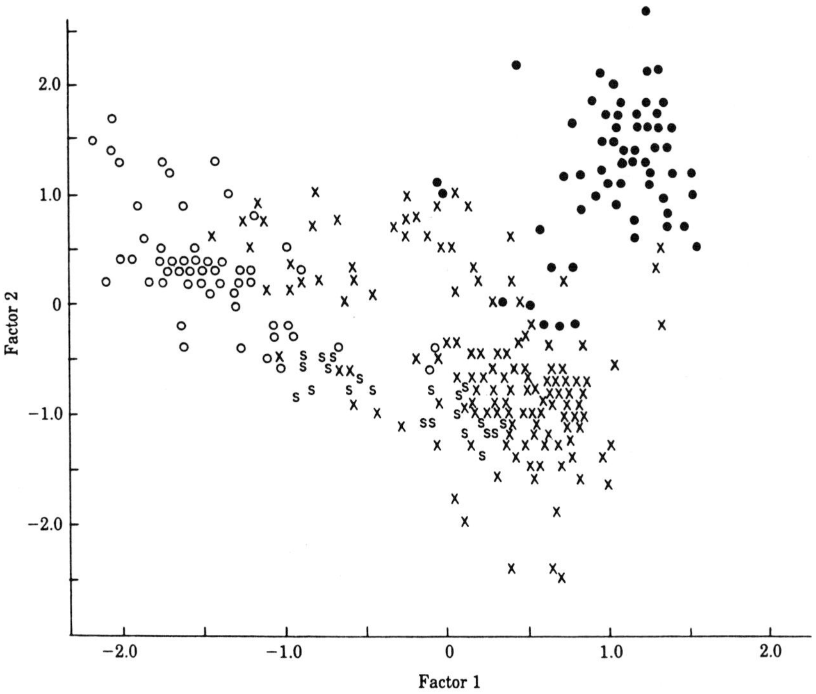

Figure 2. Three major clusters of *E. coli* and *Shigella* types based on enzyme electrophoresis (52). Each of 1705 isolates was typed at 12 enzyme loci, which distinguished 302 different combinations. These types were positioned in a multi-dimensional space by a principle components analysis, and the figure shows the first two dimensions of this space. It is clear that the strains are clustered (proximity indicates genetic similarity), though much of the information cannot be displayed in just two dimensions. A discriminant analysis based on all the data divided the *E. coli* into the three groups indicated by ○, ● and × *Shigella* isolates are indicated by s; they do not form a separate cluster, but overlap two of the *E. coli* groups. (Reproduced with permission from reference 52.)

frequency of "private alleles", unique to a single locality, is low. Thus the worldwide rate of migration appears to be relatively high.

(7) The gut flora of an individual host is made up of one or two long-established resident ETs, and many transient types that are unrelated to one another or to the residents (11).

(8) Healthy people in the same family share a significant fraction of their *E. coli* ETs; so, to a lesser extent, do their friends and their pets.

(9) Estimates of genetic distance based on ETs correlate well with measurements obtained by hybridization of the total cellular DNA. This confirms that the enzyme loci provide a reasonable sample of the overall chromosomal genome.

(10) The correlation between serotype and ET is far from perfect. Furthermore, when the same O-serogroup appears in distantly-related ETs, an examination of the lipopolysaccharide (LPS, which determines the O-serogroup) has sometimes shown consistent differences in LPS structure between the ET groups. This demonstrates convergent evolution of the LPS to common antigenicity. Conversely, strains that are otherwise very closely related often vary in serotype, suggesting that antigenicity may evolve rapidly. All in all, it appears that serotyping, although valuable in epidemiology, does not provide a reliable basis for assessing genetic relatedness.

It is apparent from this list that the analysis of genetic diversity by the relatively simple technique of enzyme electrophoresis can give us a variety of insights into the biology of bacteria, including their genetics, ecology, and taxonomy. Besides *E. coli*, a considerable number of other bacteria, mostly of medical significance, have now been investigated in this way (9). Generally, the picture is similar to that described for *E. coli*, with high genetic diversity and a strongly clonal population structure, but it is as yet too soon to say that this is universal throughout the bacterial world.

B. Selection versus Neutrality: the Significance of Variation

The discovery that enzyme electrophoretic polymorphism was ubiquitous led to a prolonged debate about the meaning of this variation (1, 2, 3, 5, 12, 13). Was the balance of different allelic forms maintained by selection, or were most of the allelic differences so trivial that they were selectively neutral? For a few well-studied enzymes there is evidence for appropriate selective differences between alleles, but there are theoretical grounds for doubting whether the majority of enzyme polymorphisms can simultaneously be stabilized by effective selection (basically, there cannot be enough selection to go round). Two approaches to resolving the question have repeatedly been tried: comparing the observed distribution of variants with theoretical expectations, and studying the biochemistry and physiology of particular polymorphic enzymes. Neither approach has provided an unchallenged general answer, but studies of bacteria have made a number of significant contributions to the debate.

When Milkman first turned to *E. coli* as a subject for enzyme electrophoresis (8), it was in the hope of providing a definitive test of the neutral theory. *E. coli* was, in Levin's words (14), a "no excuse" organism; the population size of the species was vast (much more than 10^{10}), and had presumably remained vast for the past 40 million years or more, long enough to reach equilibrium. The theory for neutral mutations predicts that in such a large population the average effective number of alleles should be more than 400 per locus. In fact, the effective number of alleles at the loci Milkman examined was between one and two, so he concluded that *E. coli* was much less polymorphic than the neutral theory predicted (8). The level of polymorphism was still substantial, though, and, occurring in a haploid species, it ruled out the possibility that heterosis (heterozygote advantage) was a selective mechanism of general importance in the maintenance of enzyme polymorphisms, which was one idea current at the time. A few years later, Levin reconsidered Milkman's arguments, and reconciled the data with the neutral theory by pointing out that periodic selection (see Section II.C) would greatly reduce the genetically effective population size, making it many orders of magnitude smaller than the actual number of cells in the species (14).

In sexually-reproducing eukaryotes, it is unusual for allelic variation at different loci to be strongly correlated, as is the case in *E. coli* (7, 10; see Section II.A), because regular recombination tends to break down this linkage disequilibrium unless the loci are very tightly linked. Since conjugative plasmids are found in natural populations, it seems likely that chromosomal recombination can occur in *E. coli*, too, but Levin has estimated that under the conditions usually prevailing in nature the rate of chromosomal recombination by conjugation may be extremely low, comparable with the mutation rate (14). If recombination were much more frequent, it would be necessary to invoke strong selection to maintain the observed linkage disequilibria. This selection would have to be a particular and unlikely kind of epistasis between the enzyme loci (10).

For certain enzymes, the effects of allelic differences on fitness have been measured directly. *E. coli* offers a number of significant advantages over eukaryotic organisms in which similar experiments have been attempted. The genetic system allows naturally-occurring allelic variants to be placed in a common genetic background (by transduction, for example), many biochemical pathways are well understood and fluxes through them can be manipulated via the growth medium, and the controlled growth of large populations in chemostats allows differences in fitness to be measured with precision. In a series of studies, Dykhuizen and Hartl tested natural alleles for a number of sugar-utilization enzymes (7). In most cases selection was not detected even when the enzyme function was made critical by supplying the appropriate limiting growth substrate. However, some alleles of 6-phosphogluconate dehydrogenase and of phosphogluconate dehydratase were relatively disfavoured on gluconate medium, although they were neutral on glucose. Since competition is more intense in a chemostat than in nature, and glucose is a widespread substrate whereas gluconate is rare, Dykhuizen and Hartl concluded that the various enzyme alleles they examined were effectively neutral, although a "potential for selection" existed in some cases where selection could be evoked by appropriate conditions. They echo a comment made by others in relation to biotype markers: great phenotypic variability is evidence for selective unimportance.

One might suppose that synonymous base substitutions—those that do not alter protein sequence—would be selectively neutral, but this is apparently not so. The usage of alternative synonymous codons is far from random, and this bias is especially strong in very highly expressed genes, indicating presumably that selection is based on the efficiency of translation. These highly biased genes also tend to have the lowest degree of synonymous substitution in interspecific comparisons between enteric bacteria (15).

One factor influencing codon usage is the overall guanine and cytosine (G + C) content of the genome, which varies among eubacteria from about 25% to 75%. Differences in (G + C) content are reflected in all parts of the genome, but are greatest in those parts that are least constrained by function (spacers between genes, and the third position of codons) (16). Presumably, genomes with divergent (G + C) contents have experienced biased mutation pressures, such that the rate of mutation from (A·T) to (G·C) is different from the reverse rate. The mechanism is unclear, as are the selective constraints that permit such differences to be maintained for long periods of evolution.

C. Periodic Selection: a Consequence of Clonal Reproduction

"Is sex necessary?" asked Muller in 1932 (17), and in reply he pointed out that, in a purely asexual population, advantageous mutations arising in different lineages would inevitably be in competition, and the fittest mutant lineage would eventually displace and eliminate other advantageous mutations. Different advantageous mutations must occur successively in the same lineage if they are to be combined. Mutations in other lineages may spread for a while but will eventually be eliminated. The lack of recombination hinders the evolutionary process of adaptation (Fig. 3).

Almost 20 years later, microbiologists arrived at essentially the same conclusion to explain their experimental data. The classic study published by Atwood, Schneider and Ryan in 1951 (18) was perhaps the first experimental population genetics of bacteria, even though its frame of reference was mutation research rather than population genetics. They showed that when a bacterial culture is serially transferred for many generations, new types with a selective advantage arise periodically and replace the existing population. The effect of this "periodic selection" is to purge the population of much of the genetic variation that would otherwise accumulate by repeated mutation, because the chances are that the new advantageous mutation will arise in a cell which has the most abundant genotype at most other genetic loci. This will protect the population from the rapid loss by mutation pressure of functions that are temporarily inessential, such as prototrophy while growing on a rich medium. From the point of view of population genetics, an important consequence of periodic selection is that the effective population size of a clonal

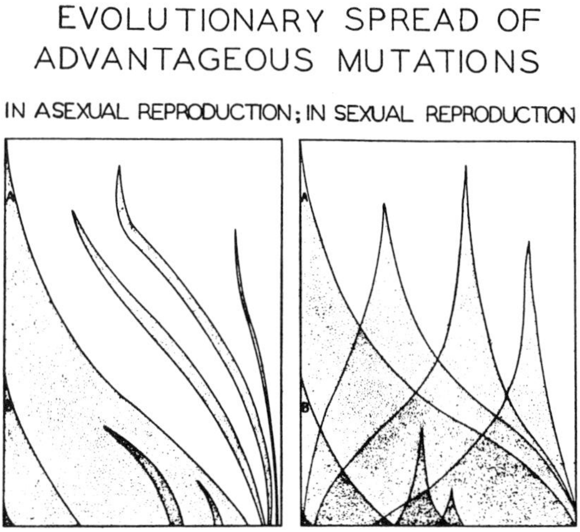

Figure 3. Muller's diagram illustrating the effect of recombination on the rate of evolution. In Muller's words (17): "Time here is the vertical dimension, progressing downwards. In the horizontal dimension a given population, stationary in total numbers, is represented. Sections of the population bearing advantageous mutant genes are darkened, proportionally to the number of such genes. In asexual organisms these genes compete and hinder one another's spread; in sexual organisms they spread through one another". As an advantageous gene spreads through an asexual population there is a purging of variation that was later called "periodic selection". (Reproduced with permission from reference 17.)

organism will be much less than the actual size (14), so that correspondingly less selectively neutral variation will be maintained, and mutations with very small advantageous effects will be quasi-neutral and liable to be lost.

"Periodic selection" was observed because no recombination occurred between lineages during the experiment. Even a small amount of recombination would be sufficient to change the situation radically, so the question of the natural rate of (chromosomal) recombination in bacteria is an important one.

III. SPECIES BOUNDARIES AND EVOLUTIONARY RELATIONSHIPS

A. The Biological Definition of Species

Population genetics is concerned with the variation within species, systematics with the differences between species. The disciplines are divided by the species boundary, and each has its definition of this. For population genetics the definition is, naturally, a genetic one; it is based on gene flow, and is summarized in Mayr's definition of a biological species: "a reproductive community of populations (reproductively isolated from others) that occupies a specific niche in nature" (19). Can this definition be applied to bacteria? Two potential, but opposite, problems are their predominantly asexual reproduction, and the apparent promiscuity of certain genetic elements.

B. Is Asexuality a Problem?

Even among eukaryotes, the gene-flow definition of species is not without problems. Indeed, it was partly a concern to encompass asexual organisms that moved Mayr to add the rider about "a specific niche in nature" to the earlier, and logically more defensible, definition in terms of reproductive isolation alone.

Asexual reproduction is widespread in plants, and this leads to taxonomic problems in certain groups, such as dandelions (*Taraxacum*). The genus *Taraxacum* includes some diploid species with normal sexual reproduction, and many polyploid forms that can reproduce asexually (20). Some of these polyploids also have some capacity for sexual reproduction; consequently, new asexual forms are continually arising by hybridization, and can then be perpetuated clonally. The result is a complex array of different genotypes, some genetically isolated, others retaining the potential to contribute to a wider gene pool. There is a great deal of variation in morphology and in chromosome structure, and prolonged careful study has led to the designation of thousands of named species, as well as broader groupings. Despite the confusion within the genus, it is clearly demarcated from related genera. Some progress has been made in unravelling its evolutionary history (20), but modern molecular techniques could add a new dimension to our understanding. The relevance of all this to bacteria is that a similar combination of processes (clonal propagation and intermittent recombination) operates in bacteria. The genetic structure of bacterial populations could, therefore, be equally complex, with some genetically isolated lineages and others more promiscuous. Until genetic variation in bacterial species is accorded the degree of attention that generations of botanists have devoted to dandelions, we will not have an adequate description of population structure, and therefore of the potential for adaptation and evolution.

Without a clear-cut genetic boundary, taxonomists sometimes disagree on the degree of difference required to differentiate species. Thus a "splitter" may name each asexual clone as a separate species, while a "lumper" will make one name serve for a whole range of forms. Such discrepancies are widespread in bacterial taxonomy. Some taxonomists of *Salmonella*, for example, have named every antigenic variant as a separate species (21), whereas the soybean symbiont "species" *Bradyrhizobium japonicum*, as defined at present, includes strains that are as divergent at the DNA level as *Salmonella* is from *E. coli*, and that also differ in many physiological and biochemical characteristics (22). To surmount this difficulty, it has been proposed that species boundaries should be drawn solely on the basis of relatedness at the DNA level. All isolates showing more than 60% DNA homology would be defined as belonging to the same species (21, 23). A problem with this approach is that DNA sequences and biological function do not necessarily evolve at a consistent relative rate (24), so that a DNA homology level that successfully divides one group of bacteria along natural discontinuities will give arbitrary results with another. A further pitfall is that accessory elements (such as plasmids) may make a significant but variable contribution to homology measurements (21).

There are problems, then, in defining bacterial species solely by similarity, whether at the phenotypic or genotypic level. A gene flow definition, on the other hand, will make sense only if bacterial species are a biological reality; if, that is, each species has a genetic cohesiveness maintained by gene flow and circumscribed by genetic barriers. However, since relatively little gene flow may suffice to hold a species together, and the barriers between species need not be absolute, isolated examples cannot prove or disprove the hypothesis that the bacterial world is divided into biological species. We need to collect data of various kinds from a diverse range of bacteria.

C. Is Promiscuity a Problem?

Examples of wide-ranging gene transfer among bacteria, particularly the spread of antibiotic resistance, have led many authors to suppose that neither biological species nor a consistent evolutionary tree can be expected. All bacteria might share "one gene pool from which any 'species' may draw genes as these are required" (25), and bacterial evolutionary relationships would therefore be represented by a network rather than a branching family tree (25, 26). By contrast, evolutionary and population geneticists have generally assumed that interspecific gene transfer has not been of major general significance as far as chromosomal genes are concerned (4, 27). Who is right?

No one disputes that accessory DNA elements—plasmids, phages, transposons, insertion sequences—may sometimes move widely (see the next section), but have chromosomal genes been constantly reshuffled? We need to consider this question on several different scales: within a species, between closely related species, and between distant species.

The question of chromosomal recombination within a species has already been considered. The existence of strong linkage disequilibria in all species so far examined demonstrates that recombination is too rare to break up these chance associations. This does not mean, however, that recombination never occurs outside the laboratory, and a number of observations can best be explained by invoking it. A plausible example involves three strains of *E. coli* that are identical in most respects but differ at each of two closely linked loci (*gnd* and *rfb*), suggesting that these genes may have been cotransferred, perhaps by transduction (a phage P2 attachment site is very close by) (10). The occasional transfer and incorporation by homologous recombination of small DNA segments, just a few genes or even part of a gene, would not prevent the accumulation of linkage disequilibria, but might be sufficient to maintain the genetic cohesiveness of a species.

We classify some bacterial isolates as *E. coli* and others as *Salmonella typhimurium*. Is the dividing line arbitrary, or does it correspond to a biological reality? Is there a continuum of intermediate types between the typical *E. coli* and the typical *Salmonella*? The answer is unequivocal: numerous phenotypic criteria and total DNA hybridization all confirm that a range of different *E. coli* isolates are all much more closely related to *E. coli* K-12 than is any *Salmonella* (21). Such a striking gap could probably not have arisen simply through a failure in sampling. In fact, similar gaps are apparent in other groups of bacteria. We must conclude that the bacteria occupy only discrete patches in the space of all conceivable genotypes. This is similar to the situation in eukaryotes, where the existence of sex and of reproductive isolation provides a ready explanation. A similar explanation in terms of gene flow can apply to bacteria only if there is sufficient recombination within species and relatively little between species. If bacteria are purely clonal, or (paradoxically) if they form a freely recombining "superorganism", then their genetic patchiness can be explained only by invoking selection for a limited number of specialized niches. In other words, it must be supposed that all *E. coli* are similar because they have a common "purpose in life", which is different from that of *Salmonella*, and if intermediates are produced they fail to survive because there is no niche for them.

The enteric bacteria form a coherent group of fairly closely related genera whose similar life-styles should provide frequent opportunities for contact. Nevertheless, the available evidence does not suggest that they indulge in uninhibited chromosomal gene swapping. Patterns of relatedness have been determined for many different proteins and nucleic acid sequences, and a fairly consistent genealogy emerges for all the genes (27). When *E. coli* DNA is transferred to *S. typhimurium* in laboratory crosses, it integrates into the chromosome many orders of magnitude less frequently than in intraspecific crosses, and this has been interpreted to mean that natural genetic exchange will be correspondingly less frequent (27). A similar effect is observed in crosses between *Rhizobium leguminosarum* and *R. meliloti* (28), but in both cases the range of independent genotypes used for intraspecific crosses was too narrow to be sure that all such crosses are more effective than interspecific ones. Further studies in this area would provide evidence for or against the reality of gene-flow barriers around species.

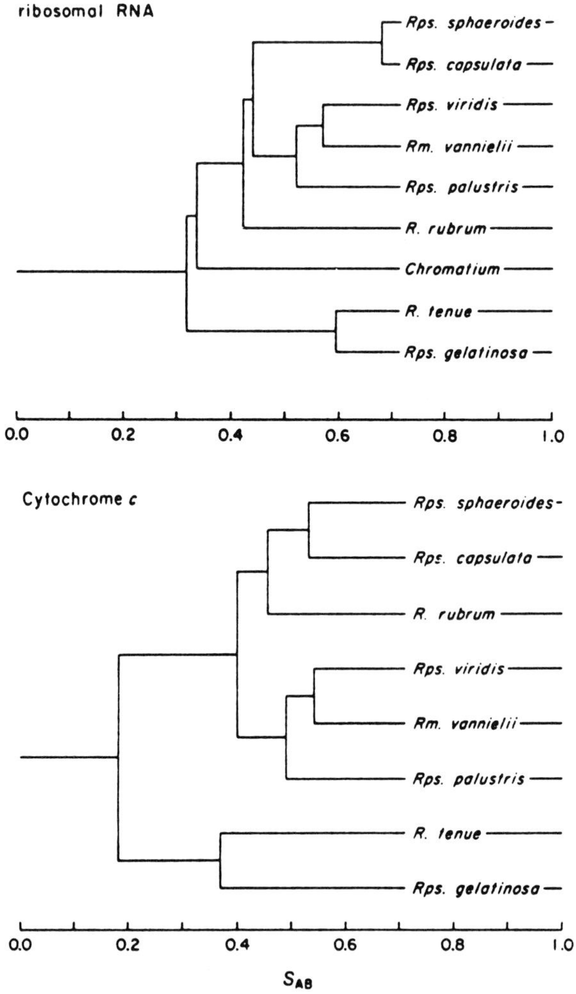

Figure 4. Evidence that evolutionary relationships have not been obscured by interspecific gene transfer. The relationship amongst various purple photosynthetic bacteria determined by 16S ribosomal RNA sequence comparisons is essentially the same as that from cytochrome *c* amino acid sequences (29). *Rps.*, *Rhodopseudomonas* (some of these species have now been assigned to other genera, Chapter 5); *R.*, *Rhodospirillum*; *Rm*, *Rhodomicrobium*. (Reproduced with permission from reference 29.)

Woese and his colleagues have devoted much effort to constructing phylogenetic trees for bacterial 16S ribosomal RNA (rRNA) sequences (4), on the assumption that these will reflect the evolutionary history of the major part of the genome. This will be true only if lateral transfer has not played a major role and bacterial evolution has been, as a consequence, predominantly treelike rather than reticulate. This view is supported by evidence that consistent phylogenies can be derived independently from different parts of the genome. For example, phylogenies derived from 16S rRNA data are essentially the same as those for purple bacteria based on cytochrome *c* protein sequences (29; Fig. 4) and for a wider range of bacteria based on nitrogenase gene sequences (30). The rapidly-growing databases of DNA sequence data will provide many additional tests.

IV. THE IMPORTANCE OF ACCESSORY ELEMENTS

A. What are Accessory Elements?

Phages, plasmids, transposons and insertion sequences make up the accessory elements of the bacterial genome, so named by analogy with the accessory chromosomes of eukaryotes. In Campbell's definition (31), they have two properties in common: they include no genes that are unconditionally required for the reproduction of the organism that harbours them, and they can over-replicate their own DNA relative to the typical chromosomal DNA of the cell. Over-replication may occur by infectious transfer to other cells, or by replicative transposition within a genome.

These properties suggested the notion that accessory elements were "selfish DNA" (32, 33). That is, the persistence of these elements need not depend on any benefit which they confer on the organism in which they reside; they could survive and spread as a simple consequence of their capacity for over-replication. Levin and Stewart have investigated mathematically the conditions under which plasmids can become established and persist in bacterial populations, and conclude that plasmids can be favoured even if they are disadvantageous to their hosts, provided they are conjugative or mobilizable at a sufficient rate (34). However, they conclude that these conditions are so stringent that in practice plasmids are unlikely to spread unless they confer a net benefit to their hosts. Similar conclusions have recently been reached for bacterial transposons (35), and in the case of the insertion sequence IS*50* there is direct experimental evidence that it confers an advantage, although the mechanism is unknown (36).

Plasmids and transposons usually, and phages sometimes, carry genes that are not necessary for the survival or spread of the accessory element itself. These genes are not a random sample of the genome, but are typically "genes that are needed occasionally rather than continually under natural conditions" (31). Examples are genes conferring resistance to antibiotics and other toxic substances, utilization of unusual substrates, production of bacteriocins, or traits involved in pathogenesis or symbiosis. Conversely, the genes on chromosomes typically provide "housekeeping" functions that are frequently required. However, many exceptions are found; genes for symbiotic nitrogen fixation are chromosomal in *Bradyrhizobium*, for example, although the homologous genes are plasmid-borne in *Rhizobium* (Chapter 19). Since transposons can shuttle between chromosome and plasmids (37), and can pick up DNA, it would appear that in principle any chromosomal gene could move into the accessory element fraction of the genome (31).

B. Consequences for Population Structure and Evolution

Although it appears that horizontal transfer of chromosomal genes has not been so extensive as to obscure the outlines of phylogeny, the transfer of accessory elements and of those genes that are their specialized cargoes could play an important role in bacterial adaptation: there could be a genetic "commonwealth" of clones in which the various strains, species, and genera in a habitat can exchange genes on accessory elements, allowing the population to "exploit as 'resource data' the grouped genetic experience of whole microbial ecosystems" (38). In the same vein, Bennett and Richmond ask, "If resistance to antibiotics, for example, can be relatively readily acquired from among a few members of a population, what is the advantage of maintaining large numbers of resistant bacteria when no antibiotic is in the environment?" (39). One answer is that if a population's response to a sudden and lethal assault depends on the survival of rare individuals that happen to possess resistance or to acquire it in the nick of time, then the resulting drastic reduction in population size will not just be a short-term setback but will also purge the population of other potentially-valuable rare genes. If antibiotics constituted an infrequent but recurring challenge, one would expect resistance genes to be ubiquitous, but perhaps inducible or cryptic (40; see Section V). In fact, the result of suddenly exposing a range of pathogens to high levels of antibiotics, through clinical and veterinary use, was the spread of antibiotic resistance plasmids and the transposons which they carry (41). Many of these have a particularly broad host range, and closely related elements have been isolated from a range of different species. This response should probably not be regarded as arising from a pre-existing "adaptation" on the part of the bacteria concerned, but could be described as an opportunistic expansion of their territory on the part of the resistance plasmids or transposons.

The example of the spread of antibiotic resistance, which has coloured so much thinking, is probably not typical of the role and importance of most accessory elements, although it illustrates the major significance that even rare transfer events can have if accompanied by strong positive selection. Most plasmids do not appear to confer any such dramatic benefit on their hosts, and yet they persist. Far from being cosmopolitan, many plasmids appear to be restricted in natural distribution to certain genotypes within a species (37). Most conjugative plasmids have evolved the means to repress conjugative ability, suggesting that unbridled transfer is not advantageous even to the plasmid itself. If such plasmids are normally repressed for transfer, the rate of genetic exchange may be low in nature.

V. EXPERIMENTAL EVOLUTION

Bacteria can be grown in large populations through many rapid generations under controlled conditions, so they can serve as models to study evolutionary processes, and their amenability to genetic and biochemical analysis means that the resulting changes can be described in some detail. In a sufficiently large population, allele frequencies are not subject to random fluctuations caused by sampling, so fitness differences can be estimated very accurately from changes in frequencies, as in the studies of enzyme alleles by Dykhuizen and Hartl (7) discussed in Section II.B. In large populations rare mutational events will occur, and this allows several types of evolutionary event to be simulated. Selection for growth on a novel medium, for example, has been used successfully to obtain mutant forms of enzymes with altered substrate specificities (42). In other circumstances, similar selection has led to the activation by mutation of cryptic genes whose existence was previously unsuspected. For example, none of 71 natural isolates of *E. coli* was able to utilize the β-glucoside sugars cellobiose, salicin and arbutin, but all except five yielded spontaneous mutants that could utilize one or more of them (40). The mutations could be assigned to at least four different operons, and detailed studies of one of these have shown that activation is due to the insertion of a mobile genetic element, IS*1* or IS*5*, upstream from the structural genes. Cryptic genes appear to be quite widespread, and to be just one mutational event away from the functional state. It is supposed that they must have been active in the fairly recent past (in evolutionary terms), otherwise they would be expected to accumulate other mutations that would render them permanently nonfunctional.

"Now, *here*, you see, it takes all the running *you* can do, to keep in the same place", said the Red Queen to Alice. Van Valen developed this theme into the Red Queen Hypothesis: as each organism improves through evolution it makes life worse for others (1). Real organisms do not live in an unvarying and inanimate environment; they have to interact with other organisms, some of which are antagonistic. As predators, parasites and pathogens pursue their victims, the two sides engage in a running evolutionary battle of genetic defences and countermeasures. The great antigenic diversity displayed by some pathogenic bacteria is presumably the outcome of just such an arms race against the immune responses of their hosts, but bacteria can be victims as well as aggressors, and one system that is readily amenable to experimental study is the coevolution of bacteria and virulent phages. Lenski and Levin have examined the interaction in chemostats of *E. coli* B with phage T4 and with other phages, both in theory and in practice (43). They conclude that coliphages cannot mutate fast enough to keep up with resistance in the host because the mutations required in the phage are much more specific, and therefore rarer. They expect "that natural communities of coliform bacteria and virulent coliphage are dominated by bacterial clones resistant to all co-occurring virulent phage" (43).

Arms races can occur within species, too, and a possible example is provided by bacteriocins, which are molecules secreted by one bacterium and toxic to other, closely related, bacteria. However, the evolutionary significance of such interactions has not yet been determined. The restriction and modification of DNA by specific nucleases and methylases is another phenomenon with potential, but largely uninvestigated, consequences for population genetics. Although the primary function of such systems is probably defence against phages, they may also serve to reduce effective gene transfer. The converse could also be argued, though, because the cohesive DNA ends generated by restriction nucleases may enhance recombination of short stretches of the genome (44).

In complex natural communities there are many different interactions between organisms, but much of the history of microbiology has depended on the study of genetically uniform strains in isolation. It is intriguing to discover, therefore, that an isolated strain can evolve into a diversified community simply through prolonged laboratory culture. When populations of *E. coli*, initiated with a single clone, were maintained in glucose-limited chemostats for many hundreds of generations, genetic polymorphisms developed and persisted (45). The distinct component clones had diverged to fill different ecological niches: they differed in maximum specific growth rate and in the transport of glucose, and incompletely respired metabolites were excreted by some clones and utilized by other, minority clones.

VI. THE PLANNED RELEASE OF NOVEL ORGANISMS

When Pandora's box was opened, all manner of evils flew out to plague the world. So runs one version of the Greek myth, but others tell us that it was blessings that escaped. Today, there are similar divergences of opinion about the planned release of microbes produced by genetic engineering. Organisms with novel combinations of genes could have significant applications in enhancing crop plant growth, protecting crops against disease and damage, extracting minerals, cleaning up environmental pollution, and so on, but is it possible that they might have disastrous unforeseen effects on natural ecosystems? This is not, of course, a question that can be answered in general terms; each case, or class of cases, must be considered in its turn. Our ability to predict the consequences of a release depends on an understanding of the ecology and population genetics of the organism and the ecosystem concerned (46, 47).

For many years, selected (but not engineered) forms of bacteria have been released without ill effect. Indeed, when strains of *Rhizobium* have been introduced into soils with pre-existing populations of the same species, the problem has usually been that the introduction fails to become established (48). However, this kind of experience should not lead us to suppose that organisms in nature are always so perfectly adapted to their immediate environment that any kind of genetic change is bound to render an organism unfit to compete. This argument has frequently been advanced by genetic engineers to justify their sanguineness, but as a general rule it is wrong: in the simplistic models of population geneticists and ecologists, populations may be devised which reach a steady state that cannot be invaded by a novel organism, but the real world is in a constant state of flux, and organisms in nature are therefore seldom optimal (49). The history of exotic introductions of higher plants and animals provides many examples of organisms that thrive in habitats very different from those in which they evolved.

It is, of course, possible to make specific changes to an organism that will certainly make it unfit to compete in the field, and such disabled organisms are widely used for genetic engineering in the laboratory, but they would usually be unsuitable for planned release since persistence and multiplication in the field, at least for a limited period, would normally be required. Since general theory cannot predict the consequences of any particular release, each case will require an experimental assessment of the fitness of the engineered organism, under appropriate conditions, and its effect on any indigenous population. Genetic variation in this indigenous population needs to be assessed, both to ensure that the introduced organism can be distinguished from all wild-types, and to detect interactions between them.

The least predictable potential interaction is the exchange of genes (47, 50). The probability that the engineered DNA will be transferred out of the introduced organism can often by minimized by a suitable choice of genetic construct, but since it can never be eliminated it would be prudent to consider whether the DNA might be a hazard in any plausible recipient. It is probably more difficult to keep out incoming genetic material that may confer new properties on the introduced organism. Since many pathogenesis determinants are encoded on transmissible plasmids, for example, it must be anticipated that the released organism, if of an appropriate species, may acquire them: the engineered DNA should not have the potential to enhance pathogenicity.

Descriptive population genetics may be of direct help in the choice of suitable organisms to minimize the potential for gene flow. It is clear from the enzyme electrophoretic studies of *E. coli*, for example, that certain chromosomal genotypes are much more frequently associated with pathogenesis than others (10), and studies on the distribution of plasmids in natural populations indicate that the host range of certain plasmids may in practice be restricted even within a bacterial species.

Clearly, the risks associated with gene transfer cannot be assessed without a much more definite knowledge of the rates and pathways of genetic exchange in natural bacterial populations. The issues are essentially population genetics questions, and provide a sense of urgency and topicality to the emerging field of bacterial population genetics.

References

(1) Futuyma, D. J. (1979). "Evolutionary Biology", Sinauer, Sunderland, Mass.

(2) Selander, R. K. (1976). Genetic variation in natural populations. *In* "Molecular Evolution" (F. J. Ayala, ed.), pp. 21–45. Sinauer, Sunderland, Mass.

(3) Lewontin, R. C. (1985). Population genetics. *Ann. Rev. Genet.* **19**, 81–102.

(4) Woese, C. R. (1987). Bacterial evolution. *Microbiol. Rev.* **51**, 221–271.

(5) Lewontin, R. C. (1974). "The Genetic Basis of Evolutionary Change". Columbia University Press, New York.

(6) Aquadro, C. F., Desse, S. F., Bland, M. M., Langley, C. H. and Laurie-Ahlberg, C. C. (1986). Molecular population genetics of the alcohol dehydrogenase gene region of *Drosophila melanogaster. Genetics* **114**, 1165–1190.

(7) Hartl, D. L. and Dykhuizen, D. E. (1984). The population genetics of *Escherichia coli. Ann. Rev. Genet.* **18**, 31–68.

(8) Milkman, R. (1973). Electrophoretic variation in *Escherichia coli* from natural sources. *Science* **182**, 1024–1026.

(9) Selander, R. K., Caugant, D. A., Ochman, H., Musser, J. M., Gilmour, M. N. and Whittam, T. S. (1986). Methods of multilocus enzyme electrophoresis for bacterial population genetics and systematics. *Appl. Envir. Microbiol.* **51**, 873–884.

(10) Selander, R. K., Caugant, D. A. and Whittam, T. S. (1987). Genetic structure and variation in natural populations of *Escherichia coli. In* "*Escherichia coli* and *Salmonella typhimurium.* Cellular and Molecular Biology" (F. C. Neidhardt, J. L. Ingraham, K. Brooks Low, B. Magasanik, M. Schaechter and H. E. Umbarger, eds.), Vol. 2, pp. 1625–1648, American Society for Microbiology, Washington.

(11) Caugant, D. A., Levin, B. R. and Selander, R. K. (1981). Genetic diversity and temporal variation in the *Escherichia coli* population of a human host. *Genetics* **98**, 467–490.

(12) Kimura, M. (1983). "The Neutral Theory of Molecular Evolution". Cambridge University Press, Cambridge.

(13) Ohta, T. and Aoki, K., eds. (1985). "Population Genetics and Molecular Evolution". Springer-Verlag, Berlin.

(14) Levin, B. R. (1981). Periodic selection, infectious gene exchange and the genetic structure of *E. coli* populations. *Genetics* **99**, 1–23.
(15) Sharp, P. M. and Li, W. H. (1987). The rate of synonymous substitution in enterobacterial genes is inversely related to codon usage bias. *Mol. Biol. Evol.* **4**, 222–230.
(16) Muto, A. and Osawa, S. (1987). The guanine and cytosine content of genomic DNA and bacterial evolution. *Proc. Natl. Acad. Sci. USA* **84**, 166–169.
(17) Muller, H. J. (1931). Some genetic aspects of sex. *Amer. Natur.* **66**, 118–138.
(18) Atwood, K. C., Schneider, L. K. and Ryan, F. J. (1951). Periodic selection in *Escherichia coli*. *Genetics* **37**, 146–155.
(19) Mayr, E. (1982). "The Growth of Biological Thought". Belknap Press, Cambridge, Mass.
(20) Richards, A. J. (1973). The origin of *Taraxacum* agamospecies. *Bot. J. Linn. Soc.* **66**, 189–211.
(21) Brenner, D. J. and Falkow, S. (1971). Molecular relationships among members of the Enterobacteriaceae. *Adv. Genet.* **16**, 81–118.
(22) Hollis, A. B., Kloos, W. E. and Elkan, G. H. (1981). DNA:DNA hybridization studies of *Rhizobium japonicum* and related *Rhizobiaceae*. *J. Gen. Microbiol.* **123**, 215–222.
(23) Johnson, J. L. (1984). Bacterial classification III: Nucleic acids in bacterial classification. *In* "Bergey's Manual of Systematic Bacteriology" (N. R. Krieg, ed.), Vol. 1, pp. 8–11, Williams and Wilkins, Baltimore, Maryland.
(24) Avise, J. C. (1983). Protein variation and phylogenetic reconstruction. *In* "Protein Polymorphism: Adaptive and Taxonomic Significance" (G. S. Oxford and D. Rollinson, eds.), pp. 103–130. Academic Press, London.
(25) Hedges, R. W. (1972). The pattern of evolutionary change in bacteria. *Heredity* **28**, 39–48.
(26) Sonea, S. and Panisset, M. (1983). "A New Bacteriology". Jones and Bartlett, Boston, Massachusetts.
(27) Ochman, H. and Wilson, A. C. (1987). Evolutionary history of enteric bacteria. *In* "*Escherichia coli* and *Salmonella typhimurium*. Cellular and Molecular Biology" (F. C. Neidhardt, J. L. Ingraham, K. Brooks Low, B. Magasanik, M. Schaechter and H. E. Umbarger, eds.), Vol. 2, pp. 1649–1654, American Society for Microbiology, Washington, D.C.
(28) Kondorosi, A., Vincze, E., Johnston, A. W. B. and Beringer, J. E. (1980). A comparison of three *Rhizobium* linkage maps. *Molec. Gen. Genet.* **178**, 403–408.
(29) Woese, C. R., Gibson, J. and Fox, G. E. (1980). Do genealogical patterns in purple photosynthetic bacteria reflect interspecific gene transfer? *Nature* **283**, 212–214.
(30) Hennecke, H., Kaluza, K., Thöny, B., Fuhrmann, M., Ludwig, W. and Stackebrandt, E. (1985). Concurrent evolution of nitrogenase genes and 16S rRNA in *Rhizobium* species and other nitrogen fixing bacteria. *Arch. Microbiol.* **142**, 342–348.
(31) Campbell, A. (1981). Evolutionary significance of accessory DNA elements in bacteria. *Ann. Rev. Microbiol.* **35**, 55–83.
(32) Doolittle, W. F. and Sapienza, C. (1980). Selfish genes, the phenotype paradigm and genome evolution. *Nature* **284**, 601–603.
(33) Orgel, L. E. and Crick, F. H. C. (1980). Selfish DNA: the ultimate parasite. *Nature* **284**, 604–607.
(34) Levin, B. R. and Stewart, F. M. (1979). The population biology of bacterial plasmids: *a priori* conditions for the existence of mobilizable nonconjugative factors. *Genetics* **94**, 425–443.
(35) Condit, R., Stewart, F. M. and Levin, B. R. (1988). The population biology of bacterial transposons: *a priori* conditions for maintenance as parasitic DNA. *Am. Natur.* (In press.)
(36) Hartl, D. L., Dykhuizen, D. E., Miller, R. D., Green, L. and de Framond, J. (1983). Transposable element IS*50* improves growth rate of *E. coli* cells without transposition. *Cell* **35**, 503–510.
(37) Hartl, D. L., Medhora, M., Green, L. and Dykhuizen, D. E. (1986). The evolution of DNA sequences in *Escherichia coli*. *Phil. Trans. R. Soc. Lond. B.* **312**, 191–204.

(38) Reanney, D. C. (1978). Coupled evolution: adaptive interactions among the genomes of plasmids, viruses and cells. *Int. Rev. Cytol.* Suppl. **8**, 1–68.

(39) Bennett, P. M. and Richmond, M. H. (1978). Plasmids and their possible influence on bacterial evolution. *In* "The Bacteria", Vol. VI (I. C. Gunsalus and R. Y. Stanier, eds.), pp. 1–69. Academic Press.

(40) Hall, B. G. and Betts, P. W. (1987). Cryptic genes for cellobiose utilization in natural isolates of *Escherichia coli. Genetics* **115**, 431–439.

(41) Wiedemann, B., Bennett, P. M., Linton, A. H., Sköld, O. and Speller, D. C. E., eds. (1986). "Evolution, Ecology and Epidemiology of Antibiotic Resistance". *J. Antimicrobial Chemotherapy* **18**, Suppl. C, pp. 1–261.

(42) Clarke, P. H. (1974). The evolution of enzymes for the utilization of novel substrates. *Soc. Gen. Microbiol. Symp.* **24**, 183–217.

(43) Lenski, R. E. and Levin, B. R. (1985). Constraints on the coevolution of bacteria and virulent phage: a model, some experiments, and predictions for natural communities. *Am. Natur.* **125**, 585–602.

(44) Chang, S. and Cohen, S. N. (1977). *In vivo* site-specific genetic recombination promoted by the *Eco*RI restriction endonuclease. *Proc. Natl. Acad. Sci. USA* **74**, 4811–4815.

(45) Helling, R. B., Vargas, C. N. and Adams, J. (1987). Evolution of *Escherichia coli* during growth in a constant environment. *Genetics* **116**, 349–358.

(46) Fiksel, J. and Covello, V. T., eds. (1986). "Biotechnology Risk Assessment. Issues and Methods for Environmental Introductions". Pergamon Press, New York.

(47) Halvorson, H. O., Pramer, D. and Rogul, M., eds. (1985). "Engineered Organisms in the Environment: Scientific Issues". American Society for Microbiology, Washington D.C.

(48) Stacey, G. (1985). The *Rhizobium* experience. *In* "Engineered Organisms in the Environment: Scientific Issues" (H. O. Halvorson, D. Pramer and M. Rogul, eds.), pp. 109–121. American Society for Microbiology, Washington D.C.

(49) Regal, P. J. (1985). The ecology of evolution: implications of the individualistic paradigm. *In* "Engineered Organisms in the Environment: Scientific Issues" (H. O. Halvorson, D. Pramer and M. Rogul, eds), pp. 11–19. American Society for Microbiology, Washington D.C.

(50) Strauss, H. S., Hattis, D., Page, G., Harrison, K., Vogel, S. and Caldart, C. (1986). Genetically-engineered microorganisms: II. Survival, multiplication and genetic transfer. *Recombinant DNA Technical Bulletin* **9**, 69–87.

(51) Young, J. P. W. (1985). *Rhizobium* population genetics: enzyme polymorphism in isolates from peas, clover, beans and lucerne grown at the same site. *J. Gen. Microbiol.* **131**, 2399–2408.

(52) Whittam, T. S., Ochman, H. and Selander, R. K. (1983). Multilocus genetic structure in natural populations of *Escherichia coli. Proc. Natl. Acad. Sci. USA* **80**, 1751–1755.

Index

D

E

F

G

H

Q

R

S

U

V

W

X

Y